SCHAUM'S OUTLINE OF

THEORY AND PROBLEMS

OF

COLLEGE PHYSICS

Tenth Edition

FREDERICK J. BUECHE, Ph.D.
Distinguished Professor at Large
University of Dayton

EUGENE HECHT, Ph.D.
Professor of Physics
Adelphi University

SCHAUM'S OUTLINE SERIES

McGRAW-HILL

New York Chicago San Francisco
Lisbon London Madrid Mexico City Milan
New Delhi San Juan Seoul Singapore
Sydney Toronto

EUGENE HECHT is a full-time member of the Physics Department of Adelphi University in New York where he takes great pleasure in teaching; not long ago the students elected him Professor of the Year. He has authored nine books, including *Optics*, 4th edition, published by Addison Wesley, which has been the leading text in the field, worldwide, for almost three decades. Professor Hecht has also written *Schaum's Outline of Optics* and two other major works, *Physics: Alg/Trig*, 3rd edition, and *Physics: Calculus*, 2nd edition, both published by Brooks/Cole. These modern, innovative introductory texts are used in the United States and abroad. As a member of the Editorial Board he helped to design and create the *Encyclopedia of Modern Optic*, published by Elsevier (2005). His book about the American ceramist G. E. Ohr, *The Mad Potter of Biloxi*, published by Abbeville, won the 1989 Art Book of the Year Award. Professor Hecht has lectured on both art and physics at museums, galleries, conferences and universities throughout the world. He is presently on the National Council of the new George E. Ohr Museum of Art, designed by Frank Gehry, and is co-author of the catalogue for, and co-curator of, the inaugural show. In recent years Professor Hecht has published several papers on the Special Theory of Relativity, the history of ideas, and on foundational issues in physics. He spends most of his time studying physics, writing about art, and working towards his fourth degree black belt in *Tae Kwan Do*.

FREDERICK J. BUECHE (deceased), lately Distinguished Professor at Large, University of Dayton, received his Ph.D. in physics from Cornell University. He published nearly 100 research papers dealing with the physics of high polymers. In addition, he was the author of a graduate-level text in macromolecular physics. His major effort, however, was in physics teaching. In 1965 his general physics text, *Principles of Physics*, was published; it is now in its fifth edition and is widely used. Dr. Bueche was the author of five other introductory-level physics texts and numerous workbooks and study guides.

The McGraw-Hill Companies

Schaum's Outline of Theory and Problems of
COLLEGE PHYSICS

6 7 8 9 0 RHR 0

ISBN 0-07-144814-4

Contents

Preface

The introductory physics course, variously known as "general physics" or "college physics", is usually a two-semester in-depth survey of classical topics capped with some selected material from modern physics. Indeed, the name "college physics" has become a euphemism for introductory physics without calculus. *Schaum's Outline of College Physics* was designed to uniquely complement just such a course, whether given in high school or in college. The requisite mathematical knowledge includes basic algebra, some trigonometry, and a tiny bit of vector analysis, much of which can be learned as the experience progresses. It is assumed, however, that the reader already has a modest understanding of algebra. Appendix B is a general review of trigonometry that serves nicely. Even so, the necessary ideas are developed in place, as needed. And the same is true of the rudimentary vector analysis that's required—it too is taught as the situation requires.

In some ways learning physics is unlike learning most other disciplines. Physics has a special vocabulary that constitutes a language of its own, a language immediately transcribed into a symbolic form that is analyzed and extended with mathematical logic and precision. Words like energy, momentum, current, flux, interference, capacitance, and so forth, have very special scientific meanings. These must be learned promptly and accurately because the discipline builds layer upon layer; unless you know exactly what velocity is, you cannot know what acceleration or momentum are, and without them you cannot know what force is, and on and on. Each chapter in this book begins with a concise summary of the important ideas, definitions, relationships, laws, rules, and equations that are associated with the topic under discussion. All of this material constitutes the conceptual framework of the discourse, and its mastery is certainly challenging in and of itself, but there's more to physics than the mere recitation of its principles.

Every physicist who has ever tried to teach this marvelous subject has heard the universal student lament, "I understand everything; I just can't do the problems." Nonetheless most teachers believe that the "doing" of problems is the crucial culmination of the entire experience, it's the ultimate proof of understanding and competence. The conceptual machinery of definitions and rules and laws all come together in the process of problem solving as nowhere else. Moreover, insofar as the problems reflect the realities of our world, the student learns a skill of immense practical value. This is no easy task; carrying out the analysis of even a moderately complex problem requires extraordinary intellectual vigilance and unflagging attention to detail above and beyond just "knowing how to do it." Like playing a musical instrument, the student must learn the basics and then practice, practice, practice. A single missed note in a sonata is overlookable; a single error in a calculation, however, can propagate throughout the entire effort producing an answer that's completely wrong. Getting it right is what this book is all about.

In this new edition we have rearranged the first several chapters to bring them into harmony with the organization of today's introductory textbooks. In order to facilitate the learning process and increase student confidence we have added a number of easy single-concept problems. Moreover we have scrutinized every problem in the book, adding clarifying comments where needed, and extending and making more accessible the accompanying solutions wherever appropriate.

The level of difficulty of each problem is now specified by the designations [I], [II], or [III]. Single-concept straightforward problems are characterized by a [I] immediately following their number. A [II] signifies a somewhat more complicated but still manageable problem. Level [III] problems are designed to be challenging.

If you have any comments about this edition, suggestions for the next edition, or favorite problems you'd like to share, send them to E. Hecht, Adelphi University, Physics Department, Garden City, NY 11530.

Freeport, NY EUGENE HECHT

Chapter 1

Speed, Displacement, and Velocity: An Introduction to Vectors

A SCALAR QUANTITY, or **scalar**, is one that has nothing to do with spatial direction. Many physical concepts such as length, time, temperature, mass, density, charge, and volume are scalars; each has a scale or size, but no associated direction. The number of students in a class, the quantity of sugar in a jar, and the cost of a house are familiar scalar quantities.

Scalars are specified by ordinary numbers and add and subtract in the usual way. Two candies in one box plus seven in another give nine candies total.

DISTANCE (l): Get in a vehicle and travel a distance, some length in space, which we'll symbolize by the letter l. Suppose the tripmeter subsequently reads 100 miles (i.e., 161 kilometers); that's how far you went along whatever path you took, with no particular regard for hills or turns. Similarly, the bug in Fig. 1-1 walked a distance l measured along a winding route; l is also called the **path-length**, and it's a scalar quantity. (Incidentally, most people avoid using d for distance because it's widely used in the representation of derivatives.)

AVERAGE SPEED (v_{av}) is a measure of how fast a thing travels in space, and it too is a scalar quantity. Imagine an object that takes a time t to travel a distance l. The ***average speed*** during that interval is defined as

$$\textit{Average speed} = \frac{\text{total distance traveled}}{\text{time elapsed}}$$

$$v_{av} = \frac{l}{t}$$

The everyday units of speed are miles per hour, but in scientific work we use kilometers per hour (km/h) or, better yet, meters per second (m/s). As we'll learn presently, speed is part of the more inclusive concept of velocity and that's why we use the letter v. A problem may concern itself with the average speed of an object, but it can also treat the special case of a ***constant speed*** v, since then $v_{av} = v = l/t$ (see Problem 1.3).

You may also see this definition written as $v_{av} = \Delta l/\Delta t$, where the symbol Δ means 'the change in." That notation just underscores that we are dealing with intervals of time (Δt) and space (Δl). If we plot a curve of **distance versus time**, and look at any two points P_i and P_f on it, their separation in space (Δl) is the *rise*, and in time (Δt) is the *run*. Thus $\Delta l/\Delta t$ is the *slope* of the line drawn from the initial location, P_i, to the final location, P_f. *The slope is the average speed during that particular interval* (see Problem 1.5). Keep in mind that distance traveled, as indicated, for example, by an odometer in a car, is always positive and never decreases, consequently the graph of l versus t is always positive, and never decreases.

INSTANTANEOUS SPEED (v): Thus far we've defined "average speed," but we often want to know the speed of an object at a specific time, say, 10 s after 1:00. Similarly, we might ask for the speed of something NOW. That's a new concept called the ***instantaneous speed***, but we can define it building on the idea of average speed. What we need is the average speed determined over a vanishingly tiny time interval centered on the desired instant. Formally, that's stated as

$$v = \lim_{\Delta t \to 0} \left[\frac{\Delta l}{\Delta t} \right]$$

Instantaneous speed (or just speed, for short) is the limiting value of the average speed ($\Delta l/\Delta t$) determined as the interval over which the averaging takes place (Δt) approaches zero. This mathematical expression becomes especially important because it leads to the calculus and the idea of the derivative. To keep the math simple we won't worry about the details, for us it's just the general concept that should be understood. In the next chapter we'll develop equations for the instantaneous speed of an object at any specific time.

Graphically, the slope of a line tangent to the distance versus time curve at any point (i.e., at any particular time) is the instantaneous speed at that time.

A VECTOR QUANTITY is a physical concept that is inherently directional and can be specified completely only if both its **magnitude** (i.e., size) and direction are provided. Many physical concepts such as displacement, velocity, acceleration, force, and momentum are vector quantities. In general, a *vector* (which stands for a specific amount of some vector quantity) is depicted as a directed line segment, and pictorially represented by an arrow (drawn to scale) whose magnitude and direction determine the vector. In printed material vectors are usually symbolically presented in boldface type (e.g. **F** for force). When written by hand it's common to distinguish a vector by just putting an arrow over the appropriate symbol (e.g., $\vec{F}$). For the sake of maximum clarity we'll combine the two and use $\vec{\mathbf{F}}$.

THE DISPLACEMENT of an object from one location to another is a vector quantity. As shown in Fig. 1-1 the displacement of the bug in going from P_1 to point P_2 is specified by the vector $\vec{\mathbf{s}}$ (the symbol *s* comes from the century-old usage corresponding to the "space" between two points). If the straight-line distance from P_1 to P_2 is, say, 2.0 m, we simply draw $\vec{\mathbf{s}}$ to be any convenient length and label it 2.0 m. In any case, $\vec{\mathbf{s}} = 2.0\,\text{m}$ — 10° NORTH OF EAST.

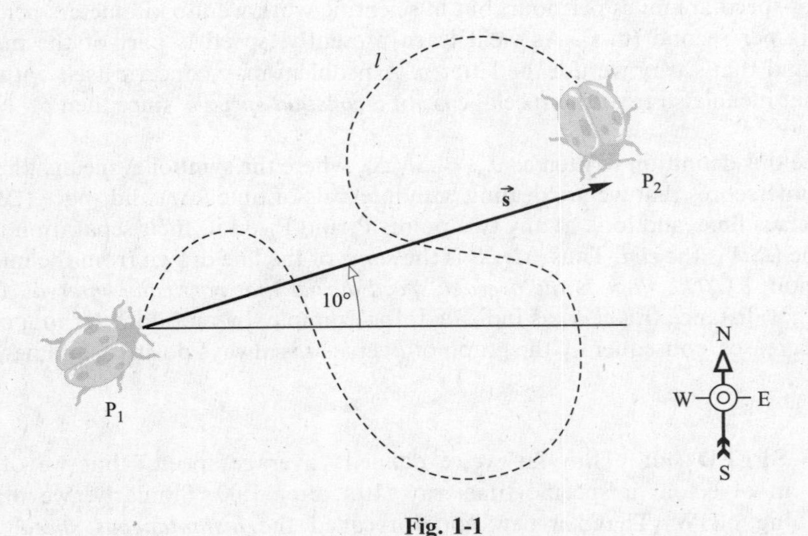

Fig. 1-1

VELOCITY is a vector quantity that embraces both the speed and the direction of motion. If an object undergoes a vector displacement $\vec{s}$ in a time interval t, then

$$\textit{Average velocity} = \frac{\text{vector displacement}}{\text{time taken}}$$

$$\vec{v}_{av} = \frac{\vec{s}}{t}$$

The direction of the velocity vector is the same as that of the displacement vector. The units of velocity (and speed) are those of distance divided by time, such as m/s or km/h.

INSTANTANEOUS VELOCITY is the average velocity evaluated for a time interval that approaches zero. Thus, if an object undergoes a displacement $\Delta\vec{s}$ in a time Δt, then for that object the instantaneous velocity is

$$\vec{v} = \lim_{\Delta t \to 0} \frac{\Delta\vec{s}}{\Delta t}$$

where the notation means that the ratio $\Delta\vec{s}/\Delta t$ is to be evaluated for a time interval Δt that approaches zero.

THE ADDITION OF VECTORS: The concept of "vector" is not completely defined until we establish some rules of behavior. For example, how do several vectors (displacements, forces, whatever) add to one another? The bug in Fig. 1-2 walks from P_1 to P_2, pauses and then goes on to P_3. It experiences two displacements $\vec{s}_1$ and $\vec{s}_2$ which combine to yield a net displacement $\vec{s}$. Here $\vec{s}$ is called the ***resultant*** or sum of the two constituent displacements and it is the physical equivalent of them taken together $\vec{s} = \vec{s}_1 + \vec{s}_2$.

THE TIP-TO-TAIL (OR POLYGON) METHOD: The two vectors in Fig. 1-2 show us how to graphically add two (or more) vectors. Simply place the tail of the second ($\vec{s}_2$) at the tip of the first ($\vec{s}_1$); the resultant then goes from the starting point, P_1 (the tail of $\vec{s}_1$), to the final point, P_2

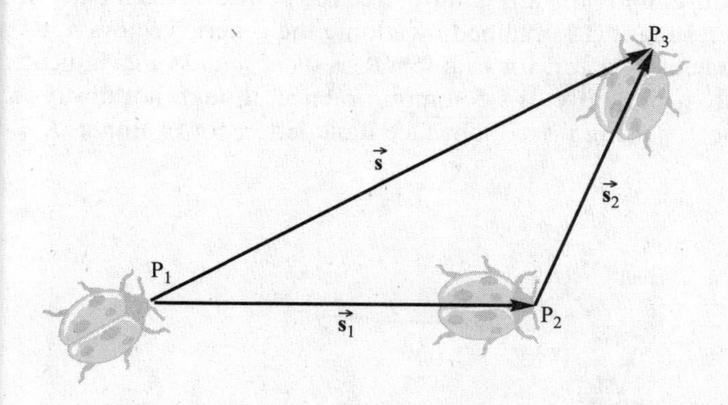

Fig. 1-2

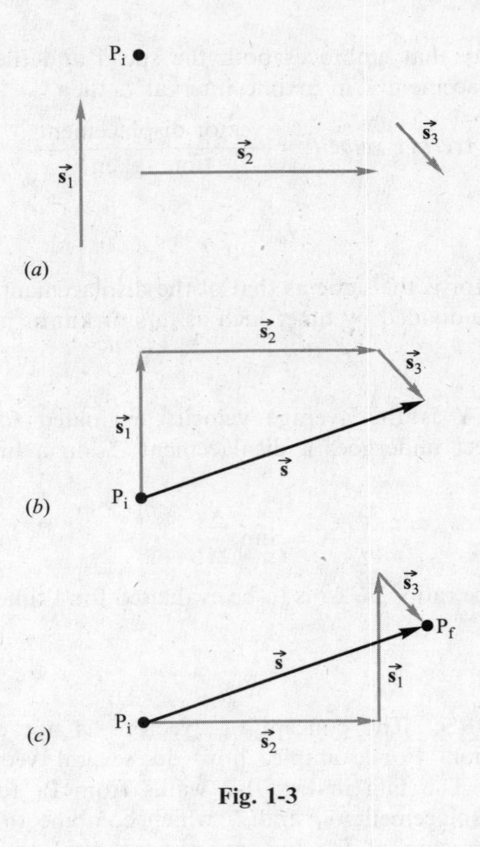

Fig. 1-3

(the tip of $\vec{s}_2$). Figure 1-3(*a*) is more general; it shows an initial starting point P_i and three displacement vectors. If we tip-to-tail those three displacements *in any order* [Fig. 1-3(*b*) and (*c*)] we'll arrive at the same final point P_f, and the same resultant $\vec{s}$. In other words

$$\vec{s} = \vec{s}_1 + \vec{s}_2 + \vec{s}_3 = \vec{s}_2 + \vec{s}_1 + \vec{s}_3 \text{ etc.}$$

As long as the bug starts at P_i and walks the three displacements, in any sequence, it will end up at P_f.

The same tip-to-tail procedure holds for any kind of vector, be it displacement, velocity, force, or anything else. Accordingly, the resultant ($\vec{\mathbf{R}}$) obtained by adding the generic vectors $\vec{\mathbf{A}}$, $\vec{\mathbf{B}}$, and $\vec{\mathbf{C}}$ is shown in Fig. 1-4. The size or **magnitude** of a vector, for example $\vec{\mathbf{R}}$, is its *absolute value* indicated symbolically as $|\vec{\mathbf{R}}|$; we'll see how to calculate it presently. It's common practice, though not always a good idea, to represent the magnitude of a vector using just a light face italic letter, for example, $R = |\vec{\mathbf{R}}|$.

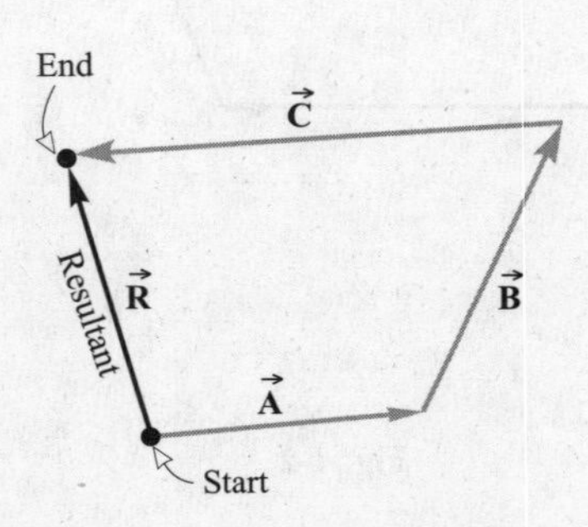

Fig. 1-4

PARALLELOGRAM METHOD for adding two vectors: The resultant of two vectors acting at any angle may be represented by the diagonal of a parallelogram. The two vectors are drawn as the sides of the parallelogram and the resultant is its diagonal, as shown in Fig. 1-5. The direction of the resultant is away from the origin of the two vectors.

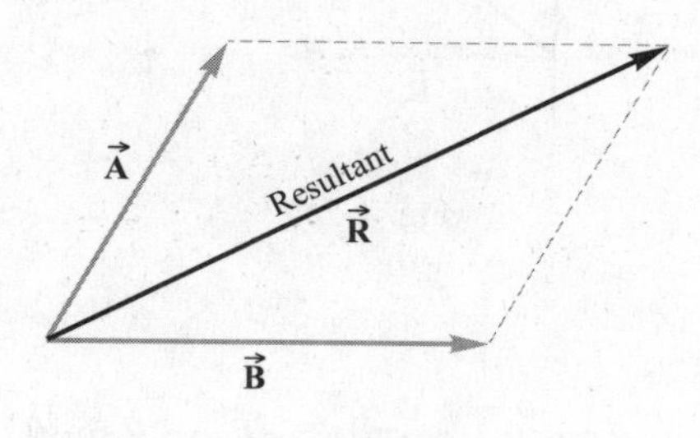

Fig. 1-5

SUBTRACTION OF VECTORS: To subtract a vector $\vec{\mathbf{B}}$ from a vector $\vec{\mathbf{A}}$, reverse the direction of $\vec{\mathbf{B}}$ and add individually to vector $\vec{\mathbf{A}}$, that is, $\vec{\mathbf{A}} - \vec{\mathbf{B}} = \vec{\mathbf{A}} + (-\vec{\mathbf{B}})$.

THE TRIGONOMETRIC FUNCTIONS are defined in relation to a right angle. For the right triangle shown in Fig. 1-6, by definition

$$\sin\theta = \frac{\text{opposite}}{\text{hypotenuse}} = \frac{B}{C}, \quad \cos\theta = \frac{\text{adjacent}}{\text{hypotenuse}} = \frac{A}{C}, \quad \tan\theta = \frac{\text{opposite}}{\text{adjacent}} = \frac{B}{A}$$

We often use these in the forms

$$B = C\sin\theta \qquad A = C\cos\theta \qquad B = A\tan\theta$$

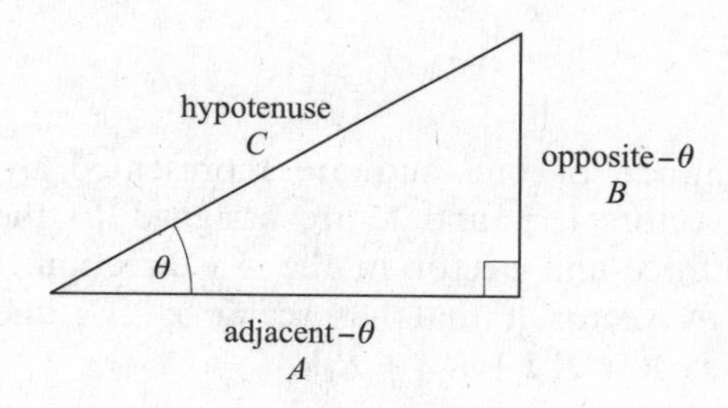

Fig. 1-6

A COMPONENT OF A VECTOR is its effective value in a given direction. For example, the x-component of a displacement is the displacement parallel to the x-axis caused by the given displacement. A vector in three dimensions may be considered as the resultant of its component vectors resolved along any three *mutually perpendicular* directions. Similarly, a vector in two dimensions may be resolved into two component vectors acting along any two mutually perpendicular directions. Figure 1-7 shows the vector $\vec{\mathbf{R}}$ and its x and y vector components, $\vec{\mathbf{R}}_x$ and $\vec{\mathbf{R}}_y$, which have magnitudes

$$|\vec{\mathbf{R}}_x| = |\vec{\mathbf{R}}|\cos\theta \qquad \text{and} \qquad |\vec{\mathbf{R}}_y| = |\vec{\mathbf{R}}|\sin\theta$$

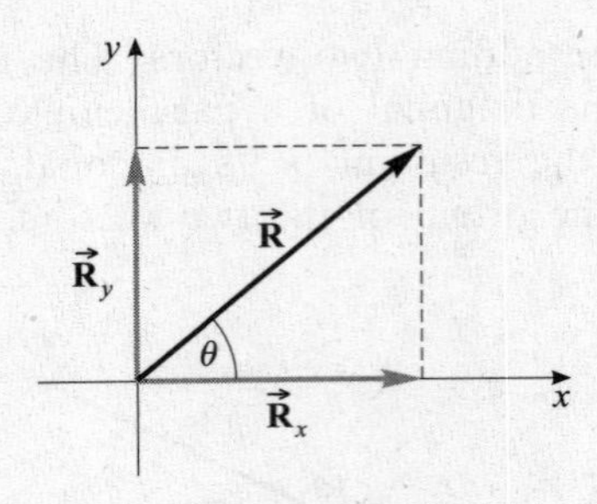

Fig. 1-7

or equivalently

$$R_x = R\cos\theta \qquad \text{and} \qquad R_y = R\sin\theta$$

COMPONENT METHOD FOR ADDING VECTORS: Each vector is resolved into its x-, y-, and z-components, with negatively directed components taken as negative. The scalar x-component R_x of the resultant $\vec{\mathbf{R}}$ is the algebraic sum of all the scalar x-components. The scalar y- and z-components of the resultant are found in a similar way. With the components known, the magnitude of the resultant is given by

$$R = \sqrt{R_x^2 + R_y^2 + R_z^2}$$

In two dimensions, the angle of the resultant with the x-axis can be found from the relation

$$\tan\theta = \frac{R_y}{R_x}$$

UNIT VECTORS have a magnitude of one and are represented by a boldface symbol topped with a caret. The special unit vectors $\hat{\mathbf{i}}$, $\hat{\mathbf{j}}$, and $\hat{\mathbf{k}}$ are assigned to the x-, y-, and z-axes, respectively. A vector $3\hat{\mathbf{i}}$ represents a three-unit vector in the $+x$-direction, while $-5\hat{\mathbf{k}}$ represents a five-unit vector in the $-z$-direction. A vector $\vec{\mathbf{R}}$ that has scalar x-, y-, and z-components R_x, R_y, and R_z, respectively, can be written as $\vec{\mathbf{R}} = R_x\hat{\mathbf{i}} + R_y\hat{\mathbf{j}} + R_z\hat{\mathbf{k}}$.

Solved Problems

1.1 [I] A toy train moves along a winding track at an average speed of 0.25 m/s. How far will it travel in 4.00 minutes?

The defining equation is $v_{av} = l/t$. Here l is in meters, and t is in seconds, so the first thing to do is convert 4.00 min into seconds: (4.00 min)(60.0 s/min) = 240 s. Solving the equation for l,

$$l = v_{av}t = (0.25\,\text{m/s})(240\,\text{s})$$

Since the speed has only two significant figures, $l = 60\,\text{m}$.

1.2 [I] A student driving a car travels 10.0 km in 30.0 min. What was her average speed?

The defining equation is $v_{av} = l/t$. Here l is in kilometers, and t is in minutes, so the first thing to do is convert 10.0 km to meters and then 30.0 min into seconds: (10.0 km)(1000 m/km) = 10.0×10^3 m and (30.0 min)(60.0 s/min) = 1800 s. We need to solve for v_{av}, giving the numerical answer to three significant figures:

$$v_{av} = \frac{l}{t} = \frac{10.0 \times 10^3 \text{ m}}{1800 \text{ s}} = 5.56 \text{ m/s}$$

1.3 [I] Rolling along across the machine shop at a constant speed of 4.25 m/s, a robot covers a distance of 17.0 m. How long did that journey take?

Since the speed is constant the defining equation is $v = l/t$. Multiply both sides of this expression by t and then divide both by v:

$$t = \frac{l}{v} = \frac{170.0 \text{ m}}{4.25 \text{ m/s}} = 4.00 \text{ s}$$

1.4 [I] Change the speed 0.200 cm/s to units of kilometers per year.

$$0.200 \frac{\text{cm}}{\text{s}} = \left(0.200 \frac{\cancel{\text{cm}}}{\cancel{\text{s}}}\right)\left(10^{-5} \frac{\text{km}}{\cancel{\text{cm}}}\right)\left(3600 \frac{\cancel{\text{s}}}{\cancel{\text{h}}}\right)\left(24 \frac{\cancel{\text{h}}}{\cancel{\text{d}}}\right)\left(365 \frac{\cancel{\text{d}}}{\text{y}}\right) = 63.1 \frac{\text{km}}{\text{y}}$$

1.5 [I] A car travels along a road and its odometer readings are plotted against time in Fig. 1-8. Find the instantaneous speed of the car at points A and B. What is the car's average speed?

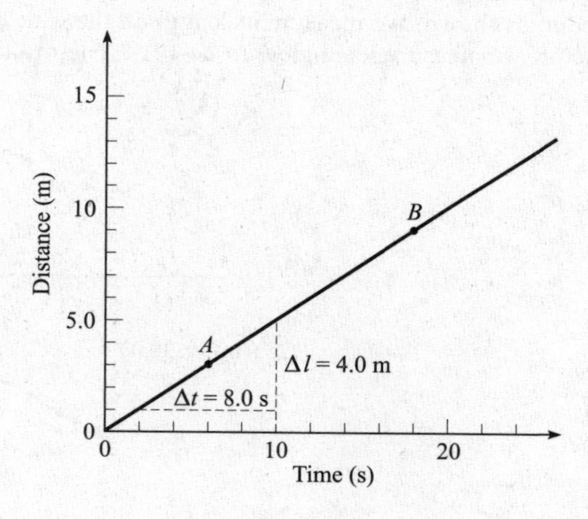

Fig. 1-8

Because the speed is given by the slope $\Delta l/\Delta t$ of the tangent line, we take a tangent to the curve at point A. The tangent line is the curve itself in this case. For the triangle shown at A, we have

$$\frac{\Delta l}{\Delta t} = \frac{4.0 \text{ m}}{8.0 \text{ s}} = 0.50 \text{ m/s}$$

This is also the speed at point B and at every other point on the straight-line graph. It follows that $v = 0.50 \text{ m/s} = v_{av}$.

1.6 [I] A kid stands 6.00 m from the base of a flagpole which is 8.00 m tall. Determine the magnitude of the displacement of the brass eagle on top of the pole with respect to the youngster's feet.

The geometry corresponds to a 3-4-5 right triangle (i.e,, $3 \times 2 - 4 \times 2 - 5 \times 2$). Thus the hypotenuse, which is the 5-side, must be 10.0 m long, and that's the magnitude of the displacement.

1.7 [II] A runner makes one lap around a 200-m track in a time of 25 s. What were the runner's (*a*) average speed and (*b*) average velocity?

(*a*) From the definition,

$$\text{Average speed} = \frac{\text{distance traveled}}{\text{time taken}} = \frac{200 \text{ m}}{25 \text{ s}} = 8.0 \text{ m/s}$$

(*b*) Because the run ended at the starting point, the displacement vector from starting pont to end point has zero length. Since $\vec{v}_{av} = \vec{s}/t$,

$$|\vec{v}_{av}| = \frac{0 \text{ m}}{25 \text{ s}} = 0 \text{ m/s}$$

1.8 [I] Using the graphical method, find the resultant of the following two displacements: 2.0 m at 40° and 4.0 m at 127°, the angles being taken relative to the $+x$-axis, as is customary. Give your answer to two significant figures. (See Appendix A on significant figures.)

Choose x- and y-axes as shown in Fig. 1-9 and lay out the displacements to scale, tip to tail from the origin. Notice that all angles are measured from the $+x$-axis. The resultant vector $\vec{s}$ points from starting point to end point as shown. We measure its length on the scale diagram to find its magnitude, 4.6 m. Using a protractor, we measure its angle θ to be 101°. The resultant displacement is therefore 4.6 m at 101°.

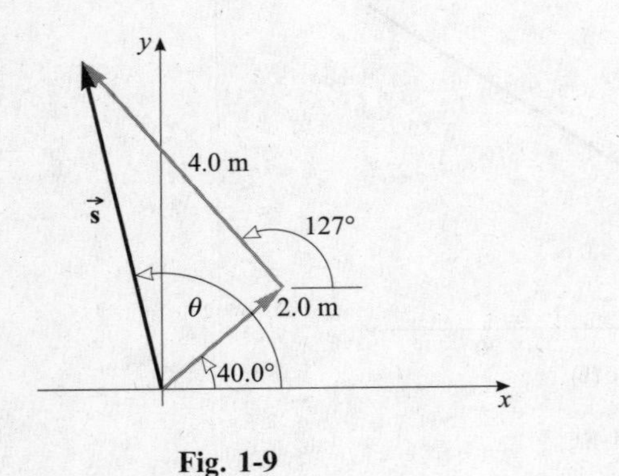

Fig. 1-9

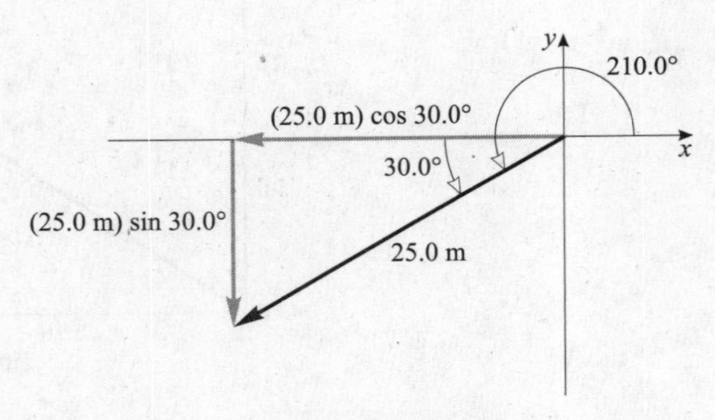

Fig. 1-10

1.9 [I] Find the x- and y-components of a 25.0-m displacement at an angle of 210.0°.

The vector displacement and its components are shown in Fig. 1-10. The scalar components are

$$x\text{-component} = -(25.0 \text{ m}) \cos 30.0° = -21.7 \text{ m}$$
$$y\text{-component} = -(25.0 \text{ m}) \sin 30.0° = -12.5 \text{ m}$$

Notice in particular that each component points in the negative coordinate direction and must therefore be taken as negative.

1.10 [II] Solve Problem 1.8 by use of rectangular components.

We resolve each vector into rectangular components as shown in Fig. 1-11(*a*) and (*b*). (Place a cross-hatch symbol on the original vector to show that it is replaced by its components.) The resultant has scalar components of

$$s_x = 1.53 \text{ m} - 2.41 \text{ m} = -0.88 \text{ m} \qquad s_y = 1.29 \text{ m} + 3.19 \text{ m} = 4.48 \text{ m}$$

Notice that components pointing in the negative direction must be assigned a negative value.

The resultant is shown in Fig. 1.11(*c*); there, we see that

$$s = \sqrt{(0.88 \text{ m})^2 + (4.48 \text{ m})^2} = 4.6 \text{ m} \qquad \tan\phi = \frac{4.48 \text{ m}}{0.88 \text{ m}}$$

and $\phi = 79°$, from which $\theta = 180° - \phi = 101°$. Hence $\vec{s} = 4.6$ m—101° FROM +*X*-AXIS; remember vectors must have their directions stated explicitly.

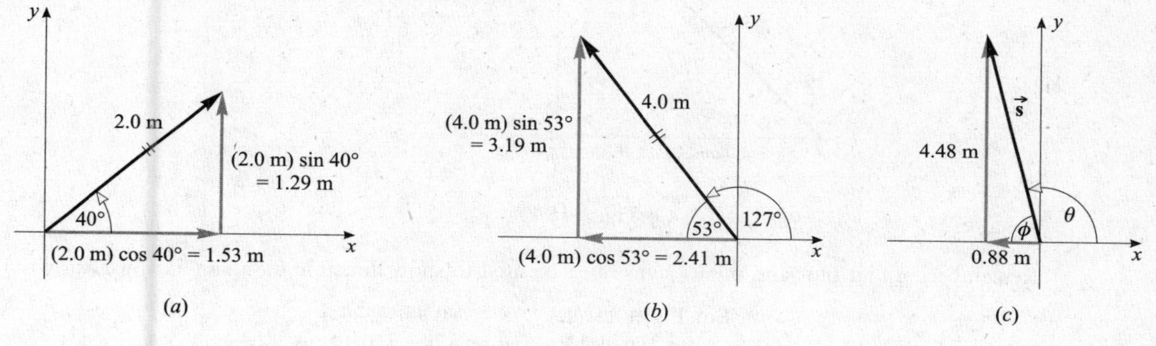

Fig. 1-11

1.11 [II] Add the following two displacement vectors using the parallelogram method: 30 m at 30° and 20 m at 140°. Remember that numbers like 30 m and 20 m have two significant figures.

The vectors are drawn with a common origin in Fig. 1-12(*a*). We construct a parallelogram using them as sides, as shown in Fig. 1-12(*b*). The resultant $\vec{s}$ is then represented by the diagonal. By measurement, we find that $\vec{s}$ is 30 m at 72°.

Fig. 1-12

1.12 [II] Express the vectors shown in Figs. 1-11(*c*), 1-13, 1-14, and 1-15 in the form $\vec{\mathbf{R}} = R_x\hat{\mathbf{i}} + R_y\hat{\mathbf{j}} + R_z\hat{\mathbf{k}}$ (leave out the units).

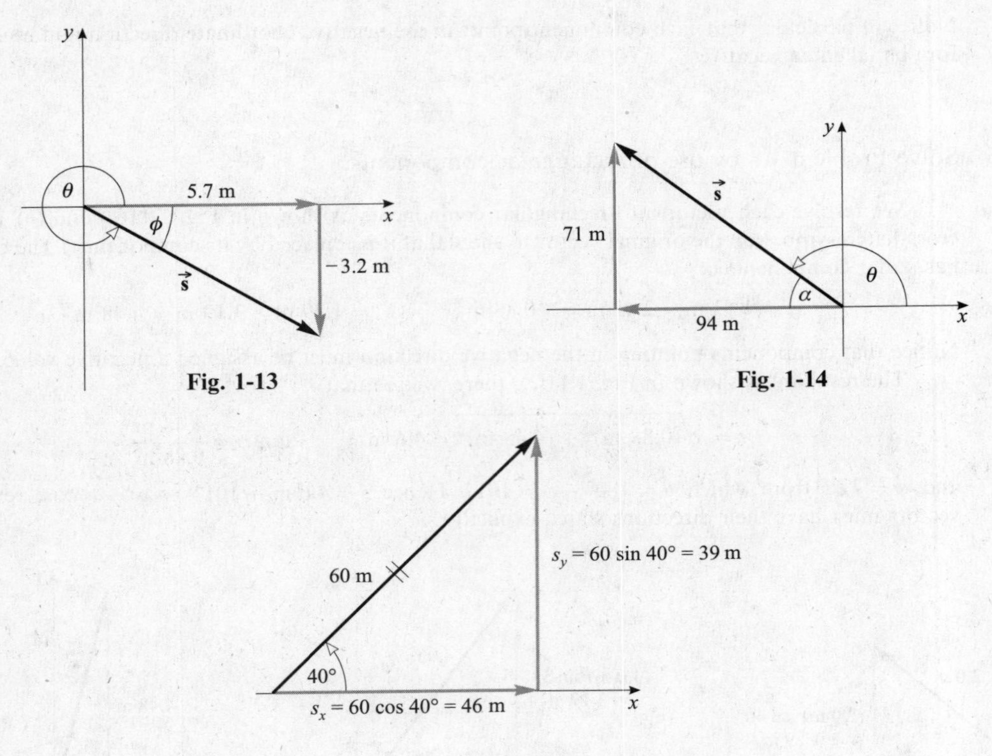

Fig. 1-13

Fig. 1-14

Fig. 1-15

Remembering that plus and minus signs must be used to show direction along an axis, we can write

For Fig. 1-11(*c*): $\vec{\mathbf{R}} = -0.88\hat{\mathbf{i}} + 4.48\hat{\mathbf{j}}$
For Fig. 1-13: $\vec{\mathbf{R}} = 5.7\hat{\mathbf{i}} - 3.2\hat{\mathbf{j}}$
For Fig. 1-14: $\vec{\mathbf{R}} = -94\hat{\mathbf{i}} + 71\hat{\mathbf{j}}$
For Fig. 1-15: $\vec{\mathbf{R}} = 46\hat{\mathbf{i}} + 39\hat{\mathbf{j}}$

1.13 [I] Perform graphically the following vector additions and subtractions, where $\vec{\mathbf{A}}$, $\vec{\mathbf{B}}$, and $\vec{\mathbf{C}}$ are the vectors shown in Fig. 1-16: (*a*) $\vec{\mathbf{A}} + \vec{\mathbf{B}}$; (*b*) $\vec{\mathbf{A}} + \vec{\mathbf{B}} + \vec{\mathbf{C}}$; (*c*) $\vec{\mathbf{A}} - \vec{\mathbf{B}}$; (*d*) $\vec{\mathbf{A}} + \vec{\mathbf{B}} - \vec{\mathbf{C}}$.

See Fig. 1-16(*a*) through (*d*). In (*c*), $\vec{\mathbf{A}} - \vec{\mathbf{B}} = \vec{\mathbf{A}} + (-\vec{\mathbf{B}})$; that is, to subtract $\vec{\mathbf{B}}$ from $\vec{\mathbf{A}}$, reverse the direction of $\vec{\mathbf{B}}$ and add it vectorially to $\vec{\mathbf{A}}$. Similarly, in (*d*), $\vec{\mathbf{A}} + \vec{\mathbf{B}} - \vec{\mathbf{C}} = \vec{\mathbf{A}} + \vec{\mathbf{B}} + (-\vec{\mathbf{C}})$, where $-\vec{\mathbf{C}}$ is equal in magnitude but opposite in direction to $\vec{\mathbf{C}}$.

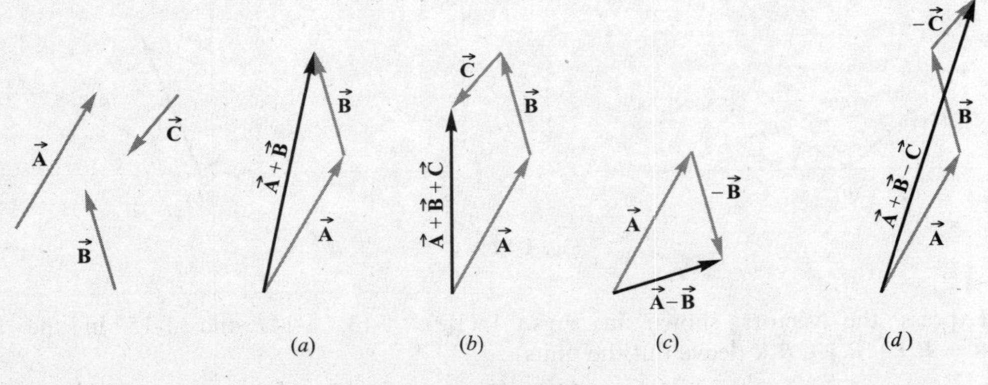

Fig. 1-16

1.14 [II] If $\vec{\mathbf{A}} = -12\hat{\mathbf{i}} + 25\hat{\mathbf{j}} + 13\hat{\mathbf{k}}$ and $\vec{\mathbf{B}} = -3\hat{\mathbf{j}} + 7\hat{\mathbf{k}}$, find the resultant when $\vec{\mathbf{A}}$ is subtracted from $\vec{\mathbf{B}}$.

From a purely mathematical approach, we have

$$\vec{\mathbf{B}} - \vec{\mathbf{A}} = (-3\hat{\mathbf{j}} + 7\hat{\mathbf{k}}) - (-12\hat{\mathbf{i}} + 25\hat{\mathbf{j}} + 13\hat{\mathbf{k}})$$
$$= -3\hat{\mathbf{j}} + 7\hat{\mathbf{k}} + 12\hat{\mathbf{i}} - 25\hat{\mathbf{j}} - 13\hat{\mathbf{k}} = 12\hat{\mathbf{i}} - 28\hat{\mathbf{j}} - 6\hat{\mathbf{k}}$$

Notice that $12\hat{\mathbf{i}} - 25\hat{\mathbf{j}} - 13\hat{\mathbf{k}}$ is simply $\vec{\mathbf{A}}$ reversed in direction. Therefore we have, in essence, reversed $\vec{\mathbf{A}}$ and added it to $\vec{\mathbf{B}}$.

1.15 [II] A boat can travel at a speed of 8 km/h in still water on a lake. In the flowing water of a stream, it can move at 8 km/h relative to the water in the stream. If the stream speed is 3 km/h, how fast can the boat move past a tree on the shore when it is traveling (*a*) upstream and (*b*) downstream?

(*a*) If the water was standing still, the boat's speed past the tree would be 8 km/h. But the stream is carrying it in the opposite direction at 3 km/h. Therefore the boat's speed relative to the tree is 8 km/h − 3 km/h = 5 km/h.

(*b*) In this case, the stream is carrying the boat in the same direction the boat is trying to move. Hence its speed past the tree is 8 km/h + 3 km/h = 11 km/h.

1.16 [III] A plane is traveling eastward at an airspeed of 500 km/h. But a 90 km/h wind is blowing southward. What are the direction and speed of the plane relative to the ground?

The plane's resultant velocity with respect to the ground, $\vec{v}_{PG}$, is the sum of two vectors, the velocity of the plane with respect to the air, $\vec{v}_{PA}$ = 500 km/h—EAST and the velocity of the air with respect to the ground, $\vec{v}_{AG}$ = 90 km/h—SOUTH. In other words, $\vec{v}_{PG} = \vec{v}_{PA} + \vec{v}_{AG}$. These component velocities are shown in Fig. 1-17. The plane's resultant speed is then

$$v_{PG} = \sqrt{(500\ \text{km/h})^2 + (90\ \text{km/h})^2} = 508\ \text{km/h}$$

The angle α is given by

$$\tan\alpha = \frac{90\ \text{km/h}}{500\ \text{km/h}} = 0.18$$

from which $\alpha = 10°$. The plane's velocity relative to the ground is 508 km/h at 10° south of east.

1.17 [III] With the same airspeed as in Problem 1.16, in what direction must the plane head in order to move due east relative to the Earth?

The sum of the plane's velocity through the air and the velocity of the wind will be the resultant velocity of the plane relative to the Earth. This is shown in the vector diagram in Fig. 1-18. Notice that, as required, the resultant velocity is eastward. Keeping in mind that the wind speed is given to two significant figures, it is seen that $\sin\theta = (90\ \text{km/h})(500\ \text{km/h})$, from which $\theta = 10°$. The plane should head 10° north of east if it is to move eastward relative to the Earth.

To find the plane's eastward speed, we note in the figure that $v_{PG} = (500\ \text{km/h})\cos\theta = 4.9 \times 10^5$ m/h.

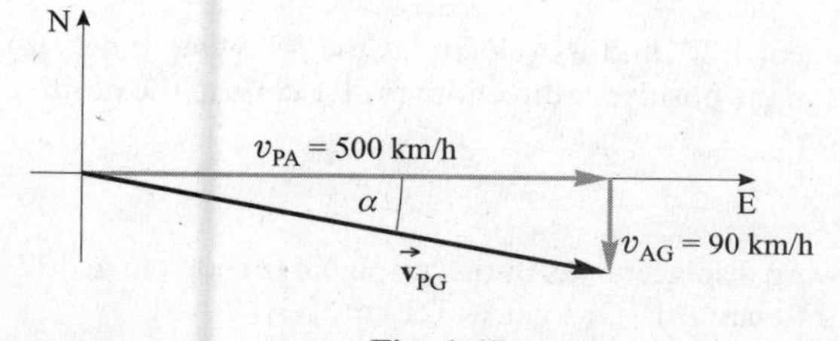

Fig. 1-17

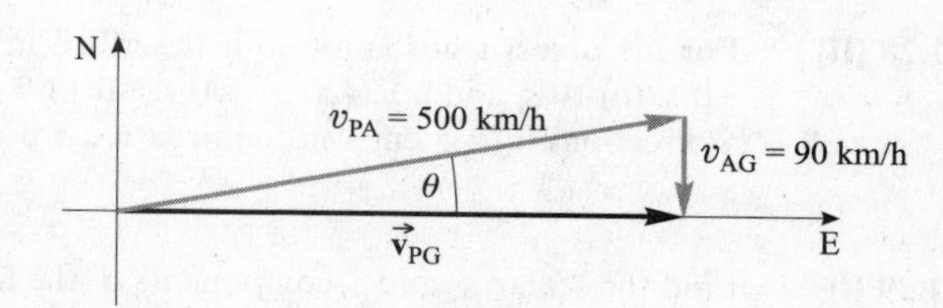

Fig. 1-18

Supplementary Problems

1.18 [I] Three kids in a parking lot launch a rocket that rises into the air along a 380-m long arc in 40 s. Determine its average speed. *Ans.* 9.5 m/s

1.19 [I] According to its computer, a robot that left its closet and traveled 1200 m, had an average speed of 20.0 m/s. How long did the trip take? *Ans.* 60.0 s

1.20 [I] A car's odometer reads 22 687 km at the start of a trip and 22 791 km at the end. The trip took 4.0 hours. What was the car's average speed in km/h and in m/s? *Ans.* 26 km/h, 7.2 m/s

1.21 [I] An auto travels at the rate of 25 km/h for 4.0 minutes, then at 50 km/h for 8.0 minutes, and finally at 20 km/h for 2.0 minutes. Find (*a*) the total distance covered in km and (*b*) the average speed for the complete trip in m/s. *Ans.* (*a*) 9.0 km; (*b*) 10.7 m/s or 11 m/s

1.22 [I] Starting from the center of town, a car travels east for 80.0 km and then turns due south for another 192 km, at which point it runs out of gas. Determine the displacement of the stopped car from the center of town. *Ans.* 208 km—67.4° SOUTH OF EAST

1.23 [II] A little turtle is placed at the origin of an xy-grid drawn on a large sheet of paper. Each grid box is 1.0 cm by 1.0 cm. The turtle walks around for a while and finally ends up at point (24, 10), that is, 24 boxes along the x-axis, and 10 boxes along the y-axis. Determine the displacement of the turtle from the origin at the point. *Ans.* 26 cm—23° ABOVE X-AXIS

1.24 [II] A bug starts at point A, crawls 8.0 cm east, then 5.0 cm south, 3.0 cm west, and 4.0 cm north to point B. (*a*) How far north and east is B from A? (*b*) Find the displacement from A to B both graphically and algebraically. *Ans.* (*a*) 1.0 cm—NORTH, 5.0 cm—EAST; (*b*) 5.10 cm—11.3° SOUTH OF EAST

1.25 [II] A runner travels 1.5 laps around a circular track in a time of 50 s. The diameter of the track is 40 m and its circumference is 126 m. Find (*a*) the average speed of the runner and (*b*) the magnitude of the runner's average velocity. Be careful here; average speed depends on the total distance traveled, whereas average velocity depends on the displacement at the end of the particular journey.
Ans. (*a*) 3.8 m/s; (*b*) 0.80 m/s

1.26 [II] During a race on an oval track, a car travels at an average speed of 200 km/h. (*a*) How far did it travel in 45.0 min? (*b*) Determine its average velocity at the end of its third lap. *Ans.* (*a*) 150 km; (*b*) zero

1.27 [II] The following data describe the position of an object along the x-axis as a function of time. Plot the data, and find the instantaneous velocity of the object at (*a*) $t = 5.0$ s, (*b*) 16 s, and (*c*) 23 s. *Ans.* (*a*) 0.018 m/s in the positive x-direction; (*b*) 0 m/s; (*c*) 0.013 m/s in the negative x-direction

t(s)	0	2	4	6	8	10	12	14	16	18	20	22	24	26	28
x(cm)	0	4.0	7.8	11.3	14.3	16.8	18.6	19.7	20.0	19.5	18.2	16.2	13.5	10.3	6.7

1.28 [III] For the object whose motion is described in Problem 1.27, find its velocity at the following times: (*a*) 3.0 s, (*b*) 10 s, and (*c*) 24 s. *Ans.* (*a*) 1.9 cm/s in the positive x-direction; (*b*) 1.1 cm/s in the positive x-direction; (*c*) 1.5 cm/s in the negative x-direction

1.29 [I] Find the scalar x- and y-components of the following displacements in the xy-plane: (*a*) 300 cm at 127° and (*b*) 500 cm at 220°. *Ans.* (*a*) −180 cm, 240 cm; (*b*) −383 cm, −321 cm

1.30 [II] Starting at the origin of coordinates, the following displacements are made in the xy-plane (that is, the displacements are coplanar): 60 mm in the $+y$-direction, 30 mm in the $-x$-direction, 40 mm at 150°, and 50 mm at 240°. Find the resultant displacement both graphically and algebraically. *Ans.* 97 mm at 158°

1.31 [II] Compute algebraically the resultant of the following coplanar displacements: 20.0 m at 30.0°, 40.0 m at 120.0°, 25.0 m at 180.0°, 42.0 m at 270.0°, and 12.0 m at 315.0°. Check your answer with a graphical solution. *Ans.* 20.1 m at 197°

1.32 [II] What displacement at 70° has an x-component of 450 m? What is its y-component? *Ans.* 1.3 km, 1.2 km

1.33 [II] What displacement must be added to a 50 cm displacement in the $+x$-direction to give a resultant displacement of 85 cm at 25°? *Ans.* 45 cm at 53°

1.34 [I] Refer to Fig. 1-19. In terms of vectors $\vec{\mathbf{A}}$ and $\vec{\mathbf{B}}$, express the vectors (*a*) $\vec{\mathbf{P}}$, (*b*) $\vec{\mathbf{R}}$, (*c*) $\vec{\mathbf{S}}$, and (*d*) $\vec{\mathbf{Q}}$. *Ans.* (*a*) $\vec{\mathbf{A}} + \vec{\mathbf{B}}$; (*b*) $\vec{\mathbf{B}}$; (*c*) $-\vec{\mathbf{A}}$; (*d*) $\vec{\mathbf{A}} - \vec{\mathbf{B}}$

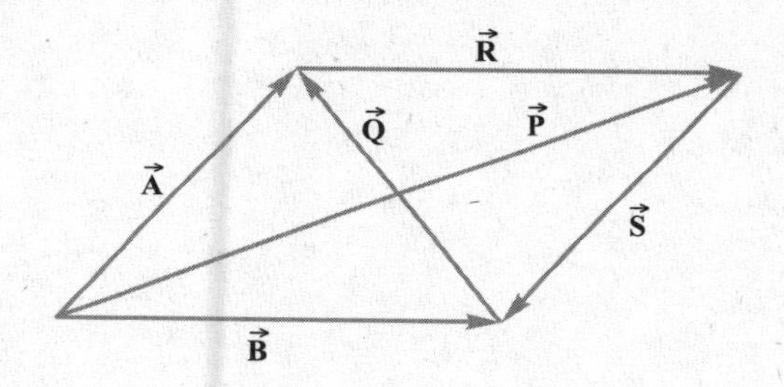

Fig. 1-19

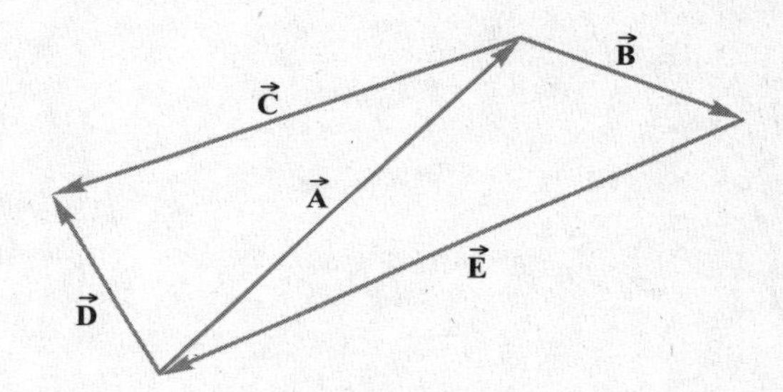

Fig. 1-20

1.35 [I] Refer to Fig. 1-20. In terms of vectors $\vec{\mathbf{A}}$ and $\vec{\mathbf{B}}$, express the vectors (*a*) $\vec{\mathbf{E}}$, (*b*) $\vec{\mathbf{D}} - \vec{\mathbf{C}}$, and (*c*) $\vec{\mathbf{E}} + \vec{\mathbf{D}} - \vec{\mathbf{C}}$. *Ans.* (*a*) $-\vec{\mathbf{A}} - \vec{\mathbf{B}}$ or $-(\vec{\mathbf{A}} + \vec{\mathbf{B}})$; (*b*) $\vec{\mathbf{A}}$; (*c*) $-\vec{\mathbf{B}}$

1.36 [II] Find (*a*) $\vec{\mathbf{A}} + \vec{\mathbf{B}} + \vec{\mathbf{C}}$, (*b*) $\vec{\mathbf{A}} - \vec{\mathbf{B}}$, and (*c*) $\vec{\mathbf{A}} - \vec{\mathbf{C}}$ if $\vec{\mathbf{A}} = 7\hat{\mathbf{i}} - 6\hat{\mathbf{j}}$, $\vec{\mathbf{B}} = -3\hat{\mathbf{i}} + 12\hat{\mathbf{j}}$, and $\vec{\mathbf{C}} = 4\hat{\mathbf{i}} - 4\hat{\mathbf{j}}$. *Ans.* (*a*) $8\hat{\mathbf{i}} + 2\hat{\mathbf{j}}$; (*b*) $10\hat{\mathbf{i}} - 18\hat{\mathbf{j}}$; (*c*) $3\hat{\mathbf{i}} - 2\hat{\mathbf{j}}$

1.37 [II] Find the magnitude and angle of $\vec{\mathbf{R}}$ if $\vec{\mathbf{R}} = 7.0\hat{\mathbf{i}} - 12\hat{\mathbf{j}}$. *Ans.* 14 at −60°

1.38 [II] Determine the displacement vector that must be added to the displacement $(25\hat{\mathbf{i}} - 16\hat{\mathbf{j}})$ m to give a displacement of 7.0 m pointing in the $+x$-direction? *Ans.* $(-18\hat{\mathbf{i}} + 16\hat{\mathbf{j}})$ m

1.39 [II] A vector $(15\hat{\mathbf{i}} - 16\hat{\mathbf{j}} + 27\hat{\mathbf{k}})$ is added to a vector $(23\hat{\mathbf{j}} - 40\hat{\mathbf{k}})$. What is the magnitude of the resultant? *Ans.* 21

1.40 [III] A truck is moving north at a speed of 70 km/h. The exhaust pipe above the truck cab sends out a trail of smoke that makes an angle of 20° east of south behind the truck. If the wind is blowing directly toward the east, what is the wind speed at that location? *Ans.* 25 km/h

1.41 [III] A ship is traveling due east at 10 km/h. What must be the speed of a second ship heading 30° east of north if it is always due north of the first ship? *Ans.* 20 km/h

1.42 [III] A boat, propelled so as to travel with a speed of 0.50 m/s in still water, moves directly across a river that is 60 m wide. The river flows with a speed of 0.30 m/s. (*a*) At what angle, relative to the straight-across direction, must the boat be pointed? (*b*) How long does it take the boat to cross the river? *Ans.* (*a*) 37° upstream; (*b*) 1.5×10^2 s

1.43 [III] A reckless drunk is playing with a gun in an airplane that is going directly east at 500 km/h. The drunk shoots the gun straight up at the ceiling of the plane. The bullet leaves the gun at a speed of 1000 km/h. According to someone standing on the Earth, what angle does the bullet make with the vertical? *Ans.* 26.6°

Chapter 2

Uniformly Accelerated Motion

ACCELERATION measures the time rate-of-change of velocity:

$$\textit{Average acceleration} = \frac{\text{change in velocity vector}}{\text{time taken}}$$

$$\vec{\mathbf{a}}_{av} = \frac{\vec{\mathbf{v}}_f - \vec{\mathbf{v}}_i}{t}$$

where $\vec{\mathbf{v}}_i$ is the initial velocity, $\vec{\mathbf{v}}_f$ is the final velocity, and t is the time interval over which the change occurred. The units of acceleration are those of velocity divided by time. Typical examples are (m/s)/s (or m/s^2) and (km/h)/s (or km/h·s). Notice that acceleration is a vector quantity. It has the direction of $\vec{\mathbf{v}}_f - \vec{\mathbf{v}}_i$, the change in velocity. It is nonetheless commonplace to speak of the magnitude of the acceleration as just the acceleration, provided there is no ambiguity.

When we concern ourselves only with accelerations tangent to the path traveled, the direction of the acceleration is known and we can write the defining equation in scalar form as

$$a_{av} = \frac{v_f - v_i}{t}$$

UNIFORMLY ACCELERATED MOTION ALONG A STRAIGHT LINE is an important situation. In this case, the *acceleration vector is constant* and lies along the line of the displacement vector, so that the directions of $\vec{\mathbf{v}}$ and $\vec{\mathbf{a}}$ can be specified with plus and minus signs. If we represent the displacement by s (positive if in the positive direction, and negative if in the negative direction), then the motion can be described with the *five equations* for uniformly accelerated motion:

$$s = v_{av}t$$

$$v_{av} = \frac{v_f + v_i}{2}$$

$$a = \frac{v_f - v_i}{t}$$

$$v_f^2 = v_i^2 + 2as$$

$$s = v_i t + \tfrac{1}{2}at^2$$

Often s is replaced by x or y, and sometimes v_f and v_i are written as v and v_0, respectively.

DIRECTION IS IMPORTANT, and a positive direction must be chosen when analyzing motion along a line. Either direction may be chosen as positive. If a displacement, velocity, or acceleration is in the opposite direction, it must be taken as negative.

GRAPHICAL INTERPRETATIONS for motion along a straight line (e.g., the x-axis) are as follows:

- A plot of *distance versus time* is always positive (i.e., the graph lies above the time axis). Such a curve never decreases (i.e., it can never have a negative slope or speed). Just think about the odometer and speedometer in a car.
- Because the displacement is a vector quantity we can only graph it against time if we limit the motion to a straight line and then use plus and minus signs to specify direction. Accordingly, it's common practice to plot *displacement along a straight line versus time* using that scheme. Such a graph representing motion along, say, the x-axis, may be either positive (plotted above the time axis) when the object is to the right of the origin ($x = 0$), or negative (plotted below the time axis) when the object is left of the origin (see Fig. 2-1). The graph can be positive and get more positive, or negative and get less negative. In both cases the curve would have a positive slope, and the object a positive velocity (it would be moving in the positive x-direction). Furthermore, the graph can be positive and get less positive, or be negative and get more negative. In both these cases the curve would have a negative slope, and the object a negative velocity (it would be moving in the negative x-direction.
- The *instantaneous velocity* of an object at a certain time is the slope of the displacement versus time graph at that time. It can be positive, negative, or zero.
- The *instantaneous acceleration* of an object at a certain time is the slope of the velocity versus time graph at that time.
- For constant-velocity motion, the x-versus-t graph is a straight line. For constant-acceleration motion, the v-versus-t graph is a straight line.

ACCELERATION DUE TO GRAVITY (g): The acceleration of a body moving only under the force of gravity is g, the gravitational (or free-fall) acceleration, which is directed vertically downward. On Earth, $g = 9.81\ \text{m/s}^2$ (i.e., $32.2\ \text{ft/s}^2$); the value varies slightly from place to place. On the Moon, the free-fall acceleration is $1.6\ \text{m/s}^2$.

VELOCITY COMPONENTS: Suppose that an object moves with a velocity $\vec{\mathbf{v}}$ at some angle θ up from the x-axis, as would initially be the case with a ball thrown into the air. That velocity then has x and y vector components (see Fig. 1-7) of $\vec{\mathbf{v}}_x$ and $\vec{\mathbf{v}}_y$. The corresponding scalar components of the velocity are

$$v_x = v\cos\theta \qquad \text{and} \qquad v_y = v\sin\theta$$

and these can turn out to be positive or negative numbers, depending on θ. As a rule, if $\vec{\mathbf{v}}$ is in the first quadrant, $v_x > 0$ and $v_y > 0$; if $\vec{\mathbf{v}}$ is in the second quadrant, $v_x < 0$ and $v_y > 0$; if $\vec{\mathbf{v}}$ is in the third quadrant, $v_x < 0$ and $v_y < 0$; finally, if $\vec{\mathbf{v}}$ is in the fourth quadrant, $v_x > 0$ and $v_y < 0$. Because these quantities have signs, and therefore implied directions along known axes, it is common to refer to them as velocities. The reader will find this usage in many texts, but it is not without pedagogical drawbacks. Instead, we shall avoid applying the term "velocity" to anything but a vector quantity (written in boldface with an arrow above) whose direction is explicitly stated. Thus for an object moving with a *velocity* $\vec{\mathbf{v}} = 100$ m/s—WEST, the *scalar value of the velocity along the x-axis* is $v_x = -100$ m/s; and the (always positive) *speed* is $v = 100$ m/s.

PROJECTILE PROBLEMS can be solved easily if air friction can be ignored. One simply considers the motion to consist of two independent parts: horizontal motion with $a = 0$ and $v_f = v_i = v_{av}$ (i.e., constant speed), and vertical motion with $a = g = 9.81\ \text{m/s}^2$ downward.

Solved Problems

2.1 [I] A robot named Fred is initially moving at 2.20 m/s along a hallway in a space terminal. It subsequently speeds up to 4.80 m/s in a time of 0.20 s. Determine the size or *magnitude* of its average acceleration along the path traveled.

The defining scalar equation is $a_{av} = (v_f - v_i)/t$. Everything is in proper SI units so we need only carry out the calculation:

$$a_{av} = \frac{4.80\ \text{m/s} - 2.20\ \text{m/s}}{0.20\ \text{s}} = 13\ \text{m/s}^2$$

Notice that the answer has two significant figures because the time has only two significant figures.

2.2 [I] A car is traveling at 20.0 m/s when the driver slams on the brakes and brings it to a straight-line stop in 4.2 s. What is the magnitude of its average acceleration?

The defining scalar equation is $a_{av} = (v_f - v_i)/t$. Note that the final speed is zero. Here the initial speed is greater than the final speed so we can expect the acceleration to be negative:

$$a_{av} = \frac{0.0\ \text{m/s} - 2.20\ \text{m/s}}{4.2\ \text{s}} = -4.76\ \text{m/s}^2$$

Because the time is provided with only two significant figures the answer is $-4.8\ \text{m/s}^2$.

2.3 [II] An object starts from rest with a constant acceleration of $8.00\ \text{m/s}^2$ along a straight line. Find (*a*) the speed at the end of 5.00 s, (*b*) the average speed for the 5-s interval, and (*c*) the distance traveled in the 5.00 s.

We are interested in the motion for the first 5.00 s. Take the direction of motion to be the $+x$-direction (that is, $s = x$). We know that $v_i = 0$, $t = 5.00$ s, and $a = 8.00\ \text{m/s}^2$. Because the motion is uniformly accelerated, the five motion equations apply.

(*a*)
$$v_{fx} = v_{ix} + at = 0 + (8.00\ \text{m/s}^2)(5.00\ \text{s}) = 40.0\ \text{m/s}$$

(*b*)
$$v_{av} = \frac{v_{ix} + v_{fx}}{2} = \frac{0 + 40.0}{2}\ \text{m/s} = 20.0\ \text{m/s}$$

(*c*) $x = v_{ix}t + \frac{1}{2}at^2 = 0 + \frac{1}{2}(8.00\ \text{m/s}^2)(5.00\ \text{s})^2 = 100\ \text{m}$ or $x = v_{av}t = (20.0\ \text{m/s})(5.00\ \text{s}) = 100\ \text{m}$

2.4 [II] A truck's speed increases uniformly from 15 km/h to 60 km/h in 20 s. Determine (*a*) the average speed, (*b*) the acceleration, and (*c*) the distance traveled, all in units of meters and seconds.

For the 20 s trip under discussion, taking $+x$ to be in the direction of motion, we have

$$v_{ix} = \left(15\ \frac{\text{km}}{\text{h}}\right)\left(1000\ \frac{\text{m}}{\text{km}}\right)\left(\frac{1}{3600}\ \frac{\text{h}}{\text{s}}\right) = 4.17\ \text{m/s}$$

$$v_{fx} = 60\ \text{km/h} = 16.7\ \text{m/s}$$

(*a*)
$$v_{av} = \tfrac{1}{2}(v_{ix} + v_{fx}) = \tfrac{1}{2}(4.17 + 16.7)\ \text{m/s} = 10\ \text{m/s}$$

(*b*)
$$a = \frac{v_{fx} - v_{ix}}{t} = \frac{(16.7 - 4.17)\ \text{m/s}}{20\ \text{s}} = 0.63\ \text{m/s}^2$$

(*c*)
$$x = v_{av}t = (10.4\ \text{m/s})(20\ \text{s}) = 208\ \text{m} = 0.21\ \text{km}$$

2.5 [II] An object's one-dimensional motion along the x-axis is graphed in Fig. 2-1. Describe its motion.

The velocity of the object at any instant is equal to the slope of the displacement–time graph at the point corresponding to that instant. Because the slope is zero from exactly $t = 0$ s to $t = 2.0$ s, the object is standing still during this time interval. At $t = 2.0$ s, the object begins to move in the $+x$-direction with constant-velocity (the slope is positive and constant). For the interval $t = 2.0$ s to $t = 4.0$ s,

$$v_{av} = \text{slope} = \frac{\text{rise}}{\text{run}} = \frac{x_f - x_i}{t_f - t_i} = \frac{3.0\text{ m} - 0\text{ m}}{4.0\text{ s} - 2.0\text{ s}} = \frac{3.0\text{ m}}{2.0\text{ s}} = 1.5\text{ m/s}$$

The average velocity is then $\vec{v}_{av} = 1.5$ m/s—POSITIVE X-DIRECTION.

During the interval $t = 4.0$ s to $t = 6.0$ s, the object is at rest; the slope of the graph is zero and x does not change for that interval.

From $t = 6.0$ s to $t = 10$ s and beyond, the object is moving in the $-x$-direction; the slope and the velocity are negative. We have

$$v_{av} = \text{slope} = \frac{x_f - x_i}{t_f - t_i} = \frac{-2.0\text{ m} - 3.0\text{ m}}{10.0\text{ s} - 6.0\text{ s}} = \frac{-5.0\text{ m}}{4.0\text{ s}} = -1.3\text{ m/s}$$

The average velocity is then $\vec{v}_{av} = 1.3$ m/s—NEGATIVE X-DIRECTION.

2.6 [II] The vertical motion of an object is graphed in Fig. 2-2. Describe its motion qualitatively, and find its instantaneous velocity at points A, B, and C.

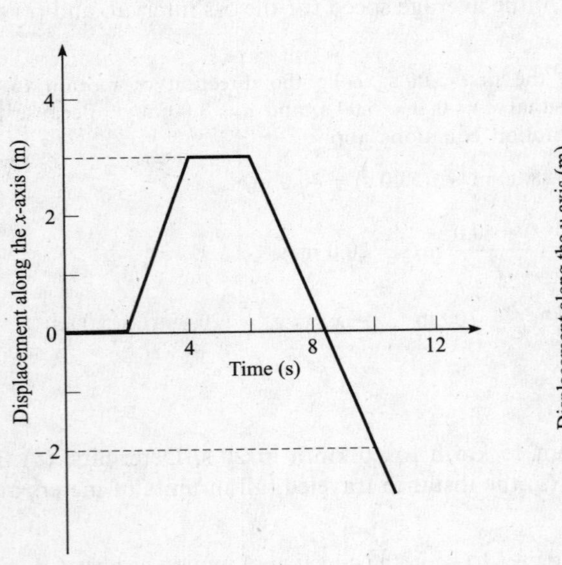

Fig. 2-1

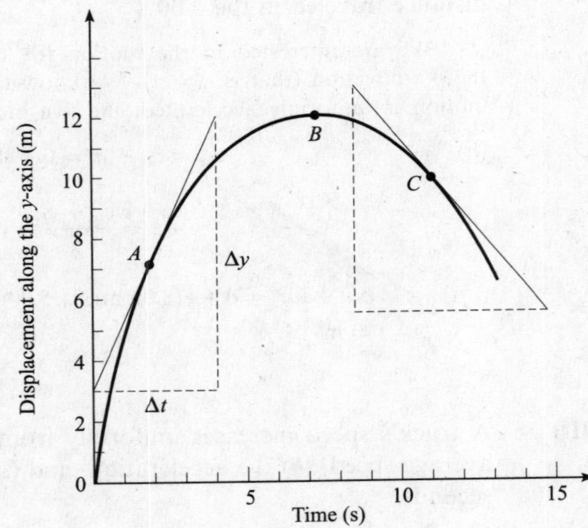

Fig. 2-2

Recalling that the instantaneous velocity is given by the slope of the graph, we see that the object is moving fastest at $t = 0$. As it rises, it slows and finally stops at B. (The slope there is zero.) Then it begins to fall back downward at ever-increasing speed.

At point A, we have

$$v_A = \text{slope} = \frac{\Delta y}{\Delta t} = \frac{12.0\text{ m} - 3.0\text{ m}}{4.0\text{ s} - 0\text{ s}} = \frac{9.0\text{ m}}{4.0\text{ s}} = 2.3\text{ m/s}$$

The velocity at A is positive, so it is in the $+y$-direction: $\vec{v}_A = 2.3$ m/s—UP. At points B and C,

$$v_B = \text{slope} = 0\,\text{m/s}$$

$$v_C = \text{slope} = \frac{\Delta y}{\Delta t} = \frac{5.5\text{ m} - 13.0\text{ m}}{15.0\text{ s} - 8.5\text{ s}} = \frac{-7.5\text{ m}}{6.5\text{ s}} = -1.2\text{ m/s}$$

Because it is negative, the velocity at C is in the $-y$-direction: $\vec{v}_C = 1.2$ m/s—DOWN. Remember that velocity is a vector quantity and direction must be specified explicitly.

2.7 [II] A ball is dropped from rest at a height of 50 m above the ground. (*a*) What is its speed just before it hits the ground? (*b*) How long does it take to reach the ground?

If we can ignore air friction, the ball is uniformly accelerated until it reaches the ground. Its acceleration is downward and is 9.81 m/s^2. Taking *down* as positive, we have for the trip:

$$y = 50.0\text{ m} \qquad a = 9.81\text{ m/s}^2 \qquad v_i = 0$$

(*a*) $$v_{fy}^2 = v_{iy}^2 + 2ay = 0 + 2(9.81\,\text{m/s}^2)(50.0\text{ m}) = 981\text{ m}^2/\text{s}^2$$

and so $v_f = 31.3$ m/s.

(*b*) From $a = (v_{fy} - v_{iy})/t$,

$$t = \frac{v_{fy} - v_{iy}}{a} = \frac{(31.3 - 0)\,\text{m/s}}{9.81\,\text{m/s}^2} = 3.19\,\text{s}$$

(We could just as well have taken *up* as positive. How would the calculation have been changed?)

2.8 [II] A skier starts from rest and slides 9.0 m down a slope in 3.0 s. In what time after starting will the skier acquire a speed of 24 m/s? Assume that the acceleration is constant and the path is straight.

We must find the skier's acceleration from the data concerning the 3.0 s trip. Taking the direction of motion as the $+x$-direction, we have $t = 3.0$ s, $v_{ix} = 0$, and $x = 9.0$ m. Then $x = v_{ix}t + \frac{1}{2}at^2$ gives

$$a = \frac{2x}{t^2} = \frac{18\,\text{m}}{(3.0\,\text{s})^2} = 2.0\,\text{m/s}^2$$

We can now use this value of a for the longer trip, from the starting point to the place where $v_{fx} = 24$ m/s. For this trip, $v_{ix} = 0$, $v_{fx} = 24$ m/s, $a = 2.0$ m/s^2. Then, from $v_f = v_i + at$,

$$t = \frac{v_{fx} - v_{ix}}{a} = \frac{24\text{ m/s}}{2.0\text{ m/s}^2} = 12\text{ s}$$

2.9 [II] A bus moving in a straight line at a speed of 20 m/s begins to slow at a constant rate of 3.0 m/s each second. Find how far it goes before stopping.

Take the direction of motion to be the $+x$-direction. For the trip under consideration, $v_i = 20$ m/s, $v_f = 0$ m/s, $a = -3.0$ m/s^2. Notice that the bus is not speeding up in the positive motion direction. Instead, it is slowing in that direction and so its acceleration is negative (a deceleration). Use

$$v_{fx}^2 = v_{ix}^2 + 2ax$$

to find

$$x = \frac{-(20\,\text{m/s})^2}{2(-3.0\,\text{m/s}^2)} = 67\,\text{m}$$

2.10 [II] A car moving along a straight road at 30 m/s slows uniformly to a speed of 10 m/s in a time of 5.0 s. Determine (*a*) the acceleration of the car and (*b*) the distance it moves during the third second.

Let us take the direction of motion to be the $+x$-direction.

(*a*) For the 5.0 s interval, we have $t = 5.0$ s, $v_{ix} = 30$ m/s, $v_f = 10$ m/s. Using $v_{fx} = v_{ix} + at$ gives

$$a = \frac{(10 - 30)\text{ m/s}}{5.0\text{ s}} = -4.0\text{ m/s}^2$$

(*b*)

$$x = (\text{distance covered in 3.0 s}) - (\text{distance covered in 2.0 s})$$
$$x = (v_{ix}t_3 + \tfrac{1}{2}at_3^2) - (v_{ix}t_2 + \tfrac{1}{2}at_2^2)$$
$$x = v_{ix}(t_3 - t_2) + \tfrac{1}{2}a(t_3^2 - t_2^2)$$

Using $v_{ix} = 30$ m/s, $a = -4.0$ m/s^2, $t_2 = 2.0$ s, $t_3 = 3.0$ s gives

$$x = (30\text{ m/s})(1.0\text{ s}) - (2.0\text{ m/s}^2)(5.0\text{ s}^2) = 20\text{ m}$$

2.11 [II] The speed of a train is reduced uniformly from 15 m/s to 7.0 m/s while traveling a distance of 90 m. (*a*) Compute the acceleration. (*b*) How much farther will the train travel before coming to rest, provided the acceleration remains constant?

Let us take the direction of motion to be the $+x$-direction.

(*a*) We have $v_{ix} = 15$ m/s, $v_{fx} = 7.0$ m/s, $x = 90$ m. Then $v_{fx}^2 = v_{ix}^2 + 2ax$ gives

$$a = -0.98\text{ m/s}^2$$

(*b*) We now have the new conditions $v_{ix} = 7.0$ m/s, $v_f = 0$, $a = -0.98$ m/s^2. Then

$$v_{fx}^2 = v_{ix}^2 + 2ax$$

gives

$$x = \frac{0 - (7.0\text{ m/s})^2}{-1.96\text{ m/s}^2} = 25\text{ m}$$

2.12 [II] A stone is thrown straight upward and it rises to a height of 20 m. With what speed was it thrown?

Let us take *up* as the positive y-direction. The stone's velocity is zero at the top of its path. Then $v_{fy} = 0$, $y = 20$ m, $a = -9.81$ m/s^2. (The minus sign arises because the acceleration due to gravity is always downward and we have taken *up* to be positive.) We use $v_{fy}^2 = v_{iy}^2 + 2ay$ to find

$$v_{iy} = \sqrt{-2(-9.81\text{ m/s}^2)(20\text{ m})} = 20\text{ m/s}$$

2.13 [II] A stone is thrown straight upward with a speed of 20 m/s. It is caught on its way down at a point 5.0 m above where it was thrown. (*a*) How fast was it going when it was caught? (*b*) How long did the trip take?

The situation is shown in Fig. 2-3. Let us take *up* as positive. Then, for the trip that lasts from the instant after throwing to the instant before catching, $v_{iy} = 20$ m/s, $y = +5.0$ m (since it is an upward displacement), $a = -9.81$ m/s^2.

(*a*) We use $v_{fy}^2 = v_{iy}^2 + 2ay$ to find

$$v_{fy}^2 = (20\text{ m/s})^2 + 2(-9.81\text{ m/s}^2)(5.0\text{ m}) = 302\text{ m}^2/\text{s}^2$$
$$v_{fy} = \pm\sqrt{302\text{ m}^2/\text{s}^2} = -17\text{ m/s}$$

We take the negative sign because the stone is moving downward, in the negative direction, at the final instant.

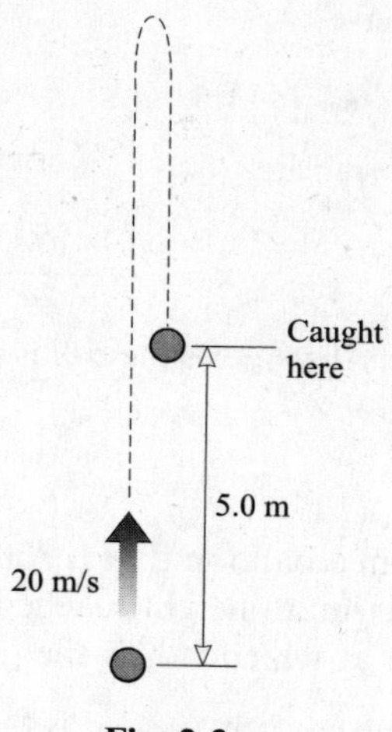

Fig. 2-3

(*b*) We use $a = (v_{fy} - v_{iy})/t$ to find

$$t = \frac{(-17.4 - 20)\ \text{m/s}}{-9.81\ \text{m/s}^2} = 3.8\ \text{s}$$

Notice that we retain the minus sign on v_{fy}.

2.14 [II] A ball that is thrown vertically upward on the Moon returns to its starting point in 4.0 s. The acceleration due to gravity there is 1.60 m/s² downward. Find the ball's original speed.

Let us take *up* as positive. For the trip from beginning to end, $y = 0$ (it ends at the same level it started at), $a = -1.60\ \text{m/s}^2$, $t = 4.0$ s. We use $y = v_{iy}t + \frac{1}{2}at^2$ to find

$$0 = v_{iy}(4.0\ \text{s}) + \tfrac{1}{2}(-1.60\ \text{m/s}^2)(4.0\ \text{s})^2$$

from which $v_{iy} = 3.2\ \text{m/s}$.

2.15 [III] A baseball is thrown straight upward on the Moon with an initial speed of 35 m/s. Compute (*a*) the maximum height reached by the ball, (*b*) the time taken to reach that height, (*c*) its velocity 30 s after it is thrown, and (*d*) when the ball's height is 100 m.

Take *up* as positive. At the highest point, the ball's velocity is zero.

(*a*) From $v_{fy}^2 = v_{iy}^2 + 2ay$ we have, since $g = 1.60\ \text{m/s}^2$ on the Moon,

$$0 = (35\ \text{m/s})^2 + 2(-1.60\ \text{m/s}^2)y \qquad \text{or} \qquad y = 0.38\ \text{km}$$

(*b*) From $v_{fy} = v_{iy} + at$ we have

$$0 = 35\ \text{m/s} + (-1.60\ \text{m/s}^2)t \qquad \text{or} \qquad t = 22\ \text{s}$$

(*c*) From $v_{fy} = v_{iy} + at$ we have

$$v_{fy} = 35\ \text{m/s} + (-1.60\ \text{m/s}^2)(30\ \text{s}) \qquad \text{or} \qquad v_{fy} = -13\ \text{m/s}$$

Because v_f is negative and we are taking *up* as positive, the velocity is directed downward. The ball is on its way down at $t = 30$ s.

(*d*) From $y = v_{iy}t + \frac{1}{2}at^2$ we have

$$100 \text{ m} = (35 \text{ m/s})t + \tfrac{1}{2}(-1.60 \text{ m/s}^2)t^2 \qquad \text{or} \qquad 0.80t^2 - 35t + 100 = 0$$

By use of the quadratic formula,

$$x = \frac{-b \pm \sqrt{b^2 - 4ac}}{2a}$$

we find $t = 3.1$ s and 41 s. At $t = 3.1$ s the ball is at 100 m and ascending; at $t = 41$ s it is at the same height but descending.

2.16 [III] A ballast bag is dropped from a balloon that is 300 m above the ground and rising at 13 m/s. For the bag, find (*a*) the maximum height reached, (*b*) its position and velocity 5.0 s after it is released, and (*c*) the time at which it hits the ground.

The initial velocity of the bag when released is the same as that of the balloon, 13 m/s upward. Let us choose *up* as positive and take $y = 0$ at the point of release.

(*a*) At the highest point, $v_f = 0$. From $v_{fy}^2 = v_{iy}^2 + 2ay$,

$$0 = (13 \text{ m/s})^2 + 2(-9.81 \text{ m/s}^2)y \qquad \text{or} \qquad y = 8.6 \text{ m}$$

The maximum height is $300 + 8.6 = 308.6$ m or 0.31 km.

(*b*) Take the end point to be its position at $t = 5.0$ s. Then, from $y = v_{iy}t + \frac{1}{2}at^2$,

$$y = (13 \text{ m/s})(5.0 \text{ s}) + \tfrac{1}{2}(-9.81 \text{ m/s}^2)(5.0 \text{ s})^2 = -57.5 \text{ m or } -58 \text{ m}$$

So its height is $300 - 58 = 242$ m. Also, from $v_{fy} = v_{iy} + at$,

$$v_{fy} = 13 \text{ m/s} + (-9.81 \text{ m/s}^2)(5.0 \text{ s}) = -36 \text{ m/s}$$

It is on its way down with a velocity of 36 m/s—DOWNWARD.

(*c*) Just as it hits the ground, the bag's displacement is -300 m. Then

$$y = v_{iy}t + \tfrac{1}{2}at^2 \qquad \text{becomes} \qquad -300 \text{ m} = (13 \text{ m/s})t + \tfrac{1}{2}(-9.81 \text{ m/s}^2)t^2$$

or $4.90t^2 - 13t - 300 = 0$. The quadratic formula gives $t = 9.3$ s and -6.6 s. Only the positive time has physical meaning, so the required answer is 9.3 s.

We could have avoided the quadratic formula by first computing v_f:

$$v_{fy}^2 = v_{iy}^2 + 2as \qquad \text{becomes} \qquad v_{fy}^2 = (13 \text{ m/s})^2 + 2(-9.81 \text{ m/s}^2)(-300 \text{ m})$$

so that $v_{fy} = \pm 77.8$ m/s. Then, using the negative value for v_{fy} (why?) in $v_{fy} = v_{iy} + at$ gives $t = 9.3$ s, as before.

2.17 [II] As shown in Fig. 2-4, a projectile is fired horizontally with a speed of 30 m/s from the top of a cliff 80 m high. (*a*) How long will it take to strike the level ground at the base of the cliff? (*b*) How far from the foot of the cliff will it strike? (*c*) With what velocity will it strike?

(*a*) The horizontal and vertical motions are independent of each other. Consider first the vertical motion. Taking *up* as positive and $y = 0$ at the top of the cliff, we have

$$y = v_{iy}t + \tfrac{1}{2}a_y t^2$$

or

$$-80 \text{ m} = 0 + \tfrac{1}{2}(-9.81 \text{ m/s}^2)t^2$$

from which $t = 4.04$ s or 4.0 s. Notice that the initial velocity had zero vertical component and so $v_i = 0$ for the vertical motion.

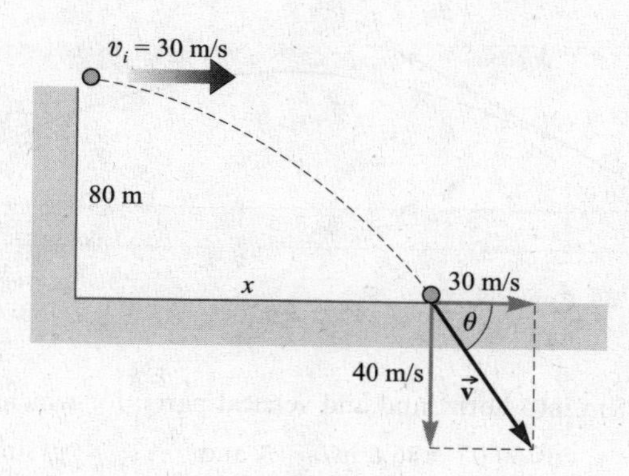

Fig. 2-4

(*b*) Now consider the horizontal motion. For it, $a = 0$ and so $v_x = v_{ix} = v_{fx} = 30$ m/s. Then, using the value of t found in (*a*), we have

$$x = v_x t = (30 \text{ m/s})(4.04 \text{ s}) = 121 \text{ m or } 0.12 \text{ km}$$

(*c*) The final velocity has a horizontal component of 30 m/s. But its vertical component at $t = 4.04$ s is given by $v_{fy} = v_{iy} + a_y t$ as

$$v_{fy} = 0 + (-9.8 \text{ m/s}^2)(4.04 \text{ s}) = -40 \text{ m/s}$$

The resultant of these two components is labeled $\vec{v}$ in Fig. 2-4; we have

$$v = \sqrt{(40 \text{ m/s})^2 + (30 \text{ m/s})^2} = 50 \text{ m/s}$$

The angle θ as shown is given by $\tan\theta = 40/30$ and is 53°. Hence, $\vec{v} = 50$ m/s—53° BELOW *X*-AXIS.

2.18 [I] A stunt flier is moving at 15 m/s parallel to the flat ground 100 m below, as shown in Fig. 2-5. How large must the distance x from plane to target be if a sack of flour released from the plane is to strike the target?

Following the same procedure as in Problem 2.17, we use $y = v_{iy}t + \frac{1}{2}a_y t^2$ to get

$$-100 \text{ m} = 0 + \tfrac{1}{2}(-9.81 \text{ m/s}^2)t^2 \qquad \text{or} \qquad t = 4.52 \text{ s}$$

Now $x = v_x t = (15 \text{ m/s})(4.52 \text{ s}) = 67.8$ m or 68 m.

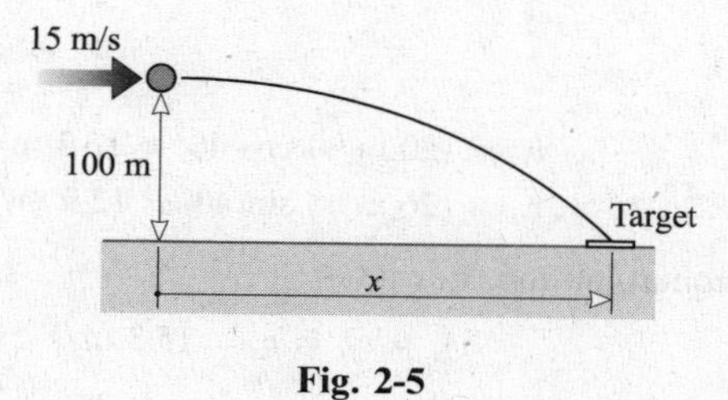

Fig. 2-5

2.19 [II] A baseball is thrown with an initial velocity of 100 m/s at an angle of 30.0° above the horizontal, as shown in Fig. 2-6. How far from the throwing point will the baseball attain its original level?

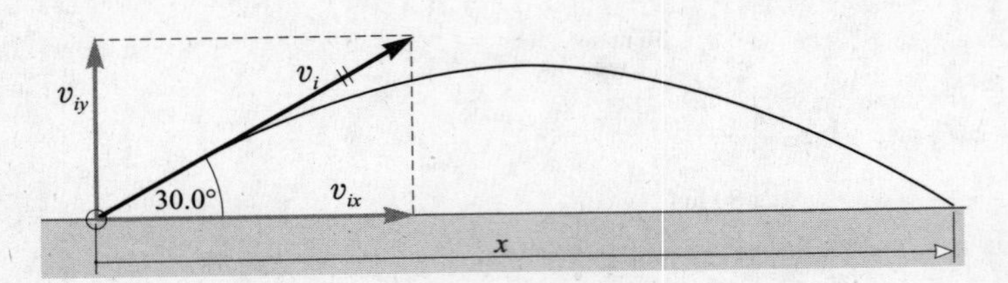

Fig. 2-6

We divide the problem into horizontal and vertical parts, for which

$$v_{ix} = v_i \cos 30.0° = 86.6 \text{ m/s} \qquad \text{and} \qquad v_{iy} = v_i \sin 30.0° = 50.0 \text{ m/s}$$

where *up* is being taken as positive.

In the vertical problem, $y = 0$ since the ball returns to its original height. Then

$$y = v_{iy}t + \tfrac{1}{2}a_y t^2 \qquad \text{or} \qquad 0 = (50.0 \text{ m/s}) + \tfrac{1}{2}(-9.81 \text{ m/s}^2)t$$

and $t = 10.2$ s.

In the horizontal problem, $v_{ix} = v_{fx} = v_x = 86.6$ m/s. Therefore,

$$x = v_x t = (86.6 \text{ m/s})(10.2 \text{ s}) = 884 \text{ m}$$

2.20 [III] As shown in Fig. 2-7, a ball is thrown from the top of one building toward a tall building 50 m away. The initial velocity of the ball is 20 m/s—40° ABOVE HORIZONTAL. How far above or below its original level will the ball strike the opposite wall?

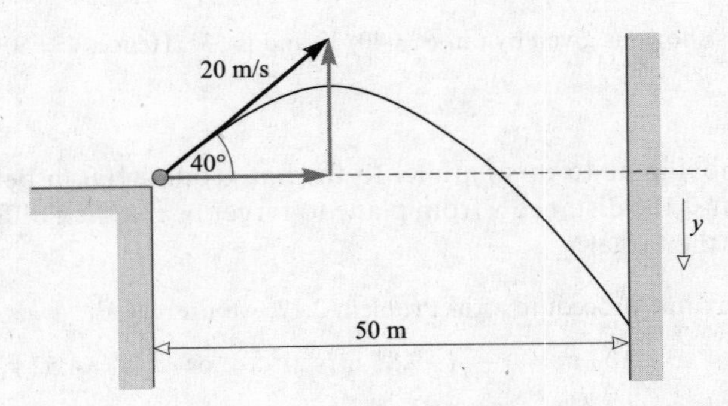

Fig. 2-7

We have

$$v_{ix} = (20 \text{ m/s}) \cos 40° = 15.3 \text{ m/s}$$
$$v_{iy} = (20 \text{ m/s}) \sin 40° = 12.9 \text{ m/s}$$

Consider first the horizontal motion. For it,

$$v_{ix} = v_{fx} = v_x = 15.3 \text{ m/s}$$

Then $x = v_x t$ gives

$$50 \text{ m} = (15.3 \text{ m/s})t \qquad \text{or} \qquad t = 3.27 \text{ s}$$

For the vertical motion, taking *down* as positive, we have

$$y = v_{iy}t + \tfrac{1}{2}a_y t^2 = (-12.9 \text{ m/s})(3.27 \text{ s}) + \tfrac{1}{2}(9.81 \text{ m/s}^2)(3.27 \text{ s})^2 = 10.3 \text{ m}$$

and to two significant figures, $y = 10$ m. Since y is positive, and since *down* is positive, the ball will hit at 10 m below the original level.

2.21 [III] (*a*) Find the range x of a gun which fires a shell with muzzle velocity v at an angle of elevation θ. (*b*) Find the angle of elevation θ of a gun which fires a shell with a muzzle velocity of 120 m/s and hits a target on the same level but 1300 m distant. (See Fig. 2-8.)

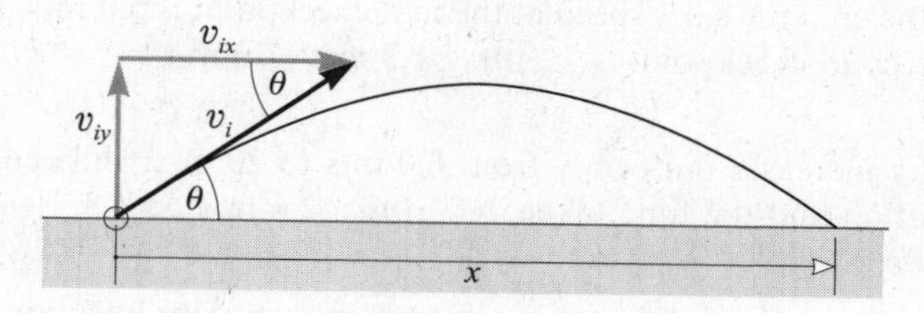

Fig. 2-8

(*a*) Let t be the time it takes the shell to hit the target. Then, $x = v_{ix}t$ or $t = x/v_{ix}$. Consider the vertical motion alone, and take *up* as positive. When the shell strikes the target,

$$\text{Vertical displacement} = 0 = v_{iy}t + \tfrac{1}{2}(-g)t^2$$

Solving this equation gives $t = 2v_{iy}/g$. But $t = x/v_{ix}$, so

$$\frac{x}{v_{ix}} = \frac{2v_{iy}}{g} \qquad \text{or} \qquad x = \frac{2v_{ix}v_{iy}}{g} = \frac{2(v_i \cos\theta)(v_i \sin\theta)}{g}$$

The formula $2 \sin\theta \cos\theta = \sin 2\theta$ can be used to simplify this. After substitution, we get

$$x = \frac{v_i^2 \sin 2\theta}{g}$$

The maximum range corresponds to $\theta = 45°$, since $\sin 2\theta$ has a maximum value of 1 when $2\theta = 90°$ or $\theta = 45°$.

(*b*) From the range equation found in (*a*), we have

$$\sin 2\theta = \frac{gx}{v_i^2} = \frac{(9.81\ \text{m/s}^2)(1300\ \text{m})}{(120\ \text{m/s})^2} = 0.885$$

Therefore, $2\theta = \arcsin 0.885 = 62°$ and so $\theta = 31°$.

Supplementary Problems

2.22 [I] For the object whose motion is plotted in Fig. 2-2, find its instantaneous velocity at the following times: (*a*) 1.0 s, (*b*) 4.0 s, and (*c*) 10 s. *Ans.* (*a*) 3.3 m/s in the positive y-direction; (*b*) 1.0 m/s in the positive y-direction; (*c*) 0.83 m/s in the negative y-direction

2.23 [I] A body with initial velocity 8.0 m/s moves along a straight line with constant acceleration and travels 640 m in 40 s. For the 40 s interval, find (*a*) the average velocity, (*b*) the final velocity, and (*c*) the acceleration. *Ans.* (*a*) 16 m/s; (*b*) 24 m/s; (*c*) 0.40 m/s^2

2.24 [I] A truck starts from rest and moves with a constant acceleration of $5.0\ m/s^2$. Find its speed and the distance traveled after 4.0 s has elapsed. *Ans.* 20 m/s, 40 m

2.25 [I] A box slides down an incline with uniform acceleration. It starts from rest and attains a speed of 2.7 m/s in 3.0 s. Find (*a*) the acceleration and (*b*) the distance moved in the first 6.0 s. *Ans.* (*a*) $0.90\ m/s^2$; (*b*) 16 m

2.26 [I] A car is accelerating uniformly as it passes two checkpoints that are 30 m apart. The time taken between checkpoints is 4.0 s, and the car's speed at the first checkpoint is 5.0 m/s. Find the car's acceleration and its speed at the second checkpoint. *Ans.* $1.3\ m/s^2$, 10 m/s

2.27 [I] An auto's velocity increases uniformly from 6.0 m/s to 20 m/s while covering 70 m in a straight line. Find the acceleration and the time taken. *Ans.* $2.6\ m/s^2$, 5.4 s

2.28 [I] A plane starts from rest and accelerates uniformly in a straight line along the ground before takeoff. It moves 600 m in 12 s. Find (*a*) the acceleration, (*b*) speed at the end of 12 s, and (*c*) the distance moved during the twelfth second. *Ans.* (*a*) $8.3\ m/s^2$; (*b*) 0.10 km/s; (*c*) 96 m

2.29 [I] A train running along a straight track at 30 m/s is slowed uniformly to a stop in 44 s. Find the acceleration and the stopping distance. *Ans.* $-0.68\ m/s^2$, 0.66 km or 6.6×10^2 m

2.30 [II] An object moving at 13 m/s slows uniformly at the rate of 2.0 m/s each second for a time of 6.0 s. Determine (*a*) its final speed, (*b*) its average speed during the 6.0 s, and (*c*) the distance moved in the 6.0 s. *Ans.* (*a*) 1.0 m/s; (*b*) 7.0 m/s; (*c*) 42 m

2.31 [I] A body falls freely from rest. Find (*a*) its acceleration, (*b*) the distance it falls in 3.0 s, (*c*) its speed after falling 70 m, (*d*) the time required to reach a speed of 25 m/s, and (*e*) the time taken to fall 300 m. *Ans.* (*a*) $9.81\ m/s^2$; (*b*) 44 m; (*c*) 37 m/s; (*d*) 2.6 s; (*e*) 7.8 s

2.32 [I] A marble dropped from a bridge strikes the water in 5.0 s. Calculate (*a*) the speed with which it strikes and (*b*) the height of the bridge. *Ans.* (*a*) 49 m/s; (*b*) 0.12 km or 1.2×10^2 m

2.33 [II] A stone is thrown straight downward with initial speed 8.0 m/s from a height of 25 m. Find (*a*) the time it takes to reach the ground and (*b*) the speed with which it strikes. *Ans.* (*a*) 1.6 s; (*b*) 24 m/s

2.34 [II] A baseball is thrown straight upward with a speed of 30 m/s. (*a*) How long will it rise? (*b*) How high will it rise? (*c*) How long after it leaves the hand will it return to the starting point? (*d*) When will its speed be 16 m/s? *Ans.* (*a*) 3.1 s; (*b*) 46 m; (*c*) 6.1 s; (*d*) 1.4 s and 4.7 s

2.35 [II] A bottle dropped from a balloon reaches the ground in 20 s. Determine the height of the balloon if (*a*) it was at rest in the air and (*b*) it was ascending with a speed of 50 m/s when the bottle was dropped. *Ans.* 2.0 km; (*b*) 0.96 km

2.36 [II] Two balls are dropped to the ground from different heights. One is dropped 1.5 s after the other, but they both strike the ground at the same time, 5.0 s after the first was dropped. (*a*) What is the difference in the heights from which they were dropped? (*b*) From what height was the first ball dropped? *Ans.* (*a*) 63 m; (*b*) 0.12 km

2.37 [II] A nut comes loose from a bolt on the bottom of an elevator as the elevator is moving up the shaft at 3.00 m/s. The nut strikes the bottom of the shaft in 2.00 s. (*a*) How far from the bottom of the shaft was the elevator when the nut fell off? (*b*) How far above the bottom was the nut 0.25 s after it fell off? *Ans.* (*a*) 13.6 m; (*b*) 14 m

2.38 [I] A marble, rolling with speed 20 cm/s, rolls off the edge of a table that is 80 cm high. (*a*) How long does it take to drop to the floor? (*b*) How far, horizontally, from the table edge does the marble strike the floor? *Ans.* (*a*) 0.40 s; (*b*) 8.1 cm

2.39 [II] A body projected upward from the level ground at an angle of 50° with the horizontal has an initial speed of 40 m/s. (*a*) How long will it take to hit the ground? (*b*) How far from the starting point will it strike? (*c*) At what angle with the horizontal will it strike? *Ans.* (*a*) 6.3 s; (*b*) 0.16 km; (*c*) 50°

2.40 [II] A body is projected downward at an angle of 30° with the horizontal from the top of a building 170 m high. Its initial speed is 40 m/s. (*a*) How long will it take before striking the ground? (*b*) How far from the foot of the building will it strike? (*c*) At what angle with the horizontal will it strike? *Ans.* (*a*) 4.2 s; (*b*) 0.15 km; (*c*) 60°

2.41 [II] A hose lying on the ground shoots a stream of water upward at an angle of 40° to the horizontal. The speed of the water is 20 m/s as it leaves the hose. How high up will it strike a wall which is 8.0 m away? *Ans.* 5.4 m

2.42 [II] A World Series batter hits a home run ball with a velocity of 40 m/s at an angle of 26° above the horizontal. A fielder who can reach 3.0 m above the ground is backed up against the bleacher wall, which is 110 m from home plate. The ball was 120 cm above the ground when hit. How high above the fielder's glove does the ball pass? *Ans.* 6.0 m

2.43 [II] Prove that a gun will shoot three times as high when its angle of elevation is 60° as when it is 30°, but the bullet will carry the same horizontal distance.

2.44 [II] A ball is thrown upward at an angle of 30° to the horizontal and lands on the top edge of a building that is 20 m away. The top edge is 5.0 m above the throwing point. How fast was the ball thrown? *Ans.* 20 m/s

2.45 [III] A ball is thrown straight upward with a speed v from a point h meters above the ground. Show that the time taken for the ball to strike the ground is $(v/g)[1 + \sqrt{1 + (2hg/v^2)}]$.

Chapter 3

Newton's Laws

THE MASS of an object is a measure of the inertia of the object. **Inertia** is the tendency of a body at rest to remain at rest, and of a body in motion to continue moving with unchanged velocity. For several centuries, physicists have found it useful to think of mass as a representation of the amount of or quantity-of-matter, but that idea is (as we have learned from Special Relativity) is no longer tenable.

THE STANDARD KILOGRAM is an object whose mass is defined to be one kilogram. The masses of other objects are found by comparison with this mass. A *gram mass* is equivalent to exactly 0.001 kg.

FORCE, in general, is the agency of change. In mechanics it is that which changes the velocity of an object. Force is a vector quantity, having magnitude and direction. An **external force** is one whose source lies outside of the system being considered.

THE NET EXTERNAL FORCE acting on an object causes the object to accelerate in the direction of that force. The acceleration is proportional to the force and inversely proportional to the mass of the object. (We now know from the Special Theory of Relativity that this statement is actually an excellent approximation applicable to all situations where the speed is appreciably less than the speed of light, c.)

THE NEWTON is the SI unit of force. One newton (1 N) is that resultant force which will give a 1 kg mass an acceleration of $1\ \text{m/s}^2$. The *pound* is 4.45 N, or alternatively a newton is about a quarter of a pound.

NEWTON'S FIRST LAW: *An object at rest will remain at rest; an object in motion will continue in motion with constant velocity, except insofar as it is acted upon by an external force.* Force is the changer of motion.

NEWTON'S SECOND LAW: As stated by Newton, the Second Law was framed in terms of the concept of momentum. This rigorously correct statement will be treated in Chapter 8. Here we focus on a less fundamental, but highly useful, variation. If the resultant (or net), force $\vec{\mathbf{F}}$ acting on an object of mass m is not zero, the object accelerates in the direction of the force. The acceleration $\vec{\mathbf{a}}$ is proportional to the force and inversely proportional to the mass of the object. With $\vec{\mathbf{F}}$ in newtons, m in kilograms, and $\vec{\mathbf{a}}$ in m/s^2, this can be written as

$$\vec{\mathbf{a}} = \frac{\vec{\mathbf{F}}}{m} \qquad \text{or} \qquad \vec{\mathbf{F}} = m\vec{\mathbf{a}}$$

The acceleration $\vec{\mathbf{a}}$ has the same direction as the resultant force $\vec{\mathbf{F}}$.

The vector equation $\vec{\mathbf{F}} = m\vec{\mathbf{a}}$ can be written in terms of components as

$$\Sigma F_x = ma_x \qquad \Sigma F_y = ma_y \qquad \Sigma F_z = ma_z$$

where the forces are the components of the external forces acting on the object.

NEWTON'S THIRD LAW: Matter *interacts* with matter – forces come in pairs. *For each force exerted on one body, there is an equal, but oppositely directed, force on some other body interacting with it.* This is often called the *Law of Action and Reaction.* Notice that the action and reaction forces act on the two different interacting objects.

THE LAW OF UNIVERSAL GRAVITATION: When two masses m and m' gravitationally interact, they attract each other with forces of equal magnitude. For point masses (or spherically symmetric bodies), the attractive force F_G is given by

$$F_G = G\frac{mm'}{r^2}$$

where r is the distance between mass centers, and where $G = 6.67 \times 10^{-11}\,\mathrm{N{\cdot}m^2/kg^2}$ when F_G is in newtons, m and m' are in kilograms, and r is in meters.

THE WEIGHT of an object (F_W) is the gravitational force acting downward on the object. On the Earth, it is the gravitational force exerted on the object by the planet. Its units are newtons (in the SI) and pounds (in the British system). Because the Earth is not a perfect uniform sphere, and moreover because it's spinning, the weight measured by a scale (often called the *effective weight*) will be very slightly different from that defined above.

RELATION BETWEEN MASS AND WEIGHT: An object of mass m falling freely toward the Earth is subject to only one force — the pull of gravity, which we call the weight F_W of the object. The object's acceleration due to F_W is the free-fall acceleration g. Therefore, $\vec{\mathbf{F}} = m\vec{a}$ provides us with the relation between $F = F_W$, $a = g$, and m; it is $F_W = mg$. Because, on average, $g = 9.81\,\mathrm{m/s^2}$ on Earth, a 1.00 kg object weighs 9.81 N (or 2.20 lb) at the Earth's surface.

THE TENSILE FORCE ($\vec{\mathbf{F}}_T$) acting on a string or chain or tendon is the applied force tending to stretch it. The magnitude of the tensile force is the **tension** (F_T).

THE FRICTION FORCE ($\vec{\mathbf{F}}_f$) is a tangential force acting on an object that opposes the sliding of that object on an adjacent surface with which it is in contact. The friction force is parallel to the surface and opposite to the direction of motion or of impending motion. Only when the applied force exceeds the maximum static friction force will an object begin to slide.

THE NORMAL FORCE ($\vec{\mathbf{F}}_N$) on an object that is being supported by a surface is the component of the supporting force that is perpendicular to the surface.

THE COEFFICIENT OF KINETIC FRICTION (μ_k) is defined for the case in which one surface is sliding across another at constant speed. It is

$$\mu_k = \frac{\text{friction force}}{\text{normal force}} = \frac{F_f}{F_N}$$

THE COEFFICIENT OF STATIC FRICTION (μ_s) is defined for the case in which one surface is just on the verge of sliding across another surface. It is

$$\mu_s = \frac{\text{maximum friction force}}{\text{normal force}} = \frac{F_f(\text{max})}{F_N}$$

where the maximum friction force occurs when the object is just on the verge of slipping but is nonetheless at rest.

DIMENSIONAL ANALYSIS: All mechanical quantities, such as acceleration and force, can be expressed in terms of three fundamental dimensions: length L, mass M, and time T. For example, acceleration is a length (a distance) divided by $(\text{time})^2$; we say it has the *dimensions* L/T^2, which we write as $[LT^{-2}]$. The dimensions of volume are $[L^3]$, and those of velocity are $[LT^{-1}]$. Because force is mass multiplied by acceleration, its dimensions are $[MLT^{-2}]$. Dimensions are helpful in checking equations, since each term of an equation must have the same dimensions. For example, the dimensions of the equation

$$s = v_i t \qquad + \tfrac{1}{2}at^2$$

are

$$[L] \rightarrow [LT^{-1}][T] + [LT^{-2}][T^2]$$

so each term has the dimensions of length. *Remember, all terms in an equation must have the same dimensions.* As examples, an equation cannot have a volume $[L^3]$ added to an area $[L^2]$, or a force $[MLT^{-2}]$ subtracted from a velocity $[LT^{-1}]$; these terms do not have the same dimensions.

MATHEMATICAL OPERATIONS WITH UNITS: In every mathematical operation, the units terms (for example, lb, cm, ft^3, mi/h, m/s^2) must be carried along with the numbers and must undergo the same mathematical operations as the numbers.

Quantities cannot be added or subtracted directly unless they have the same units (as well as the same dimensions). For example, if we are to add algebraically 5 m (length) and 8 cm (length), we must first convert m to cm or cm to m. However, quantities of any sort can be combined in multiplication or division, in which the units as well as the numbers obey the algebraic laws of squaring, cancellation, etc. Thus:

(1) $6\ \text{m}^2 + 2\ \text{m}^2 = 8\ \text{m}^2 \qquad (\text{m}^2 + \text{m}^2 \rightarrow \text{m}^2)$

(2) $5\ \text{cm} \times 2\ \text{cm}^2 = 10\ \text{cm}^3 \qquad (\text{cm} \times \text{cm}^2 \rightarrow \text{cm}^3)$

(3) $2\ \text{m}^3 \times 1500\ \dfrac{\text{kg}}{\text{m}^3} = 3000\ \text{kg} \qquad \left(\text{m}^3 \times \dfrac{\text{kg}}{\text{m}^3} \rightarrow \text{kg}\right)$

(4) $2\ \text{s} \times 3\ \dfrac{\text{km}}{\text{s}^2} = 6\ \dfrac{\text{km}}{\text{s}} \qquad \left(\text{s} \times \dfrac{\text{km}}{\text{s}^2} \rightarrow \dfrac{\text{km}}{\text{s}}\right)$

(5) $\dfrac{15\ \text{g}}{3\ \text{g/cm}^3} = 5\ \text{cm}^3 \qquad \left(\dfrac{\text{g}}{\text{g/cm}^3} \rightarrow \text{g} \times \dfrac{\text{cm}^3}{\text{g}} \rightarrow \text{cm}^3\right)$

Solved Problems

3.1 [I] Four coplanar forces act on a body at point O as shown in Fig. 3-1(a). Find their resultant graphically.

Starting from O, the four vectors are plotted in turn as shown in Fig. 3-1(b). We place the tail end of each vector at the tip end of the preceding one. The arrow from O to the tip of the last vector represents the resultant of the vectors.

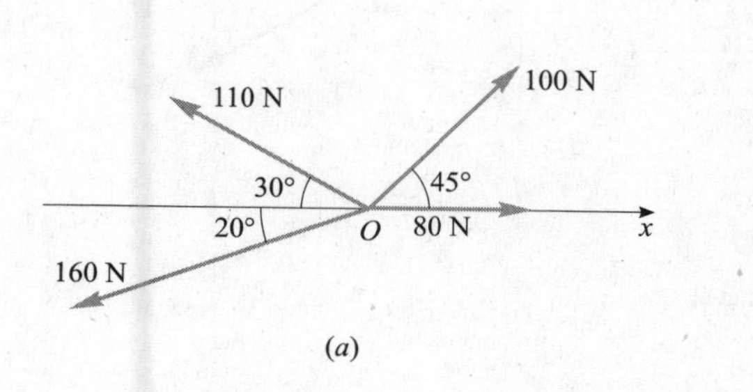

(a)

(b)

Fig. 3-1

We measure R from the scale drawing in Fig. 3-1(b) and find it to be 119 N. Angle α is measured by protractor and is found to be 37°. Hence the resultant makes an angle $\theta = 180° - 37° = 143°$ with the positive x-axis. The resultant is 119 N at 143°.

3.2 [II] The five coplanar forces shown in Fig. 3-2(a) act on an object. Find their resultant.

(1) First we find the x- and y-components of each force. These components are as follows:

Force	x-Component	y-Component
19.0 N	19.0 N	0 N
15.0 N	$(15.0\ \text{N}) \cos 60.0° = 7.50\ \text{N}$	$(15.0\ \text{N}) \sin 60.0° = 13.0\ \text{N}$
16.0 N	$-(16.0\ \text{N}) \cos 45.0° = -11.3\ \text{N}$	$(16.0\ \text{N}) \sin 45.0° = 11.3\ \text{N}$
11.0 N	$-(11.0\ \text{N}) \cos 30.0° = -9.53\ \text{N}$	$-(11.0\ \text{N}) \sin 30.0° = -5.50\ \text{N}$
22.0 N	0 N	$-22.0\ \text{N}$

Notice the + and − signs to indicate direction.

(2) The resultant $\vec{\mathbf{R}}$ has components $R_x = \Sigma F_x$ and $R_y = \Sigma F_y$, where we read ΣF_x as "the sum of all the x-force components." We then have

$$R_x = 19.0\ \text{N} + 7.50\ \text{N} - 11.3\ \text{N} - 9.53\ \text{N} + 0\ \text{N} = +5.7\ \text{N}$$
$$R_y = 0\ \text{N} + 13.0\ \text{N} + 11.3\ \text{N} - 5.50\ \text{N} - 22.0\ \text{N} = -3.2\ \text{N}$$

(3) The magnitude of the resultant is

$$R = \sqrt{R_x^2 + R_y^2} = 6.5\ \text{N}$$

(4) Finally, we sketch the resultant as shown in Fig. 3-2(b) and find its angle. We see that

$$\tan \phi = \frac{3.2\ \text{N}}{5.7\ \text{N}} = 0.56$$

from which $\phi = 29°$. Then $\theta = 360° - 29° = 331°$. The resultant is 6.5 N at 331° (or −29°) or $\vec{\mathbf{R}} = 6.5\ \text{N}$—331° FROM +$X$-AXIS.

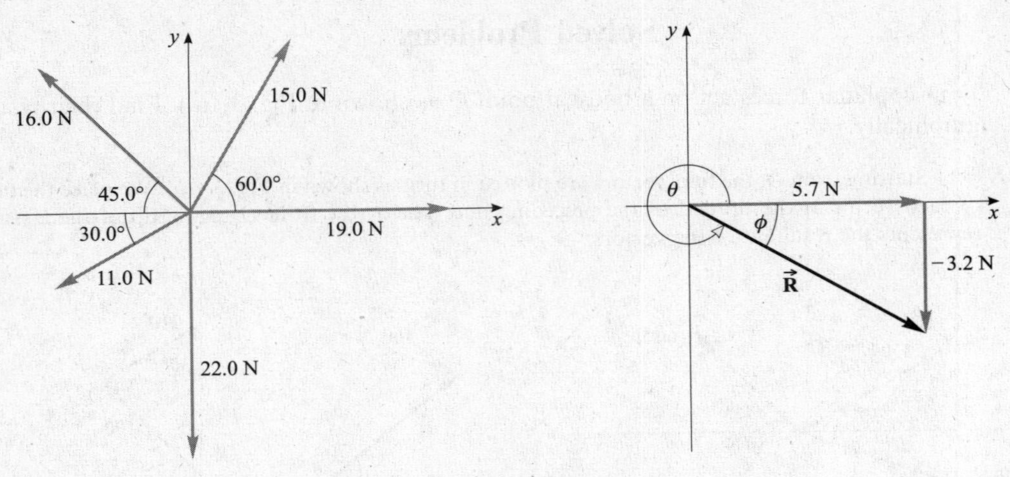

Fig. 3-2

3.3 [II] Solve Problem 3-1 by use of the component method. Give your answer for the magnitude to two significant figures.

The forces and their components are:

Force	x-Component	y-Component
80 N	80 N	0
100 N	(100 N) cos 45° = 71 N	(100 N) sin 45° = 71 N
110 N	−(110 N) cos 30° = −95 N	(110 N) sin 30° = 55 N
160 N	−(160 N) cos 20° = −150 N	−(160 N) sin 20° = −55 N

Notice the sign of each component. To find the resultant, we have

$$R_x = \Sigma F_x = 80\text{ N} + 71\text{ N} - 95\text{ N} - 150\text{ N} = -94\text{ N}$$
$$R_y = \Sigma F_y = 0 + 71\text{ N} + 55\text{ N} - 55\text{ N} = 71\text{ N}$$

The resultant is shown in Fig. 3-3; there, we see that

$$R = \sqrt{(94\text{ N})^2 + (71\text{ N})^2} = 1.2 \times 10^2\text{ N}$$

Further, $\tan\alpha = (71\text{ N})/(94\text{ N})$, from which $\alpha = 37°$. Therefore the resultant is 118 N at $180° - 37° = 143°$ or $\vec{\mathbf{R}} = 118$ N—143° FROM $+x$-AXIS.

3.4 [II] A force of 100 N makes an angle of θ with the x-axis and has a scalar y-component of 30 N. Find both the scalar x-component of the force and the angle θ. (Remember that the number 100 N has three significant figures whereas 30 N has only two.)

The data are sketched roughly in Fig. 3-4. We wish to find F_x and θ. We know that

$$\sin\theta = \frac{30\text{ N}}{100\text{ N}} = 0.30$$

$\theta = 17.46°$, and thus, to two significant figures, $\theta = 17°$. Then, using the cos θ, we have

$$F_x = (100\text{ N})\cos 17.46° = 95\text{ N}$$

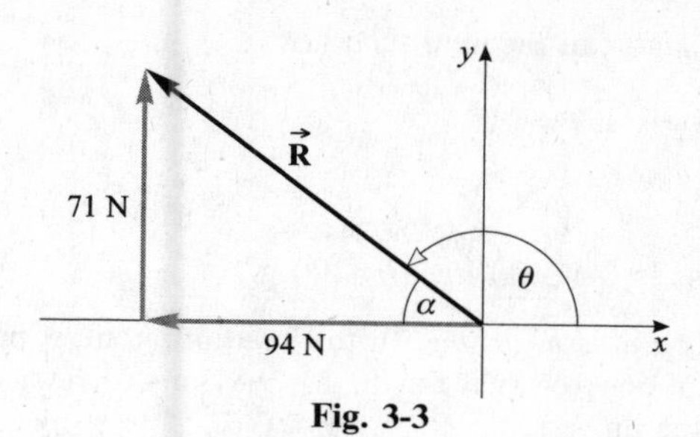

Fig. 3-3

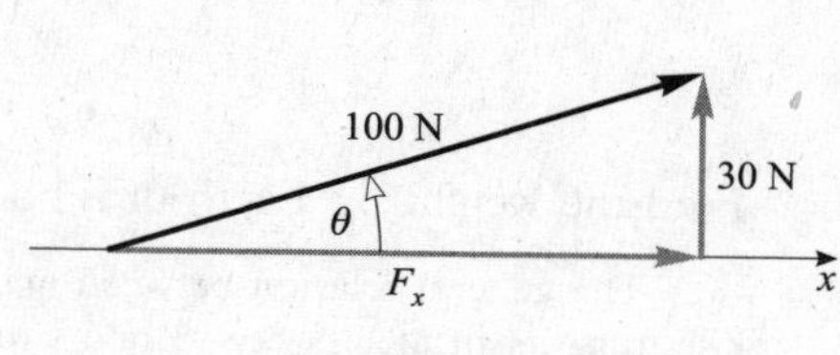

Fig. 3-4

3.5 [I] A child pulls on a rope attached to a sled with a force of 60 N. The rope makes an angle of 40° to the ground. (*a*) Compute the effective value of the pull tending to move the sled along the ground. (*b*) Compute the force tending to lift the sled vertically.

As shown in Fig. 3-5, the components of the 60 N force are 39 N and 46 N. (*a*) The pull along the ground is the horizontal component, 46 N. (*b*) The lifting force is the vertical component, 39 N.

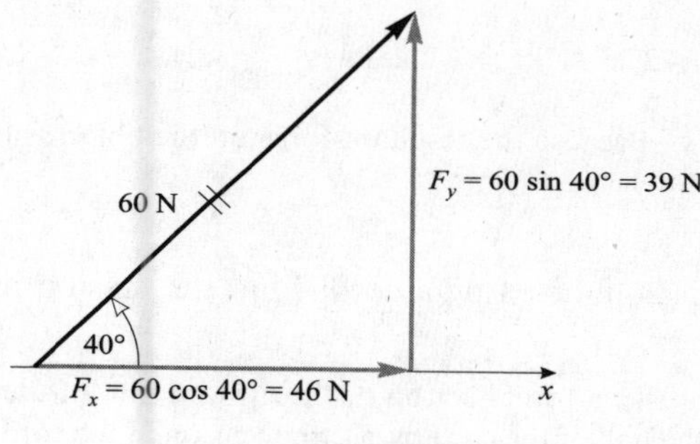

Fig. 3-5

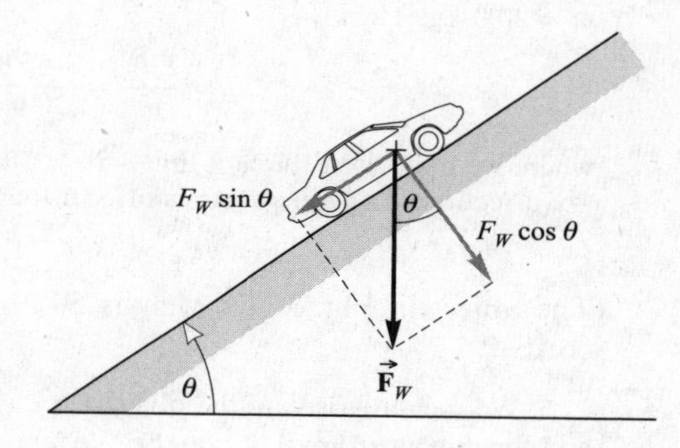

Fig. 3-6

3.6 [I] A car whose weight is F_W is on a ramp which makes an angle θ to the horizontal. How large a perpendicular force must the ramp withstand if it is not to break under the car's weight?

As shown in Fig. 3-6, the car's weight is a force $\vec{\mathbf{F}}_W$ that pulls straight down on the car. We take components of $\vec{\mathbf{F}}$ along the incline and perpendicular to it. The ramp must balance the force component $F_W \cos\theta$ if the car is not to crash through the ramp.

3.7 [II] Three forces that act on a particle are given by $\vec{\mathbf{F}}_1 = (20\hat{\mathbf{i}} - 36\hat{\mathbf{j}} + 73\hat{\mathbf{k}})$ N, $\vec{\mathbf{F}}_2 = (-17\hat{\mathbf{i}} + 21\hat{\mathbf{j}} - 46\hat{\mathbf{k}})$ N, and $\vec{\mathbf{F}}_3 = (-12\hat{\mathbf{k}})$ N. Find their resultant vector. Also find the magnitude of the resultant to two significant figures.

We know that

$$R_x = \Sigma F_x = 20\text{ N} - 17\text{ N} + 0\text{ N} = 3\text{ N}$$
$$R_y = \Sigma F_y = -36\text{ N} + 21\text{ N} + 0\text{ N} = -15\text{ N}$$
$$R_z = \Sigma F_z = 73\text{ N} - 46\text{ N} - 12\text{ N} = 15\text{ N}$$

Since $\vec{\mathbf{R}} = R_x\hat{\mathbf{i}} + R_y\hat{\mathbf{j}} + R_z\hat{\mathbf{k}}$, we find

$$\vec{\mathbf{R}} = 3\hat{\mathbf{i}} - 15\hat{\mathbf{j}} + 15\hat{\mathbf{k}}$$

To two significant figures, the three-dimensional pythagorean theorem then gives

$$R = \sqrt{R_x^2 + R_y^2 + R_z^2} = \sqrt{459} = 21 \text{ N}$$

3.8 [I] Find the weight on Earth of a body whose mass is (*a*) 3.00 kg, (*b*) 200 g.

The general relation between mass m and weight F_W is $F_W = mg$. In this relation, m must be in kilograms, g in meters per second squared, and F_W in newtons. On Earth, $g = 9.81 \text{ m/s}^2$. The acceleration due to gravity varies from place to place in the universe.

(*a*) $$F_W = (3.00 \text{ kg})(9.81 \text{ m/s}^2) = 29.4 \text{ kg·m/s}^2 = 29.4 \text{ N}$$

(*b*) $$F_W = (0.200 \text{ kg})(9.81 \text{ m/s}^2) = 1.96 \text{ N}$$

3.9 [I] A 20.0 kg object that can move freely is subjected to a resultant force of 45.0 N in the $-x$-direction. Find the acceleration of the object.

We make use of the second law in component form, $\Sigma F_x = ma_x$, with $\Sigma F_x = -45.0$ N and $m = 20.0$ kg. Then

$$a_x = \frac{\Sigma F_x}{m} = \frac{-45.0 \text{ N}}{20.0 \text{ kg}} = -2.25 \text{ N/kg} = -2.25 \text{ m/s}^2$$

where we have used the fact that $1 \text{ N} = 1 \text{ kg·m/s}^2$. Because the resultant force on the object is in the $-x$-direction, its acceleration is also in that direction.

3.10 [I] The object in Fig. 3-7(*a*) weighs 50 N and is supported by a cord. Find the tension in the cord.

We mentally isolate the object for discussion. Two forces act on it, the upward pull of the cord and the downward pull of gravity. We represent the pull of the cord by F_T, the tension in the cord. The pull of gravity, the weight of the object, is $F_W = 50$ N. These two forces are shown in the free-body diagram in Fig. 3-1(*b*).

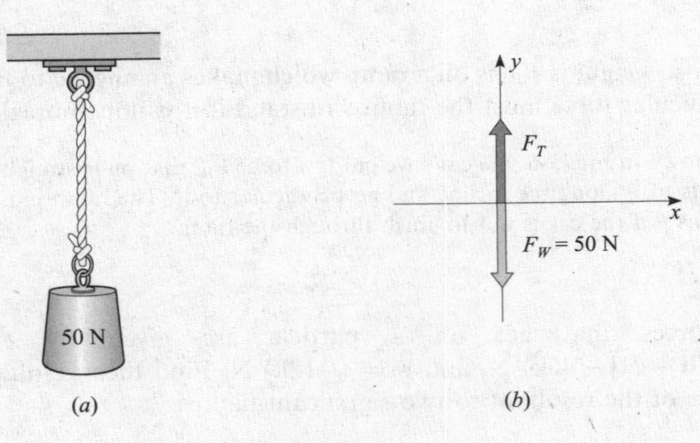

Fig. 3-7

The forces are already in component form and so we can write the first condition for equilibrium at once, taking *up* and to the *right* as positive directions:

$$\overset{+}{\rightarrow} \Sigma F_x = 0 \quad \text{becomes} \quad 0 = 0$$

$$+\uparrow \Sigma F_y = 0 \quad \text{becomes} \quad F_T - 50 \text{ N} = 0$$

from which $F_T = 50$ N. Thus, when a single vertical cord supports a body at equilibrium, the tension in the cord equals the weight of the body.

3.11 [I] A 5.0 kg object is to be given an upward acceleration of 0.30 m/s^2 by a rope pulling straight upward on it. What must be the tension in the rope?

The free-body diagram for the object is shown in Fig. 3-8. The tension in the rope is F_T, and the weight of the object is $F_W = mg = (5.0 \text{ kg})(9.81 \text{ m/s}^2) = 49.1$ N. Using $\Sigma F_y = ma_y$ with *up* taken as positive, we have

$$F_T - mg = ma_y \qquad \text{or} \qquad F_T - 49.1 \text{ N} = (5.0 \text{ kg})(0.30 \text{ m/s}^2)$$

from which $F_T = 50.6 \text{ N} = 51$ N. As a check, we notice that F_T is larger than F_W as it must be if the object is to accelerate upward.

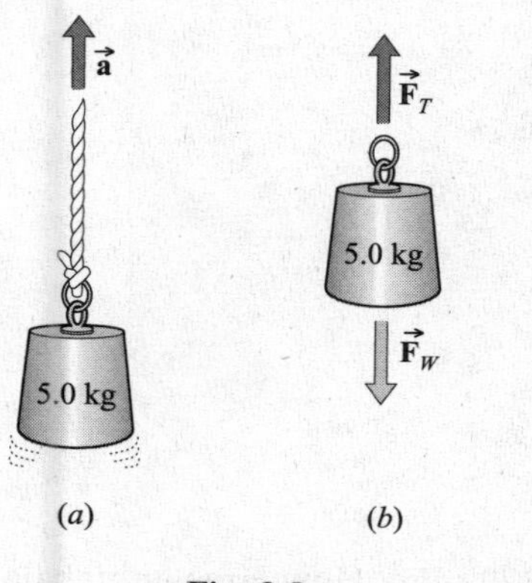

Fig. 3-8

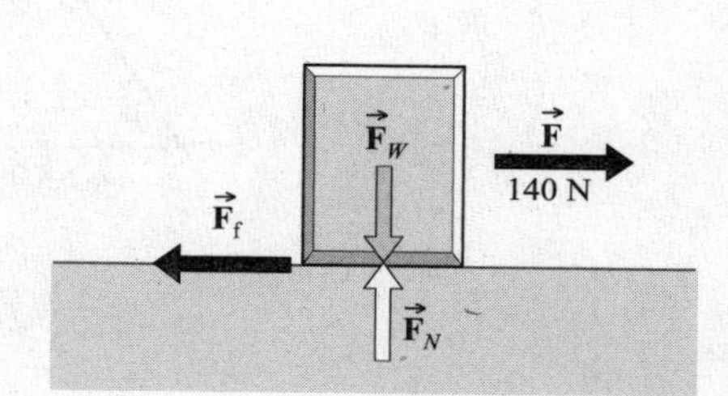

Fig. 3-9

3.12 [II] A horizontal force of 140 N is needed to pull a 60.0 kg box across the horizontal floor at constant speed. What is the coefficient of friction between floor and box? Determine it to three significant figures even though that's quite unrealistic.

The free-body diagram for the box is shown in Fig. 3-9. Because the box does not move up or down, $a_y = 0$. Therefore,

$$\Sigma F_y = ma_y \qquad \text{gives} \qquad F_N - mg = (m)(0 \text{ m/s}^2)$$

from which we find that $F_N = mg = (60.0 \text{ kg})(9.81 \text{ m/s}^2) = 588.6$ N. Further, because the box is moving horizontally at constant speed, $a_x = 0$ and so

$$\Sigma F_x = ma_x \qquad \text{gives} \qquad 140 \text{ N} - F_f = 0$$

from which the friction force is $F_f = 140$ N. We then have

$$\mu_k = \frac{F_f}{F_N} = \frac{140 \text{ N}}{588.6 \text{ N}} = 0.238$$

3.13 [II] The only force acting on a 5.0 kg object has components $F_x = 20$ N and $F_y = 30$ N. Find the acceleration of the object.

We make use of $\Sigma F_x = ma_x$ and $\Sigma F_y = ma_y$ to obtain

$$a_x = \frac{\Sigma F_x}{m} = \frac{20\ \text{N}}{5.0\ \text{kg}} = 4.0\ \text{m/s}^2$$

$$a_y = \frac{\Sigma F_y}{m} = \frac{30\ \text{N}}{5.0\ \text{kg}} = 6.0\ \text{m/s}^2$$

These components of the acceleration are shown in Fig. 3-10. From the figure, we see that

$$a = \sqrt{(4.0)^2 + (6.0)^2}\ \text{m/s}^2 = 7.2\ \text{m/s}^2$$

and $\theta = \arctan(6.0/4.0) = 56°$.

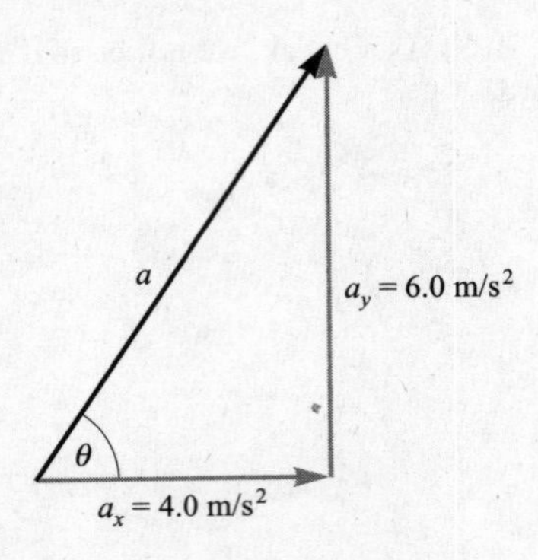

Fig. 3-10

3.14 [II] A 600 N object is to be given an acceleration of 0.70 m/s². How large an unbalanced force must act upon it?

Notice that the weight, not the mass, of the object is given. Assuming the weight was measured on the Earth, we use $F_W = mg$ to find

$$m = \frac{F_W}{g} = \frac{600\ \text{N}}{9.81\ \text{m/s}^2} = 61\ \text{kg}$$

Now that we know the mass of the object (61 kg) and the desired acceleration (0.70 m/s²), we have

$$F = ma = (61\ \text{kg})(0.70\ \text{m/s}^2) = 43\ \text{N}$$

3.15 [III] A constant force acts on a 5.0 kg object and reduces its velocity from 7.0 m/s to 3.0 m/s in a time of 3.0 s. Find the force.

We must first find the acceleration of the object, which is constant because the force is constant. Taking the direction of motion as positive, from Chapter 2 we have

$$a = \frac{v_f - v_i}{t} = \frac{-4.0\ \text{m/s}}{3.0\ \text{s}} = -1.33\ \text{m/s}^2$$

Now we can use $F = ma$ with $m = 5.0$ kg:

$$F = (5.0\ \text{kg})(-1.33\ \text{m/s}^2) = -6.7\ \text{N}$$

The minus sign indicates that the force is a retarding force, directed opposite to the motion.

3.16 [II] A 400-g block with an initial speed of 80 cm/s slides along a horizontal tabletop against a friction force of 0.70 N. (*a*) How far will it slide before stopping? (*b*) What is the coefficient of friction between the block and the tabletop?

(*a*) We take the direction of motion as positive. The only unbalanced force acting on the block is the friction force, −0.70 N. Therefore,

$$\Sigma F = ma \qquad \text{becomes} \qquad -0.70\ \text{N} = (0.400\ \text{kg})(a)$$

from which $a = -1.75\ \text{m/s}^2$. (Notice that m is always in kilograms.) To find the distance the block slides, we have $v_{ix} = 0.80$ m/s, $v_{fx} = 0$, and $a = -1.75\ \text{m/s}^2$. Then $v_{fx}^2 - v_{ix}^2 = 2ax$ gives

$$x = \frac{v_{fx}^2 - v_{ix}^2}{2a} = \frac{(0 - 0.64)\ \text{m}^2/\text{s}^2}{(2)(-1.75\ \text{m/s}^2)} = 0.18\ \text{m}$$

(*b*) Because the vertical forces on the block must cancel, the upward push of the table F_N must equal the weight mg of the block. Then

$$\mu_k = \frac{\text{friction force}}{F_N} = \frac{0.70\ \text{N}}{(0.40\ \text{kg})(9.81\ \text{m/s}^2)} = 0.18$$

3.17 [II] A 600-kg car is moving on a level road at 30 m/s. (*a*) How large a retarding force (assumed constant) is required to stop it in a distance of 70 m? (*b*) What is the minimum coefficient of friction between tires and roadway if this is to be possible? Assume the wheels are not locked, in which case we are dealing with static friction – there's no sliding.

(*a*) We must first find the car's acceleration from a motion equation. It is known that $v_{ix} = 30$ m/s, $v_{fx} = 0$, and $x = 70$ m. We use $v_{fx}^2 = v_{ix}^2 + 2ax$ to find

$$a = \frac{v_{fx}^2 - v_{ix}^2}{2x} = \frac{0 - 900\ \text{m}^2/\text{s}^2}{140\ \text{m}} = -6.43\ \text{m/s}^2$$

Now we can write

$$F = ma = (600\ \text{kg})(-6.43\ \text{m/s}^2) = -3860\ \text{N} = -3.9\ \text{kN}$$

(*b*) The force found in (*a*) is supplied as the friction force between the tires and roadway. Therefore, the magnitude of the friction force on the tires is $F_f = 3860$ N. The coefficient of friction is given by $\mu_s = F_f/F_N$, where F_N is the normal force. In the present case, the roadway pushes up on the car with a force equal to the car's weight. Therefore,

$$F_N = F_W = mg = (600\ \text{kg})(9.81\ \text{m/s}^2) = 5886\ \text{N}$$

so that

$$\mu_s = \frac{F_f}{F_N} = \frac{3860}{5886} = 0.66$$

The coefficient of friction must be at least 0.66 if the car is to stop within 70 m.

3.18 [I] An 8000-kg engine pulls a 40 000-kg train along a level track and gives it an acceleration $a_1 = 1.20\ \text{m/s}^2$. What acceleration (a_2) would the engine give to a 16 000-kg train?

For a given engine force, the acceleration is inversely proportional to the total mass. Thus

$$a_2 = \frac{m_1}{m_2} a_1 = \frac{8000\ \text{kg} + 40\,000\ \text{kg}}{8000\ \text{kg} + 16\,000\ \text{kg}}(1.20\ \text{m/s}^2) = 2.40\ \text{m/s}^2$$

3.19 [I] As shown in Fig. 3-11(a), an object of mass m is supported by a cord. Find the tension in the cord if the object is (a) at rest, (b) moving at constant velocity, (c) accelerating upward with acceleration $a = 3g/2$, and (d) accelerating downward at $a = 0.75g$.

Two forces act on the object: the tension F_T upward and the downward pull of gravity mg. They are shown in the free-body diagram in Fig. 3-11(b). We take *up* as the positive direction and write $\Sigma F_y = ma_y$ in each case.

(a)	$a_y = 0$:	$F_T - mg = ma_y = 0$	or	$F_T = mg$
(b)	$a_y = 0$:	$F_T - mg = ma_y = 0$	or	$F_T = mg$
(c)	$a_y = 3g/2$:	$F_T - mg = m(3g/2)$	or	$F_T = 2.5mg$
(d)	$a_y = -3g/4$:	$F_T - mg = m(-3g/4)$	or	$F_T = 0.25mg$

Notice that the tension in the cord is less than mg in part (d); only then can the object have a downward acceleration. Can you explain why $F_T = 0$ if $a_y = -g$?

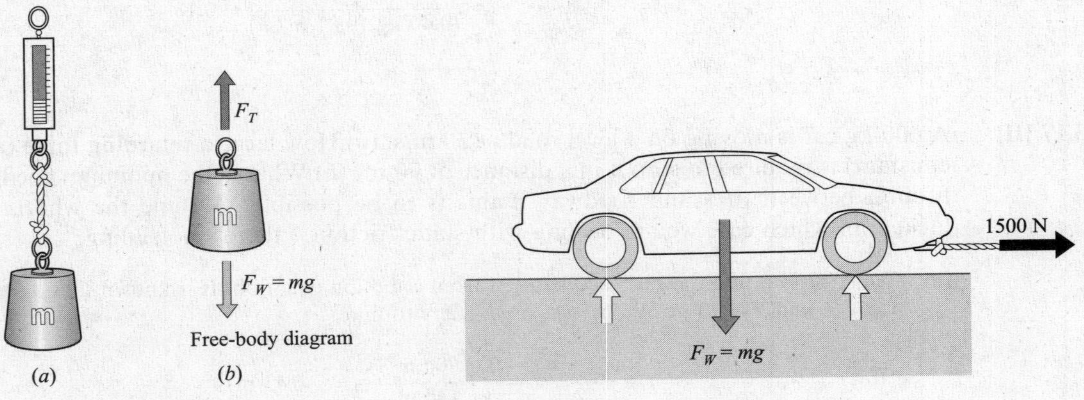

Fig. 3-11

Fig. 3-12

3.20 [I] A tow rope will break if the tension in it exceeds 1500 N. It is used to tow a 700-kg car along level ground. What is the largest acceleration the rope can give to the car? (Remember that 1500 has four significant figures; see Appendix A.)

The forces acting on the car are shown in Fig. 3-12. Only the x-directed force is of importance, because the y-directed forces balance each other. Indicating the positive direction with a + sign and a little arrow we write,

$$\underset{\rightarrow}{+}\Sigma F_x = ma_x \qquad \text{becomes} \qquad 1500\ \text{N} = (700\ \text{kg})(a)$$

from which $a = 2.14\ \text{m/s}^2$.

3.21 [I] Compute the least acceleration with which a 45-kg woman can slide down a rope if the rope can withstand a tension of only 300 N.

The weight of the woman is $mg = (45\ \text{kg})(9.81\ \text{m/s}^2) = 441\ \text{N}$. Because the rope can support only 300 N, the unbalanced downward force F on the woman must be at least $441\ \text{N} - 300\ \text{N} = 141\ \text{N}$. Her minimum downward acceleration is then

$$a = \frac{F}{m} = \frac{141\ \text{N}}{45\ \text{kg}} = 3.1\ \text{m/s}^2$$

3.22 [II] A 70-kg box is slid along the floor by a 400-N force as shown in Fig. 3-13. The coefficient of friction between the box and the floor is 0.50 when the box is sliding. Find the acceleration of the box.

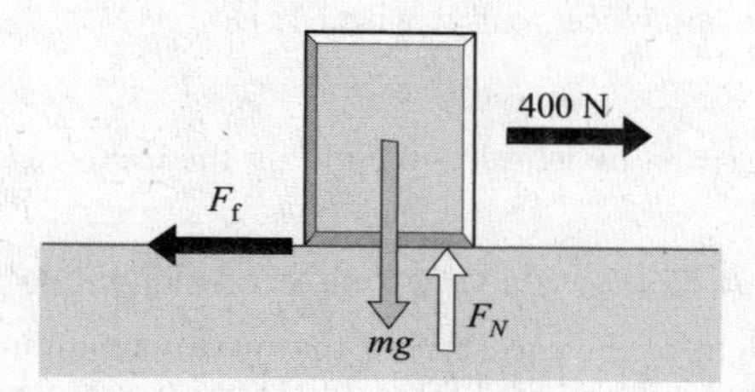

Fig. 3-13

Since the y-directed forces must balance,

$$F_N = mg = (70\text{ kg})(9.81\text{ m/s}^2) = 687\text{ N}$$

But the friction force F_f is given by

$$F_f = \mu_k F_N = (0.50)(687\text{ N}) = 344\text{ N}$$

Now write $\Sigma F_x = ma_x$ for the box, taking the direction of motion as positive:

$$400\text{ N} - 344\text{ N} = (70\text{ kg})(a) \qquad \text{or} \qquad a = 0.80\text{ m/s}^2$$

3.23 [II] Suppose, as shown in Fig. 3-14, that a 70-kg box is pulled by a 400-N force at an angle of 30° to the horizontal. The coefficient of kinetic friction is 0.50. Find the acceleration of the box.

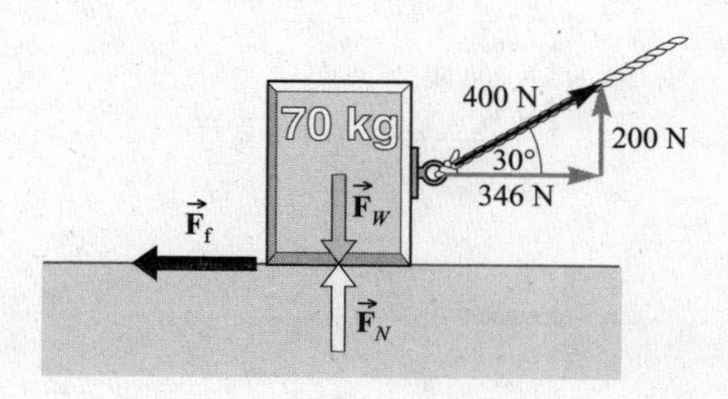

Fig. 3-14

Because the box does not move up or down, we have $\Sigma F_y = ma_y = 0$. From Fig. 3-14, we see that this equation is

$$F_N + 200\text{ N} - mg = 0$$

But $mg = (70\text{ kg})(9.81\text{ m/s}^2) = 687\text{ N}$, and it follows that $F_N = 486\text{ N}$.

We next find the friction force acting on the box:

$$F_f = \mu_k F_N = (0.50)(486\text{ N}) = 243\text{ N}$$

Now let us write $\Sigma F_x = ma_x$ for the box. It is

$$(346 - 243)\text{ N} = (70\text{ kg})(a_x)$$

from which $a_x = 1.5\text{ m/s}^2$.

3.24 [III] A car moving at 20 m/s along a horizontal road has its brakes suddenly applied and eventually comes to rest. What is the shortest distance in which it can be stopped if the friction coefficient between tires and road is 0.90? Assume that all four wheels brake identically. If the brakes don't lock the car stops via static friction.

The friction force at one wheel, call it wheel 1, is

$$F_{f1} = \mu_s F_{N1} = \mu F_{W1}$$

where F_{W1} is the weight carried by wheel 1. We obtain the total friction force F_f by adding such terms for all four wheels:

$$F_f = \mu_s F_{W1} + \mu_s F_{W2} + \mu_s F_{W3} + \mu_s F_{W4} = \mu_s(F_{W1} + F_{W2} + F_{W3} + F_{W4}) = \mu_s F_W$$

where F_W is the total weight of the car. (Notice that we are assuming optimal braking at each wheel.) This friction force is the only unbalanced force on the car (we neglect wind friction and such). Writing $F = ma$ for the car with F replaced by $-\mu_s F_W$ gives $-\mu_s F_W = ma$, where m is the car's mass and the positive direction is taken as the direction of motion. However, $F_W = mg$; so the car's acceleration is

$$a = -\frac{\mu_s F_W}{m} = -\frac{\mu_s mg}{m} = -\mu_s g = (-0.90)(9.81 \text{ m/s}^2) = -8.8 \text{ m/s}^2$$

We can find how far the car went before stopping by solving a motion problem. Knowing that $v_i = 20$ m/s, $v_f = 0$, and $a = -8.8\,\text{m/s}^2$, we find from $v_f^2 - v_i^2 = 2ax$ that

$$x = \frac{(0 - 400)\text{ m}^2/\text{s}^2}{-17.6 \text{ m/s}^2} = 23 \text{ m}$$

If the four wheels had not all been braking optimally, the stopping distance would have been longer.

3.25 [II] As shown in Fig. 3-15, a force of 400 N pushes on a 25-kg box. Starting from rest, the box achieves a velocity of 2.0 m/s in a time of 4.0 s. Find the coefficient of kinetic friction between box and floor.

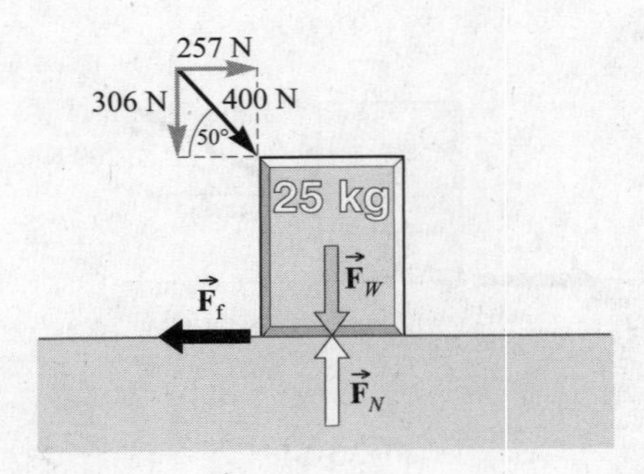

Fig. 3-15

We will need to find f by use of $F = ma$. But first we must find a from a motion problem. We know that $v_i = 0$, $v_f = 2.0$ m/s, $t = 4.0$ s. Using $v_f = v_i + at$ gives

$$a = \frac{v_f - v_i}{t} = \frac{2.0\,\text{m/s}}{4.0\,\text{s}} = 0.50 \text{ m/s}^2$$

Now we can write $\Sigma F_x = ma_x$, where $a_x = a = 0.50$ m/s^2. From Fig. 3-9, this equation becomes

$$257 \text{ N} - F_f = (25 \text{ kg})(0.50 \text{ m/s}^2) \qquad \text{or} \qquad F_f = 245 \text{ N}$$

We now wish to use $\mu = F_f/F_N$. To find F_N we write $\Sigma F_y = ma_y = 0$, since no vertical motion occurs. From Fig. 3-15,

$$F_N - 306 \text{ N} - (25)(9.81) \text{ N} = 0 \qquad \text{or} \qquad F_N = 551 \text{ N}$$

Then

$$\mu_k = \frac{F_f}{F_N} = \frac{245}{551} = 0.44$$

3.26 [I] A 200-N wagon is to be pulled up a 30° incline at constant speed. How large a force parallel to the incline is needed if friction effects are negligible?

The situation is shown in Fig. 3-16(*a*). Because the wagon moves at a constant speed along a straight line, its velocity vector is constant. Therefore the wagon is in translational equilibrium, and the first condition for equilibrium applies to it.

We isolate the wagon as the object. Three nonnegligible forces act on it: (1) the pull of gravity F_W (its weight), directed straight down; (2) the force F exerted on the wagon parallel to the incline to pull it up the incline; (3) the push F_N of the incline that supports the wagon. These three forces are shown in the free-body diagram in Fig. 3-10(*b*).

For situations involving inclines, it is convenient to take the x-axis parallel to the incline and the y-axis perpendicular to it. After taking components along these axes, we can write the first condition for equilibrium:

$$\overset{+}{\nearrow}\ \Sigma F_x = 0 \quad \text{becomes} \quad F - 0.50\ F_W = 0$$

$$\underset{+}{\nwarrow}\ \Sigma F_y = 0 \quad \text{becomes} \quad F_N - 0.87\ F_W = 0$$

Solving the first equation and recalling that $F_W = 200$ N, we find that $F = 0.50\ F_W$. The required pulling force to two significant figures is 0.10 kN.

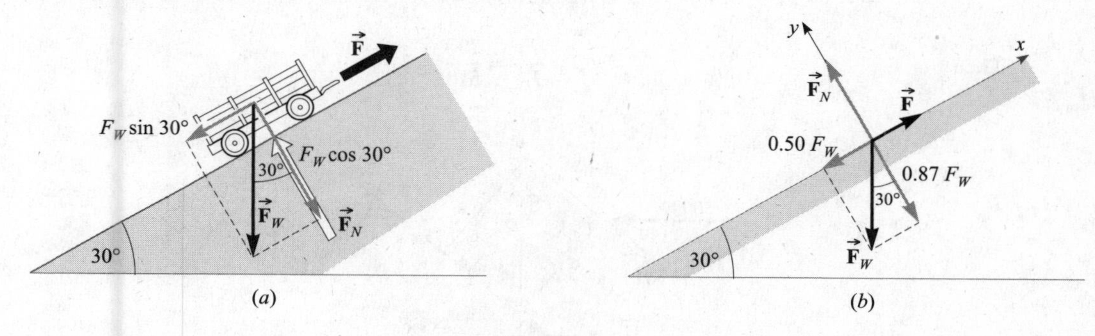

Fig. 3-16

3.27 [II] A 20-kg box sits on an incline as shown in Fig. 3-17. The coefficient of kinetic friction between box and incline is 0.30. Find the acceleration of the box down the incline.

In solving inclined-plane problems, we take x- and y-axes as shown in the figure, parallel and perpendicular to the incline. We shall find the acceleration by writing $\Sigma F_x = ma_x$. But first we must find the friction force F_f. Using the fact that $\cos 30° = 0.866$,

$$F_y = ma_y = 0 \quad \text{gives} \quad F_N - 0.87mg = 0$$

from which $F_N = (0.87)(20\ \text{kg})(9.81\ \text{m/s}^2) = 171$ N. Now we can find F_f from

$$F_f = \mu_k F_N = (0.30)(171\ \text{N}) = 51\ \text{N}$$

Writing $\Sigma F_x = ma_x$, we have

$$F_f - 0.50mg = ma_x \quad \text{or} \quad 51\ \text{N} - (0.50)(20)(9.81)\ \text{N} = (20\ \text{kg})(a_x)$$

from which $a_x = -2.35$ m/s^2. The box accelerates down the incline at 2.4 m/s^2.

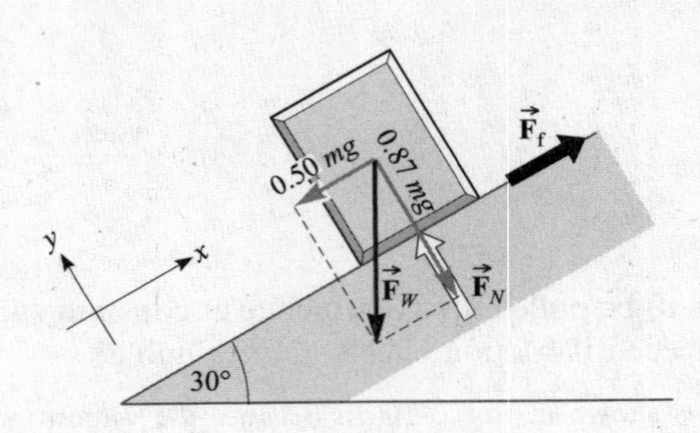

Fig. 3-17

3.28 [III] When a force of 500 N pushes on a 25-kg box as shown in Fig. 3-18, the acceleration of the box up the incline is 0.75 m/s^2. Find the coefficient of kinetic friction between box and incline.

The acting forces and their components are shown in Fig. 3-18. Notice how the x- and y-axes are taken. Since the box moves up the incline, the friction force (which always acts to retard the motion) is directed down the incline.

Let us first find F_f by writing $\Sigma F_x = ma_x$. From Fig. 3-18, using $\sin 40° = 0.643$,

$$383 \text{ N} - F_f - (0.64)(25)(9.81) \text{ N} = (25 \text{ kg})(0.75 \text{ m/s}^2)$$

from which $F_f = 207$ N.

We also need F_N. Writing $\Sigma F_y = ma_y = 0$, and using $\cos 40° = 0.766$, we get

$$F_N - 321 \text{ N} - (0.77)(25)(9.81) \text{ N} = 0 \qquad \text{or} \qquad F_N = 510 \text{ N}$$

Then

$$\mu_k = \frac{F_f}{F_N} = \frac{207}{510} = 0.41$$

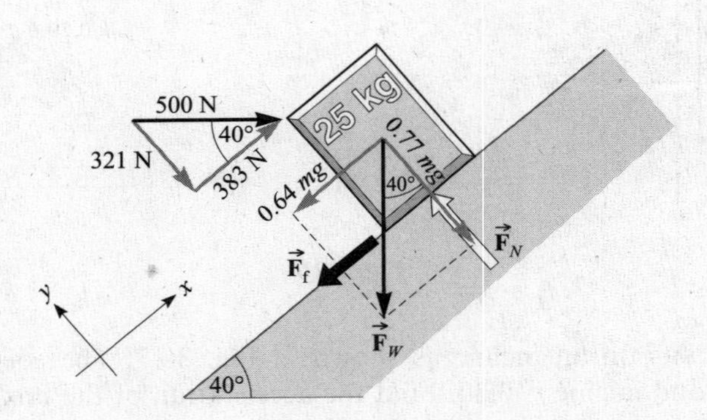

Fig. 3-18

3.29 [III] Two blocks, of masses m_1 and m_2, are pushed by a force F as shown in Fig. 3-19. The coefficient of friction between each block and the table is 0.40. (*a*) What must be the value of F if the blocks are to have an acceleration of 200 cm/s^2? How large a force does m_1 then exert on m_2? Use $m_1 = 300$ g and $m_2 = 500$ g. Remember to work in SI units.

The friction forces on the blocks are $F_{f1} = 0.4m_1 g$ and $F_{f2} = 0.4m_2 g$. We take the two blocks in combination as the object for discussion; the horizontal forces on the object from outside (i.e. the *external* forces on it) are F, F_{f1}, and F_{f2}. Although the two blocks do push on each other, the pushes are *internal* forces; they are not part of the unbalanced external force on the two-mass object. For that object,

$$\Sigma F_x = ma_x \qquad \text{becomes} \qquad F - F_{f1} - F_{f2} = (m_1 + m_2)a_x$$

(*a*) Solving for F and substituting known values, we find

$$F = 0.40\,g(m_1 + m_2) + (m_1 + m_2)a_x = 3.14\ \text{N} + 1.60\ \text{N} = 4.7\ \text{N}$$

(*b*) Now consider block m_2 alone. The forces acting on it in the x-direction are the push of block m_1 on it (which we represent by F_b) and the retarding friction force $F_{f2} = 0.4m_2g$. Then, for it,

$$\Sigma F_x = ma_x \qquad \text{becomes} \qquad F_b - F_{f2} = m_2a_x$$

We know that $a_x = 2.0\ \text{m/s}^2$ and so

$$F_b = F_{f2} + m_2a_x = 1.96\ \text{N} + 1.00\ \text{N} = 2.96\ \text{N} = 3.0\ \text{N}$$

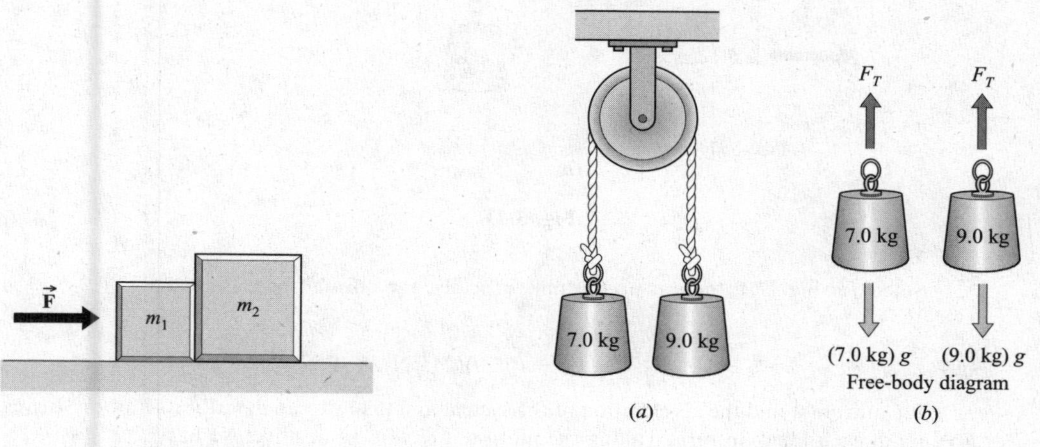

Fig. 3-19 **Fig. 3-20**

3.30 [II] A cord passing over an easily turned pulley (one that is both massless and frictionless) has a 7.0-kg mass hanging from one end and a 9.0-kg mass hanging from the other, as shown in Fig. 3-20. (This arrangement is called *Atwood's machine*.) Find the acceleration of the masses and the tension in the cord.

Because the pulley is easily turned, the tension in the cord will be the same on each side. The forces acting on each of the two masses are drawn in Fig. 3-20. Recall that the weight of an object is mg.

It is convenient in situations involving objects connected by cords to take the direction of motion as the positive direction. In the present case, we take *up* positive for the 7.0-kg mass, and *down* positive for the 9.0-kg mass. (If we do this, the acceleration will be positive for each mass. Because the cord doesn't stretch, the accelerations are numerically equal.) Writing $\Sigma F_y = ma_y$ for each mass in turn, we have

$$F_T - (7.0)(9.81)\ \text{N} = (7.0\ \text{kg})(a) \qquad \text{and} \qquad (9.0)(9.81)\ \text{N} - F_T = (9.0\ \text{kg})(a)$$

If we add these two equations, the unknown F_T drops out, giving

$$(9.0 - 7.0)(9.81)\ \text{N} = (16\ \text{kg})(a)$$

for which $a = 1.23\ \text{m/s}^2$. We can now substitute $1.23\ \text{m/s}^2$ for a in either equation and obtain $F_T = 77\ \text{N}$.

3.31 [III] In Fig. 3-21, the coefficient of kinetic friction between block A and the table is 0.20. Also, $m_A = 25$ kg, $m_B = 15$ kg. How far will block B drop in the first 3.0 s after the system is released?

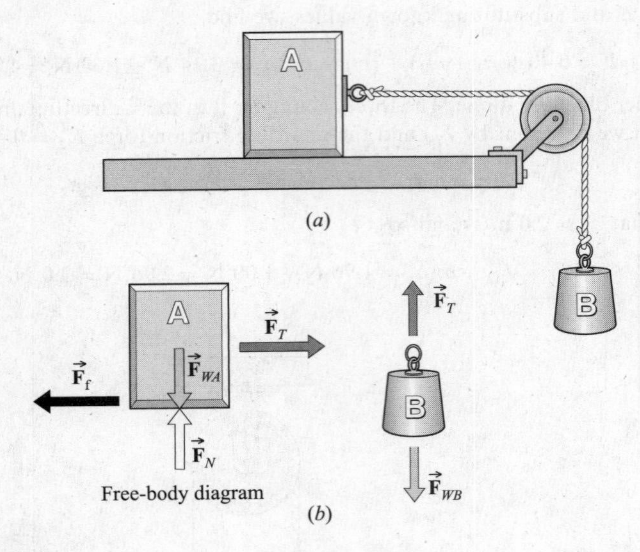

Fig. 3-21

Since, for block A, there is no motion vertically, the normal force is

$$F_N = m_A g = (25\text{ kg})(9.81\text{ m/s}^2) = 245\text{ N}$$

and

$$F_f = \mu_k F_N = (0.20)(245\text{ N}) = 49\text{ N}$$

We must first find the acceleration of the system and then we can describe its motion. Let us apply $F = ma$ to each block in turn. Taking the motion direction as positive, we have

$$F_T - F_f = m_A a \quad \text{or} \quad F_T - 49\text{ N} = (25\text{ kg})(a)$$

and

$$m_B g - F_T = m_B a \quad \text{or} \quad -F_T + (15)(9.81)\text{ N} = (15\text{ kg})(a)$$

We can eliminate F_T by adding the two equations. Then, solving for a, we find $a = 2.45\text{ m/s}^2$.

Now we can work a motion problem with $a = 2.45\text{ m/s}^2$, $v_i = 0$, $t = 3.0$ s:

$$y = v_{iy}t + \tfrac{1}{2}at^2 \quad \text{gives} \quad y = 0 + \tfrac{1}{2}(2.45\text{ m/s}^2)(3.0\text{ s})^2 = 11\text{ m}$$

as the distance B falls in the first 3.0 s.

3.32 [II] How large a horizontal force in addition to F_T must pull on block A in Fig. 3-21 to give it an acceleration of 0.75 m/s^2 *toward the left*? Assume, as in Problem 3.31, that $\mu_k = 0.20$, $m_A = 25$ kg, and $m_B = 15$ kg.

If we were to redraw Fig 3-21 for this case, we would show a force F pulling toward the left on A. In addition, the retarding friction force F_f should be reversed in direction in the figure. As in Problem 3.31, $F_f = 49$ N.

We write $F = ma$ for each block in turn, taking the direction of motion to be positive. We have

$$F - F_T - 49\text{ N} = (25\text{ kg})(0.75\text{ m/s}^2) \quad \text{and} \quad F_T - (15)(9.81)\text{ N} = (15\text{ kg})(0.75\text{ m/s}^2)$$

We solve the last equation for F_T and substitute in the previous equation. We can then solve for the single unknown F, and we find it to be 226 N or 0.23 kN.

3.33 [II] The coefficient of static friction between a box and the flat bed of a truck is 0.60. What is the maximum acceleration the truck can have along level ground if the box is not to slide?

The box experiences only one x-directed force, the friction force. When the box is on the verge of slipping, $F_f = \mu_s F_W$, where F_W is the weight of the box.

As the truck accelerates, the friction force must cause the box to have the same acceleration as the truck; otherwise, the box will slip. When the box is not slipping, $\Sigma F_x = ma_x$ applied to the box gives $F_f = ma_x$. However, if the box is on the verge of slipping, $F_f = \mu_s F_W$ so that $\mu_s F_W = ma_x$. Because $F_W = mg$, this gives

$$a_x = \frac{\mu_s mg}{m} = \mu_s g = (0.60)(9.81 \text{ m/s}^2) = 5.9 \text{ m/s}^2$$

as the maximum acceleration without slipping.

3.34 [III] In Fig. 3-22, the two boxes have identical masses of 40 kg. Both experience a sliding friction force with $\mu_k = 0.15$. Find the acceleration of the boxes and the tension in the tie cord.

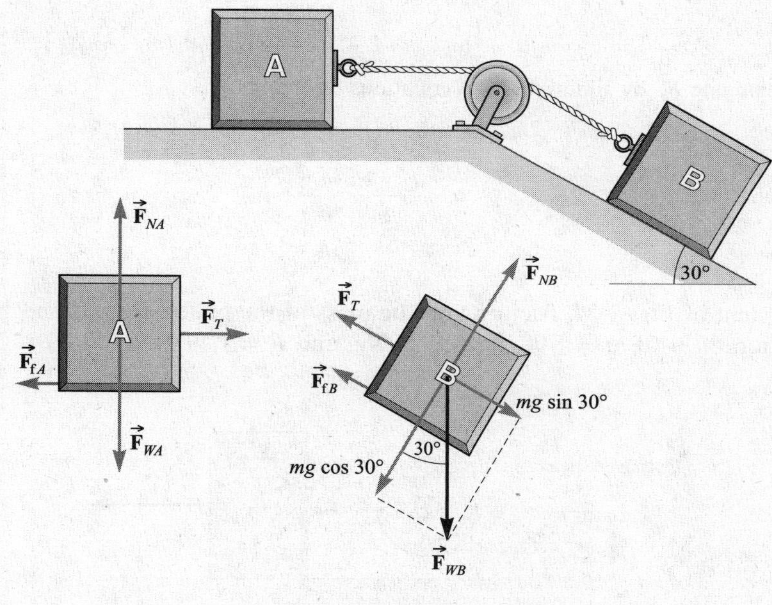

Fig. 3-22

Using $F_f = \mu_k F_N$, we find that the friction forces on the two boxes are

$$F_{fA} = (0.15)(mg) \qquad \text{and} \qquad F_{fB} = (0.15)(0.87mg)$$

But $m = 40$ kg, so $F_{fA} = 59$ N and $F_{fB} = 51$ N.

Let us now apply $\Sigma F_x = ma_x$ to each block in turn, taking the direction of motion as positive. This gives

$$F_T - 59 \text{ N} = (40 \text{ kg})(a) \qquad \text{and} \qquad 0.5mg - F_T - 51 \text{ N} = (40 \text{ kg})(a)$$

Solving these two equations for a and F_T gives $a = 1.1 \text{ m/s}^2$ and $F_T = 0.10$ kN.

3.35 [III] In the system shown in Fig. 3-23(a), force F accelerates block m_1 to the right. Find its acceleration in terms of F and the coefficient of friction μ_k at the contact surfaces.

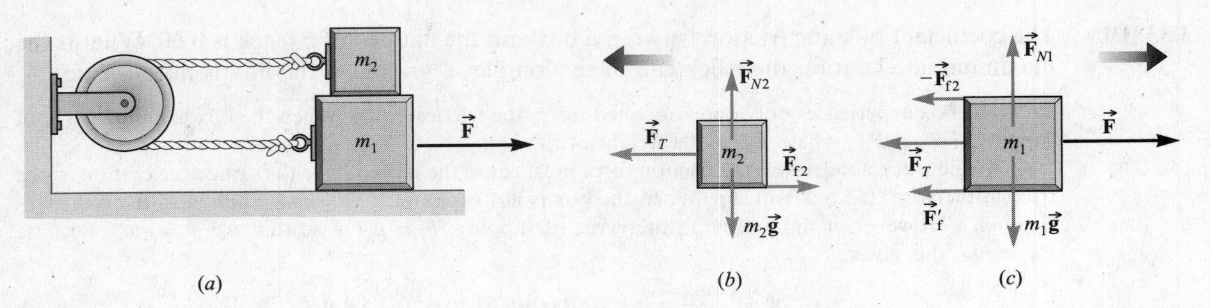

Fig. 3-23

The horizontal forces on the blocks are shown in Fig. 3-23(*b*) and (*c*). Block m_2 is pressed against m_1 by its weight $m_2 g$. This is the normal force where m_1 and m_2 are in contact, so the friction force there is $F_{f2} = \mu_k m_2 g$. At the bottom surface of m_1, however, the normal force is $(m_1 + m_2)g$. Hence, $F'_f = \mu_k(m_1 + m_2)g$. We now write $\Sigma F_x = ma_x$ for each block, taking the direction of motion as positive:

$$F_T = \mu_k m_2 g = m_2 a \qquad \text{and} \qquad F - F_T - \mu m_2 g - \mu_k(m_1 + m_2)g = m_1 a$$

We can eliminate F_T by adding the two equations to obtain

$$F - 2\mu_k m_2 g - \mu_k(m_1 + m_2)(g) = (m_1 + m_2)(a)$$

from which

$$a = \frac{F - 2\mu_k m_2 g}{m_1 + m_2} - \mu_k g$$

3.36 [II] In the system of Fig. 3-24, friction and the mass of the pulley are both negligible. Find the acceleration of m_2 if $m_1 = 300$ g, $m_2 = 500$ g, and $F = 1.50$ N.

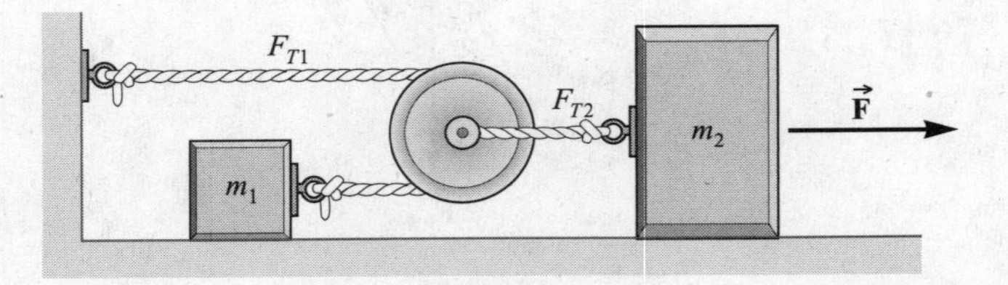

Fig. 3-24

Notice that m_1 has twice as large an acceleration as m_2. (When the pulley moves a distance d, m_1 moves a distance $2d$.) Also notice that the tension F_{T1} in the cord pulling m_1 is half F_{T2}, that in the cord pulling the pulley, because the total force on the pulley must be zero. ($F = ma$ tells us that this is so because the mass of the pulley is zero.) Writing $\Sigma F_x = ma_x$ for each mass, we have

$$F_{T1} = (m_1)(2a) \qquad \text{and} \qquad F - F_{T2} = m_2 a$$

However, we know that $F_{T1} = \frac{1}{2}F_{T2}$ and so the first equation gives $F_{T2} = 4m_1 a$. Substitution in the second equation yields

$$F = (4m_1 + m_2)(a) \qquad \text{or} \qquad a = \frac{F}{4m_1 + m_2} = \frac{1.50\ \text{N}}{1.20\ \text{kg} + 0.50\ \text{kg}} = 0.882\ \text{m/s}^2$$

3.37 [III] In Fig. 3-25, the weights of the objects are 200 N and 300 N. The pulleys are essentially frictionless and massless. Pulley P_1 has a stationary axle, but pulley P_2 is free to move up and down. Find the tensions F_{T1} and F_{T2} and the acceleration of each body.

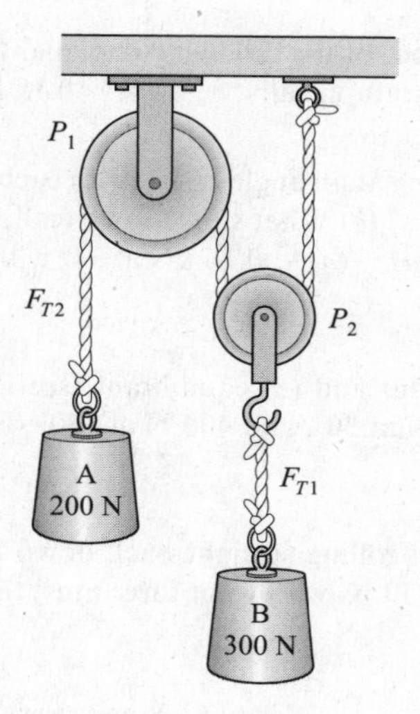

Fig. 3-25

Mass B will rise and mass A will fall. You can see this by noticing that the forces acting on pulley P_2 are $2F_{T2}$ up and F_{T1} down. Since the pulley has no mass, it can have no acceleration, and so $F_{T1} = 2F_{T2}$ (the inertialess object transmits the tension). Twice as large a force is pulling upward on B as on A.

Let a be the downward acceleration of A. Then $a/2$ is the upward acceleration of B. (Why?) We now write $\Sigma F_y = ma_y$ for each mass in turn, taking the direction of motion as positive in each case. We have

$$F_{T1} - 300 \text{ N} = (m_B)(\tfrac{1}{2}a) \qquad \text{and} \qquad 200 \text{ N} - F_{T2} = m_A a$$

But $m = F_W/g$ and so $m_A = (200/9.81)$ kg and $m_B = (300/9.81)$ kg. Further $F_{T1} = 2F_{T2}$. Substitution of these values in the two equations allows us to compute F_{T2} and then F_{T1} and a. The results are

$$F_{T1} = 327 \text{ N} \qquad F_{T2} = 164 \text{ N} \qquad a = 1.78 \text{ m/s}^2$$

3.38 [II] Compute the mass of the Earth, assuming it to be a sphere of radius 6370 km. Give your answer to three significant figures.

Let M be the mass of the Earth, and m the mass of an object on the Earth's surface. The weight of the object is equal to mg. It is also equal to the gravitational force $G(Mm)/r^2$, where r is the Earth's radius. Hence,

$$mg = G\frac{Mm}{r^2}$$

from which
$$M = \frac{gr^2}{G} = \frac{(9.81 \text{ m/s}^2)(6.37 \times 10^6 \text{ m})^2}{6.67 \times 10^{-11} \text{ N}\cdot\text{m}^2/\text{kg}^2} = 5.97 \times 10^{24} \text{ kg}$$

Supplementary Problems

3.39 [I] Two forces act on a point object as follows: 100 N at 170.0° and 100 N at 50.0°. Find their resultant. *Ans.* 100 N at 110°

3.40 [I] Compute algebraically the resultant of the following coplanar forces: 100 N at 30°, 141.4 N at 45°, and 100 N at 240°. Check your result graphically. *Ans.* 0.15 kN at 25°

3.41 [I] Two forces, 80 N and 100 N acting at an angle of 60° with each other, pull on an object. (*a*) What single force would replace the two forces? (*b*) What single force (called the *equilibrant*) would balance the two forces? Solve algebraically. *Ans.* (*a*) $\vec{R}$: 0.16 kN at 34° with the 80 N force; (*b*) $-\vec{R}$: 0.16 kN at 214° with the 80 N force

3.42 [I] Find algebraically the (*a*) resultant and (*b*) equilibrant (see Problem 1.26) of the following coplanar forces: 300 N at exactly 0°, 400 N at 30°, and 400 N at 150°. *Ans.* (*a*) 0.50 kN at 53°; (*b*) 0.50 kN at 233°

3.43 [I] A child is holding a wagon from rolling straight back down a driveway that is inclined at 20° to the horizontal. If the wagon weighs 150 N, with what force must the child pull on the handle if the handle is parallel to the incline? *Ans.* 51 N

3.44 [II] Repeat Problem 3.43 if the handle is at an angle of 30° above the incline. *Ans.* 59 N

3.45 [I] Once ignited, a small rocket motor on a spacecraft exerts a constant force of 10 N for 7.80 s. During the burn the rocket causes the 100-kg craft to accelerate uniformly. Determine that acceleration. *Ans.* 0.10 m/s^2

3.46 [II] Typically, a bullet leaves a standard 45-caliber pistol (5.0-in. barrel) at a speed of 262 m/s. If it takes 1 ms to traverse the barrel, determine the average acceleration experienced by the 16.2-g bullet within the gun and then compute the average force exerted on it. *Ans.* 3×10^5 m/s^2; 0.4×10^2 N

3.47 [I] A force acts on a 2-kg mass and gives it an acceleration of 3 m/s^2. What acceleration is produced by the same force when acting on a mass of (*a*) 1 kg? (*b*) 4 kg? (*c*) How large is the force? *Ans.* (*a*) 6 m/s^2; (*b*) 2 m/s^2; (*c*) 6 N

3.48 [I] An object has a mass of 300 g. (*a*) What is its weight on Earth? (*b*) What is its mass on the Moon? (*c*) What will be its acceleration on the Moon when a 0.500 N resultant force acts on it? *Ans.* (*a*) 2.94 N; (*b*) 0.300 kg; (*c*) 1.67 m/s^2

3.49 [I] A horizontal cable pulls a 200-kg cart along a horizontal track. The tension in the cable is 500 N. Starting from rest, (*a*) How long will it take the cart to reach a speed of 8.0 m/s? (*b*) How far will it have gone? *Ans.* (*a*) 3.2 s; (*b*) 13 m

3.50 [II] A 900-kg car is going 20 m/s along a level road. How large a constant retarding force is required to stop it in a distance of 30 m? (*Hint*: First find its deceleration.) *Ans.* 6.0 kN

3.51 [II] A 12.0-g bullet is accelerated from rest to a speed of 700 m/s as it travels 20.0 cm in a gun barrel. Assuming the acceleration to be constant, how large was the accelerating force? (*Be careful of units.*) *Ans.* 14.7 kN

3.52 [II] A 20-kg crate hangs at the end of a long rope. Find its acceleration (magnitude and direction) when the tension in the rope is (*a*) 250 N, (*b*) 150 N, (*c*) zero, (*d*) 196 N. *Ans.* (*a*) 2.7 m/s^2 up; (*b*) 2.3 m/s^2 down; (*c*) 9.8 m/s^2 down; (*d*) zero

3.53 [II] A 5.0-kg mass hangs at the end of a cord. Find the tension in the cord if the acceleration of the mass is (*a*) 1.5 m/s^2 up, (*b*) 1.5 m/s^2 down, (*c*) 9.8 m/s^2 down. *Ans.* (*a*) 57 N; (*b*) 42 N; (*c*) zero

3.54 [II] A 700-N man stands on a scale on the floor of an elevator. The scale records the force it exerts on whatever is on it. What is the scale reading if the elevator has an acceleration of (*a*) 1.8 m/s^2 up? (*b*) 1.8 m/s^2 down? (*c*) 9.8 m/s^2 down? *Ans.* (*a*) 0.83 kN; (*b*) 0.57 kN; (*c*) zero

3.55 [II] Using the scale described in Problem 3.54, a 65.0 kg astronaut weighs himself on the Moon, where $g = 1.60\ m/s^2$. What does the scale read? *Ans.* 104 N

3.56 [II] A cord passing over a frictionless, massless pulley has a 4.0-kg object tied to one end and a 12-kg object tied to the other. Compute the acceleration and the tension in the cord. *Ans.* 4.9 m/s^2, 59 N

3.57 [II] An elevator starts from rest with a constant upward acceleration. It moves 2.0 m in the first 0.60 s. A passenger in the elevator is holding a 3.0-kg package by a vertical string. What is the tension in the string during the accelerating process? *Ans.* 63 N

3.58 [II] Just as her parachute opens, a 60-kg parachutist is falling at a speed of 50 m/s. After 0.80 s has passed, the chute is fully open and her speed has dropped to 12.0 m/s. Find the average retarding force exerted upon the chutist during this time if the deceleration is uniform. *Ans.* $2850\ N + 588\ N = 3438\ N = 3.4\ kN$

3.59 [II] A 300-g mass hangs at the end of a string. A second string hangs from the bottom of that mass and supports a 900-g mass. (*a*) Find the tension in each string when the masses are accelerating upward at 0.700 m/s^2. (*b*) Find the tension in each string when the acceleration is 0.700 m/s^2 downward. *Ans.* (*a*) 12.6 N and 9.45 N; (*b*) 10.9 N and 8.19 N

3.60 [II] A 20-kg wagon is pulled along the level ground by a rope inclined at 30° above the horizontal. A friction force of 30 N opposes the motion. How large is the pulling force if the wagon is moving with (*a*) constant speed and (*b*) an acceleration of 0.40 m/s^2? *Ans.* (*a*) 35 N; (*b*) 44 N

3.61 [II] A 12-kg box is released from the top of an incline that is 5.0 m long and makes an angle of 40° to the horizontal. A 60-N friction force impedes the motion of the box. (*a*) What will be the acceleration of the box and (*b*) how long will it take to reach the bottom of the incline? *Ans.* (*a*) 1.3 m/s^2; (*b*) 2.8 s

3.62 [II] For the situation outlined in Problem 3.61, what is the coefficient of friction between box and incline? *Ans.* 0.67

3.63 [II] An inclined plane makes an angle of 30° with the horizontal. Find the constant force, applied parallel to the plane, required to cause a 15-kg box to slide (*a*) up the plane with acceleration 1.2 m/s^2 and (*b*) down the incline with acceleration 1.2 m/s^2. Neglect friction forces. *Ans.* (*a*) 92 N; (*b*) 56 N

3.64 [II] A horizontal force F is exerted on a 20-kg box to slide it up a 30° incline. The friction force retarding the motion is 80 N. How large must F be if the acceleration of the moving box is to be (*a*) zero and (*b*) 0.75 m/s^2? *Ans.* (*a*) 0.21 kN; (*b*) 0.22 kN

3.65 [II] An inclined plane making an angle of 25° with the horizontal has a pulley at its top. A 30-kg block on the plane is connected to a freely hanging 20-kg block by means of a cord passing over the pulley. Compute the distance the 20-kg block will fall in 2.0 s starting from rest. Neglect friction. *Ans.* 2.9 m

3.66 [III] Repeat Problem 3.65 if the coefficient of friction between block and plane is 0.20. *Ans.* 0.74 m

3.67 [III] A horizontal force of 200 N is required to cause a 15-kg block to slide up a 20° incline with an acceleration of 25 cm/s^2. Find (*a*) the friction force on the block and (*b*) the coefficient of friction. *Ans.* (*a*) 0.13 kN; (*b*) 0.65

3.68 [II] Find the acceleration of the blocks in Fig. 3-26 if friction forces are negligible. What is the tension in the cord connecting them? *Ans.* 3.3 m/s^2, 13 N

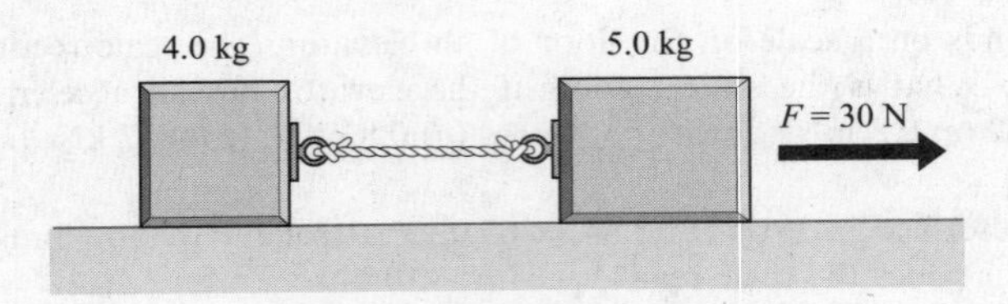

Fig. 3-26

3.69 [II] Repeat Problem 3.68 if the coefficient of kinetic friction between the blocks and the table is 0.30. *Ans.* 0.39 m/s^2, 13 N

3.70 [III] How large a force F is needed in Fig. 3-27 to pull out the 6.0-kg block with an acceleration of 1.50 m/s^2 if the coefficient of friction at its surfaces is 0.40? *Ans.* 48 N

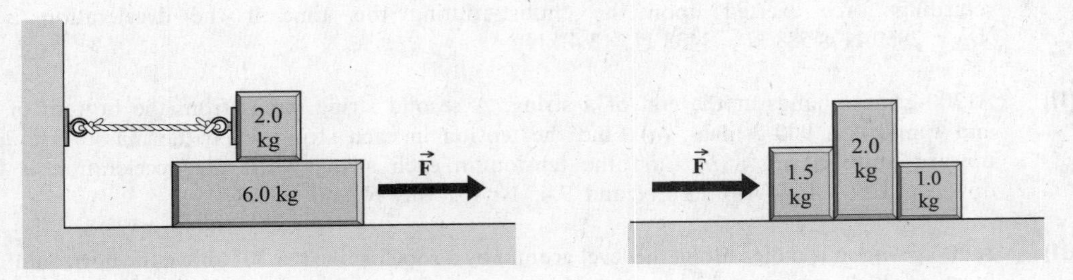

Fig. 3-27 **Fig. 3-28**

3.71 [III] In Fig. 3-28, how large a force F is needed to give the blocks an acceleration of 3.0 m/s^2 if the coefficient of kinetic friction between blocks and table is 0.20? How large a force does the 1.50-kg block then exert on the 2.0-kg block? *Ans.* 22 N, 15 N

3.72 [III] (*a*) What is the smallest force parallel to a 37° incline needed to keep a 100-N weight from sliding down the incline if the coefficients of static and kinetic friction are both 0.30? (*b*) What parallel force is required to keep the weight moving up the incline at constant speed? (*c*) If the parallel pushing force is 94 N, what will be the acceleration of the object? (*d*) If the object in (*c*) starts from rest, how far will it move in 10 s? *Ans.* (*a*) 36 N; (*b*) 84 N; (*c*) 0.98 m/s^2 up the plane; (*d*) 49 m

3.73 [III] A 5.0-kg block rests on a 30° incline. The coefficient of static friction between the block and the incline is 0.20. How large a horizontal force must push on the block if the block is to be on the verge of sliding (*a*) up the incline and (*b*) down the incline? *Ans.* (*a*) 43 N; (*b*) 16.6 N

3.74 [III] Three blocks with masses 6.0 kg, 9.0 kg, and 10 kg are connected as shown in Fig. 3-29. The coefficient of friction between the table and the 10-kg block is 0.20. Find (*a*) the acceleration of the system and (*b*) the tension in the cord on the left and in the cord on the right. *Ans.* (*a*) 0.39 m/s^2; (*b*) 61 N, 85 N

3.75 [II] The Earth's radius is about 6370 km. An object that has a mass of 20 kg is taken to a height of 160 km above the Earth's surface. (*a*) What is the object's mass at this height? (*b*) How much does the object weigh (i.e., how large a gravitational force does it experience) at this height? *Ans.* (*a*) 20 kg; (*b*) 0.19 kN

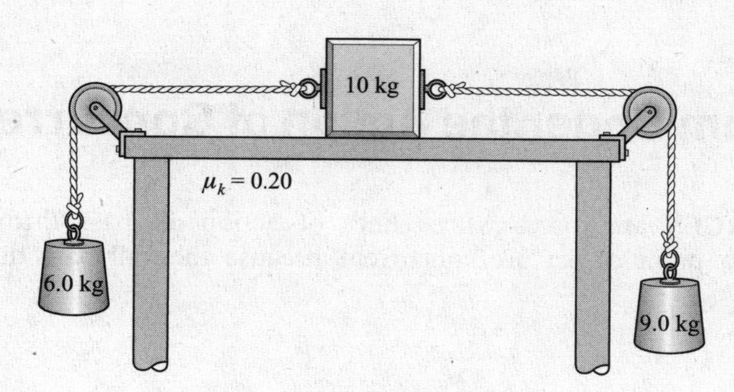

Fig. 3-29

3.76 [II] The radius of the Earth is about 6370 km, while that of Mars is about 3440 km. If an object weighs 200 N on Earth, what would it weigh, and what would be the acceleration due to gravity, on Mars? The mass of Mars is 0.11 that of Earth. *Ans.* 75 N, 3.7 m/s^2

Chapter 4

Equilibrium Under the Action of Concurrent Forces

CONCURRENT FORCES are forces whose lines of action all pass through a common point. The forces acting on a point object are concurrent because they all pass through the same point, the point object.

AN OBJECT IS IN EQUILIBRIUM under the action of concurrent forces provided it is not accelerating.

THE FIRST CONDITION FOR EQUILIBRIUM is the requirement that $\Sigma \vec{\mathbf{F}} = 0$ or, in component form, that

$$\Sigma F_x = \Sigma F_y = \Sigma F_z = 0$$

That is, the resultant of all external forces acting on the object must be zero. This condition is sufficient for equilibrium when the external forces are concurrent. A second condition must also be satisfied if an object is to be in equilibrium under nonconcurrent forces; it is discussed in Chapter 5.

PROBLEM SOLUTION METHOD (CONCURRENT FORCES):

(1) Isolate the object for discussion.
(2) Show the forces acting on the isolated object in a diagram (the *free-body diagram*).
(3) Find the rectangular components of each force.
(4) Write the first condition for equilibrium in equation form.
(5) Solve for the required quantities.

THE WEIGHT OF AN OBJECT ($\vec{\mathbf{F}}_W$) is essentially the force with which gravity pulls downward upon it.

THE TENSILE FORCE ($\vec{\mathbf{F}}_T$) acting on a string or cable or chain (or indeed, on any structural member) is the applied force tending to stretch it. The scalar magnitude of the tensile force is the *tension* (F_T).

THE FRICTION FORCE ($\vec{\mathbf{F}}_f$) is a tangential force acting on an object that opposes the sliding of that object across an adjacent surface with which it is in contact. The friction force is parallel to the surface and opposite to the direction of motion or of impending motion.

THE NORMAL FORCE ($\vec{\mathbf{F}}_N$) on an object that is being supported by a surface is the component of the supporting force that is perpendicular to the surface.

PULLEYS: When a system of several frictionless light-weight pulleys has a single continuous rope wound around it, the tension *in each length of the rope* equals the force applied to the end of the rope (F) by some external agency. Thus when the load is supported by N lengths of this rope, the net force delivered to the load, the output force, is NF. Often the pulley attached to the load moves with the load and we need only count up the number of lengths of rope (N) acting on that pulley to determine the output force.

Solved Problems

4.1 [II] In Fig. 4-1(*a*), the tension in the horizontal cord is 30 N as shown. Find the weight of the object.

The tension in cord 1 is equal to the weight of the object hanging from it. Therefore $F_{T1} = F_W$, and we wish to find F_{T1} or F_W.

Notice that the unknown force F_{T1} and the known force of 30 N both pull on the knot at point P. It therefore makes sense to isolate the knot at P as our object. The free-body diagram showing the forces on the knot is drawn as in Fig. 4-1(*b*). The force components are also shown there.

We next write the first condition for equilibrium for the knot. From the free-body diagram,

$$\overset{+}{\rightarrow} \Sigma F_x = 0 \quad \text{becomes} \quad 30\ \text{N} - F_{T2}\cos 40^\circ = 0$$

$$+\uparrow \Sigma F_y = 0 \quad \text{becomes} \quad F_{T2}\sin 40^\circ - F_W = 0$$

Solving the first equation for F_{T2} gives $F_{T2} = 39.2$ N. Substituting this value in the second equation gives $F_W = 25$ N as the weight of the object.

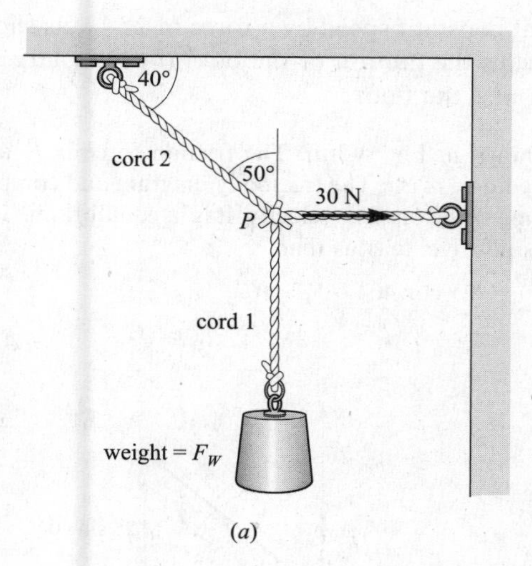

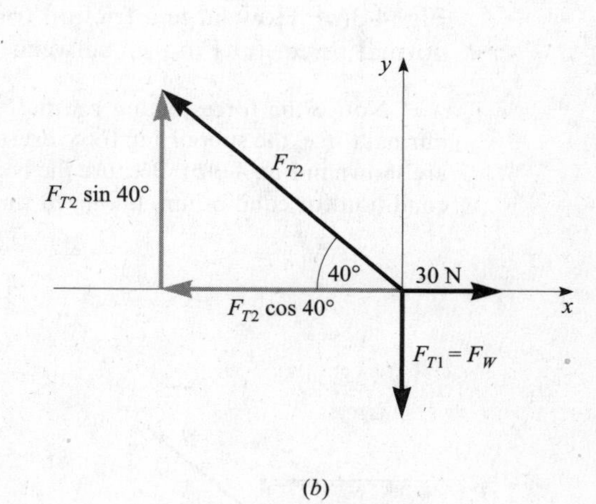

Fig. 4-1

4.2 [II] A rope extends between two poles. A 90-N boy hangs from it as shown in Fig. 4-2(*a*). Find the tensions in the two parts of the rope.

We label the two tensions F_{T1} and F_{T2}, and isolate the rope at the boy's hands as the object. The free-body diagram for the object is shown in Fig. 4-2(*b*).

After resolving the forces into their components as shown, we can write the first condition for equilibrium:

$$\overset{+}{\rightarrow} \Sigma F_x = 0 \quad \text{becomes} \quad F_{T2} \cos 5.0^\circ - F_{T1} \cos 10^\circ = 0$$

$$+\uparrow \Sigma F_y = 0 \quad \text{becomes} \quad F_{T2} \sin 5.0^\circ + F_{T1} \sin 10^\circ - 90\text{ N} = 0$$

When we evaluate the sines and cosines, these equations become

$$0.996F_{T2} - 0.985F_{T1} = 0 \quad \text{and} \quad 0.087F_{T2} + 0.174F_{T1} - 90 = 0$$

Solving the first for F_{T2} gives $F_{T2} = 0.990F_{T1}$. Substituting this in the second equation gives

$$0.086F_{T1} + 0.174F_{T1} - 90 = 0$$

from which $F_{T1} = 0.35$ kN. Then, because $F_{T2} = 0.990F_{T1}$, we have $F_{T2} = 0.34$ kN.

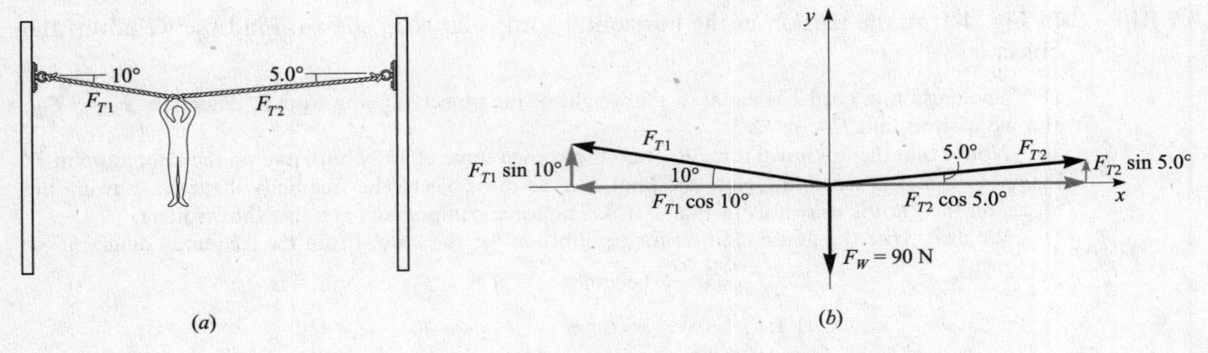

Fig. 4-2

4.3 [II] A 50-N box is slid straight across the floor at constant speed by a force of 25 N, as shown in Fig. 4-3(*a*). How large a friction force impedes the motion of the box? (*b*) How large is the normal force? (*c*) Find μ_k between the box and the floor.

Notice the forces acting on the box, as shown in Fig. 4-3(*a*). The friction force is F_f and the normal force, the supporting force exerted by the floor, is F_N. The free-body diagram and components are shown in Fig. 4-3(*b*). Because the box is moving with constant velocity, it is in equilibrium. The first condition for equilibrium, taking to the right as positive, tells us that

$$\overset{+}{\rightarrow} \Sigma F_x = 0 \quad \text{or} \quad 25 \cos 40^\circ - F_f = 0$$

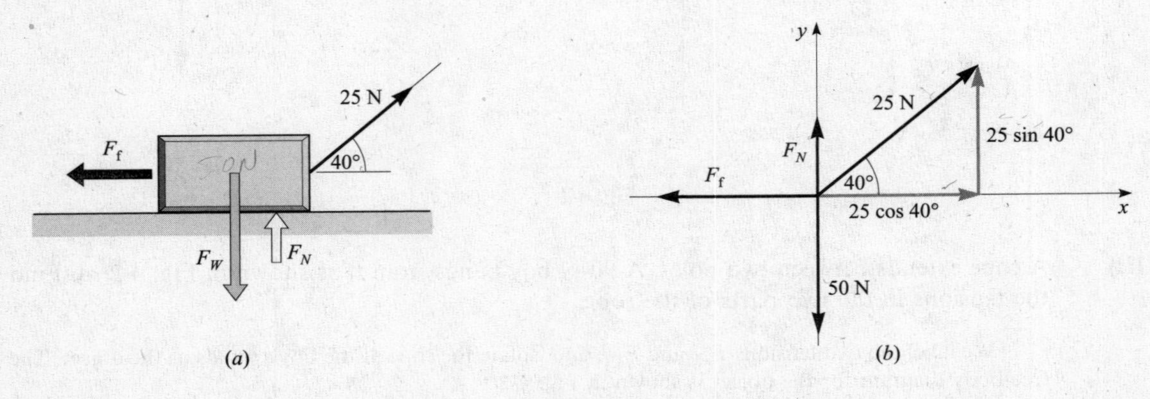

Fig. 4-3

(*a*) We can solve for the friction force F_f at once to find that $F_f = 19.2\,\text{N}$, or to two significant figures, $F_f = 19\,\text{N}$.

(*b*) To find F_N we use the fact that

$$+\uparrow \Sigma F_y = 0 \qquad \text{or} \qquad F_N + 25 \sin 40^\circ - 50 = 0$$

Solving gives the normal force as $F_N = 33.9$ N or, to two significant figures, $F_N = 34$ N.

(*c*) From the definition of μ_k, we have

$$\mu_k = \frac{F_f}{F_N} = \frac{19.2\,\text{N}}{33.9\,\text{N}} = 0.57$$

4.4 [II] Find the tensions in the ropes shown in Fig. 4-4(*a*) if the supported object weighs 600 N.

Let us select as our object the knot at A because we know one force acting on it. The weight pulls down on it with a force of 600 N, and so the free-body diagram for the knot is as shown in Fig. 4-4(*b*). Applying the first condition for equilibrium to that diagram, we have

$$\overset{+}{\rightarrow} \Sigma F_x = 0 \qquad \text{or} \qquad F_{T2} \cos 60^\circ - F_{T1} \cos 60^\circ = 0$$

$$+\uparrow \Sigma F_y = 0 \qquad \text{or} \qquad F_{T1} \sin 60^\circ + F_{T2} \sin 60^\circ - 600 = 0$$

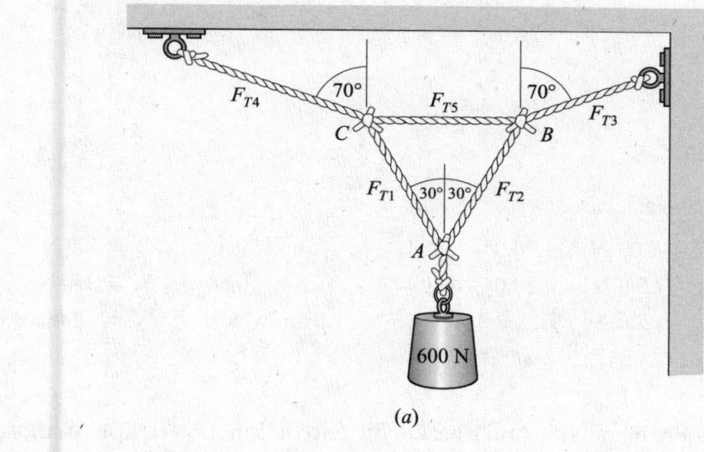

(*a*)

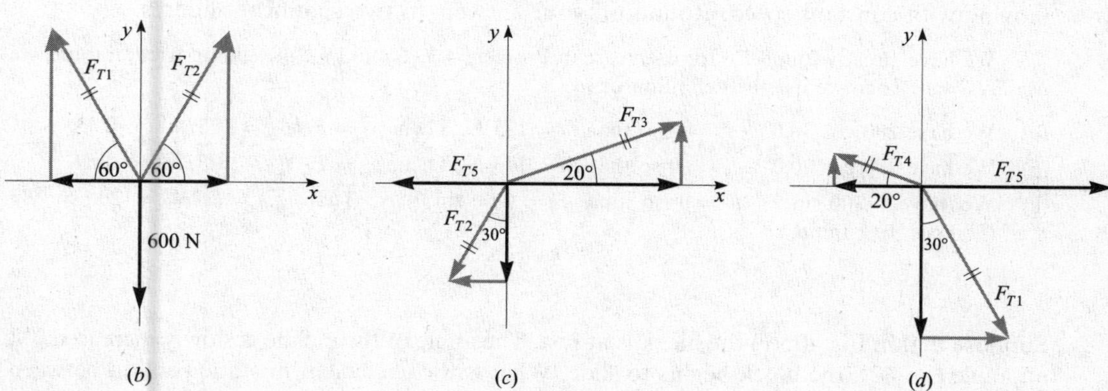

(*b*) (*c*) (*d*)

Fig. 4-4

The first equation yields $F_{T1} = F_{T2}$. (We could have inferred this from the symmetry of the system. Also symmetry, $F_{T3} = F_{T4}$.) Substitution of F_{T1} for F_{T2} in the second equation gives $F_{T1} = 346$ N, and so $F_{T2} = 346$ N also.

Let us now isolate knot B as our object. Its free-body diagram is shown in Fig. 4-4(*c*). We have already found that $F_{T2} = 346$ N or 0.35 kN and so the equilibrium equations are

$$\overset{+}{\rightarrow}\Sigma F_x = 0 \quad \text{or} \quad F_{T3}\cos 20^\circ - F_{T5} - 346 \sin 30^\circ = 0$$

$$+\uparrow \Sigma F_y = 0 \quad \text{or} \quad F_{T3}\sin 20^\circ - 346\cos 30^\circ = 0$$

The last equation yields $F_{T3} = 877$ N or 0.88 kN. Substituting this in the prior equation gives $F_{T5} = 651$ N or 0.65 kN. As stated previously from symmetry $F_{T4} = F_{T3} = 877$ N or 0.88 kN. How could you have found F_{T4} without recourse to symmetry? (*Hint*: See Fig. 4.4(*d*).)

4.5 [I] Each of the objects in Fig. 4-5 is in equilibrium. Find the normal force F_N in each case.

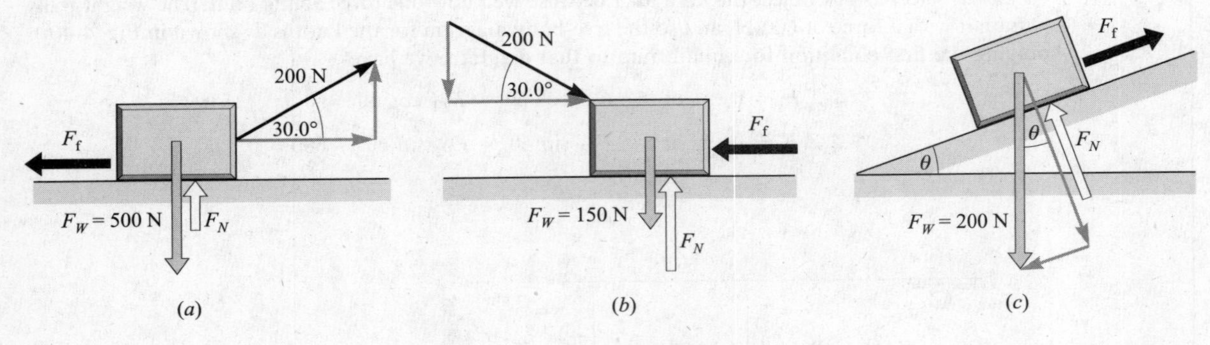

Fig. 4-5

We apply $\Sigma F_y = 0$ in each case.

(*a*) $F_N + (200 \text{ N})\sin 30.0^\circ - 500 = 0$ from which $F_N = 400$ N

(*b*) $F_N - (200 \text{ N})\sin 30.0^\circ - 150 = 0$ from which $F_N = 250$ N

(*c*) $F_N - (200 \text{ N})\cos\theta = 0$ from which $F_N = (200\cos\theta)$ N

4.6 [I] For the situations of Problem 4.5, find the coefficient of kinetic friction if the object is moving with constant speed. Round off your answers to two significant figures.

We have already found F_N for each case in Problem 4.5. To find F_f, the sliding-friction force, we use $\Sigma F_x = 0$. Then we use the definition of μ_k.

(*a*) We have $200\cos 30.0^\circ - F_f = 0$ so that $F_f = 173$ N. Then, $\mu_k = F_f/F_N = 173/400 = 0.43$.

(*b*) We have $200\cos 30.0^\circ - F_f = 0$ so that $F_f = 173$ N. Then, $\mu_k = F_f/F_N = 173/250 = 0.69$.

(*c*) We have $-200\sin\theta + F_f = 0$ so that $F_f = (200\sin\theta)$ N. Then, $\mu_k = F_f/F_N = (200\sin\theta)/(200\cos\theta) = \tan\theta$.

4.7 [II] Suppose that in Fig. 4-5(*c*) the block is at rest. The angle of the incline is slowly increased. At an angle $\theta = 42^\circ$, the block begins to slide. What is the coefficient of static friction between the block and the incline? (The block and surface are not the same as in Problems 4.5 and 4.6.)

At the instant the block begins to slide, the friction has its critical value. Therefore, $\mu_s = F_f/F_N$ at that instant. Following the method of Problems 4.5 and 4.6, we have

$$F_N = F_W \cos\theta \qquad \text{and} \qquad F_f = F_W \sin\theta$$

Therefore, when sliding just starts,

$$\mu_s = \frac{F_f}{F_N} = \frac{F_W \sin\theta}{F_W \cos\theta} = \tan\theta$$

But θ was found by experiment to be 42°. Therefore, $\mu_s = \tan 42° = 0.90$.

4.8 [II] Pulled by the 8.0-N block shown in Fig. 4-6(*a*), the 20-N block slides to the right at a constant velocity. Find μ_k between the block and the table. Assume the pulley to be frictionless.

Because it is moving at a constant velocity, the 20-N block is at equilibrium. Since the pulley is frictionless, the tension in the continuous rope is the same on both sides of the pulley. Thus, we have $F_{T1} = F_{T2} = 8.0$ N.

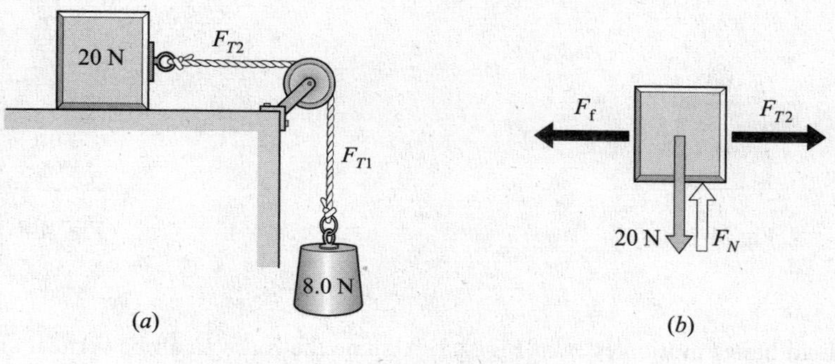

Fig. 4-6

Looking at the free-body diagram in Fig. 4-6(*b*) and recalling that the block is at equilibrium, we have

$$\overset{+}{\rightarrow}\Sigma F_x = 0 \qquad \text{or} \qquad F_f = F_{T2} = 8.0 \text{ N}$$

$$+\uparrow \Sigma F_y = 0 \qquad \text{or} \qquad F_N = 20 \text{ N}$$

Then, from the definition of μ_k,

$$\mu_k = \frac{F_f}{F_N} = \frac{8.0\text{ N}}{20\text{ N}} = 0.40$$

Supplementary Problems

4.9 [I] The load in Fig. 4.7 is hanging at rest. Take the ropes to all be vertical and the pulleys to be weightless and frictionless. (*a*) How many segments of rope support the combination of the lower pulley and load? (*b*) What is the tension in the rope wound around the pulleys? (*c*) How much force is the man exerting? (*d*) How much force acts downward on the hook in the ceiling? *Ans.* (*a*) 2; (*b*) 100 N; (*c*) 100 N; (*d*) 300 N

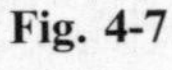

Fig. 4-7

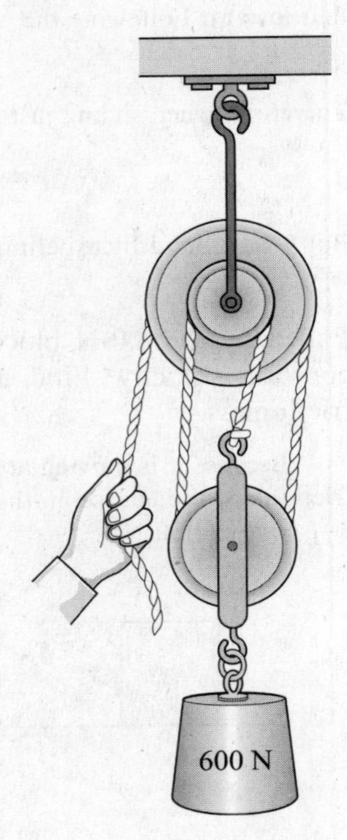

Fig. 4-8

4.10 [I] A 600-N load hangs motionlessly in Fig. 4.8. Assume the ropes to all be vertical and the pulleys to be weightless and frictionless. (*a*) What is the tension in the bottom hook attached, via a ring, to the load? (*b*) How many lengths of rope support the movable pulley? (*c*) What is the tension in the long rope? (*d*) How much force does the man apply? (*e*) How much force acts downward on the ceiling? *Ans.* (*a*) 600 N; (*b*) 3; (*c*) 200 N; (*d*) 200 N; (*e*) 800 N

4.11 [I] For the situation shown in Fig. 4-9, find the values of F_{T1} and F_{T2} if the object's weight is 600 N. *Ans.* 503 N, 783 N

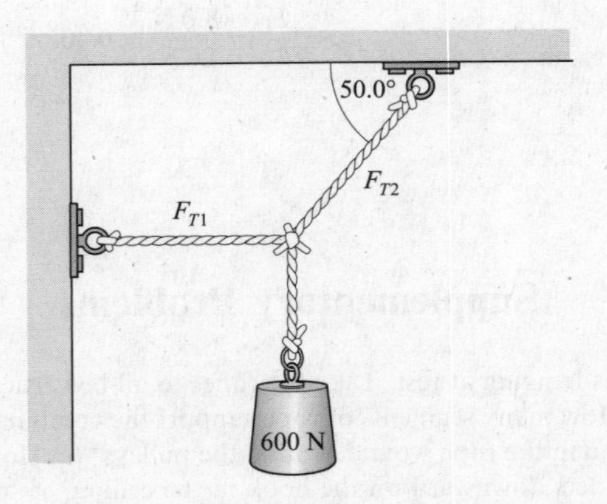

Fig. 4-9

4.12 [I] The following coplanar forces pull on a ring: 200 N at 30.0°, 500 N at 80.0°, 300 N at 240°, and an unknown force. Find the magnitude and direction of the unknown force if the ring is to be in equilibrium. *Ans.* 350 N at 252°

4.13 [II] In Fig. 4-10, the pulleys are frictionless and the system hangs in equilibrium. If F_{W3}, the weight of the object on the right, is 200 N, what are the values of F_{W1} and F_{W2}? *Ans.* 260 N, 150 N

4.14 [II] Suppose F_{W1} in Fig. 4-10 is 500 N. Find the values of F_{W2} and F_{W3} if the system is to hang in equilibrium as shown. *Ans.* 288 N, 384 N

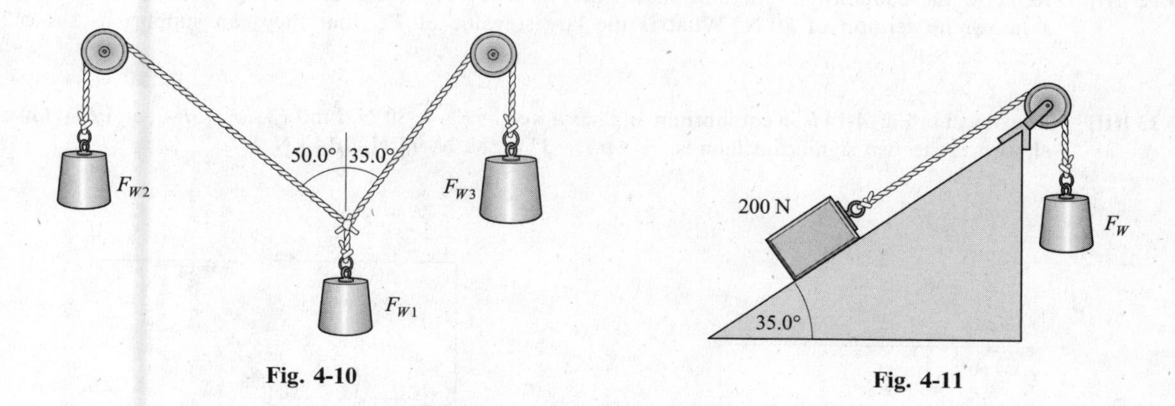

Fig. 4-10 Fig. 4-11

4.15 [I] If in Fig. 4-11 the friction between the block and the incline is negligible, how much must the object on the right weigh if the 200-N block is to remain at rest? *Ans.* 115 N

4.16 [II] The system in Fig. 4-11 remains at rest when $F_W = 220$ N. What are the magnitude and direction of the friction force on the 200-N block? *Ans.* 105 N down the incline

4.17 [II] Find the normal force acting on the block in each of the equilibrium situations shown in Fig. 4-12. *Ans.* (*a*) 34 N; (*b*) 46 N; (*c*) 91 N

4.18 [II] The block shown in Fig. 4-12(*a*) slides with constant speed under the action of the force shown. (*a*) How large is the retarding friction force? (*b*) What is the coefficient of kinetic friction between the block and the floor? *Ans.* (*a*) 12 N; (*b*) 0.34

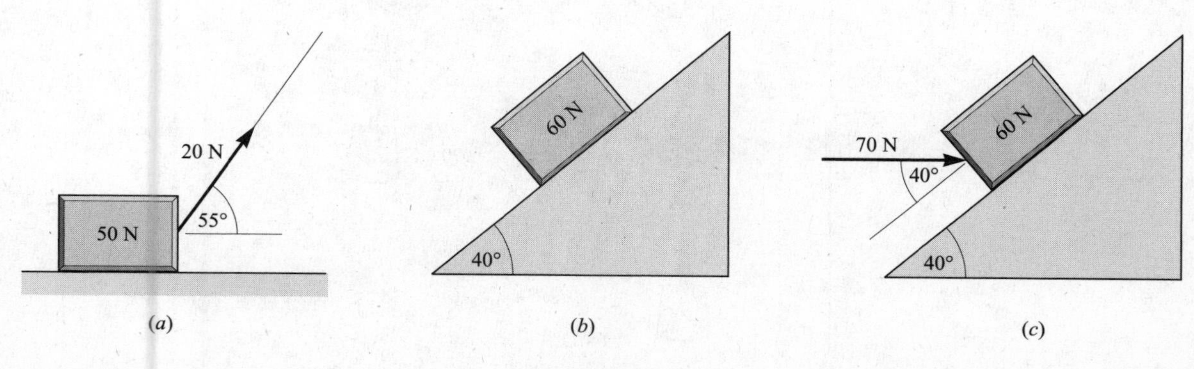

Fig. 4-12

4.19 [II] The block shown in Fig. 4-12(*b*) slides at a constant speed down the incline. (*a*) How large is the friction force that opposes its motion? (*b*) What is the coefficient of sliding (kinetic) friction between the block and the plane? *Ans.* (*a*) 39 N; (*b*) 0.84

4.20 [II] The block in Fig. 4-12(*c*) just begins to slide up the incline when the pushing force shown is increased to 70 N. (*a*) What is the critical static friction force on it? (*b*) What is the value of the coefficient of static friction? *Ans.* (*a*) 15 N; (*b*) 0.17

4.21 [II] If $F_W = 40$ N in the equilibrium situation shown in Fig. 4-13, find F_{T1} and F_{T2}. *Ans.* 58 N, 31 N

4.22 [III] Refer to the equilibrium situation shown in Fig. 4-13. The cords are strong enough to withstand a maximum tension of 80 N. What is the largest value of F_W that they can support as shown? *Ans.* 55 N

4.23 [III] The object in Fig. 4-14 is in equilibrium and has a weight $F_W = 80$ N. Find F_{T1}, F_{T2}, F_{T3}, and F_{T4}. Give all answers to two significant figures. *Ans.* 37 N, 88 N, 77 N, 0.14 kN

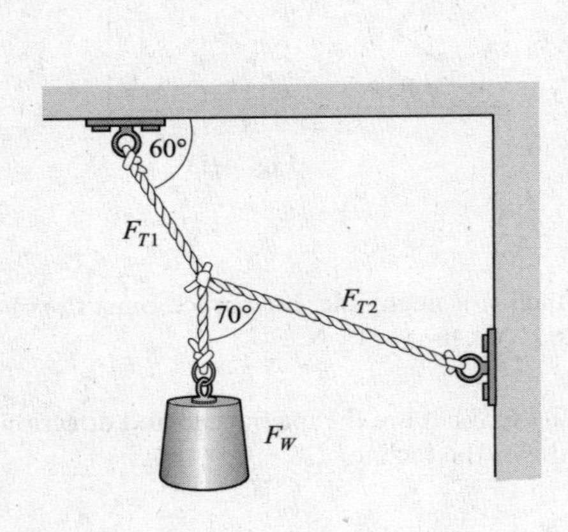

Fig. 4-13

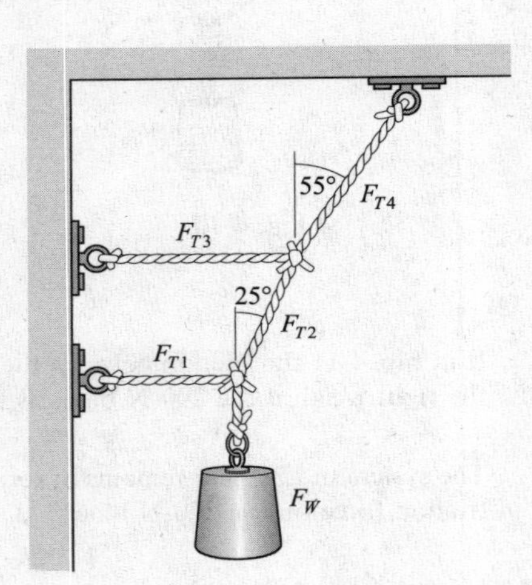

Fig. 4-14

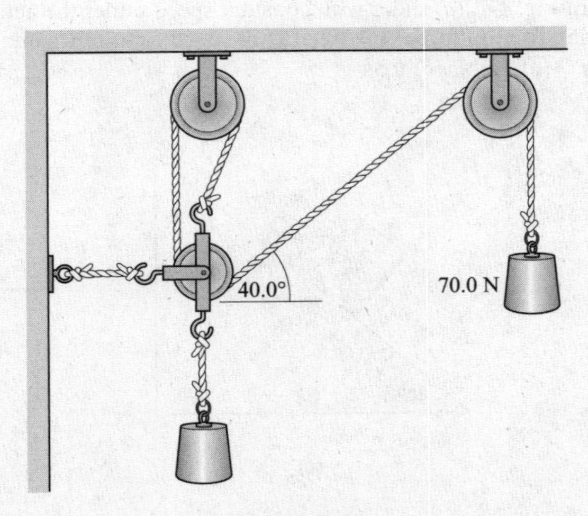

Fig. 4-15

4.24 [III] The pulleys shown in Fig. 4-15 have negligible weight and friction. What is the value of F_W if the system is at equilibrium? *Ans.* 185 N

4.25 [III] In Fig. 4-16, the system is in equilibrium. (*a*) What is the maximum value that F_W can have if the friction force on the 40-N block cannot exceed 12.0 N? (*b*) What is the coefficient of static friction between the block and the tabletop? *Ans.* (*a*) 6.9 N; (*b*) 0.30

4.26 [III] The system in Fig. 4-16 is just on the verge of slipping. If $F_W = 8.0$ N, what is the coefficient of static friction between the block and tabletop? *Ans.* 0.35

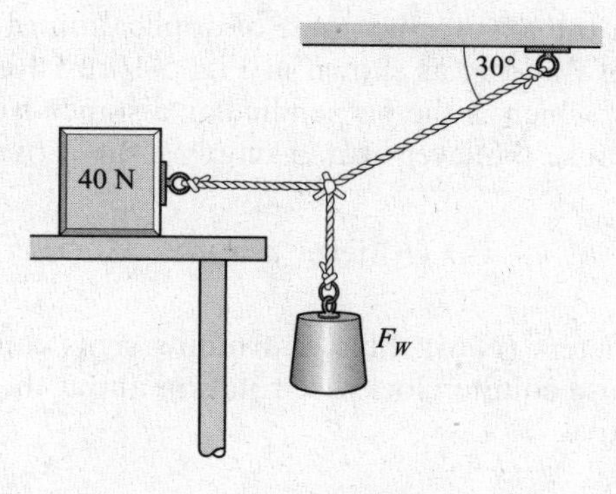

Fig. 4-16

Chapter 5

Equilibrium of a Rigid Body Under Coplanar Forces

THE TORQUE (τ) about an axis, due to a force, is a measure of the effectiveness of the force in producing rotation about that axis. It is defined in the following way:

$$\text{Torque} = \tau = rF \sin \theta$$

where r is the radial distance from the axis to the point of application of the force, and θ is the acute angle between the lines-of-action of $\vec{r}$ and $\vec{F}$, as shown in Fig. 5-1(*a*). Often this definition is written in terms of the *lever arm* of the force, which is the perpendicular distance from the axis to the line of the force, as shown in Fig. 5-1(*b*). Because the lever arm is simply $r \sin \theta$, the torque becomes

$$\tau = (F)(\text{lever arm})$$

The units of torque are newton-meters (N·m). Plus and minus signs can be assigned to torques; for example, a torque that tends to cause counterclockwise rotation about the axis is positive, whereas one causing clockwise rotation is negative.

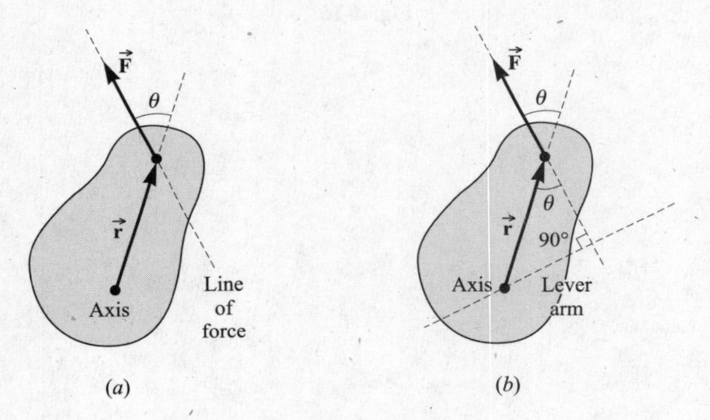

Fig. 5-1

THE TWO CONDITIONS FOR EQUILIBRIUM of a rigid object under the action of *coplanar forces* are

(1) The *first* or *force condition*: The vector sum of all forces acting on the body must be zero:

$$\Sigma F_x = 0 \qquad \Sigma F_y = 0$$

where the plane of the coplanar forces is taken to be the *xy*-plane.

(2) The *second* or *torque condition*: Take an axis perpendicular to the plane of the coplanar forces. Call the torques that tend to cause clockwise rotation about the axis negative, and counterclockwise torques positive; then the sum of all the torques acting on the object must be zero:

$$\circlearrowleft\!(+)\; \Sigma \tau = 0$$

THE CENTER OF GRAVITY of an object is the point at which the entire weight of the object may be considered concentrated; i.e., the line-of-action of the weight passes through the center of gravity. A single vertically upward directed force, equal in magnitude to the weight of the object and applied through its center of gravity, will keep the object in equilibrium.

THE POSITION OF THE AXIS IS ARBITRARY: If the sum of the torques is zero about one axis for a body that obeys the force condition, it is zero about all other axes parallel to the first. We can choose the axis in such a way that the line of an unknown force passes through the intersection of the axis and the plane of the forces. The angle θ between $\vec{r}$ and $\vec{F}$ is then zero; hence, that particular unknown force exerts zero torque and therefore does not appear in the torque equation.

Solved Problems

5.1 [I] Find the torque about axis A (which is perpendicular to the page) in Fig. 5-2 due to each of the forces shown.

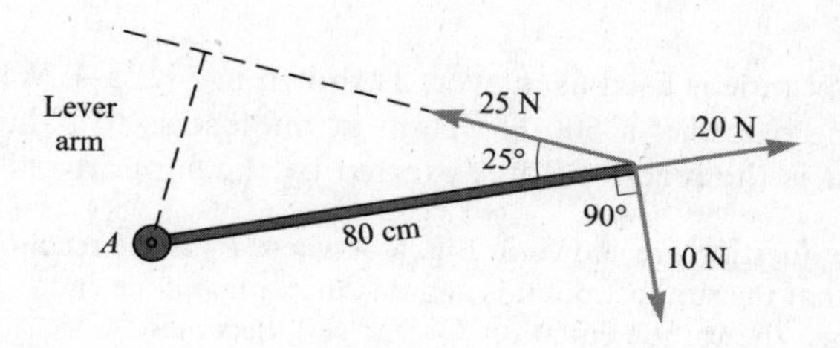

Fig. 5-2

We use $\tau = rF \sin\theta$, recalling that clockwise torques are negative while counterclockwise torques are positive. The torques due to the three forces are

For 10 N: $\tau = -(0.80\text{ m})(10\text{ N})(\sin 90°) = -8.0\text{ N·m}$

For 25 N: $\tau = +(0.80\text{ m})(25\text{ N})(\sin 25°) = +8.5\text{ N·m}$

For 20 N: $\tau = \pm(0.80\text{ m})(20\text{ N})(\sin 0°) = 0$

The line of the 20-N force goes through the axis and so $\theta = 0°$ for it. Or, put another way, because the line of the force passes through the axis, its lever arm is zero. Either way, the torque is zero for this (and any) force whose line passes through the axis.

5.2 [II] A uniform metal beam of length L weighs 200 N and holds a 450-N object as shown in Fig. 5-3. Find the magnitudes of the forces exerted on the beam by the two supports at its ends. Assume the lengths are exact.

Rather than draw a separate free-body diagram, we show the forces on the object being considered (the beam) in Fig. 5-3. Because the beam is uniform, its center of gravity is at its geometric center. Thus the weight of the beam (200 N) is shown acting at the beam's center. The forces F_1 and F_2 are exerted on the beam by the supports. Because there are no x-directed forces acting on the beam, we have only two equations to write for this equilibrium situation: $\Sigma F_y = 0$ and $\Sigma\tau = 0$.

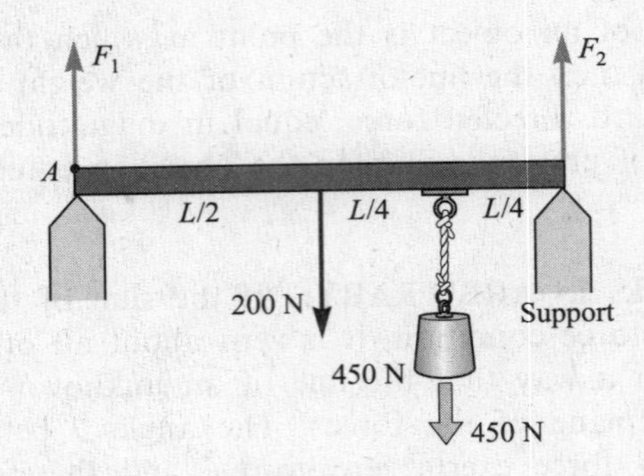

Fig. 5-3

$$+\uparrow \Sigma F_y = 0 \qquad \text{becomes} \qquad F_1 + F_2 - 200\text{ N} - 450\text{ N} = 0$$

Before the torque equation is written, an axis must be chosen. We choose it at A, so that the unknown force F_1 will pass through it and exert no torque. The torque equation is then

$$\circlearrowleft\!\!+ \Sigma\tau = -(L/2)(200\text{ N})(\sin 90°) - (3L/4)(450\text{ N})(\sin 90°) + LF_2 \sin 90° = 0$$

Dividing through the equation by L and solving for F_2, we find that $F_2 = 438$ N.

To find F_1 we substitute the value of F_2 in the force equation, obtaining $F_1 = 212$ N.

5.3 [II] A uniform, 100-N pipe is used as a lever, as shown in Fig. 5-4. Where must the fulcrum (the support point) be placed if a 500-N weight at one end is to balance a 200-N weight at the other end? What is the reaction force exerted by the support on the pipe?

The forces in question are shown in Fig. 5-4, where F_R is the reaction force of the support on the pipe. We assume that the support point is at a distance x from one end. Let us take the axis to be at the support point. Then the torque equation, $\circlearrowleft\!\!+ \Sigma\tau = 0$, becomes

$$+(x)(200\text{ N})(\sin 90°) + (x - L/2)(100\text{ N})(\sin 90°) - (L - x)(500\text{ N})(\sin 90°) = 0$$

This simplifies to

$$(800\text{ N})(x) = (550\text{ N})(L)$$

and so $x = 0.69L$. The support should be placed 0.69 of the way from the lighter-loaded end.

To find the load F_R held by the support, we use $+\uparrow \Sigma F_y = 0$, which gives

$$-200\text{ N} - 100\text{ N} - 500\text{ N} = 0$$

from which $F_R = 800$ N.

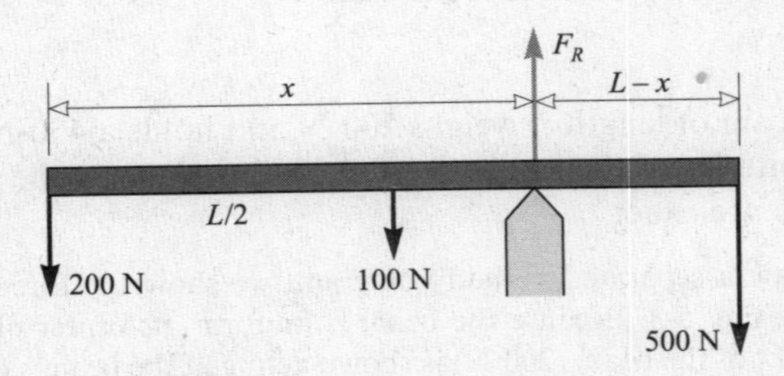

Fig. 5-4

5.4 [II] Where must a 0.80-kN object be hung on a uniform, horizontal, rigid 100-N pole so that a girl pushing up at one end supports one-third as much as a woman pushing up at the other end?

The situation is shown in Fig. 5-5. We represent the force exerted by the girl as F, and that by the woman as $3F$. Take the axis point at the left end. Then the torque equation becomes

$$-(x)(800\,\text{N})(\sin 90^\circ) - (L/2)(100\,\text{N})(\sin 90^\circ) + (L)(F)(\sin 90^\circ) = 0$$

A second equation we can write is $\Sigma F_y = 0$, or

$$3F - 800\,\text{N} - 100\,\text{N} + F = 0$$

from which $F = 225$ N. Substitution of this value in the torque equation gives

$$(800\,\text{N})(x) = (225\,\text{N})(L) - (100\,\text{N})(L/2)$$

from which $x = 0.22L$. The load should be hung 0.22 of the way from the woman to the girl.

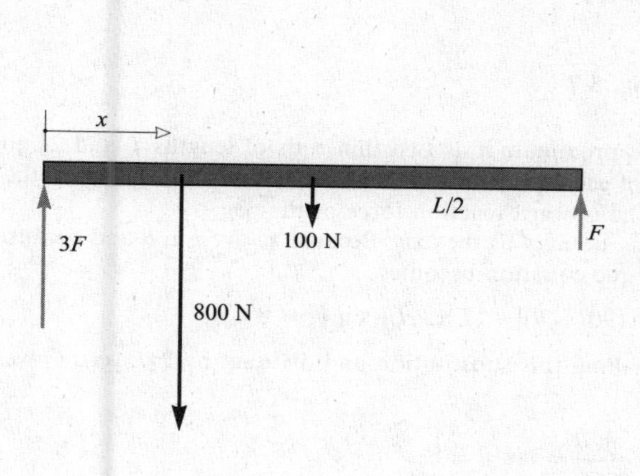

Fig. 5-5

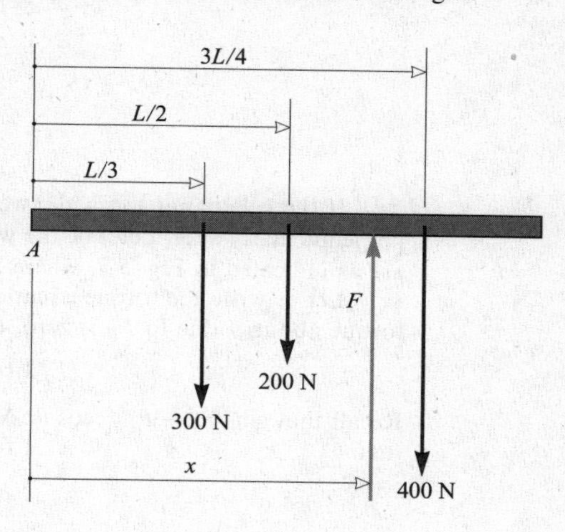

Fig. 5-6

5.5 [II] A uniform, 0.20-kN board of length L has two objects hanging from it: 300 N at exactly $L/3$ from one end, and 400 N at exactly $3L/4$ from the same end. What single additional force acting on the board will cause the board to be in equilibrium?

The situation is shown in Fig. 5-6, where F is the force we wish to find. For equilibrium, $\Sigma F_y = 0$ and so

$$F = 400\text{ N} + 200\text{ N} + 300\text{ N} = 900\text{ N}$$

Because the board is to be in equilibrium, we are free to choose the axis anywhere. Choose it at point A. Then $\Sigma\tau = 0$ gives

$$+(x)(F)(\sin 90^\circ) - (3L/4)(400\text{ N})(\sin 90^\circ) - (L/2)(200\text{ N})(\sin 90^\circ) - (L/3)(300\text{ N})(\sin 90^\circ) = 0$$

Using $F = 900$ N, we find that $x = 0.56L$. The required force is 0.90 kN upward at $0.56L$ from the left end.

5.6 [II] The right-angle rule (or square) shown in Fig. 5-7 hangs at rest from a peg as shown. It is made of a uniform metal sheet. One arm is L cm long, while the other is $2L$ cm long. Find (to two significant figures) the angle θ at which it will hang.

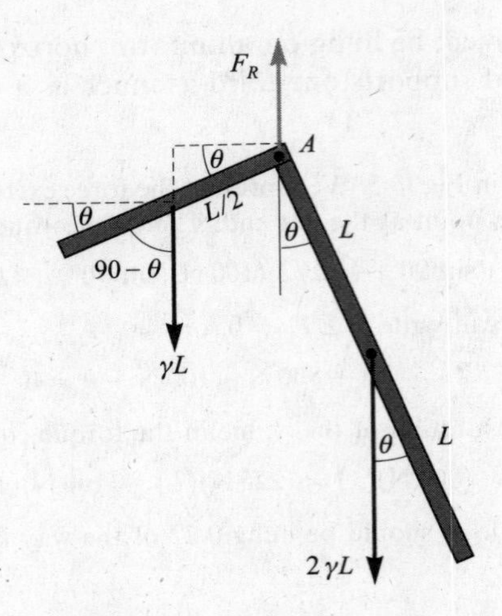

Fig. 5-7

If the rule is not too wide, we can approximate it as two thin rods of lengths L and $2L$ joined perpendicularly at A. Let γ be the weight of each centimeter of rule. Then the forces acting on the rule are as indicated in Fig. 5-7, where F_R is the upward reaction force of the peg.

Let us write the torque equation using point A as the axis. Because $\tau = rF \sin \theta$ and because the torque about A due to F_R is zero, the torque equation becomes

$$+(L/2)(\gamma L)[\sin(90° - \theta)] - (L)(2\gamma L)(\sin \theta) = 0$$

Recall that $\sin(90° - \theta) = \cos \theta$. After making this substitution and dividing by $2\gamma L^2 \cos \theta$, we find that

$$\frac{\sin \theta}{\cos \theta} = \tan \theta = \frac{1}{4}$$

which yields $\theta = 14°$.

5.7 [II] Consider the situation shown in Fig. 5-8(*a*). The uniform 0.60-kN beam is hinged at P. Find the tension in the tie rope and the components of the reaction force exerted by the hinge on the beam. Give your answers to two significant figures.

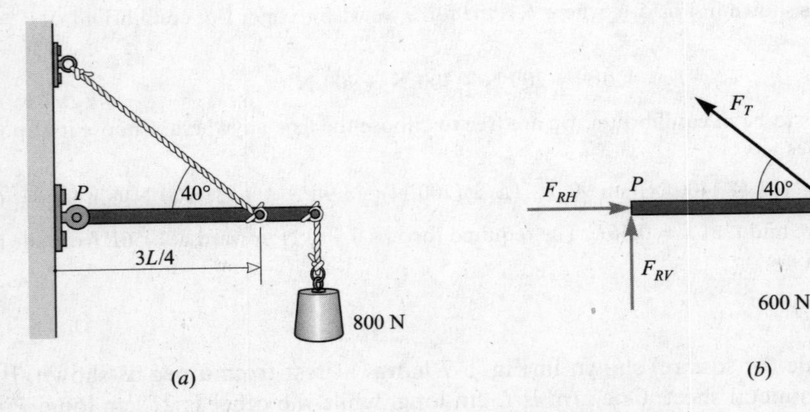

Fig. 5-8

The reaction forces acting on the beam are shown in Fig. 5-8(*b*), where the force exerted by the hinge is represented by its components, F_{RH} and F_{RV}. The torque equation about P as axis is

$$+(3L/4)(F_T)(\sin 40°) - (L)(800\text{ N})(\sin 90°) - (L/2)(600\text{ N})(\sin 90°) = 0$$

(We take the axis at P because then F_{RH} and F_{RV} do not appear in the torque equation.) Solution of this equation yields $F_T = 2280$ N or to two significant figures $F_T = 2.3$ kN.

To find F_{RH} and F_{RV} we write

$$\overset{+}{\rightarrow}\Sigma F_x = 0 \quad \text{or} \quad -F_T \cos 40° + F_{RH} = 0$$

$$+\uparrow \Sigma F_y = 0 \quad \text{or} \quad F_T \sin 40° + F_{RV} - 600 - 800 = 0$$

Since we know F_T, these equations give $F_{RH} = 1750$ N or 1.8 kN and $F_{RV} = 65.6$ N or 66 N.

5.8 [II] A uniform, 0.40-kN boom is supported as shown in Fig. 5-9(*a*). Find the tension in the tie rope and the force exerted on the boom by the pin at P.

The forces acting on the boom are shown in Fig. 5-9(*b*). Take the pin as axis. The torque equation is then

$$+(3L/4)(F_T)(\sin 50°) - (L/2)(400\text{ N})(\sin 40°) - (L)(2000\text{ N})(\sin 40°) = 0$$

from which $F_T = 2460$ N or 2.5 kN. We now write:

$$\overset{+}{\rightarrow}\Sigma F_x = 0 \quad \text{or} \quad F_{RH} - F_T = 0$$

and so $F_{RH} = 2.5$ kN. Also

$$\Sigma F_y = 0 \quad \text{or} \quad F_{RV} - 2000\text{ N} - 400\text{ N} = 0$$

and so $F_{RV} = 2.4$ kN. F_{RV} and F_{RH} are the components of the reaction force at the pin. The magnitude of this force is

$$\sqrt{(2400)^2 + (2460)^2} = 3.4\text{ kN}$$

The tangent of the angle it makes with the horizontal is $\tan\theta = 2400/2460$, and so $\theta = 44°$.

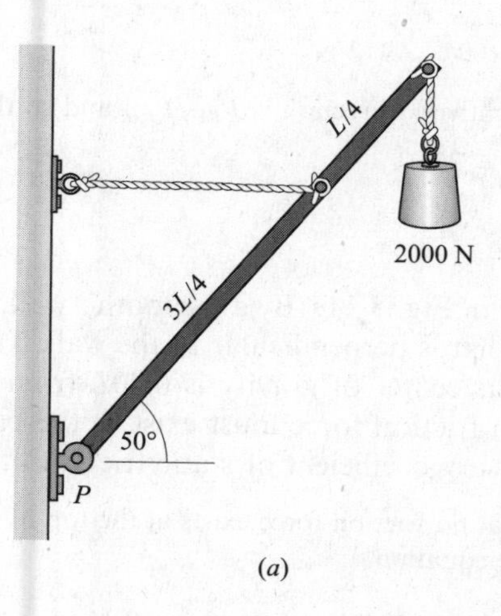

(*a*)

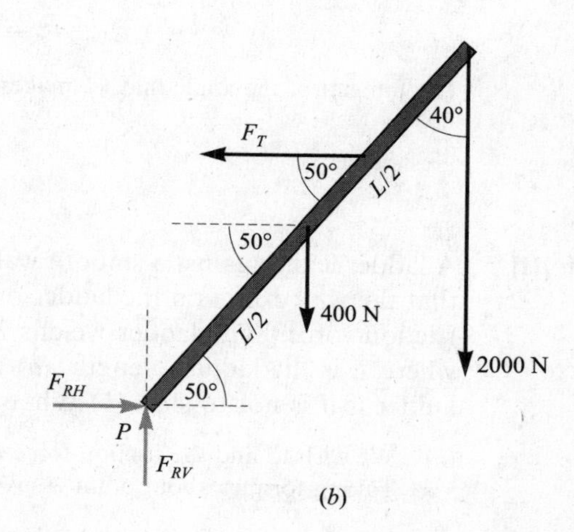

(*b*)

Fig. 5-9

5.9 [II] As shown in Fig. 5-10, hinges A and B hold a uniform, 400-N door in place. If the upper hinge happens to support the entire weight of the door, find the forces exerted on the door at both hinges. The width of the door is exactly $h/2$, where h is the distance between the hinges.

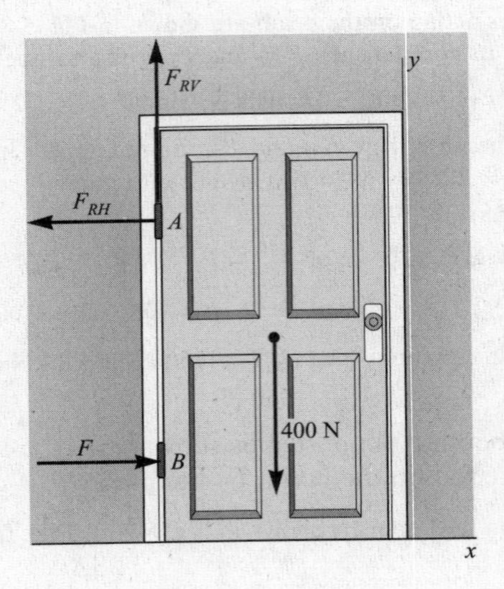

Fig. 5-10

The forces acting on the door are shown in Fig. 5-10. Only a horizonal force acts at B, because the upper hinge is assumed to support the door's weight. Let us take torques about point A as axis:

$$\circlearrowleft^{+} \Sigma \tau = 0 \qquad \text{becomes} \qquad +(h)(F)(\sin 90.0^\circ) - (h/4)(400\text{ N})(\sin 90.0^\circ) = 0$$

from which $F = 100$ N. We also have

$$\xrightarrow{+} \Sigma F_x = 0 \qquad \text{or} \qquad F - F_{RH} = 0$$

$$+\uparrow \Sigma F_y = 0 \qquad \text{or} \qquad F_{RV} - 400\text{ N} = 0$$

We find from these that $F_{RH} = 100$ N and $F_{RV} = 400$ N.

For the resultant reaction force F_R on the hinge at A, we have

$$F_R = \sqrt{(400)^2 + (100)^2} = 412\text{ N}$$

The tangent of the angle that $\vec{\mathbf{F}}_R$ makes with the negative x-direction is F_{RV}/F_{RH} and so the angle is

$$\arctan 4.00 = 76.0^\circ$$

5.10 [II] A ladder leans against a smooth wall, as shown in Fig. 5-11. (By a "smooth" wall, we mean that the wall exerts on the ladder only a force that is perpendicular to the wall. There is no friction force.) The ladder weighs 200 N and its center of gravity is $0.40L$ from the base, where L is the ladder's length. (*a*) How large a friction force must exist at the base of the ladder if it is not to slip? (*b*) What is the necessary coefficient of static friction?

(*a*) We wish to find the friction force F_f. Notice that no friction force exists at the top of the ladder. Taking torques about point A gives the torque equation

$$\circlearrowleft^{+} \Sigma \tau_A = -(0.40L)(200\text{ N})(\sin 40^\circ) + (L)(F_{N2})(\sin 50^\circ) = 0$$

Solving gives $F_{N2} = 67.1$ N. We can also write

$$\Sigma F_x = 0 \qquad \text{or} \qquad F_f - F_{N2} = 0$$
$$\Sigma F_y = 0 \qquad \text{or} \qquad F_{N1} - 200 = 0$$

and so $F_f = 67$ N and $F_{N1} = 0.20$ kN.

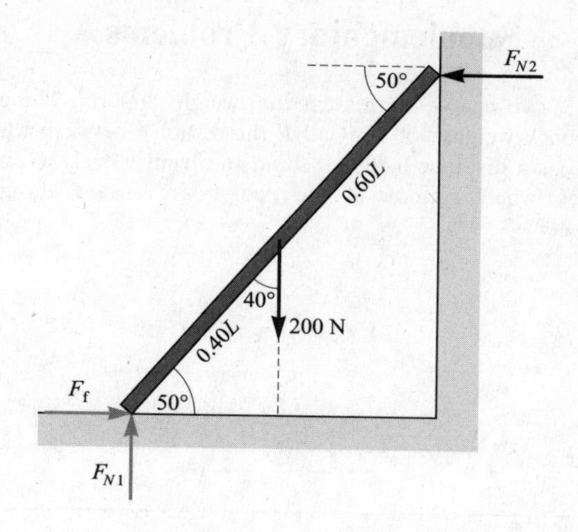

Fig. 5-11

(*b*)

$$\mu_s = \frac{F_f}{F_{N1}} = \frac{67.1}{200} = 0.34$$

5.11 [III] For the situation shown in Fig. 5-12(*a*), find F_{T1}, F_{T2}, and F_{T3}. The boom is uniform and weighs 800 N.

Let us first apply the force condition to point A. The appropriate free-body diagram is shown in Fig. 5-12(*b*). We then have

$$F_{T2} \cos 50.0° - 2000 \text{ N} = 0 \qquad \text{and} \qquad F_{T1} - F_{T2} \sin 50.0° = 0$$

From the first of these we find $F_{T2} = 3.11$ kN; then the second equation gives $F_{T1} = 2.38$ kN.

Let us now isolate the boom and apply the equilibrium conditions to it. The appropriate free-body diagram is shown in Fig. 5-12(*c*). The torque equation, for torques taken about point C, is

$$\circlearrowleft\!\!+ \Sigma\, \tau_c = +(L)(F_{T3})(\sin 20.0°) - (L)(3110 \text{ N})(\sin 90.0°) - (L/2)(800 \text{ N})(\sin 40.0°) = 0$$

Solving for F_{T3}, we find it to be 9.84 kN. If it were required, we could find F_{RH} and F_{RV} by using the x- and y-force equations.

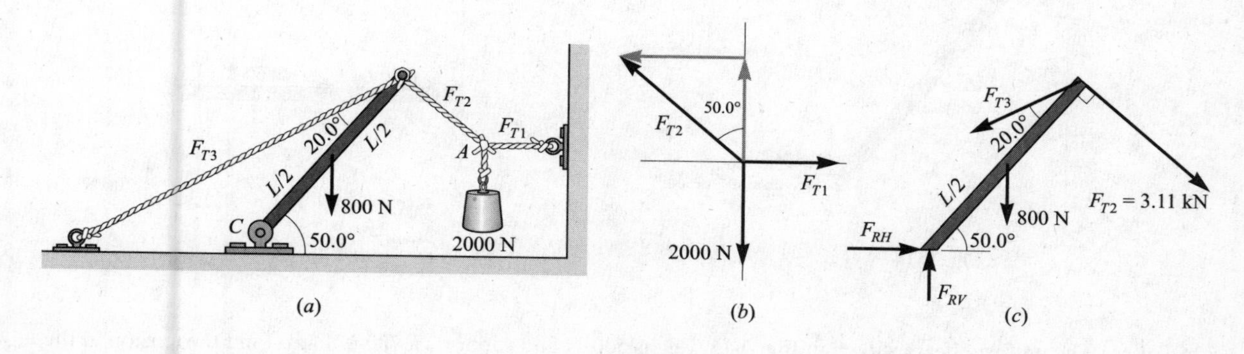

Fig. 5-12

Supplementary Problems

5.12 [II] As shown in Fig. 5-13, two people sit in a car that weighs 8000 N. The person in front weighs 700 N, while the one in the back weighs 900 N. Call L the distance between the front and back wheels. The car's center of gravity is a distance $0.400L$ behind the front wheels. How much force does each front wheel and each back wheel support if the people are seated along the centerline of the car? *Ans.* 2.09 kN, 2.71 kN

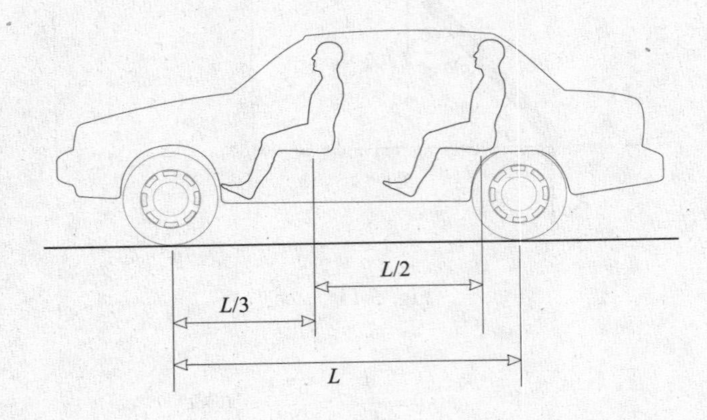

Fig. 5-13

5.13 [I] Two people, one at each end of a uniform beam that weighs 400 N, hold the beam at an angle of 25.0° to the horizontal. How large a vertical force must each person furnish to the beam? *Ans.* 200 N

5.14 [II] Repeat Problem 5.13 if a 140-N child sits on the beam at a point one-fourth of the way along the beam from its lower end. *Ans.* 235 N, 305 N

5.15 [II] As shown in Fig. 5-14, the uniform, 1600-N beam is hinged at one end and held by a tie rope at the other. Determine the tension F_T in the rope and the force components at the hinge. *Ans.* $F_T = 0.67$ kN, $F_{RH} = 0.67$ kN, $F_{RV} = 1.6$ kN

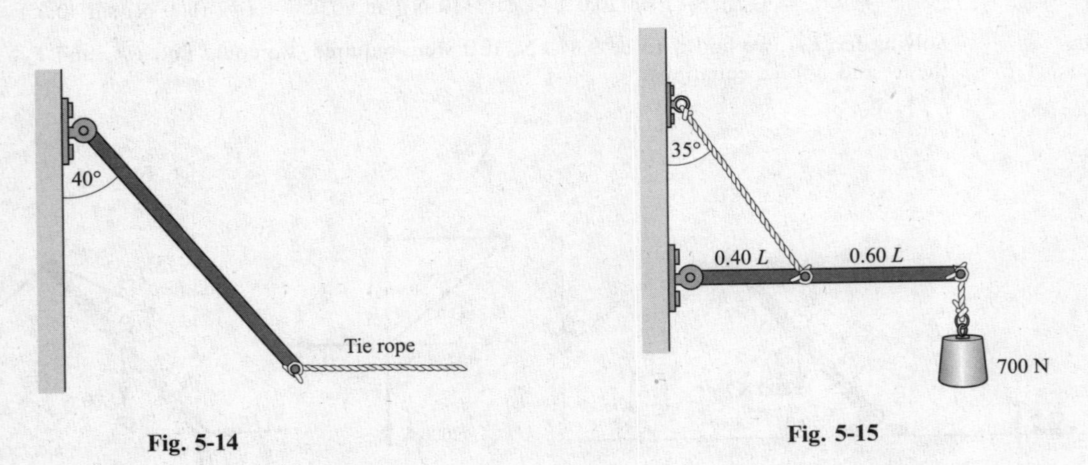

Fig. 5-14 **Fig. 5-15**

5.16 [II] The uniform beam shown in Fig. 5-15 weighs 500 N and supports a 700-N load. Find the tension in the tie rope and the force of the hinge on the beam. *Ans.* 2.9 kN, 2.0 kN at 35° below the horizontal

5.17 [II] The arm shown in Fig. 5-16 supports a 4.0-kg sphere. The mass of the hand and forearm together is 3.0 kg and its weight acts at a point 15 cm from the elbow. Determine the force exerted by the biceps muscle. *Ans.* 0.43 kN

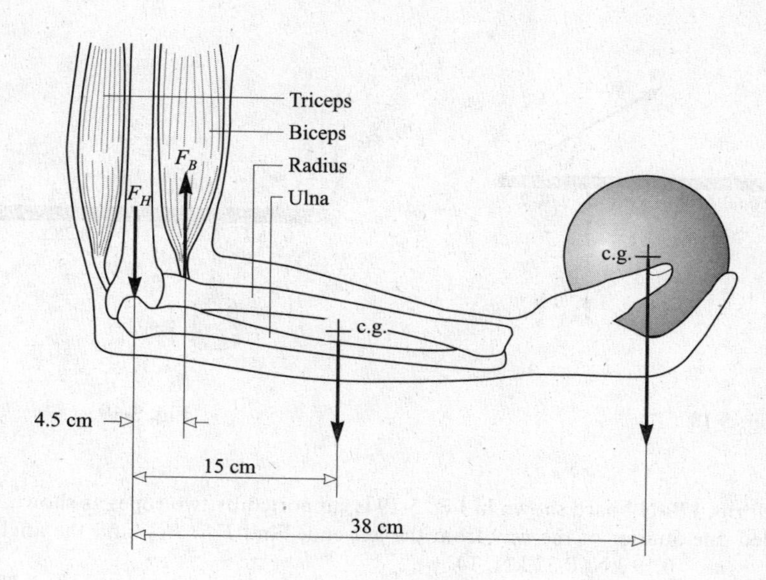

Fig. 5-16

5.18 [II] The mobile shown in Fig. 5-17 hangs at equilibrium. It consists of objects held by vertical strings. Object 3 weighs 1.40 N, while each of the identical uniform horizontal bars weighs 0.50 N. Find (*a*) the weights of objects 1 and 2, and (*b*) the tension in the upper string. *Ans.* (*a*) 1.5 N, 1.4 N; (*b*) 5.3 N

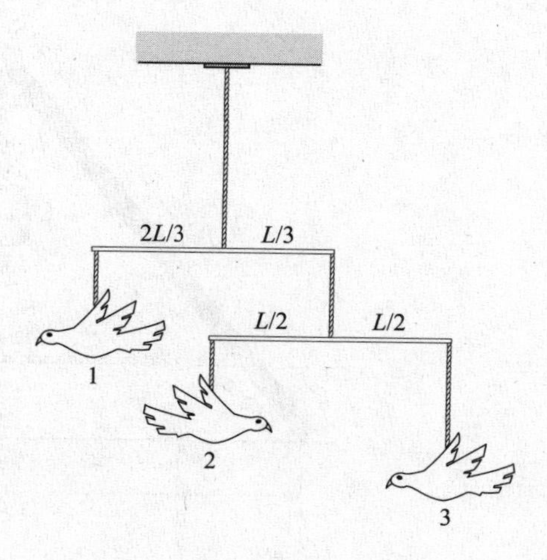

Fig. 5-17

5.19 [II] The hinges of a uniform door which weighs 200 N are 2.5 m apart. One hinge is a distance d from the top of the door, while the other is a distance d from the bottom. The door is 1.0 m wide. The weight of the door is supported by the lower hinge. Determine the forces exerted by the hinges on the door. *Ans.* The horizontal force at the upper hinge is 40 N. The force at the lower hinge is 0.20 kN at 79° above the horizontal.

5.20 [III] The uniform bar shown in Fig. 5-18 weighs 40 N and is subjected to the forces shown. Find the magnitude, location, and direction of the force needed to keep the bar in equilibrium. *Ans.* 0.11 kN, $0.68L$ from right end, at 49°

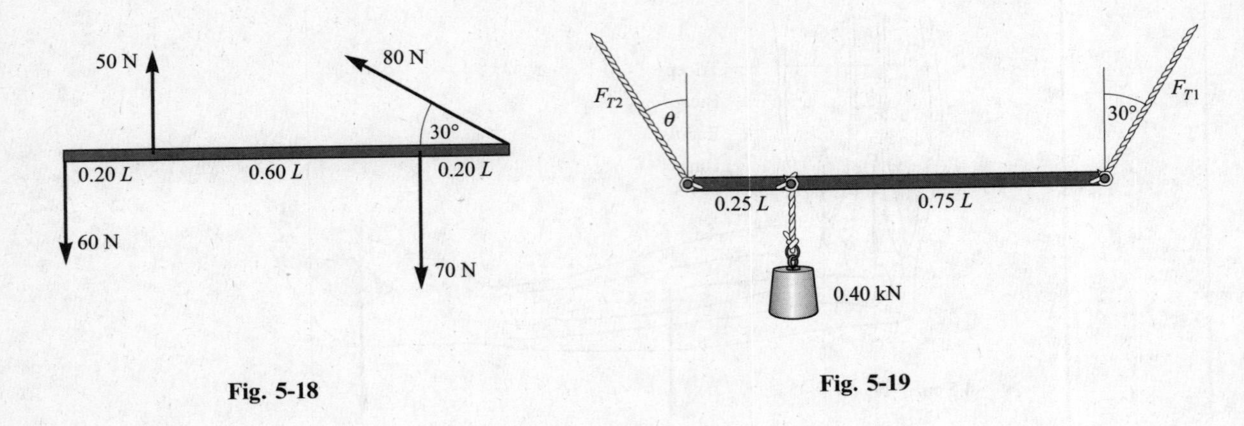

Fig. 5-18

Fig. 5-19

5.21 [III] The uniform, 120-N board shown in Fig. 5-19 is supported by two ropes as shown. A 0.40-kN weight is suspended one-quarter of the way from the left end. Find F_{T1}, F_{T2}, and the angle θ made by the left rope. *Ans.* 0.19 kN, 0.37 kN, 14°

5.22 [III] The foot of a ladder rests against a wall and its top is held by a tie rope, as shown in Fig. 5-20. The ladder weighs 100 N, and its center of gravity is 0.40 of its length from the foot. A 150-N child hangs from a rung that is 0.20 of the length from the top. Determine the tension in the tie rope and the components of the force on the foot of the ladder. *Ans.* $F_T = 0.12$ kN, $F_{RH} = 0.12$ kN, $F_{RV} = 0.25$ kN

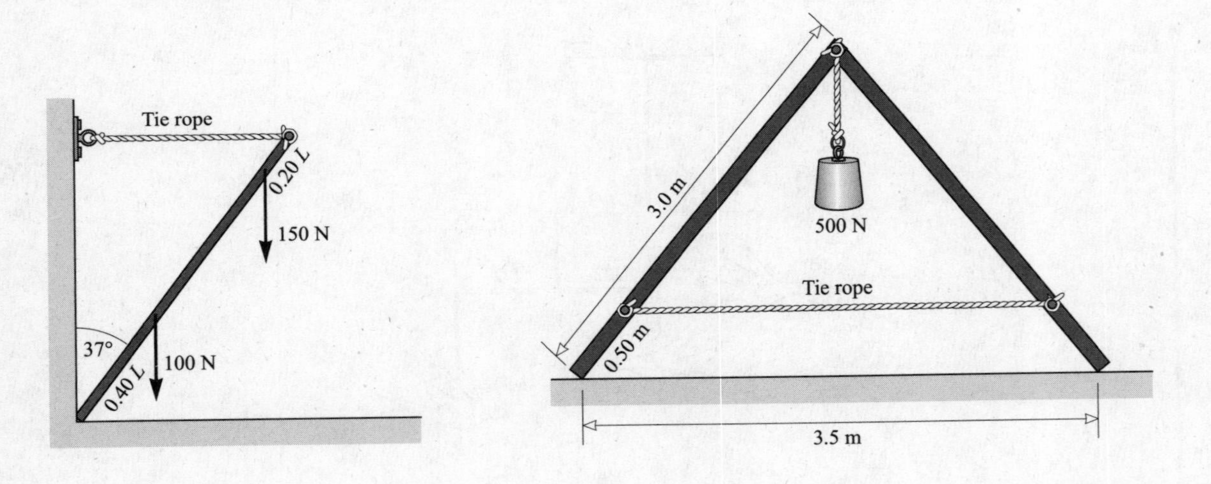

Fig. 5-20

Fig. 5-21

5.23 [III] A truss is made by hinging two uniform, 150-N rafters as shown in Fig. 5-21. They rest on an essentially frictionless floor and are held together by a tie rope. A 500-N load is held at their apex. Find the tension in the tie rope. *Ans.* 0.28 kN

5.24 [III] A 900-N lawn roller is to be pulled over a 5.0-cm high curb as shown in Fig. 5-22. The radius of the roller is 25 cm. What minimum pulling force is needed if the angle θ made by the handle is (*a*) 0° and (*b*) 30°? (*Hint*: Find the force needed to keep the roller balanced against the edge of the curb, just clear of the ground.) *Ans.* (*a*) 0.68 kN; (*b*) 0.55 kN

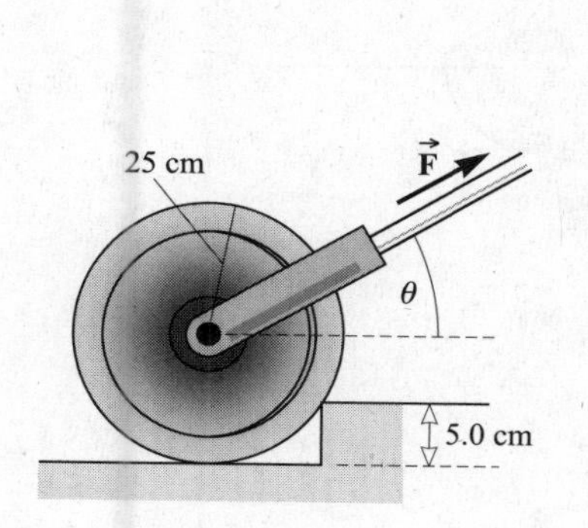

Fig. 5-22

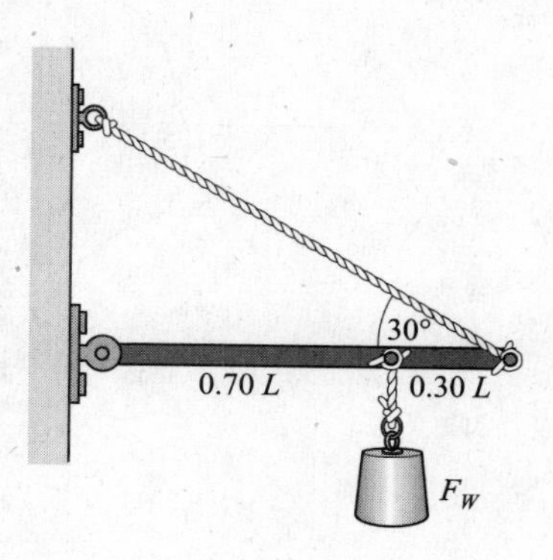

Fig. 5-23

5.25 [II] In Fig. 5-23, the uniform beam weighs 500 N. If the tie rope can support 1800 N, what is the maximum value the load F_W can have? *Ans.* 0.93 kN

5.26 [III] The beam in Fig. 5-24 has negligible weight. If the system hangs in equilibrium when $F_{W1} = 500$ N, what is the value of F_{W2}? *Ans.* 0.64 kN

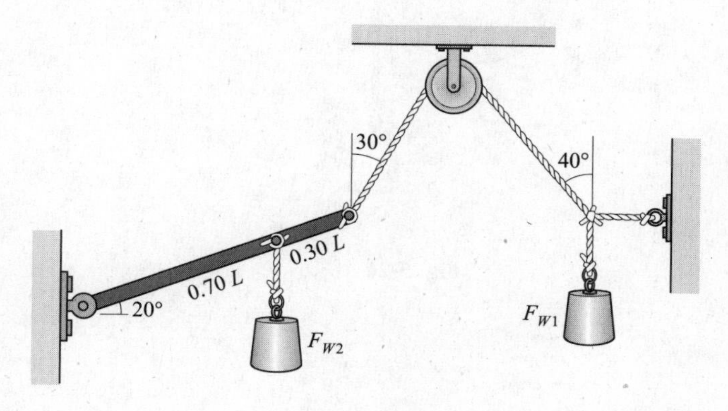

Fig. 5-24

5.27 [III] Repeat Problem 5.26, but now find F_{W1} if F_{W2} is 500 N. The beam weighs 300 N and is uniform. *Ans.* 0.56 kN

5.28 [III] An object is subjected to the forces shown in Fig. 5-25. What single force F applied at a point on the x-axis will balance these forces? (First find its components, and then find the force.) Where on the x-axis should the force be applied? *Ans.* $F_x = 232$ N, $F_y = -338$ N; $F = 410$ N at $-55.5°$; at $x = 2.14$ m

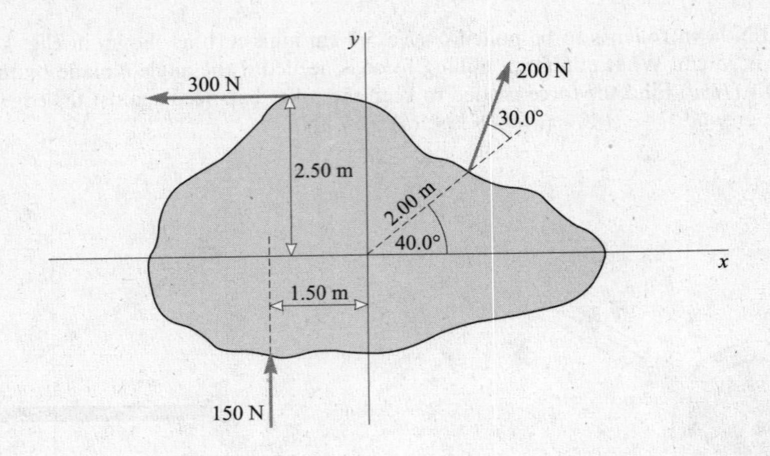

Fig. 5-25

5.29 [III] The solid uniform disk of radius b shown in Fig. 5-26 can turn freely on an axle through its center. A hole of diameter D is drilled through the disk; its center is a distance r from the axle. The weight of the material drilled out is F_{Wh}. Find the weight F_W of an object hung from a string wound on the disk that will hold the disk at equilibrium in the position shown. *Ans.* $F_W = F_{Wh}(r/b)\cos\theta$

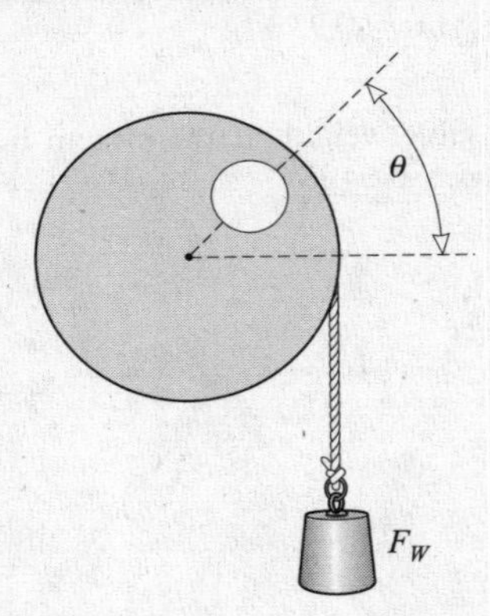

Fig. 5-26

Chapter 6

Work, Energy, and Power

THE WORK (W) done by a force is defined as the product of that force times the parallel distance over which it acts. Consider the simple case of straight-line motion shown in Fig. 6-1, where a force $\vec{\mathbf{F}}$ acts on a body that simultaneously undergoes a vector displacement $\vec{\mathbf{s}}$. The component of $\vec{\mathbf{F}}$ in the direction of $\vec{\mathbf{s}}$ is $F \cos \theta$. The work W done by the force $\vec{\mathbf{F}}$ is defined to be the component of $\vec{\mathbf{F}}$ in the direction of the displacement, multiplied by the displacement:

$$W = (F \cos \theta)(s) = Fs \cos \theta$$

Notice that θ is the angle between the force and displacement vectors. Work is a scalar quantity.

If $\vec{\mathbf{F}}$ and $\vec{\mathbf{s}}$ are in the same direction, $\cos \theta = \cos 0° = 1$ and $W = Fs$. But, if $\vec{\mathbf{F}}$ and $\vec{\mathbf{s}}$ are in opposite directions, then $\cos \theta = \cos 180° = -1$ and $W = -Fs$; the work is negative. Forces such as friction often slow the motion of an object and are then opposite in direction to the displacement. Such forces usually do negative work. Inasmuch as the friction force opposes the motion of an object the work done in overcoming friction (along any path, curved or straight) equals the product of F_f and the path-length traveled. Thus, if an object is dragged against friction, back to the point where the journey started, work is done even if the net displacement is zero.

Work is the transfer of energy from one entity to another by way of the action of a force applied over a distance. *The point of application of the force must move if work is to be done.*

THE UNIT OF WORK in the SI is the *newton-meter*, called the *joule* (J). One joule is the work done by a force of 1 N when it displaces an object 1 m in the direction of the force. Other units sometimes used for work are the *erg*, where 1 erg $= 10^{-7}$ J, and the *foot-pound* (ft·lb), where 1 ft·lb $= 1.355$ J.

ENERGY (E) is a measure of the change imparted to a system. It can be mechanically transferred to an object when a force does work on that object. The amount of energy given to an object via the action of a force over a distance equals the work done. Further, when an object does work, it gives up an amount of energy equal to the work it does. Because change can be effectuated in many different ways there are a variety of forms of energy. All forms of energy, including work, have the same units, joules. Energy is a scalar quantity. An object that is capable of doing work possesses energy.

KINETIC ENERGY (KE) is the energy possessed by an object because it is in motion. If an object of mass m is moving with a speed v, it has translational KE given by

$$\text{KE} = \tfrac{1}{2}mv^2$$

When m is in kg and v is in m/s, the units of KE are joules.

GRAVITATIONAL POTENTIAL ENERGY (PE_G) is the energy possessed by an object because of the gravitational interaction. In falling through a vertical distance h, a mass m can do work in the amount mgh. We define the PE_G of an object relative to an arbitrary zero level, often the Earth's surface. If the object is at a height h above the zero (or reference) level, its

$$\text{PE}_G = mgh$$

where g is the acceleration due to gravity. Notice that mg is the weight of the object. The units of PE_G are joules when m is in kg, g is in m/s^2, and h is in m.

THE WORK-ENERGY THEOREM: When work is done on a point mass or a rigid body, and there is no change in PE, the energy imparted can only appear as KE. Insofar as a body is not totally rigid, however, energy can be transferred to its parts and the work done on it will not precisely equal its change in KE.

CONSERVATION OF ENERGY: Energy can neither be created nor destroyed, but only transformed from one kind to another. (Mass can be regarded as one form of energy. Ordinarily, the conversion of mass into energy, and vice versa, predicted by the Special Theory of Relativity can be ignored. This subject is treated in Chapter 41.)

POWER (P) is the time rate of doing work:

$$\text{Average power} = \frac{\text{work done by a force}}{\text{time taken to do this work}} = \text{force} \times \text{speed}$$

where the speed is measured in the direction of the force applied to the object. More generally, power is the rate of transfer of energy. In the SI, the unit of power is the *watt* (W), and $1\text{ W} = 1\text{ J/s}$.

Another unit of power often used is the *horsepower*: $1\text{ hp} = 746\text{ W}$. Generally speaking, **power is the rate at which energy is transferred.**

THE KILOWATT-HOUR is a unit of energy. If a force is doing work at a rate of 1 kilowatt (which is 1000 J/s), then in 1 hour it will do 1 kW·h of work:

$$1\text{ kW}\cdot\text{h} = 3.6 \times 10^6\text{ J} = 3.6\text{ MJ}$$

Solved Problems

6.1 [I] In Fig. 6-1, assume that the object is being pulled along the ground by a 75-N force directed 28° above the horizontal. How much work does the force do in pulling the object 8.0 m?

The work done is equal to the product of the displacement, 8.0 m, and the component of the force that is parallel to the displacement, (75 N)(cos 28°). Thus,

$$W = (75\text{ N})(\cos 28°)(8.0\text{ m}) = 0.53\text{ kJ}$$

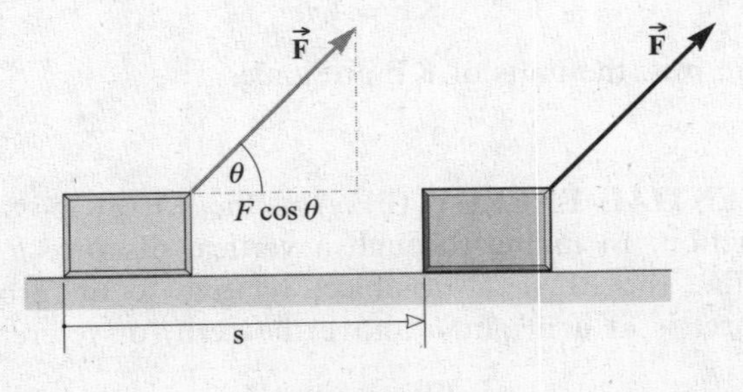

Fig. 6-1

6.2 [I] A block moves up a 30° incline under the action of certain forces, three of which are shown in Fig. 6-2. $\vec{\mathbf{F}}_1$ is horizontal and of magnitude 40 N. $\vec{\mathbf{F}}_2$ is normal to the plane and of magnitude 20 N. $\vec{\mathbf{F}}_3$ is parallel to the plane and of magnitude 30 N. Determine the work done by each force as the block (and point of application of each force) moves 80 cm up the incline.

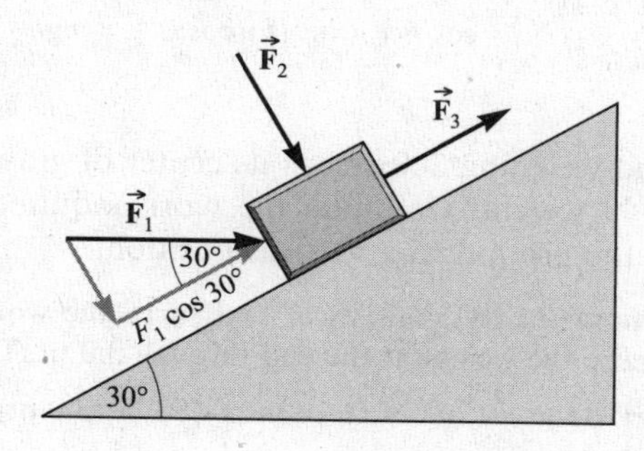

Fig. 6-2

The component of $\vec{\mathbf{F}}_1$ along the direction of the displacement is

$$F_1 \cos 30° = (40\ \text{N})(0.866) = 34.6\ \text{N}$$

Hence the work done by $\vec{\mathbf{F}}_1$ is (34.6 N)(0.80 m) = 28 J. (Notice that the distance must be expressed in meters.)

Because it has no component in the direction of the displacement, $\vec{\mathbf{F}}_2$ does no work.

The component of $\vec{\mathbf{F}}_3$ in the direction of the displacement is 30 N. Hence the work done by $\vec{\mathbf{F}}_3$ is (30 N)(0.80 m) = 24 J.

6.3 [II] A 300-g object slides 80 cm along a horizontal tabletop. How much work is done in overcoming friction between the object and the table if the coefficient of kinetic friction is 0.20?

We first find the friction force. Since the normal force equals the weight of the object,

$$F_f = \mu_k F_N = (0.20)(0.300\ \text{kg})(9.81\ \text{m/s}^2) = 0.588\ \text{N}$$

The work done overcoming friction is $F_f s \cos\theta$. Because the friction force is opposite in direction to the displacement, $\theta = 180°$. Therefore,

$$\text{Work} = F_f s \cos 180° = (0.588\ \text{N})(0.80\ \text{m})(-1) = -0.47\ \text{J}$$

The work is negative because the friction force slows the object; it decreases the object's kinetic energy.

6.4 [I] How much work is done against gravity in lifting a 3.0-kg object through a vertical distance of 40 cm?

An external force is needed to lift an object. If the object is lifted at constant speed, the lifting force must equal the weight of the object. The work done by the lifting force is what we refer to as *work done against gravity*. Because the lifting force is mg, where m is the mass of the object, we have

$$\text{Work} = (mg)(h)(\cos\theta) = (3.0\ \text{kg} \times 9.81\ \text{N})(0.40\ \text{m})(1) = 12\ \text{J}$$

In general, the work done against gravity in lifting an object of mass m through a vertical distance h is mgh.

6.5 [I] How much work is done on an object by the force that supports it as the object is lowered at a constant speed through a vertical distance h? How much work does the gravitational force on the object do in this same process?

The supporting force is mg, where m is the mass of the object. It is directed upward while the displacement is downward. Hence the work it does is

$$Fs \cos \theta = (mg)(h)(\cos 180°) = -mgh$$

The force of gravity acting on the object is also mg, but it is directed downward in the same direction as the displacement. The work done on the object by the force of gravity is therefore

$$Fs \cos \theta = (mg)(h)(\cos 0°) = mgh$$

6.6 [II] A ladder 3.0 m long and weighing 200 N has its center of gravity 120 cm from the bottom. At its top end is a 50-N weight. Compute the work required to raise the ladder from a horizontal position on the ground to a vertical position.

The work done (against gravity) consists of two parts, the work to raise the center of gravity 1.20 m and the work to raise the weight at the end through 3.0 m. Therefore

$$\text{Work done} = (200\ \text{N})(1.20\ \text{m}) + (50\ \text{N})(3.0\ \text{m}) = 0.39\ \text{kJ}$$

6.7 [II] Compute the work done against gravity by a pump that discharges 600 liters of fuel oil into a tank 20 m above the pump's intake. One cubic centimeter of fuel oil has a mass of 0.82 g. One liter is 1000 cm^3.

The mass lifted is

$$(600\ \text{liters})\left(1000\ \frac{\text{cm}^3}{\text{liter}}\right)\left(0.82\ \frac{\text{g}}{\text{cm}^3}\right) = 492\,000\ \text{g} = 492\ \text{kg}$$

The lifting work is then

$$\text{Work} = (mg)(h) = (492\ \text{kg} \times 9.81\ \text{m/s}^2)(20\ \text{m}) = 96\ \text{kJ}$$

6.8 [I] A 2.0-kg mass falls 400 cm. (*a*) How much work was done on it by the gravitational force? (*b*) How much PE_G did it lose?

Gravity pulls with a force mg on the object, and the displacement is 4 m in the direction of the force. The work done by gravity is therefore

$$(mg)(4.00\ \text{m}) = (2.0\ \text{kg} \times 9.81\ \text{N})(4.00\ \text{m}) = 78\ \text{J}$$

The change in PE_G of the object is $mgh_f - mgh_i$, where h_i and h_f are the initial and final heights of the object above the reference level. We then have

$$\text{Change in PE}_\text{G} = mgh_f - mgh_i = mg(h_f - h_i) = (2.0\ \text{kg} \times 9.81\ \text{N})(-4.0\ \text{m}) = -78\ \text{J}$$

The loss in PE_G is 78 J.

6.9 [II] A force of 1.50 N acts on a 0.20-kg cart so as to accelerate it along an air track. The track and force are horizontal and in line. How fast is the cart going after acceleration from rest through 30 cm, if friction is negligible?

The work done by the force causes, and is equal to, the increase in KE of the cart. Therefore,

$$\text{Work done} = (\text{KE})_\text{end} - (\text{KE})_\text{start} \quad \text{or} \quad Fs \cos 0° = \tfrac{1}{2}mv_f^2 - 0$$

Substituting gives

$$(1.50\ \text{N})(0.30\ \text{m}) = \tfrac{1}{2}(0.20\ \text{kg})v_f^2$$

from which $v_f = 2.1\ \text{m/s}$.

6.10 [II] A 0.50-kg block slides across a tabletop with an initial velocity of 20 cm/s and comes to rest in a distance of 70 cm. Find the average friction force that retarded its motion.

The KE of the block is decreased because of the slowing action of the friction force. That is,

$$\text{Change in KE of block} = \text{work done on block by friction force}$$

$$\tfrac{1}{2}mv_f^2 - \tfrac{1}{2}mv_i^2 = F_f s \cos\theta$$

Because the friction force on the block is opposite in direction to the displacement, $\cos\theta = -1$. Using $v_f = 0$, $v_i = 0.20$ m/s, and $s = 0.70$ m, we find

$$0 - \tfrac{1}{2}(0.50\ \text{kg})(0.20\ \text{m/s})^2 = (F_f)(0.70\ \text{m})(-1)$$

from which $F_f = 0.014$ N.

6.11 [II] A car going 15 m/s is brought to rest in a distance of 2.0 m as it strikes a pile of dirt. How large an average force is exerted by seatbelts on a 90-kg passenger as the car is stopped?

We assume the seatbelts stop the passenger in 2.0 m. The force F they apply acts through a distance of 2.0 m and decreases the passenger's KE to zero. So

$$\text{Change in KE of passenger} = \text{work done by } F$$

$$0 - \tfrac{1}{2}(90\ \text{kg})(15\ \text{m/s}^2) = (F)(2.0\ \text{m})(-1)$$

where $\cos\theta = -1$ because the restraining force on the passenger is opposite in direction to the displacement. Solving, we find $F = 5.1$ kN.

6.12 [II] A projectile is shot upward from the earth with a speed of 20 m/s. Using energy considerations, how high is the projectile when its speed is 8.0 m/s? Ignore air friction.

Because the projectile's energy is conserved, we have

$$\text{Change in KE} + \text{change in PE}_G = 0$$

$$\tfrac{1}{2}mv_f^2 - \tfrac{1}{2}mv_i^2 + (mg)(h_f - h_i) = 0$$

We wish to find $h_f - h_i$. After a little algebra, we obtain

$$h_f - h_i = -\frac{v_f^2 - v_i^2}{2g} = -\frac{(8.0\ \text{m/s})^2 - (20\ \text{m/s})^2}{2(9.81\ \text{m/s}^2)} = 17\ \text{m}$$

6.13 [II] In an Atwood machine (see Problem 3.30) the two masses are 800 g and 700 g. The system is released from rest. How fast is the 800-g mass moving after it has fallen 120 cm?

The 700-g mass rises 120 cm while the 800-g mass falls 120 cm, so the net change in PE_G is

$$\text{Change in PE}_G = (0.70\ \text{kg})(9.81\ \text{m/s}^2)(1.20\ \text{m}) - (0.80\ \text{kg})(9.81\ \text{m/s}^2)(1.20\ \text{m}) = -1.18\ \text{J}$$

which is a loss in PE_G. Because energy is conserved, the KE of the masses must increase by 1.18 J. Therefore,

$$\text{Change in KE} = 1.18\ \text{J} = \tfrac{1}{2}(0.70\ \text{kg})(v_f^2 - v_i^2) + \tfrac{1}{2}(0.80\ \text{kg})(v_f^2 - v_i^2)$$

The system started from rest, so $v_i = 0$. We solve the above equation for v_f and find $v_f = 1.25$ m/s.

6.14 [II] Figure 6-3 shows a bead sliding on a wire. If friction forces are negligible and the bead has a speed of 200 cm/s at A, what will be its speed (*a*) at point B? (*b*) At point C?

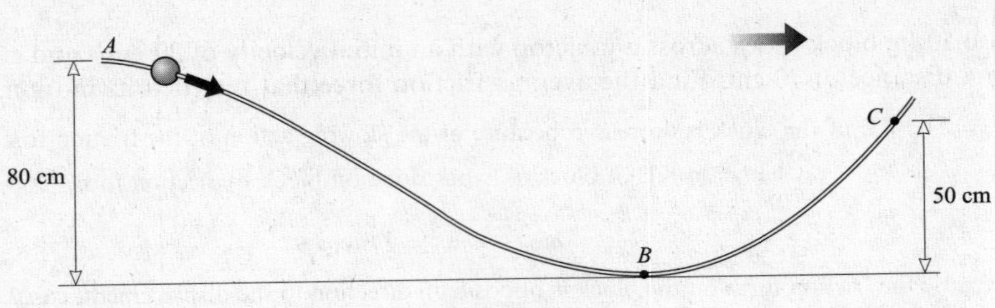

Fig. 6-3

We know the energy of the bead is conserved, so we can write

$$\text{Change in KE} + \text{change in PE}_G = 0$$

$$\tfrac{1}{2}mv_f^2 - \tfrac{1}{2}mv_i^2 + mg(h_f - h_i) = 0$$

(*a*) Here, $v_i = 2.0$ m/s, $h_i = 0.80$ m, and $h_f = 0$. Using these values, while noticing that m cancels out, gives $v_f = 4.4$ m/s.

(*b*) Here, $v_i = 2.0$ m/s, $h_i = 0.80$ m, and $h_f = 0.50$ m. Using these values gives $v_f = 3.1$ m/s.

6.15 [II] Suppose the bead in Fig. 6-3 has a mass of 15 g and a speed of 2.0 m/s at *A*, and it stops as it reaches point *C*. The length of the wire from *A* to *C* is 250 cm. How large an average friction force opposed the motion of the bead?

When the bead moves from *A* to *C*, it experiences a change in its total energy: it loses both KE and PE_G. This total energy change is equal to the work done on the bead by the friction force. Therefore,

$$\text{Change in PE}_G + \text{change in KE} = \text{work done by friction force}$$

$$mg(h_C - h_A) + \tfrac{1}{2}m(v_C^2 - v_A^2) = F_f s \cos\theta$$

Notice that $\cos\theta = -1$, $v_C = 0$, $v_A = 2.0$ m/s, $h_C - h_A = -0.30$ m, $s = 2.50$ m, and $m = 0.015$ kg. Using these values, we find that $F_f = 0.030$ N.

6.16 [II] A 1200-kg car is coasting down a 30° hill as shown in Fig. 6-4. At a time when the car's speed is 12 m/s, the driver applies the brakes. What constant force *F* (parallel to the road) must result if the car is to stop after traveling 100 m?

The change in total energy of the car ($\text{KE} + \text{PE}_G$) is equal to the work done on it by the braking force *F*. This work is $Fs \cos 180°$ because *F* retards the car's motion. We have

$$\tfrac{1}{2}m(v_f^2 - v_i^2) + mg(h_f - h_i) = Fs(-1)$$

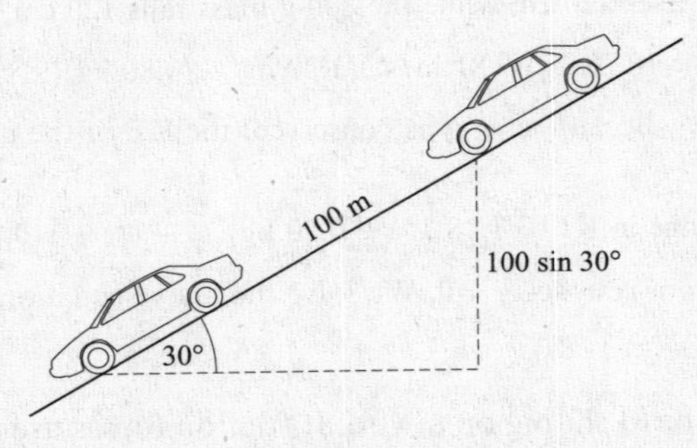

Fig. 6-4

where $m = 1200$ kg, $v_f = 0$, $v_i = 12$ m/s, $h_f - h_i = (100 \text{ m}) \sin 30°$ and $s = 100$ m
With these values, the equation yields $F = 6.7$ kN.

6.17 [II] A ball at the end of a 180-cm long string swings as a pendulum as shown in Fig. 6-5. The ball's speed is 400 cm/s as it passes through its lowest position. (*a*) To what height h above this position will it rise before stopping? (*b*) What angle does the pendulum then make to the vertical?

(*a*) The pull of the string on the ball is always perpendicular to the ball's motion, and therefore does no work on the ball. Consequently, the ball's total energy remains constant; it loses KE but gains an equal amount of PE_G. That is,

$$\text{Change in KE} + \text{change in PE}_\text{G} = 0$$

$$\tfrac{1}{2}mv_f^2 - \tfrac{1}{2}mv_i^2 + mgh = 0$$

Since $v_f = 0$ and $v_i = 4.00$ m/s, we find $h = 0.816$ m as the height to which the ball rises.

(*b*) From Fig. 6-5,

$$\cos\theta = \frac{L-h}{L} = 1 - \frac{0.816}{1.80}$$

which gives $\theta = 56.9°$.

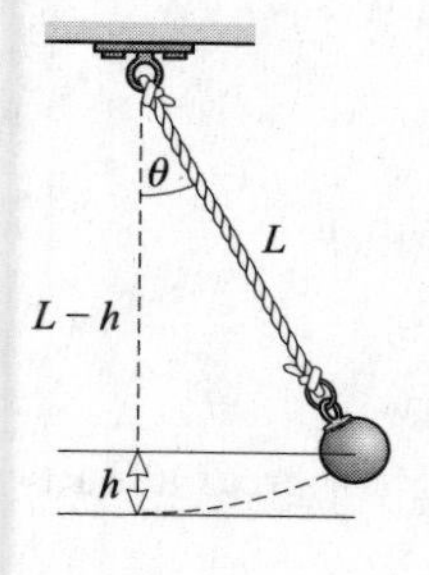

Fig. 6-5

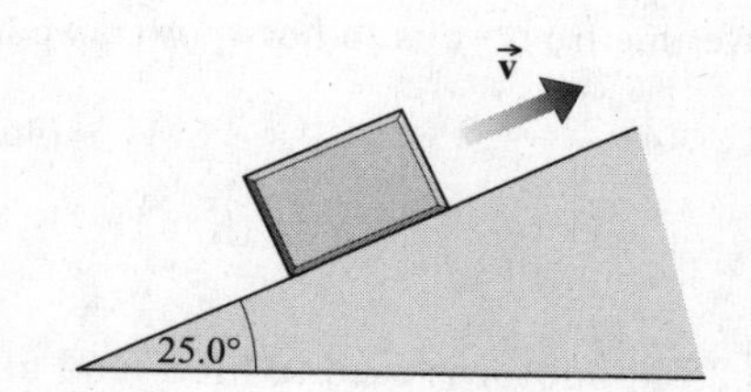

Fig. 6-6

6.18 [II] A 500-g block is shot up the incline in Fig. 6-6 with an initial speed of 200 cm/s. How far up the incline will it go if the coefficient of friction between it and the incline is 0.150?

We first find the friction force on the block using

$$F_f = \mu F_N = \mu(mg \cos 25.0°)$$

$$F_f = 0.667 \text{ N}$$

As the block slides up the incline a distance D, it rises a distance $D \sin 25.0°$. Because the change in energy of the block equals the work done on it by the friction force, we have

$$\text{Change in KE} + \text{change in PE}_\text{G} = F_f D \cos 180°$$

$$\tfrac{1}{2}m(v_f^2 - v_i^2) + mg(D \sin 25.0°) = -F_f D$$

The friction force opposes the motion, it's down the incline, while the displacement is up the incline, hence the work it does is negative.

We know $v_i = 2.00$ m/s and $v_f = 0$. Notice that the mass of the block could be canceled out in this case (but only because F_f is given in terms of it). Substitution gives $D = 0.365$ m.

6.19 [II] A 60 000-kg train is being pulled up a 1.0 percent grade (it rises 1.0 m for each horizontal 100 m) by a drawbar pull of 3.0 kN. The friction force opposing the motion of the train is 4.0 kN. The train's initial speed is 12 m/s. Through what horizontal distance s will the train move before its speed is reduced to 9.0 m/s?

The height the train rises in traveling a horizontal distance s is $0.010s$. The change in total energy of the train is due to the work of the friction force (which is negative) and the drawbar pull:

$$\text{Change in KE} + \text{change in PE}_G = W_{\text{drawbar}} + W_{\text{friction}}$$

$$\tfrac{1}{2}m(v_f^2 - v_i^2) + mg(0.010s) = (3000\ \text{N})(s)(1) + (4000\ \text{N})(s)(-1)$$

from which $s = 275\ \text{m} = 0.28\ \text{km}$.

6.20 [III] An advertisement claims that a certain 1200-kg car can accelerate from rest to a speed of 25 m/s in a time of 8.0 s. What average power must the motor produce to cause this acceleration? Give your answer in both watts and horsepower. Ignore friction losses.

The work done in accelerating the car is given by

$$\text{Work done} = \text{change in KE} = \tfrac{1}{2}m(v_f^2 - v_i^2) = \tfrac{1}{2}mv_f^2$$

The time taken for this work to be performed is 8.0 s. Therefore to two significant figures,

$$\text{Power} = \frac{\text{work}}{\text{time}} = \frac{\tfrac{1}{2}(1200\ \text{kg})(25\ \text{m/s})^2}{8.0\ \text{s}} = 46\,875\ \text{W} = 47\ \text{kW}$$

Converting from watts to horsepower, we have

$$\text{Power} = (46\,875\ \text{W})\left(\frac{1\ \text{hp}}{746\ \text{W}}\right) = 63\ \text{hp}$$

6.21 [III] A 0.25-hp motor is used to lift a load at the rate of 5.0 cm/s. How great a load can it raise at this constant speed?

We assume the power *output* of the motor to be 0.25 hp = 186.5 W. In 1.0 s, the load mg is lifted a distance of 0.050 m. Therefore,

$$\text{Work done in 1.0 s} = (\text{weight})(\text{height change in 1.0 s}) = (mg)(0.050\ \text{m})$$

By definition, power = work/time, so that

$$186.5\ \text{W} = \frac{(mg)(0.050\ \text{m})}{1.0\ \text{s}}$$

Using $g = 9.81\ \text{m/s}^2$, we find that $m = 381$ kg. The motor can lift a load of about 0.38×10^3 kg at this speed.

6.22 [III] Repeat Problem 6.20 if the data apply to a car going up a 20° incline.

Work must be done to lift the car as well as to accelerate it:

$$\begin{aligned}\text{Work done} &= \text{change in KE} + \text{change in PE}_G \\ &= \tfrac{1}{2}m(v_f^2 - v_i^2) + mg(h_f - h_i)\end{aligned}$$

where $h_f - h_i = s \sin 20°$ and s is the total distance the car travels along the incline in the 8 s under consideration. Knowing $v_i = 0$, $v_f = 25$ m/s, and $t = 8.0$ s, we have

$$s = v_{av}t = \tfrac{1}{2}(v_i + v_f)t = 100\ \text{m}$$

Then $\quad$ Work done $= \frac{1}{2}(1200 \text{ kg})(625 \text{ m}^2/\text{s}^2) + (1200 \text{ kg})(9.81 \text{ m/s}^2)(100 \text{ m})(\sin 20°) = 777.6 \text{ kJ}$

from which
$$\text{Power} = \frac{778 \text{ kJ}}{8.0 \text{ s}} = 97 \text{ kW} = 0.13 \times 10^3 \text{ hp}$$

6.23 [III] In unloading grain from the hold of a ship, an elevator lifts the grain through a distance of 12 m. Grain is discharged at the top of the elevator at a rate of 2.0 kg each second, and the discharge speed of each grain particle is 3.0 m/s. Find the minimum-horsepower motor that can elevate grain in this way.

The power output of the motor is

$$\text{Power} = \frac{\text{change in KE} + \text{change in PE}_G}{\text{time taken}} = \frac{\frac{1}{2}m(v_f^2 - v_i^2) + mgh}{t}$$
$$= \frac{m}{t}\left[\tfrac{1}{2}(9.0 \text{ m}^2/\text{s}^2) + (9.81 \text{ m/s}^2)(12 \text{ m})\right]$$

The mass transported per second, m/t, is 2.0 kg/s. Using this value gives the power as 0.24 kW.

Supplementary Problems

6.24 [I] A force of 3.0 N acts through a distance of 12 m in the direction of the force. Find the work done. *Ans.* 36 J

6.25 [I] A 4.0-kg object is lifted 1.5 m. (*a*) How much work is done against the Earth's gravity? (*b*) Repeat if the object is lowered instead of lifted. *Ans.* (*a*) 59 J; (*b*) −59 J

6.26 [I] A uniform rectangular marble slab is 3.4 m long and 2.0 m wide. It has a mass of 180 kg. If it is originally lying on the flat ground, how much work is needed to stand it on end? *Ans.* 3.0 kJ

6.27 [I] How large a force is required to accelerate a 1300-kg car from rest to a speed of 20 m/s in a horizontal distance of 80 m? *Ans.* 3.3 kN

6.28 [I] A 1200-kg car going 30 m/s applies its brakes and skids to rest. If the friction force between the sliding tires and the pavement is 6000 N, how far does the car skid before coming to rest? *Ans.* 90 m

6.29 [I] A proton ($m = 1.67 \times 10^{-27}$ kg) that has a speed of 5.0×10^6 m/s passes through a metal film of thickness 0.010 mm and emerges with a speed of 2.0×10^6 m/s. How large an average force opposed its motion through the film? *Ans.* 1.8×10^{-9} N

6.30 [I] A 200-kg cart is pushed slowly up an incline. How much work does the pushing force do in moving the cart up to a platform 1.5 m above the starting point if friction is negligible? *Ans.* 2.9 kJ

6.31 [II] Repeat Problem 6.30 if the distance along the incline to the platform is 7.0 m and a friction force of 150 N opposes the motion. *Ans.* 4.0 kJ

6.32 [II] A 50 000-kg freight car is pulled 800 m up along a 1.20 percent grade at constant speed. (*a*) Find the work done against gravity by the drawbar pull. (*b*) If the friction force retarding the motion is 1500 N, find the total work done. *Ans.* (*a*) 4.70 MJ; (*b*) 5.90 MJ

6.33 [II] A 60-kg woman walks up a flight of stairs that connects two floors 3.0 m apart. (*a*) How much lifting work is done by the woman? (*b*) By how much does the woman's PE_G change? *Ans.* (*a*) 1.8 kJ; (*b*) 1.8 kJ

6.34 [II] A pump lifts water from a lake to a large tank 20 m above the lake. How much work against gravity does the pump do as it transfers 5.0 m^3 of water to the tank? One cubic meter of water has a mass of 1000 kg. *Ans.* 9.8×10^5 J

6.35 II] Just before striking the ground, a 2.00-kg mass has 400 J of KE. If friction can be ignored, from what height was it dropped? *Ans.* 20.0 m

6.36 [II] A 0.50-kg ball falls past a window that is 1.50 m in vertical length. (*a*) How much did the KE of the ball increase as it fell past the window? (*b*) If its speed was 3.0 m/s at the top of the window, what was its speed at the bottom? *Ans.* (*a*) 7.4 J; (*b*) 6.2 m/s

6.37 [II] At sea level a nitrogen molecule in the air has an average translational KE of 6.2×10^{-21} J. Its mass is 4.7×10^{-26} kg. (*a*) If such a molecule could shoot straight up without striking other air molecules, how high would it rise? (*b*) What is that molecule's initial upward speed? *Ans.* 14 km; (*b*) 0.51 km/s

6.38 [II] The coefficient of sliding friction between a 900-kg car and the pavement is 0.80. If the car is moving at 25 m/s along level pavement when it begins to skid to a stop, how far will it go before coming to rest? *Ans.* 40 m

6.39 [II] Consider the simple pendulum shown in Fig. 6-7. (*a*) If it is released from point *A*, what will be the speed of the ball as it passes through point C? (*b*) What is the ball's speed at point *B*? *Ans.* (*a*) 3.8 m/s; (*b*) 3.4 m/s

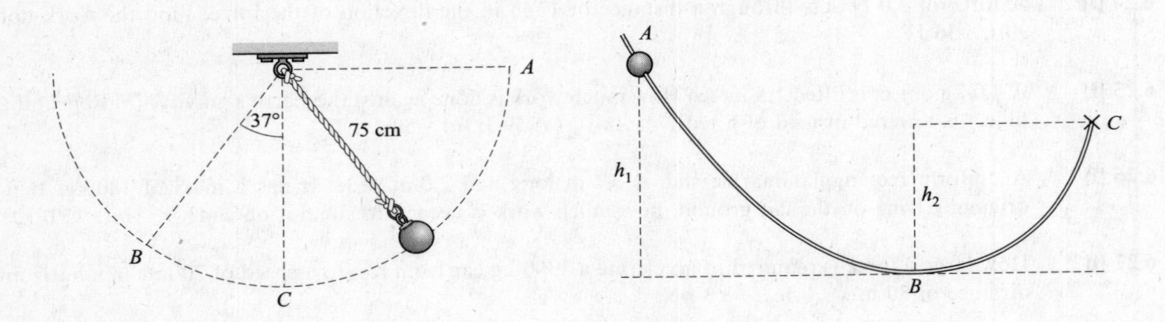

Fig. 6-7

Fig. 6-8

6.40 [II] A 1200-kg car coasts from rest down a driveway that is inclined 20° to the horizontal and is 15 m long. How fast is the car going at the end of the driveway if (*a*) friction is negligible and (*b*) a friction force of 3000 N opposes the motion? *Ans.* (*a*) 10 m/s; (*b*) 5.1 m/s

6.41 [II] The driver of a 1200-kg car notices that the car slows from 20 m/s to 15 m/s as it coasts a distance of 130 m along level ground. How large a force opposes the motion? *Ans.* 0.81 kN

6.42 [II] A 2000-kg elevator rises from rest in the basement to the fourth floor, a distance of 25 m. As it passes the fourth floor, its speed is 3.0 m/s. There is a constant frictional force of 500 N. Calculate the work done by the lifting mechanism. *Ans.* 0.51 MJ

6.43 [II] Figure 6-8 shows a bead sliding on a wire. How large must height h_1 be if the bead, starting at rest at *A*, is to have a speed of 200 cm/s at point *B*? Ignore friction. *Ans.* 20.4 cm

6.44 [II] In Fig. 6-8, $h_1 = 50.0$ cm, $h_2 = 30.0$ cm, and the length along the wire from A to C is 400 cm. A 3.00-g bead released at A coasts to point C and stops. How large an average friction force opposed its motion?
Ans. 1.47 mN

6.45 [III] In Fig. 6-8, $h_1 = 200$ cm, $h_2 = 150$ cm, and at A the 3.00-g bead has a downward speed along the wire of 800 cm/s. (*a*) How fast is the bead moving as it passes point B if friction is negligible? (*b*) How much energy did the bead lose to friction work if it rises to a height of 20.0 cm above C after it leaves the wire?
Ans. (*a*) 10.2 m/s; (*b*) 105 mJ

6.46 [I] Calculate the average horsepower required to raise a 150-kg drum to a height of 20 m in a time of 1.0 minute. *Ans.* 0.66 hp

6.47 [I] Compute the power output of a machine that lifts a 500-kg crate through a height of 20.0 m in a time of 60.0 s. *Ans.* 1.63 kW

6.48 [I] An engine expends 40.0 hp in propelling a car along a level track at a constant speed of 15.0 m/s. How large is the total retarding force acting on the car? *Ans.* 1.99 kN

6.49 [II] A 1000-kg auto travels up a 3.0 percent grade at 20 m/s. Find the horsepower required, neglecting friction. *Ans.* 7.9 hp

6.50 [II] A 900-kg car whose motor delivers a maximum power of 40.0 hp to its wheels can maintain a steady speed of 130 km/h on a horizontal roadway. How large is the friction force that impedes its motion at this speed? *Ans.* 826 N

6.51 [II] Water flows from a reservoir at the rate of 3000 kg/min, to a turbine 120 m below. If the efficiency of the turbine is 80 percent, compute the horsepower output of the turbine. Neglect friction in the pipe and the small KE of the water leaving the turbine. *Ans.* 63 hp

6.52 [II] Find the mass of the largest box that a 40-hp engine can pull along a level road at 15 m/s if the friction coefficient between road and box is 0.15. *Ans.* 1.4×10^3 kg

6.53 [II] A 1300-kg car is to accelerate from rest to a speed of 30.0 m/s in a time of 12.0 s as it climbs a 15.0° hill. Assuming uniform acceleration, what minimum horsepower is needed to accelerate the car in this way?
Ans. 132 hp

Chapter 7

Simple Machines

A MACHINE is any device by which the magnitude, direction, or method of application of a force is changed so as to achieve some advantage. Examples of simple machines are the lever, inclined plane, pulley, crank and axle, and jackscrew.

THE PRINCIPLE OF WORK that applies to a continuously operating machine is as follows:

$$\text{Work input} = \text{useful work output} + \text{work to overcome friction}$$

In machines that operate for only a short time, some of the input work may be used to store energy within the machine. An internal spring might be stretched, or a movable pulley might be raised, for example.

MECHANICAL ADVANTAGE: The **actual mechanical advantage** (AMA) of a machine is

$$\text{AMA} = \text{force ratio} = \frac{\text{force exerted by machine on load}}{\text{force used to operate machine}}$$

The **ideal mechanical advantage** (IMA) of a machine is

$$\text{IMA} = \text{distance ratio} = \frac{\text{distance moved by input force}}{\text{distance moved by load}}$$

Because friction is always present, the AMA is always less than the IMA. In general, both the AMA and IMA are greater than one.

THE EFFICIENCY of a machine is

$$\text{Efficiency} = \frac{\text{work output}}{\text{work input}} = \frac{\text{power output}}{\text{power input}}$$

The efficiency is also equal to the ratio AMA/IMA.

Solved Problems

7.1 [I] In a particular hoist system, the load is lifted 10 cm for each 70 cm of movement of the rope that operates the device. What is the smallest input force that could possibly lift a 5.0-kN load?

The most advantageous situation possible is that in which all the input work is used to lift the load, i.e., in which friction and other loss mechanisms are negligible. In that case,

$$\text{Work input} = \text{lifting work}$$

If the load is lifted a distance s, the lifting work is (5.0 kN)(s). The input force F, however, must work through a distance $7.0s$. The above equation then becomes

$$(F)(7.0s) = (5.0\ \text{kN})(s)$$

which gives $F = 0.71$ kN as the smallest possible force required.

7.2 [III] A hoisting machine lifts a 3000-kg load a height of 8.00 m in a time of 20.0 s. The power supplied to the engine is 18.0 hp. Compute (*a*) the work output, (*b*) the power output and power input, and (*c*) the efficiency of the engine and hoist system.

(*a*) $$\text{Work output} = (\text{lifting force})\ (\text{height}) = (3000 \times 9.81\ \text{N})(8.00\ \text{m}) = 235\ \text{kJ}$$

(*b*) $$\text{Power output} = \frac{\text{work output}}{\text{time taken}} = \frac{235\ \text{kJ}}{20.0\ \text{s}} = 11.8\ \text{kW}$$

$$\text{Power input} = (18.0\ \text{hp})\left(\frac{0.746\ \text{kW}}{1\ \text{hp}}\right) = 13.4\ \text{kW}$$

(*c*) $$\text{Efficiency} = \frac{\text{power output}}{\text{power input}} = \frac{11.8\ \text{kW}}{13.4\ \text{kW}} = 0.881 = 88.1\%$$

or $$\text{Efficiency} = \frac{\text{work output}}{\text{work input}} = \frac{235\ \text{kJ}}{(13.4\ \text{kJ/s})(20.0\ \text{s})} = 0.877 = 87.7\%$$

The efficiency is 88%; the differences arise from the rounding off process.

7.3 [II] What power in kW is supplied to a 12.0-hp motor having an efficiency of 90.0 percent when it is delivering its full rated output?

From the definition of efficiency,

$$\text{Power input} = \frac{\text{power output}}{\text{efficiency}} = \frac{(12.0\ \text{hp})(0.746\ \text{kW/hp})}{0.900} = 9.95\ \text{kW}$$

7.4 [II] For the three levers shown in Fig. 7-1, determine the vertical forces F_1, F_2, and F_3 required to support the load $F_W = 90$ N. Neglect the weights of the levers. Also find the IMA, AMA, and efficiency for each system.

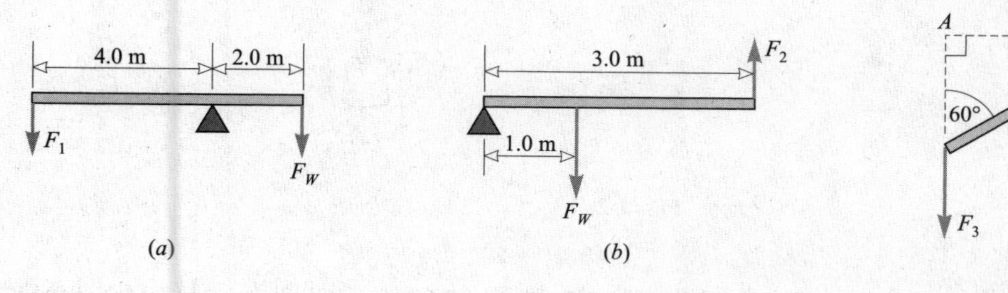

Fig. 7-1

In each case, we take torques about the fulcrum point as axis. If we assume that the lifting is occurring slowly at constant speed, then the systems are in equilibrium; the clockwise torques balance the counterclockwise torques. (Recall that torque $= rF \sin\theta$.)

Clockwise torque = counterclockwise torque

(*a*) $(2.0\ \text{m})(90\ \text{N})(1) = (4.0\ \text{m})(F_1)(1)$ from which $F_1 = 45$ N

(*b*) $(1.0\ \text{m})(90\ \text{N})(1) = (3.0\ \text{m})(F_2)(1)$ from which $F_2 = 30$ N

(*c*) $(2.0\ \text{m})(90\ \text{N})(1) = (5.0\ \text{m})(F_3)\sin 60^\circ$ from which $F_3 = 42$ N

To find the IMA of the system in Fig. 7-1(*a*), we notice that the load moves only half as far as the input force, and so

$$\text{IMA} = \text{distance ratio} = 2.0$$

Similarly, in Fig. 7-1(*b*). IMA $= 3/1 = 3$. In Fig. 7-1(*c*), however, the lever arm is $(5.0\ \text{m})\sin 60^\circ = 4.33$ m and so the distance ratio is $4.33/2 = 2.16$. To summarize,

	Lever (*a*)	Lever (*b*)	Lever (*c*)
IMA	2.0	3.0	2.2
AMA	$\dfrac{90\ \text{N}}{45\ \text{N}} = 2.0$	$\dfrac{90\ \text{N}}{30\ \text{N}} = 3.0$	$\dfrac{90\ \text{N}}{41.6\ \text{N}} = 2.2$
Eff.	1.0	1.0	1.0

The efficiencies are 1.0 because we have neglected friction at the fulcrums.

7.5 [II] Determine the force F required to lift a 100-N load F_W with each of the pulley systems shown in Fig. 7-2. Neglect friction and the weights of the pulleys.

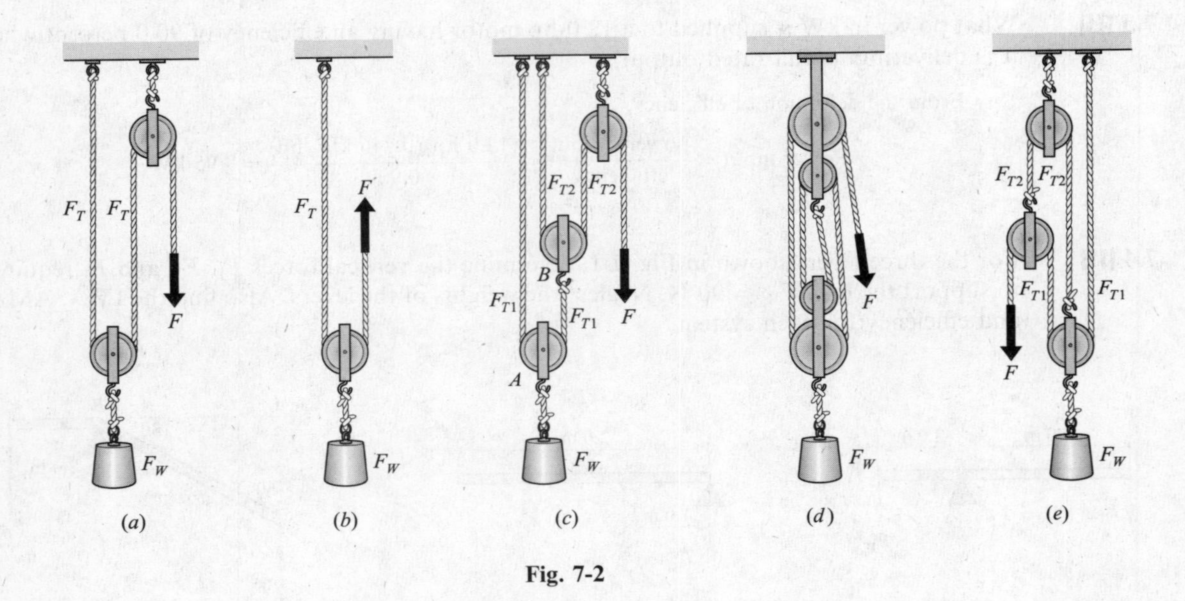

Fig. 7-2

(*a*) Load F_W is supported by two ropes; each rope exerts an upward pull of $F_T = \frac{1}{2}F_W$. Because the rope is continuous and the pulleys are frictionless, $F_T = F$. Then

$$F = F_T = \tfrac{1}{2}F_W = \tfrac{1}{2}(100\ \text{N}) = 50\ \text{N}$$

(*b*) Here, too, the load is supported by the tensions in two ropes, F_T and F, where $F_T = F$. Then

$$F_T + F = F_W \qquad \text{or} \qquad F = \tfrac{1}{2}F_W = 50\ \text{N}$$

(*c*) Let F_{T1} and F_{T2} be tensions around pulleys A and B, respectively. Pulley A is in equilibrium, so

$$F_{T1} + F_{T1} - F_W = 0 \qquad \text{or} \qquad F_{T1} = \tfrac{1}{2}F_W$$

Pulley B, too, is in equilibrium, so

$$F_{T2} + F_{T2} - F_{T1} = 0 \qquad \text{or} \qquad F_{T2} = \tfrac{1}{2}F_{T1} = \tfrac{1}{4}F_W$$

But $F = F_{T2}$ and so $F = \frac{1}{4}F_W = 25$ N.

(*d*) Four ropes, each with the same tension F_T, support the load F_W. Therefore,

$$4F_{T1} = F_W \qquad \text{and so} \qquad F = F_{T1} = \tfrac{1}{4}F_W = 25\ \text{N}$$

(*e*) We see at once $F = F_{T1}$. Because the pulley on the left is in equilibrium, we have

$$F_{T2} - F_{T1} - F = 0$$

But $F_{T1} = F$ and so $F_{T2} = 2F$. The pulley on the right is also in equilibrium, and so

$$F_{T1} + F_{T2} + F_{T1} - F_W = 0$$

Recalling that $F_{T1} = F$ and that $F_{T2} = 2F$ gives $4F = F_W$, so $F = 25$ N.

7.6 [II] Using the wheel and axle shown in Fig. 7-3, a 400-N load can be raised by a force of 50 N applied to the rim of the wheel. The radii of the wheel (R) and axle (r) are 85 cm and 6.0 cm, respectively. Determine the IMA, AMA, and efficiency of the machine.

We know that in one turn of the wheel-axle system, a length of cord equal to the circumference of the wheel or axle will be wound or unwound.

$$\text{IMA} = \frac{\text{distance moved by } F}{\text{distance moved by } F_W} = \frac{2\pi R}{2\pi r} = \frac{85\ \text{cm}}{6.0\ \text{cm}} = 14.2 = 14$$

$$\text{AMA} = \text{force ratio} = \frac{400\ \text{N}}{50\ \text{N}} = 8.0$$

$$\text{Efficiency} = \frac{\text{AMA}}{\text{IMA}} = \frac{8.0}{14.2} = 0.56 = 56\%$$

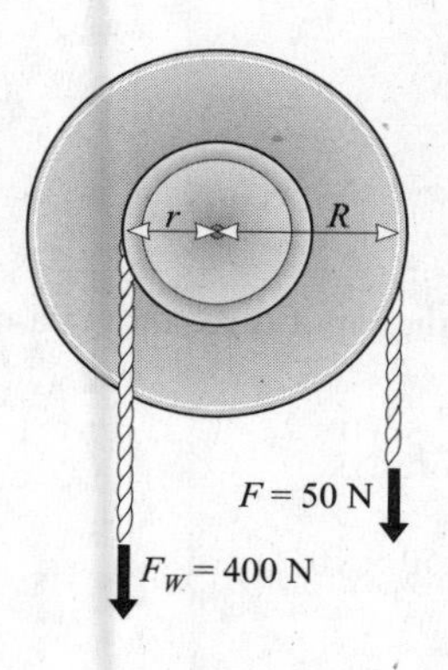

Fig. 7-3

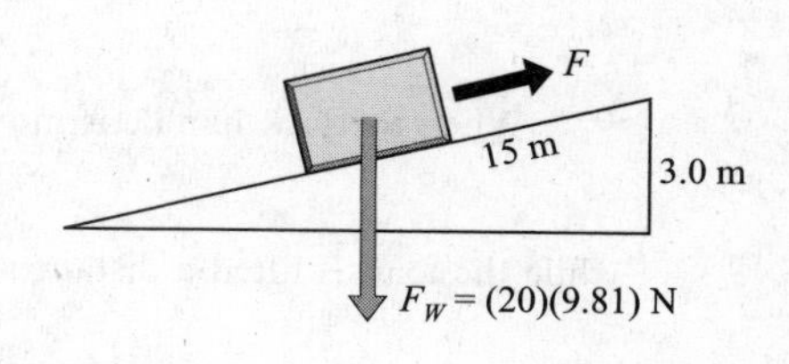

Fig. 7-4

7.7 [II] The inclined plane shown in Fig. 7-4 is 15 m long and rises 3.0 m. (*a*) What minimum force F parallel to the plane is required to slide a 20-kg box up the plane if friction is neglected? (*b*) What is the IMA of the plane? (*c*) Find the AMA and efficiency if a 64-N force is actually required.

(*a*) There are several ways to approach this. Let us consider energy. Since there is no friction, the work done by the pushing force, $(F)(15\ \text{m})$, must equal the lifting work done, $(20\ \text{kg})(9.81\ \text{m/s}^2)(3.0\ \text{m})$. Equating these two expressions and solving for F gives $F = 39$ N.

(*b*)
$$\text{IMA} = \frac{\text{distance moved by } F}{\text{distance } F_W \text{ is lifted}} = \frac{15\ \text{m}}{3.0\ \text{m}} = 5.0$$

(*c*)
$$\text{AMA} = \text{force ratio} = \frac{F_W}{F} = \frac{196 \text{ N}}{64 \text{ N}} = 3.06 = 3.1$$

$$\text{Efficiency} = \frac{\text{AMA}}{\text{IMA}} = \frac{3.06}{5.0} = 0.61 = 61\%$$

Or, as a check,

$$\text{Efficiency} = \frac{\text{work output}}{\text{work input}} = \frac{(F_W)(3.0 \text{ m})}{(F)(15 \text{ m})} = 0.61 = 61\%$$

7.8 [III] As shown in Fig. 7-5, a jackscrew has a lever arm of 40 cm and a pitch of 5.0 mm. If the efficiency is 30 percent, what horizontal force F applied perpendicularly at the end of the lever arm is required to lift a load F_W of 270 kg?

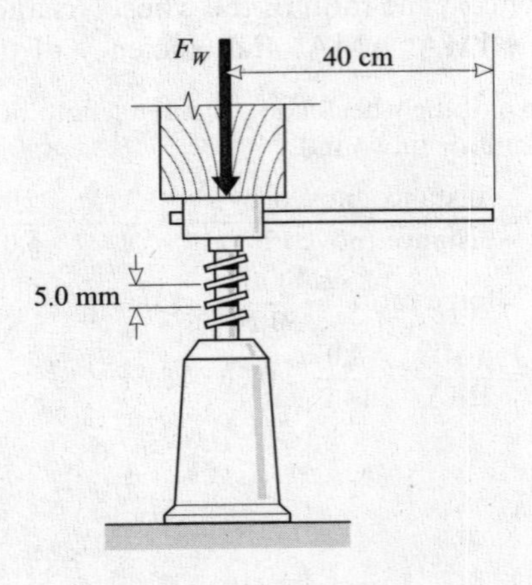

Fig. 7-5

When the jack handle is moved around one complete circle, the input force moves a distance

$$2\pi r = 2\pi(0.40 \text{ m})$$

while the load is lifted a distance of 0.0050 m. The IMA is therefore

$$\text{IMA} = \text{distance ratio} = \frac{2\pi(0.40 \text{ m})}{0.0050 \text{ m}} = 0.50 \times 10^3$$

Since efficiency = AMA/IMA, we have

$$\text{AMA} = (\text{efficiency})(\text{IMA}) = (0.30)(502) = 0.15 \times 10^3$$

But AMA = (load lifted)/(input force) and so

$$F = \frac{\text{load lifted}}{\text{AMA}} = \frac{(270 \text{ kg})(9.81 \text{ m/s}^2)}{151} = 18 \text{ N}$$

7.9 [III] A differential pulley (chain hoist) is shown in Fig. 7-6. Two toothed pulleys of radii $r = 10$ cm and $R = 11$ cm are fastened together and turn on the same axle. A continuous chain passes over the smaller (10 cm) pulley, then around the movable pulley at the bottom, and finally around the 11 cm pulley. The operator exerts a downward force F on the chain to lift the load F_W. (*a*) Determine the IMA. (*b*) What is the efficiency of the machine if an applied force of 50 N is required to lift a load of 700 N?

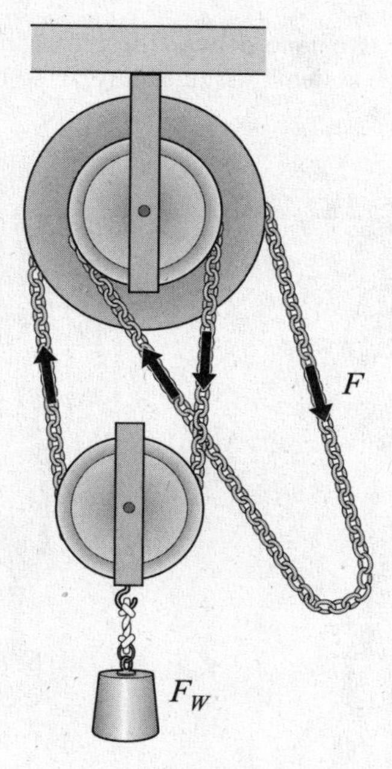

Fig. 7-6

(*a*) Suppose that the force F moves down a distance sufficient to cause the upper rigid system of pulleys to turn one revolution. Then the smaller upper pulley unwinds a length of chain equal to its circumference, $2\pi r$, while the larger upper pulley winds a length $2\pi R$. As a result, the chain supporting the lower pulley is shortened by a length $2\pi R - 2\pi r$. The load F_W is lifted half this distance, or

$$\tfrac{1}{2}(2\pi R - 2\pi r) = \pi(R - r)$$

when the input force moves a distance $2\pi R$. Therefore,

$$\text{IMA} = \frac{\text{distance moved by } F}{\text{distance moved by } F_W} = \frac{2\pi R}{\pi(R - r)} = \frac{2R}{R - r} = \frac{22 \text{ cm}}{1.0 \text{ cm}} = 22$$

(*b*) From the data,

$$\text{AMA} = \frac{\text{load lifted}}{\text{input force}} = \frac{700 \text{ N}}{50 \text{ N}} = 14$$

and

$$\text{Efficiency} = \frac{\text{AMA}}{\text{IMA}} = \frac{14}{22} = 0.64 = 64\%$$

Supplementary Problems

7.10 [I] A motor furnishes 120 hp to a device that lifts a 5000-kg load to a height of 13.0 m in a time of 20 s. Find the efficiency of the machine. *Ans.* 36%

7.11 [I] Refer back to Fig. 7-2(*d*). If a force of 200 N is required to lift a 50-kg load, find the IMA, AMA, and efficiency for the system. *Ans.* 4, 2.5, 61%

7.12 [II] In Fig. 7-7, the 300-N load is balanced by a force F in both systems. Assuming efficiencies of 100 percent, how large is F in each system? Assume all ropes to be vertical. *Ans.* (*a*) 100 N; (*b*) 75.0 N

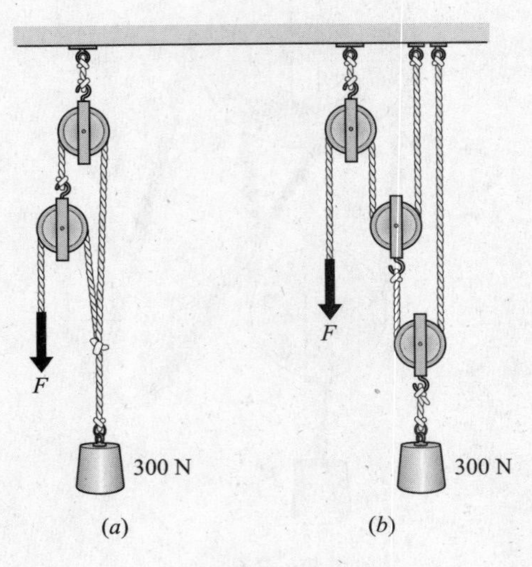

Fig. 7-7

7.13 [II] With a certain machine, the applied force moves 3.3 m to raise a load 8.0 cm. Find the (*a*) IMA and (*b*) AMA if the efficiency is 60 percent. What load can be lifted by an applied force of 50 N if the efficiency is (*c*) 100 percent and (*d*) 60 percent? *Ans.* (*a*) 41; (*b*) 25; (*c*) 2.1 kN; (*d*) 1.2 kN

7.14 [II] With a wheel and axle, a force of 80 N applied to the rim of the wheel can lift a load of 640 N. The diameters of the wheel and axle are 36 cm and 4.0 cm, respectively. Determine the AMA, IMA, and efficiency of the machine. *Ans.* 8.0, 9.0, 89%

7.15 [II] A certain hydraulic jack in a gas station lifts a 900-kg car a distance of 0.25 cm when a force of 150 N pushes a piston through a distance of 20 cm. Find the IMA, AMA, and efficiency. *Ans.* 80, 59, 74%

7.16 [II] The screw of a certain press has a pitch of 0.20 cm. The diameter of the wheel to which a tangential turning force F is applied is 55 cm. If the efficiency is 40 percent, how large must F be to produce a force of 12 kN in the press? *Ans.* 35 N

7.17 [II] The diameters of the two upper pulleys of a chain hoist (Fig. 7-6) are 18 cm and 16 cm. If the efficiency of the hoist is 45 percent, what force is required to lift a 400-kg crate? *Ans.* 0.48 kN

Chapter 8

Impulse and Momentum

THE LINEAR MOMENTUM ($\vec{\mathbf{p}}$) of a body is the product of its mass (m) and velocity ($\vec{\mathbf{v}}$):

$$\text{Linear momentum} = (\text{mass of body})\,(\text{velocity of body})$$
$$\vec{\mathbf{p}} = m\vec{\mathbf{v}}$$

Momentum is a vector quantity whose direction is that of the velocity. The units of momentum are kg·m/s in the SI.

AN IMPULSE is the product of a force ($\vec{\mathbf{F}}$) and the time interval (Δt) over which the force acts:

$$\text{Impulse} = (\text{force})\,(\text{length of time the force acts})$$

Impulse is a vector quantity whose direction is that of the force. Its units are N·s in the SI.

AN IMPULSE CAUSES A CHANGE IN MOMENTUM: The change of momentum produced by an impulse is equal to the impulse in both magnitude and direction. Thus, if a constant force $\vec{\mathbf{F}}$ acting for a time Δt on a body of mass m changes its velocity from an initial value $\vec{\mathbf{v}}_i$ to a final value $\vec{\mathbf{v}}_f$, then

$$\text{Impulse} = \text{change in momentum}$$
$$\vec{\mathbf{F}}\,\Delta t = m(\vec{\mathbf{v}}_f - \vec{\mathbf{v}}_i)$$

Newton's Second Law, as he gave it, is $\vec{\mathbf{F}} = \Delta\vec{\mathbf{p}}/\Delta t$ from which it follows that $\vec{\mathbf{F}}\,\Delta t = \Delta\vec{\mathbf{p}}$. Moreover, $\vec{\mathbf{F}}\,\Delta t = \Delta(m\vec{\mathbf{v}})$ and if m is constant $\vec{\mathbf{F}}\,\Delta t = m(\vec{\mathbf{v}}_f - \vec{\mathbf{v}}_i)$.

CONSERVATION OF LINEAR MOMENTUM: If the net external force acting on a system of objects is zero, the vector sum of the momenta of the objects will remain constant.

IN COLLISIONS AND EXPLOSIONS, the vector sum of the momenta just before the event equals the vector sum of the momenta just after the event. The vector sum of the momenta of the objects involved does not change during the collision or explosion.

Thus, when two bodies of masses m_1 and m_2 collide,

$$\text{Total momentum before impact} = \text{total momentum after impact}$$
$$m_1\vec{\mathbf{u}}_1 + m_2\vec{\mathbf{u}}_2 = m_1\vec{\mathbf{v}}_1 + m_2\vec{\mathbf{v}}_2$$

where $\vec{\mathbf{u}}_1$ and $\vec{\mathbf{u}}_2$ are the velocities before impact, and $\vec{\mathbf{v}}_1$ and $\vec{\mathbf{v}}_2$ are the velocities after. In one dimension, in component form,

$$m_1u_{1x} + m_2u_{2x} = m_1v_{1x} + m_2v_{2x}$$

and similarly for the y- and z-components. Remember that vector quantities are always boldfaced and velocity is a vector. On the other hand, u_{1x}, u_{2x}, v_{1x}, and v_{2x} are the scalar values of the velocities (they can be positive or negative). A positive direction is initally selected and vectors pointing opposite to this have negative numerical scalar values.

A PERFECTLY ELASTIC COLLISION is one in which the sum of the translational KEs of the objects is not changed during the collision. In the case of two bodies,

$$\tfrac{1}{2}m_1u_1^2 + \tfrac{1}{2}m_2u_2^2 = \tfrac{1}{2}m_1v_1^2 + \tfrac{1}{2}m_2v_2^2$$

COEFFICIENT OF RESTITUTION: For any collision between two bodies in which the bodies move only along a single straight line (e.g., the x-axis), a **coefficient of restitution** e is defined. It is a pure number given by

$$e = \frac{v_{2x} - v_{1x}}{u_{1x} - u_{2x}}$$

where u_{1x} and u_{2x} are values before impact, and v_{1x} and v_{2x} are values after impact. Notice that $|u_{1x} - u_{2x}|$ is the relative speed of approach and $|v_{2x} - v_{1x}|$ is the relative speed of recession.

For a perfectly elastic collision, $e = 1$. For inelastic collisions, $e < 1$. If the bodies stick together after collision, $e = 0$.

THE CENTER OF MASS of an object (of mass m) is the single point that moves in the same way as a point mass (of mass m) would move when subjected to the same external forces that act on the object. That is, if the resultant force acting on an object (or system of objects) of mass m is $\vec{\mathbf{F}}$, the acceleration of the center of mass of the object (or system) is given by $\vec{\mathbf{a}}_{\text{cm}} = \vec{\mathbf{F}}/m$.

If the object is considered to be composed of tiny masses m_1, m_2, m_3, and so on, at coordinates (x_1, y_1, z_1), (x_2, y_2, z_2), and so on, then the coordinates of the center of mass are given by

$$x_{\text{cm}} = \frac{\Sigma\, x_i m_i}{\Sigma\, m_i} \qquad y_{\text{cm}} = \frac{\Sigma\, y_i m_i}{\Sigma\, m_i} \qquad z_{\text{cm}} = \frac{\Sigma\, z_i m_i}{\Sigma\, m_i}$$

where the sums extend over all masses composing the object. In a uniform gravitational field, the center of mass and the center of gravity coincide.

Solved Problems

8.1 [II] An 8.0-g bullet is fired horizontally into a 9.00-kg cube of wood, which is at rest, and sticks in it. The cube is free to move and has a speed of 40 cm/s after impact. Find the initial velocity of the bullet.

Consider the system (cube + bullet). The velocity, and hence the momentum, of the cube before impact is zero. Take the bullet's initial motion to be positive in the positive x-direction. The momentum conservation law tells us that

Momentum of system before impact = momentum of system after impact

(momentum of bullet) + (momentum of cube) = (momentum of bullet + cube)

$$m_B v_{Bx} + m_C v_{Cx} = (m_B + m_C)v_x$$

$$(0.0080\ \text{kg})v_{Bx} + 0 = (9.008\ \text{kg})(0.40\ \text{m/s})$$

Solving gives $v_{Bx} = 0.45$ km/s and so $\vec{\mathbf{v}}_B = 0.45$ km/s — POSITIVE X-DIRECTION.

8.2 [II] A 16-g mass is moving in the $+x$-direction at 30 cm/s while a 4.0-g mass is moving in the $-x$-direction at 50 cm/s. They collide head on and stick together. Find their velocity after the collision.

Let the 16-g mass be m_1 and the 4.0-g mass be m_2. Take the $+x$-direction to be positive. That means that the velocity of the 4.0-g mass has a scalar value of $v_{2x} = -50$ cm/s. We apply the law of conservation of momentum to the system consisting of the two masses:

$$\text{Momentum before impact} = \text{momentum after impact}$$
$$m_1 v_{1x} + m_2 v_{2x} = (m_1 + m_2)v_x$$
$$(0.016\text{ kg})(0.30\text{ m/s}) + (0.0040\text{ kg})(-0.50\text{ m/s}) = (0.020\text{ kg})v_x$$
$$v_x = +0.14\text{ m/s}$$

(Notice that the 4.0-g mass has negative momentum.) Hence, $\vec{v} = 0.14$ m/s—POSITIVE X-DIRECTION.

8.3 [I] A 2.0-kg brick is moving at a speed of 6.0 m/s. How large a force F is needed to stop the brick in a time of 7.0×10^{-4} s?

Let us solve this by use of the impulse equation:

$$\text{Impulse on brick} = \text{change in momentum of brick}$$
$$F\,\Delta t = mv_f - mv_i$$
$$F(7.0 \times 10^{-4}\text{ s}) = 0 - (2.0\text{ kg})(6.0\text{ m/s})$$

from which $F = -1.7 \times 10^4$ N. The minus sign indicates that the force opposes the motion.

8.4 [II] A 15-g bullet moving at 300 m/s passes through a 2.0 cm thick sheet of foam plastic and emerges with a speed of 90 m/s. Assuming that the speed change takes place uniformly, what average force impeded the bullet's motion through the plastic?

Use the impulse equation to find the force F on the bullet as it takes a time Δt to pass through the plastic. Taking the initial direction of motion to be positive,

$$F\Delta t = mv_f - mv_i$$

We can find Δt by assuming uniform deceleration and using $x = v_{av}t$, where $x = 0.020$ m and $v_{av} = \frac{1}{2}(v_i + v_f) = 195$ m/s. This gives $\Delta t = 1.026 \times 10^{-4}$ s. Then

$$(F)(1.026 \times 10^{-4}\text{ s}) = (0.015\text{ kg})(90\text{ m/s}) - (0.015\text{ kg})(300\text{ m/s})$$

which gives $F = -3.1 \times 10^4$ N as the average retarding force. How could this problem have been solved using $F = ma$ instead of the impulse equation? By using energy methods?

8.5 [II] The nucleus of an atom has a mass of 3.80×10^{-25} kg and is at rest. The nucleus is radioactive and suddenly ejects a particle of mass 6.6×10^{-27} kg and speed 1.5×10^7 m/s. Find the recoil speed of the nucleus that is left behind.

Take the direction of the ejected particle as positive. We are given, $m_{ni} = 3.80 \times 10^{-25}$ kg, $m_p = 6.6 \times 10^{-27}$ kg, $m_{nf} = m_{ni} - m_p = 3.73 \times 10^{-25}$ kg, and $v_{pf} = 1.5 \times 10^7$ m/s; find the final speed of the nucleus, v_{nf}. The momentum of the system is conserved during the explosion.

$$\text{Momentum before} = \text{momentum after}$$
$$0 = m_{nf}v_{nf} + m_p v_{pf}$$
$$0 = (3.73 \times 10^{-25}\text{ kg})(v_{nf}) + (6.6 \times 10^{-27}\text{ kg})(1.5 \times 10^7\text{ m/s})$$

Solving gives

$$-v_{nf} = \frac{(6.6 \times 10^{-27}\ \text{kg})(1.5 \times 10^7\ \text{m/s})}{3.73 \times 10^{-25}\ \text{kg}} = \frac{10.0 \times 10^{-20}}{3.73 \times 10^{-25}} = 2.7 \times 10^5\ \text{m/s}$$

The fact that this is negative tells us that the velocity vector of the nucleus points in the negative direction, opposite to the velocity of the particle.

8.6 [II] A 0.25-kg ball moving in the $+x$-direction at 13 m/s is hit by a bat. Its final velocity is 19 m/s in the $-x$-direction. The bat acts on the ball for 0.010 s. Find the average force F exerted on the ball by the bat.

We have $v_i = 13$ m/s and $v_f = -19$ m/s. Taking the initial direction of motion as positive, the impulse equation then gives

$$F\,\Delta t = mv_f - mv_i$$

$$F(0.010\ \text{s}) = (0.25\ \text{kg})(-19\ \text{m/s}) - (0.25\ \text{kg})(13\ \text{m/s})$$

from which $F = -0.80$ kN.

8.7 [II] Two girls (masses m_1 and m_2) are on roller skates and stand at rest, close to each other and face to face. Girl-1 pushes squarely against girl-2 and sends her moving backward. Assuming the girls move freely on their skates, write an expression for the speed with which girl-1 moves.

We take the two girls to comprise the system under consideration. The problem states that girl-2 moves "backward," so let that be the negative direction; therefore the "forward" direction is positive. There is no resultant external force on the system (the push of one girl on the other is an internal force), and so momentum is conserved:

$$\text{Momentum before} = \text{momentum after}$$

$$0 = m_1 v_1 + m_2 v_2$$

from which

$$v_1 = -\frac{m_2}{m_1} v_2$$

Girl-1 recoils with this speed. Notice that if m_2/m_1 is very large, v_1 is much larger than v_2. The velocity of girl-1, $\vec{v}_1$, points in the positive forward direction. The velocity of girl-2, $\vec{v}_2$, points in the negative backward direction. If we put numbers into the equation, v_2 would have to be negative and v_1 would come out positive.

8.8 [II] As shown in Fig. 8-1, a 15-g bullet is fired horizontally into a 3.000-kg block of wood suspended by a long cord. The bullet sticks in the block. Compute the speed of the bullet if the impact causes the block to swing 10 cm above its initial level.

Consider first the collision of block and bullet. During the collision, momentum is conserved, so

$$\text{Momentum just before} = \text{momentum just after}$$

$$(0.015\ \text{kg})v + 0 = (3.015\ \text{kg})V$$

where v is the initial speed of the bullet, and V is the speed of block and bullet just after collision.

We have two unknowns in this equation. To find another equation, we can use the fact that the block swings 10 cm high. If we let $\text{PE}_G = 0$ at the initial level of the block, energy conservation gives

$$\text{KE just after collision} = \text{final PE}_G$$

$$\tfrac{1}{2}(3.015\ \text{kg})V^2 = (3.015\ \text{kg})(9.81\ \text{m/s}^2)(0.10\ \text{m})$$

From this we find $V = 1.40$ m/s. Substituting this in the previous equation gives $v = 0.28$ km/s for the speed of the bullet.

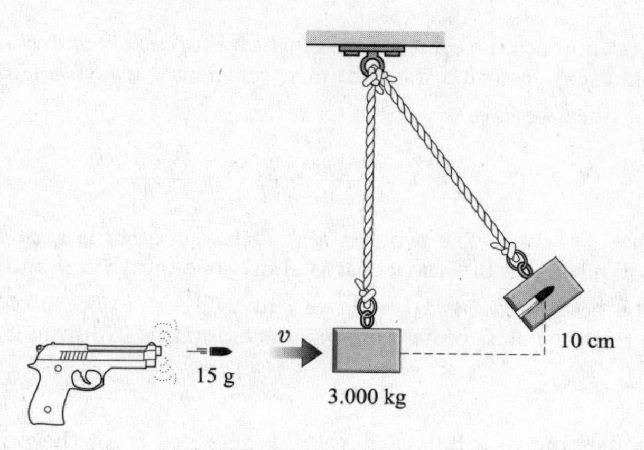

Fig. 8-1

Notice that we cannot write the conservation of energy equation $\frac{1}{2}mv^2 = (m + M)gh$, where $m = 0.015$ kg and $M = 3.000$ kg because energy is lost (through friction) in the collision process.

8.9 [I] Three masses are placed on the x-axis: 200 g at $x = 0$, 500 g at $x = 30$ cm, and 400 g at $x = 70$ cm. Find their center of mass.

$$x_{cm} = \frac{\Sigma x_i m_i}{\Sigma m_i} = \frac{(0)(0.20 \text{ kg}) + (0.30 \text{ m})(0.50 \text{ kg}) + (0.70 \text{ m})(0.40 \text{ kg})}{(0.20 + 0.50 + 0.40) \text{ kg}} = 0.39 \text{ m}$$

The y- and z-coordinates of the mass center are zero.

8.10 [II] A system consists of the following masses in the xy-plane: 4.0 kg at coordinates ($x = 0$, $y = 5.0$ m), 7.0 kg at (3.0 m, 8.0 m), and 5.0 kg at (−3.0 m, −6.0 m). Find the position of its center of mass.

$$x_{cm} = \frac{\Sigma x_i m_i}{\Sigma m_i} = \frac{(0)(4.0 \text{ kg}) + (3.0 \text{ m})(7.0 \text{ kg}) + (-3.0 \text{ m})(5.0 \text{ kg})}{(4.0 + 7.0 + 5.0) \text{ kg}} = 0.38 \text{ m}$$

$$y_{cm} = \frac{\Sigma y_i m_i}{\Sigma m_i} = \frac{(5.0 \text{ m})(4.0 \text{ kg}) + (8.0 \text{ m})(7.0 \text{ kg}) + (-6.0 \text{ m})(5.0 \text{ kg})}{16 \text{ kg}} = 2.9 \text{ m}$$

and $z_{cm} = 0$.

8.11 [II] Two identical railroad cars sit on a horizontal track, with a distance D between their centers. By means of a cable between them, a winch on one is used to pull the two together. (*a*) Describe their relative motion. (*b*) Repeat the analysis if the mass of one car is now three times that of the other.

The forces due to the cable on the two cars are internal forces for the two-car system. The net external force on the system is zero, and so its center of mass does not move, even though each car moves toward the other. Taking the origin of coordinates at the mass center, we have

$$x_{cm} = 0 = \frac{\Sigma m_i x_i}{\Sigma m_i} = \frac{m_1 x_1 + m_2 x_2}{m_1 + m_2}$$

where x_1 and x_2 are the positions of the centers of the two cars.

(*a*) If $m_1 = m_2$, this equation becomes

$$0 = \frac{x_1 + x_2}{2} \qquad \text{or} \qquad x_1 = -x_2$$

The two cars approach the center of mass, which is originally midway between the two cars (that is, $D/2$ from each), in such a way that their centers are always equidistant from it.

(*b*) If $m_1 = 3m_2$, then we have

$$0 = \frac{3m_2x_1 + m_2x_2}{3m_2 + m_2} = \frac{3x_1 + x_2}{4}$$

from which $x_1 = -x_2/3$. The two cars approach each other in such a way that the mass center remains motionless and the heavier car is always one-third as far away from it as the lighter car.

Originally, because $|x_1| + |x_2| = D$, we had $x_2/3 + x_2 = D$. So m_2 was originally a distance $x_2 = 3D/4$ from the mass center, and m_1 was a distance $D/4$ from it.

8.12 [III] A pendulum consisting of a ball of mass m is released from the position shown in Fig. 8-2 and strikes a block of mass M. The block slides a distance D before stopping under the action of a steady friction force $0.20Mg$. Find D if the ball rebounds to an angle of 20°.

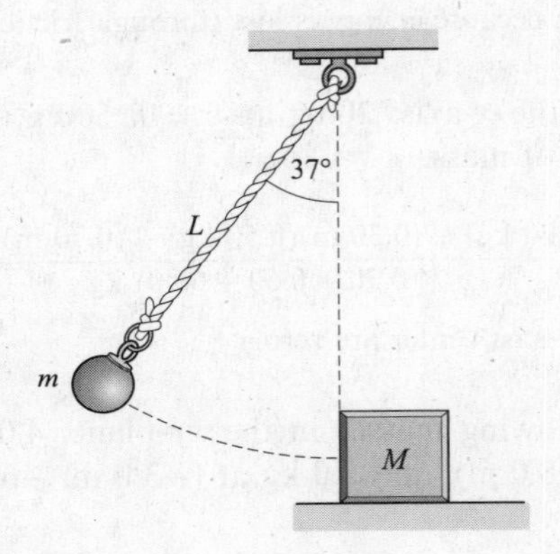

Fig. 8-2

The pendulum ball falls through a height $(L - L\cos 37°) = 0.201L$ and rebounds to a height $(L - L\cos 20°) = 0.060\,3L$. Because $(mgh)_{\text{top}} = (\frac{1}{2}mv^2)_{\text{bottom}}$ for the ball, its speed at the bottom is $v = \sqrt{2gh}$. Thus, just before it hits the block the ball has a speed equal to $\sqrt{2g(0.201L)}$. Since the ball rises up to a height of $0.060\,3L$ after the collision it must have rebounded with an initial speed of $\sqrt{2g(0.060\,3L)}$.

Although KE is not conserved in the collision, momentum is. Therefore, for the collision,

$$\text{Momentum just before} = \text{momentum just after}$$

$$m\sqrt{2g(0.201L)} + 0 = -m\sqrt{2g(0.060\,3L)} + MV$$

where V is the velocity of the block just after the collision. (Notice the minus sign on the momentum of the rebounding ball.) Solving this equation, we find

$$V = \frac{m}{M}0.981\sqrt{gL}$$

The block uses up its translational KE doing work against friction as it slides a distance D. Therefore,

$$\tfrac{1}{2}MV^2 = F_f D \qquad \text{or} \qquad \tfrac{1}{2}M(0.963gL)\left(\frac{m}{M}\right)^2 = (0.2Mg)(D)$$

from which $D = 2.4(m/M)^2 L$.

8.13 [II] Two balls of equal mass approach the coordinate origin, one moving downward along the $+y$-axis at 2.00 m/s and the other moving to the right along the $-x$-axis at 3.00 m/s. After they collide, one ball moves out to the right along the $+x$-axis at 1.20 m/s. Find the scalar x and y velocity components of the other ball.

Take *up* and to the *right* as positive. Momentum is conserved in the collision, so we can write

$$(\text{momentum before})_x = (\text{momentum after})_x$$

or

$$m(3.00\ \text{m/s}) + 0 = m(1.20\ \text{m/s}) + mv_x$$

and

$$(\text{momentum before})_y = (\text{momentum after})_y$$

or

$$0 + m(-2.00\ \text{m/s}) = 0 + mv_y$$

(Why the minus sign?) Solving, we find that $v_x = 1.80$ m/s and $v_y = -2.00$ m/s.

8.14 [III] A 7500-kg truck traveling at 5.0 m/s east collides with a 1500-kg car moving at 20 m/s in a direction 30° south of west. After collision, the two vehicles remain tangled together. With what speed and in what direction does the wreckage begin to move?

The original momenta are shown in Fig. 8-3(*a*), while the final momentum $M\vec{v}$ is shown in Fig. 8-3(*b*). Momentum must be conserved in both the north and east directions. Therefore,

$$(\text{momentum before})_{\text{East}} = (\text{momentum after})_{\text{East}}$$

$$(7500\ \text{kg})(5.0\ \text{m/s}) - (1500\ \text{kg})[(20\ \text{m/s})\cos 30°] = Mv_{\text{E}}$$

where $M = 7500\ \text{kg} + 1500\ \text{kg} = 9000$ kg, and v_{E} is the scalar eastward component of the velocity of the wreckage [see Fig. 8-3(*b*)].

$$(\text{momentum before})_{\text{North}} = (\text{momentum after})_{\text{North}}$$

$$(7500\ \text{kg})(0) - (1500\ \text{kg})[(20\ \text{m/s})\sin 30°] = Mv_{\text{N}}$$

The first equation gives $v_{\text{E}} = 1.28$ m/s, and the second gives $v_{\text{N}} = -1.67$ m/s. The resultant is

$$v = \sqrt{(1.67\ \text{m/s})^2 + (1.28\ \text{m/s})^2} = 2.1\ \text{m/s}$$

The angle θ in Fig. 8-3(*b*) is

$$\theta = \arctan\left(\frac{1.67}{1.28}\right) = 53°$$

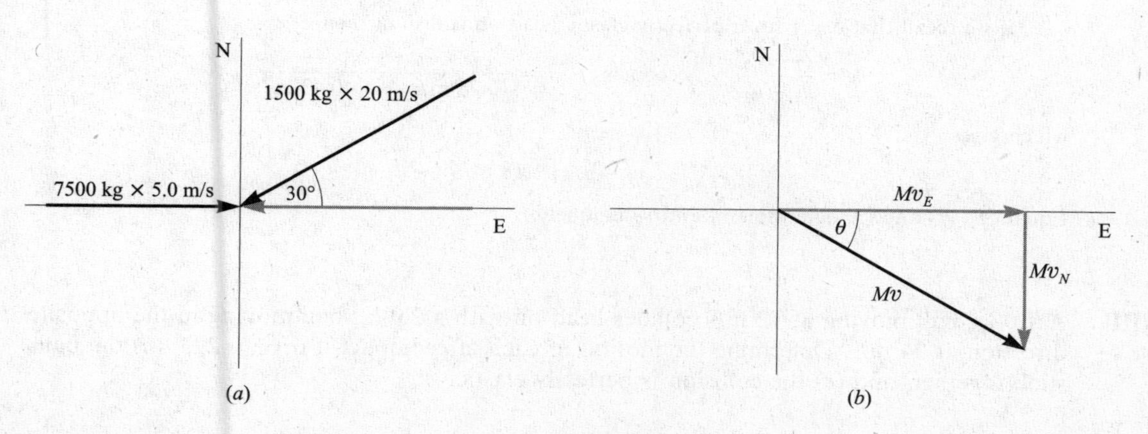

Fig. 8-3

8.15 [III] Two identical balls collide head-on. The initial velocity of one is 0.75 m/s—EAST, while that of the other is 0.43 m/s—WEST. If the collision is perfectly elastic, what is the final velocity of each ball?

Since the collision is head-on, all motion takes place along a straight line. Take east as positive and call the mass of each ball m. Momentum is conserved in a collision, so we can write

$$\text{Momentum before} = \text{momentum after}$$
$$m(0.75\ \text{m/s}) + m(-0.43\ \text{m/s}) = mv_1 + mv_2$$

where v_1 and v_2 are the final values. This equation simplifies to

$$0.32\ \text{m/s} = v_1 + v_2 \qquad (1)$$

Because the collision is assumed to be perfectly elastic, KE is also conserved. Thus,

$$\text{KE before} = \text{KE after}$$
$$\tfrac{1}{2}m(0.75\ \text{m/s})^2 + \tfrac{1}{2}m(0.43\ \text{m/s})^2 = \tfrac{1}{2}mv_1^2 + \tfrac{1}{2}mv_2^2$$

This equation can be simplified to

$$0.747 = v_1^2 + v_2^2 \qquad (2)$$

We can solve for v_2 in Eq. (*1*) to get $v_2 = 0.32 - v_1$ and substitute this in Eq. (*2*). This yields

$$0.747 = (0.32 - v_1)^2 + v_1^2$$

from which
$$2v_1^2 - 0.64v_1 - 0.645 = 0$$

Using the quadratic formula, we find that

$$v_1 = \frac{0.64 \pm \sqrt{(0.64)^2 + 5.16}}{4} = 0.16 \pm 0.59\,\text{m/s}$$

from which $v_1 = 0.75$ m/s or -0.43 m/s. Substitution back into Eq. (*1*) gives $v_2 = -0.43$ m/s or 0.75 m/s.

Two choices for answers are available:

$$(v_1 = 0.75\ \text{m/s},\ v_2 = -0.43\ \text{m/s}) \qquad \text{and} \qquad (v_1 = -0.43\ \text{m/s},\ v_2 = 0.75\ \text{m/s})$$

We must discard the first choice because it implies that the balls continue on unchanged; that is to say, no collision occurred. The correct answer is therefore $v_1 = -0.43$ m/s and $v_2 = 0.75$ m/s, which tells us that in a perfectly elastic, head-on collision between equal masses, the two bodies simply exchange velocities. Hence $\vec{v}_1 = 0.43$ m/s—WEST and $\vec{v}_2 = 0.75$ m/s—EAST.

Alternative Method

If we recall that $e = 1$ for a perfectly elastic head-on collision, then

$$e = \frac{v_2 - v_1}{u_1 - u_2} \qquad \text{becomes} \qquad 1 = \frac{v_2 - v_1}{(0.75\ \text{m/s}) - (-0.43\ \text{m/s})}$$

which gives

$$v_2 - v_1 = 1.18\ \text{m/s} \qquad (3)$$

Equations (*1*) and (*3*) determine v_1 and v_2 uniquely.

8.16 [III] A 1.0-kg ball moving at 12 m/s collides head-on with a 2.0-kg ball moving in the opposite direction at 24 m/s. Determine the motion of each after impact if (*a*) $e = 2/3$, (*b*) the balls stick together, and (*c*) the collision is perfectly elastic.

In all three cases momentum is conserved, and so we can write

$$\text{Momentum before} = \text{momentum after}$$
$$(1.0 \text{ kg})(12 \text{ m/s}) + (2.0 \text{ kg})(-24 \text{ m/s}) = (1.0 \text{ kg})v_1 + (2.0 \text{ kg})v_2$$

which becomes

$$-36 \text{ m/s} = v_1 + 2v_2$$

(*a*) When $e = 2/3$,

$$e = \frac{v_2 - v_1}{u_1 - u_2} \qquad \text{becomes} \qquad \frac{2}{3} = \frac{v_2 - v_1}{(12 \text{ m/s}) - (-24 \text{ m/s})}$$

from which $24 \text{ m/s} = v_2 - v_1$. Combining this with the momentum equation found above gives $v_2 = -4.0 \text{ m/s}$ and $v_1 = -28 \text{ m/s}$.

(*b*) In this case $v_1 = v_2 = v$ and so the momentum equation becomes

$$3v = -36 \text{ m/s} \qquad \text{or} \qquad v = -12 \text{ m/s}$$

(*c*) Here $e = 1$, so

$$e = \frac{v_2 - v_1}{u_1 - u_2} \qquad \text{becomes} \qquad 1 = \frac{v_2 - v_1}{(12 \text{ m/s}) - (-24 \text{ m/s})}$$

from which $v_2 - v_1 = 36 \text{ m/s}$. Adding this to the momentum equation gives $v_2 = 0$. Using this value for v_2 then gives $v_1 = -36 \text{ m/s}$.

8.17 [III] A ball is dropped from a height h above a tile floor and rebounds to a height of $0.65h$. Find the coefficient of restitution between ball and floor.

The initial and final velocities of the floor, u_1 and v_1, are zero. Therefore,

$$e = \frac{v_2 - v_1}{u_1 - u_2} = -\frac{v_2}{u_2}$$

But we can write equations for the interchange of PE_G and KE both before and after the bounce:

$$mgh = \tfrac{1}{2}mu_2^2 \qquad \text{and} \qquad mg(0.65h) = \tfrac{1}{2}mv_2^2$$

Therefore, taking *down* as positive, we have $u_2 = \sqrt{2gh}$ and $v_2 = -\sqrt{1.30gh}$. Substitution gives

$$e = \frac{\sqrt{1.30gh}}{\sqrt{2gh}} = \sqrt{0.65} = 0.81$$

8.18 [III] The two balls shown in Fig. 8-4 collide off center and bounce off each other as shown. (*a*) What is the final velocity of the 500-g ball if the 800-g ball has a speed of 15 cm/s after the collision? (*b*) Is the collision perfectly elastic?

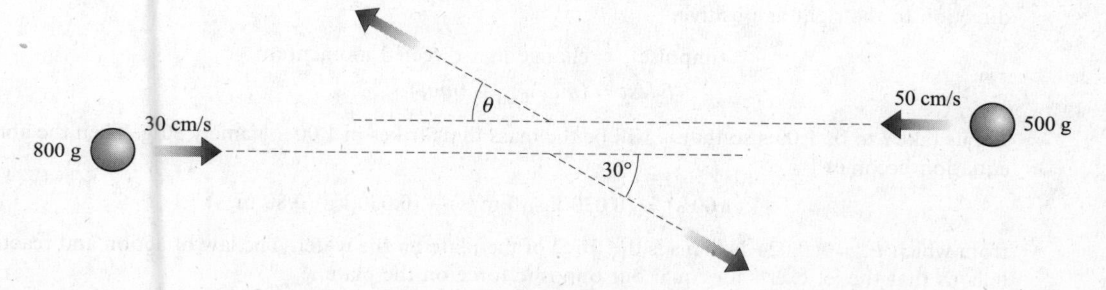

Fig. 8-4

(*a*) Take motion to the right as positive. From the law of conservation of momentum,

$$(\text{momentum before})_x = (\text{momentum after})_x$$
$$(0.80\ \text{kg})(0.30\ \text{m/s}) + (0.50\ \text{kg})(-0.5\ \text{m/s}) = (0.80\ \text{kg})[(0.15\ \text{m/s})\cos 30°] + (0.50\ \text{kg})v_x$$

from which $v_x = -0.228$ m/s. Taking motion upward as positive,

$$(\text{momentum before})_y = (\text{momentum after})_y$$
$$0 = (0.80\ \text{kg})[-(0.15\ \text{m/s})\sin 30°] + (0.50\ \text{kg})v_y$$

from which $v_y = 0.120$ m/s. Then

$$v = \sqrt{v_x^2 + v_y^2} = \sqrt{(-0.228\ \text{m/s})^2 + (0.120\ \text{m/s})^2} = 0.26\ \text{m/s}$$

and $\vec{v} = 0.26$ m/s—RIGHT.

Also, for the angle θ shown in Fig. 8-4,

$$\theta = \arctan\left(\frac{0.120}{0.228}\right) = 28°$$

(*b*)
$$\text{Total KE before} = \tfrac{1}{2}(0.80\ \text{kg})(0.30\ \text{m/s})^2 + \tfrac{1}{2}(0.50\ \text{kg})(0.50\ \text{m/s})^2 = 0.099\ \text{J}$$
$$\text{Total KE after} = \tfrac{1}{2}(0.80\ \text{kg})(0.15\ \text{m/s})^2 + \tfrac{1}{2}(0.50\ \text{kg})(0.26\ \text{m/s})^2 = 0.026\ \text{J}$$

Because KE is lost in the collision, it is not perfectly elastic.

8.19 [II] What force is exerted on a stationary flat plate held perpendicular to a jet of water as shown in Fig. 8-5? The horizontal speed of the water is 80 cm/s, and 30 mL of the water hits the plate each second. Assume the water moves parallel to the plate after striking it. One milliliter (mL) of water has a mass of 1.00 g.

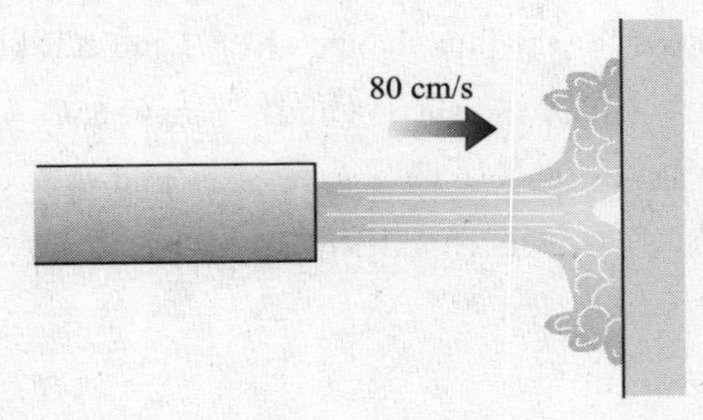

Fig. 8-5

The plate exerts an impulse on the water and changes its horizontal momentum. Taking the direction to the right as positive,

$$(\text{impulse})_x = \text{change in } x\text{-directed momentum}$$
$$F_x\,\Delta t = (mv_x)_{\text{final}} - (mv_x)_{\text{initial}}$$

Let us take t to be 1.00 s so that m will be the mass that strikes in 1.00 s, namely 30 g. Then the above equation becomes

$$F_x\,(1.00\ \text{s}) = (0.030\ \text{kg})(0\ \text{m/s}) - (0.030\ \text{kg})(0.80\ \text{m/s})$$

from which $F_x = -0.024$ N. This is the force of the plate on the water. The law of action and reaction tells us that the jet exerts an equal but opposite force on the plate.

8.20 [III] A rocket standing on its launch platform points straight upward. Its engines are activated and eject gas at a rate of 1500 kg/s. The molecules are expelled with a speed of 50 km/s. How much mass can the rocket initially have if it is slowly to rise because of the thrust of the engines?

Because the motion of the rocket itself is negligible in comparison to the speed of the expelled gas, we can assume the gas to be accelerated from rest to a speed of 50 km/s. The impulse required to provide this acceleration to a mass m of gas is

$$F\,\Delta t = mv_f - mv_i = m(50\,000\text{ m/s}) - 0$$

from which

$$F = (50\,000\text{ m/s})\frac{m}{\Delta t}$$

But we are told that the mass ejected per second ($m/\Delta t$) is 1500 kg/s, and so the force exerted on the expelled gas is

$$F = (50\,000\text{ m/s})(1500\text{ kg/s}) = 75\text{ MN}$$

An equal but opposite reaction force acts on the rocket, and this is the upward thrust on the rocket. The engines can therefore support a weight of 75 MN, so the maximum mass the rocket could have is

$$M_{\text{rocket}} = \frac{\text{weight}}{g} = \frac{75 \times 10^6\text{ N}}{9.81\text{ m/s}^2} = 7.7 \times 10^6\text{ kg}$$

Supplementary Problems

8.21 [I] Typically, a tennis ball hit during a serve travels away at about 51 m/s. If the ball is at rest mid-air when struck, and it has a mass of 0.058 kg, what is the change in its momentum on leaving the racket? *Ans.* 3.0 kg·m/s

8.22 [I] During a soccer game a ball (of mass 0.425 kg), which is initially at rest, is kicked by one of the players. The ball moves off at a speed of 26 m/s. Given that the impact lasted for 8.0 ms, what was the average force exerted on the ball? *Ans.* 1.4 kN

8.23 [II] A 40 000-kg freight car is coasting at a speed of 5.0 m/s along a straight track when it strikes a 30 000-kg stationary freight car and couples to it. What will be their combined speed after impact? *Ans.* 2.9 m/s

8.24 [I] An empty 15 000-kg coal car is coasting on a level track at 5.00 m/s. Suddenly 5000 kg of coal is dumped into it from directly above it. The coal initially has zero horizontal velocity with respect to the ground. Find the final speed of the car. *Ans.* 3.75 m/s.

8.25 [II] Sand drops at a rate of 2000 kg/min from the bottom of a stationary hopper onto a belt conveyer moving horizontally at 250 m/min. Determine the force needed to drive the conveyer, neglecting friction. *Ans.* 139 N

8.26 [II] Two bodies of masses 8 kg and 4 kg move along the x-axis in opposite directions with velocities of 11 m/s—POSITIVE x-DIRECTION and 7 m/s—NEGATIVE x-DIRECTION, respectively. They collide and stick together. Find their velocity just after collision. *Ans.* 5 m/s—POSITIVE x-DIRECTION

8.27 [II] A 1200-kg gun mounted on wheels shoots an 8.00-kg projectile with a muzzle velocity of 600 m/s at an angle of 30.0° above the horizontal. Find the horizontal recoil speed of the gun. *Ans.* 3.46 m/s

8.28 [I] Three masses are placed on the y-axis: 2 kg at $y = 300$ cm, 6 kg at $y = 150$ cm, and 4 kg at $y = -75$ cm. Find their center of mass. *Ans.* $y = 1$ m

8.29 [II] Four masses are positioned in the xy-plane as follows: 300 g at ($x = 0$, $y = 2.0$ m), 500 g at (-2.0 m, -3.0 m), 700 g at (50 cm, 30 cm), and 900 g at (-80 cm, 150 cm). Find their center of mass. *Ans.* $x = -0.57$ m, $y = 0.28$ m

8.30 [II] A ball of mass m sits at the coordinate origin when it explodes into two pieces that shoot along the x-axis in opposite directions. When one of the pieces (which has mass $0.270m$) is at $x = 70$ cm, where is the other piece? (*Hint*: What happens to the mass center?) *Ans.* at $x = -26$ cm

8.31 [II] A ball of mass m at rest at the coordinate origin explodes into three equal pieces. At a certain instant, one piece is on the x-axis at $x = 40$ cm and another is at $x = 20$ cm, $y = -60$ cm. Where is the third piece at that instant? *Ans.* at $x = -60$ cm, $y = 60$ cm

8.32 [II] A 2.0-kg block of wood rests on a long tabletop. A 5.0-g bullet moving horizontally with a speed of 150 m/s is shot into the block and sticks in it. The block then slides 270 cm along the table and stops. (*a*) Find the speed of the block just after impact. (*b*) Find the friction force between block and table. *Ans.* (*a*) 0.37 m/s; (*b*) 0.052 N

8.33 [II] A 2.0-kg block of wood rests on a tabletop. A 7.0-g bullet is shot straight up through a hole in the table beneath the block. The bullet lodges in the block, and the block flies 25 cm above the tabletop. How fast was the bullet going initially? *Ans.* 0.64 km/s

8.34 [III] A 6000-kg truck traveling north at 5.0 m/s collides with a 4000-kg truck moving west at 15 m/s. If the two trucks remain locked together after impact, with what speed and in what direction do they move immediately after the collision? *Ans.* 6.7 m/s at 27° north of west

8.35 [I] What average resisting force must act on a 3.0-kg mass to reduce its speed from 65 cm/s to 15 cm/s in 0.20 s? *Ans.* 7.5 N

8.36 [II] A 7.00-g bullet moving horizontally at 200 m/s strikes and passes through a 150-g tin can sitting on a post. Just after impact, the can has a horizontal speed of 180 cm/s. What was the bullet's speed after leaving the can? *Ans.* 161 m/s

8.37 [III] Two balls of equal mass, moving with speeds of 3 m/s, collide head-on. Find the speed of each after impact if (*a*) they stick together, (*b*) the collision is perfectly elastic, (*c*) the coefficient of restitution is 1/3. *Ans.* (*a*) 0 m/s; (*b*) each rebounds at 3 m/s; (*c*) each rebounds at 1 m/s

8.38 [III] A 90-g ball moving at 100 cm/s collides head-on with a stationary 10-g ball. Determine the speed of each after impact if (*a*) they stick together, (*b*) the collision is perfectly elastic, (*c*) the coefficient of restitution is 0.90. *Ans.* (*a*) 90 cm/s; (*b*) 80 cm/s, 1.8 m/s; (*c*) 81 cm/s, 1.7 m/s

8.39 [II] A ball is dropped onto a horizontal floor. It reaches a height of 144 cm on the first bounce, and 81 cm on the second bounce. Find (*a*) the coefficient of restitution between the ball and floor and (*b*) the height it attains on the third bounce. *Ans.* (*a*) 0.75; (*b*) 46 cm

8.40 [II] Two identical balls undergo a collision at the origin of coordinates. Before collision their scalar velocity components are ($u_x = 40$ cm/s, $u_y = 0$) and ($u_x = -30$ cm/s, $u_y = 20$ cm/s). After collision, the first ball is standing still. Find the scalar velocity components of the second ball. *Ans.* $v_x = 10$ cm/s, $v_y = 20$ cm/s

8.41 [II] Two identical balls traveling parallel to the x-axis have speeds of 30 cm/s and are oppositely directed. They collide off center perfectly elastically. After the collision, one ball is moving at an angle of 30°

above the $+x$-axis. Find its speed and the velocity of the other ball. *Ans.* 30 cm/s, 30 cm/s at 30° below the $-x$-axis (opposite to the first ball)

8.42 [II] (*a*) What minimum thrust must the engines of a 2.0×10^5 kg rocket have if the rocket is to be able to rise from the Earth when aimed straight upward? (*b*) If the engines eject gas at the rate of 20 kg/s, how fast must the gaseous exhaust be moving as it leaves the engines? Neglect the small change in the mass of the rocket due to the ejected fuel. *Ans.* (*a*) 20×10^5 N; (*b*) 98 km/s

Chapter 9

Angular Motion in a Plane

ANGULAR DISPLACEMENT (θ) is usually expressed in radians, in degrees, or in revolutions:

$$1 \text{ rev} = 360^\circ = 2\pi \text{ rad} \qquad \text{or} \qquad 1 \text{ rad} = 57.3^\circ$$

One radian is the angle subtended at the center of a circle by an arc equal in length to the radius of the circle. Thus an angle θ in radians is given in terms of the arc length l it subtends on a circle of radius r by

$$\theta = \frac{l}{r}$$

The radian measure of an angle is a dimensionless number. Radians, like degrees, are not a physical unit – the radian is not expressable in meters, kilograms, or seconds. Nonetheless, we will use the abbreviation rad to remind us that we are working with radians. As we'll soon see, "rad" does not always carry through the equations in a consistent fashion. We'll have to remove it and insert it as needed.

THE ANGULAR SPEED (ω) of an object whose axis of rotation is fixed is the rate at which its angular coordinate, the angular displacement θ, changes with time. If θ changes from θ_i to θ_f in a time t, then the **average angular speed** is

$$\omega_{av} = \frac{\theta_f - \theta_i}{t}$$

The units of ω_{av} are exclusively rad/s. Since each complete turn or cycle of a revolving system carries it through 2π rad

$$\omega = 2\pi f$$

where f is the **frequency** in revolutions per second, rotations per second, or cycles per second. Accordingly, ω is also called the **angular frequency**. We can associate a direction with ω and thereby create a vector quantity $\vec{\omega}$. Thus if the fingers of the right hand curve around in the direction of rotation, the thumb points along the axis of rotation in the direction of $\vec{\omega}$, the **angular velocity** vector.

THE ANGULAR ACCELERATION (α) of an object whose axis of rotation is fixed is the rate at which its angular speed changes with time. If the angular speed changes uniformly from ω_i to ω_f in a time t, then the **angular acceleration** is constant and

$$\alpha = \frac{\omega_f - \omega_i}{t}$$

The units of α are typically rad/s^2, rev/min^2, and such. It is possible to associate a direction with $\Delta\omega$, and therefore with α, thereby specifying the angular acceleration vector $\vec{\alpha}$, but we will have no need to do so here.

EQUATIONS FOR UNIFORMLY ACCELERATED ANGULAR MOTION are exactly analogous to those for uniformly accelerated linear motion. In the usual notation we have:

Linear	Angular
$v_{av} = \frac{1}{2}(v_i + v_f)$	$\omega_{av} = \frac{1}{2}(\omega_i + \omega_f)$
$s = v_{av}t$	$\theta = \omega_{av}t$
$v_f = v_i + at$	$\omega_f = \omega_i + \alpha t$
$v_f^2 = v_i^2 + 2as$	$\omega_f^2 = \omega_i^2 + 2\alpha\theta$
$s = v_i t + \frac{1}{2}at^2$	$\theta = \omega_i t + \frac{1}{2}\alpha t^2$

Taken alone, the second of these equations is just the definition of average speed, so it is valid whether the acceleration is constant or not.

RELATIONS BETWEEN ANGULAR AND TANGENTIAL QUANTITIES: When a disk of radius r rotates about a fixed central axis, a point on the rim of the disk is described in terms of the circumferential distance l it has moved, its tangential speed v, and its tangential acceleration a_T. These quantities are related to the angular quantities θ, ω, and α, which describe the rotation of the wheel, through the relations

$$l = r\theta \qquad v = r\omega \qquad a_T = r\alpha$$

provided radian measure is used for θ, ω, and α.

By simple reasoning, l can be shown to be the distance traveled by a point on a belt wound around a portion of a rotating wheel, or the distance a wheel would roll (without slipping) if free to do so. In such cases, v and a_T refer to the tangential speed and acceleration of a point on the belt, or of the center of the wheel, where r is the radius of the wheel. This can be seen in Fig. 9-1 which depicts a rolling wheel uniformly accelerating at an angular rate α (without slipping). The motion of the wheel can be thought of as composed of a simultaneous rotation about its center O, and a translation of O to O''. The point initially touching the ground (A), is in effect rotated into A' through an angle θ, and translated into A'' over a distance $l_O = r\theta$, which is also the distance O translates. Seen by someone standing still A moves along a cycloid (the dotted curve) to its position at A''. The speed at which O translates at any instant is $v_O = r\omega$, where ω is the angular speed at that instant. The linear (or tangential) acceleration of O, which is constant since α is contant, is $a_{TO} = r\alpha$.

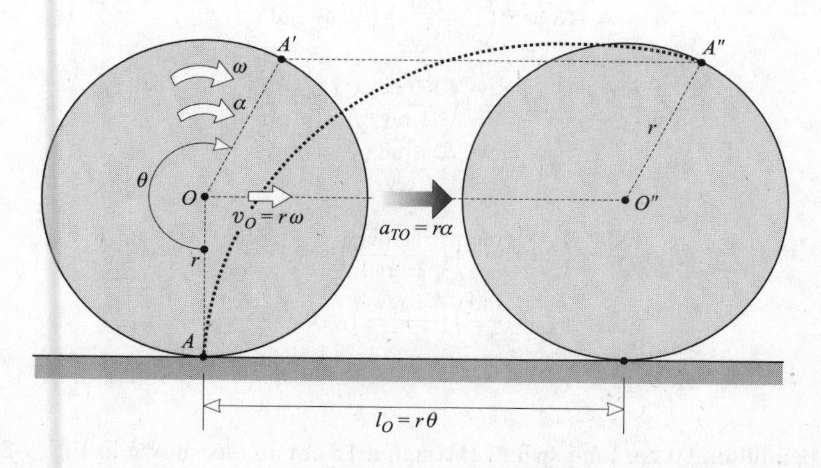

Fig. 9-1

CENTRIPETAL ACCELERATION (a_C): A point mass m moving with constant speed v around a circle of radius r is undergoing acceleration. Although the magnitude of its linear velocity is not changing, the direction of the velocity is continually changing. This change in velocity gives rise to an acceleration a_C of the mass directed toward the center of the circle. We call this acceleration the **centripetal acceleration**; its magnitude is given by

$$a_C = \frac{(\text{tangential speed})^2}{\text{radius of circular path}} = \frac{v^2}{r}$$

where v is the speed of the mass around the perimeter of the circle.

Because $v = r\omega$, we also have $a_C = r\omega^2$, where ω must be in rad/s. Notice that the word "acceleration" is commonly used in physics as either a scalar or a vector quantity. Fortunately, there's usually no ambiguity.

THE CENTRIPETAL FORCE ($\vec{F}_C$) is the force that must act on a mass m moving in a circular path of radius r to give it the required centripetal acceleration v^2/r. From $F = ma$, we have

$$F_C = \frac{mv^2}{r} = mr\omega^2$$

where $\vec{F}_C$ is directed toward the center of the circular path. Centripetal force is not a new kind of force; it's just the name given to whatever force (be it gravity, the tension in a string, magnetism, friction, etc.) that causes an object to move (off it's straight-line inertial path) along an arc.

Solved Problems

9.1 [I] Express each of the following in terms of other angular measures: (*a*) 28°, (*b*) $\frac{1}{4}$ rev/s, (*c*) 2.18 rad/s^2.

(*a*)
$$28° = (28 \text{ deg})\left(\frac{1 \text{ rev}}{360 \text{ deg}}\right) = 0.078 \text{ rev}$$
$$= (28 \text{ deg})\left(\frac{2\pi \text{ rad}}{360 \text{ deg}}\right) = 0.49 \text{ rad}$$

(*b*)
$$\frac{1}{4}\frac{\text{rev}}{\text{s}} = \left(0.25\ \frac{\text{rev}}{\text{s}}\right)\left(\frac{360 \text{ deg}}{1 \text{ rev}}\right) = 90\ \frac{\text{deg}}{\text{s}}$$
$$= \left(0.25\ \frac{\text{rev}}{\text{s}}\right)\left(\frac{2\pi \text{ rad}}{1 \text{ rev}}\right) = \frac{\pi}{2}\ \frac{\text{rad}}{\text{s}}$$

(*c*)
$$2.18\frac{\text{rad}}{\text{s}^2} = \left(2.18\ \frac{\text{rad}}{\text{s}^2}\right)\left(\frac{360 \text{ deg}}{2\pi \text{ rad}}\right) = 125\ \frac{\text{deg}}{\text{s}^2}$$
$$= \left(2.18\ \frac{\text{rad}}{\text{s}^2}\right)\left(\frac{1 \text{ rev}}{2\pi \text{ rad}}\right) = 0.347\ \frac{\text{rev}}{\text{s}^2}$$

9.2 [I] The bob of a pendulum 90 cm long swings through a 15-cm arc, as shown in Fig. 9-2. Find the angle θ, in radians and in degrees, through which it swings.

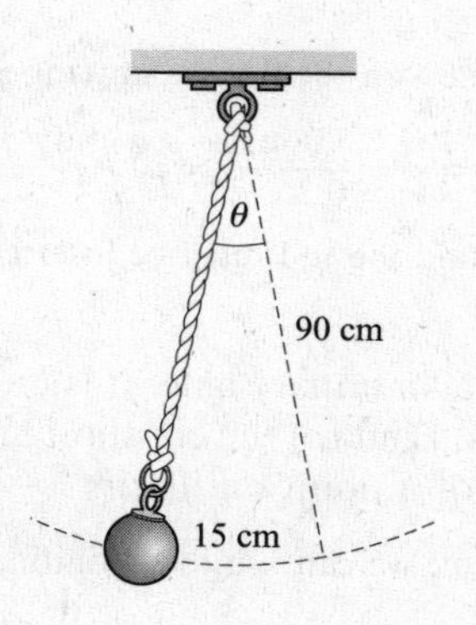

Fig. 9-2

Recall that $l = r\theta$ applies only to angles in radian measure. Therefore, in radians

$$\theta = \frac{l}{r} = \frac{0.15\ \text{m}}{0.90\ \text{m}} = 0.167\ \text{rad} = 0.17\ \text{rad}$$

Then in degrees $$\theta = (0.167\ \text{rad})\left(\frac{360\ \text{deg}}{2\pi\ \text{rad}}\right) = 9.6^\circ$$

9.3 [I] A fan turns at a rate of 900 rpm (i.e., rev/min). (*a*) Find the angular speed of any point on one of the fan blades. (*b*) Find the tangential speed of the tip of a blade if the distance from the center to the tip is 20.0 cm.

(*a*) $$f = 900\frac{\text{rev}}{\text{min}} = 15.0\frac{\text{rev}}{\text{s}}$$

and since $\omega = 2\pi f$

$$\omega = 94.2\frac{\text{rad}}{\text{s}}$$

for all points of the fan blade.

(*b*) The tangential speed is $r\omega$, where ω must be in rad/s. Therefore,

$$v = r\omega = (0.200\ \text{m})(94.2\ \text{rad/s}) = 18.8\ \text{m/s}$$

Notice that the rad does not carry through the equations properly – we insert it and delete it as needed.

9.4 [I] A belt passes over a wheel of radius 25 cm, as shown in Fig. 9-3. If a point on the belt has a speed of 5.0 m/s, how fast is the wheel turning?

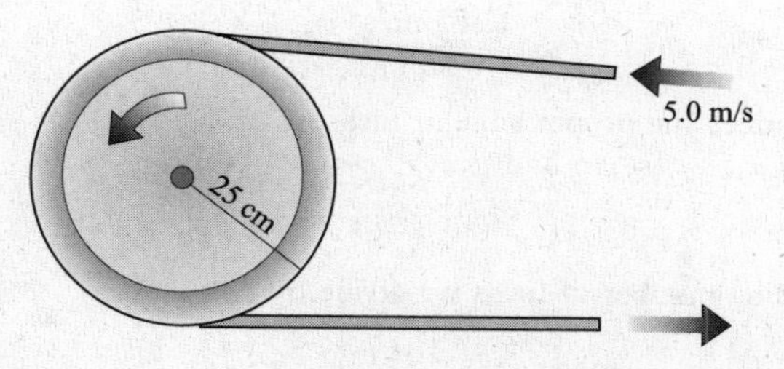

Fig. 9-3

A point on the wheel's circumference (i.e., on the belt) is moving at a linear speed $v = r\omega$. Hence,

$$\omega = \frac{v}{r} = \frac{5.0 \text{ m/s}}{0.25 \text{ m}} = 20 \ \frac{\text{rad}}{\text{s}}$$

As a rule, ω comes out in units of s^{-1} and the rad must be inserted ad hoc.

9.5 [I] A wheel of 40-cm radius rotates on a stationary axle. It is uniformly speeded up from rest to a speed of 900 rpm in a time of 20 s. Find (*a*) the constant angular acceleration of the wheel and (*b*) the tangential acceleration of a point on its rim.

(*a*) Because the acceleration is constant, we can use the definition $\alpha = (\omega_f - \omega_i)/t$ to get

$$\alpha = \frac{\left(2\pi\frac{\text{rad}}{\text{rev}}\right)\left(\frac{900 \text{ rev}}{60 \text{ s}}\right) - \left(2\pi\frac{\text{rad}}{\text{rev}}\right)\left(0\frac{\text{rev}}{\text{s}}\right)}{20 \text{ s}} = 4.7 \ \frac{\text{rad}}{\text{s}^2}$$

(*b*) Then $$a_T = r\alpha = (0.40 \text{ m})\left(4.7\frac{\text{rad}}{\text{s}^2}\right) = 1.88 \ \frac{\text{m}}{\text{s}^2} = 1.9 \text{ m/s}^2$$

9.6 [II] A pulley of 5.0-cm radius, on a motor, is turning at 30 rev/s and slows down uniformly to 20 rev/s in 2.0 s. Calculate (*a*) the angular acceleration of the motor, (*b*) the angle through which it turns in this time, and (*c*) the length of belt it winds in this time.

Because the pulley is decelerating we can anticipate that α will be negative:

(*a*) $$\alpha = \frac{\omega_f - \omega_i}{t} = 2\pi\frac{(20 - 30) \text{ rad/s}}{2.0 \text{ s}} = -10\pi \text{ rad/s}^2$$

And to two significant figures,

(*b*) $$\theta = \omega_{av}t = \tfrac{1}{2}(\omega_f + \omega_i)t = \tfrac{1}{2}(100\pi \text{ rad/s})(2.0 \text{ s}) = 100\pi \text{ rad} = 1.0 \times 10^2\pi \text{ rad}$$

(*c*) With $\theta = 314$ rad

$$l = r\theta = (0.050 \text{ m})(314 \text{ rad}) = 16 \text{ m}$$

9.7 [II] A car has wheels of radius 30 cm. It starts from rest and (without slipping) accelerates uniformly to a speed of 15 m/s in a time of 8.0 s. Find the angular acceleration of its wheels and the number of rotations one wheel makes in this time.

Remember that the center of the rolling wheel accelerates tangentially at the same rate as does a point on its circumference. We know that $a_T = (v_f - v_i)/t$, and so

$$a_T = \frac{15 \text{ m/s}}{8.0 \text{ s}} = 1.875 \text{ m/s}^2$$

Then $a_T = r\alpha$ gives

$$\alpha = \frac{a_T}{r} = \frac{1.875 \text{ m/s}^2}{0.30 \text{ m}} = 6.2 \text{ rad/s}^2$$

Notice that we must introduce the proper angular measure, radians.

Now we can use $\theta = \omega_i t + \frac{1}{2}\alpha t^2$ to find

$$\theta = 0 + \tfrac{1}{2}(6.2 \text{ rad/s}^2)(8.0 \text{ s})^2 = 200 \text{ rad}$$

and to get the corresponding number of turns we divide by 2π,

$$(200 \text{ rad})\left(\frac{1 \text{ rev}}{2\pi \text{ rad}}\right) = 32 \text{ rev}$$

9.8 [II] The spin-drier of a washing machine revolving at 900 rpm slows down uniformly to 300 rpm while making 50 revolutions. Find (*a*) the angular acceleration and (*b*) the time required to turn through these 50 revolutions.

We easily find that 900 rev/min $= 15.0$ rev/s $= 30.0\pi$ rad/s and 300 rev/min $= 5.00$ rev/s $= 10.0\pi$ rad/s.

(*a*) From $\omega_f^2 = \omega_i^2 + 2\alpha\theta$, we have

$$\alpha = \frac{\omega_f^2 - \omega_i^2}{2\theta} = \frac{(10.0\pi\ \text{rad/s})^2 - (30.0\pi\ \text{rad/s})^2}{2(100\pi\ \text{rad})} = -4.0\pi\ \text{rad/s}^2$$

(*b*) Because $\omega_{av} = \frac{1}{2}(\omega_i + \omega_f) = 20.0\pi$ rad/s, $\theta = \omega_{av}t$ yields

$$t = \frac{\theta}{\omega_{av}} = \frac{100\pi\ \text{rad}}{20.0\pi\ \text{rad/s}} = 5.0\ \text{s}$$

9.9 [II] A 200-g object is tied to the end of a cord and whirled in a horizontal circle of radius 1.20 m at a constant 3.0 rev/s. Assume that the cord is horizontal, i.e., that gravity can be neglected. Determine (*a*) the acceleration of the object and (*b*) the tension in the cord.

(*a*) The object is not accelerating tangentially to the circle but is undergoing a radial, or centripetal, acceleration given by

$$a_C = \frac{v^2}{r} = r\omega^2$$

where ω must be in rad/s. Since 3.0 rev/s $= 6.0\pi$ rad/s,

$$a_C = (6.0\pi\ \text{rad/s})^2(1.20\ \text{m}) = 426\ \text{m/s}^2 = 0.43\ \text{km/s}^2$$

(*b*) To cause the acceleration found in (*a*), the cord must pull on the 0.200-kg mass with a centripetal force given by

$$F_C = ma_C = (0.200\ \text{kg})(426\ \text{m/s}^2) = 85\ \text{N}$$

This is the tension in the cord.

9.10 [II] What is the maximum speed at which a car can round a curve of 25-m radius on a level road if the coefficient of static friction between the tires and road is 0.80?

The radial force required to keep the car in the curved path (the centripetal force) is supplied by the force of friction between the tires and the road. If the mass of the car is m, then the maximum friction force (which is the centripetal force) equals $\mu_s F_N$ or $0.80mg$; this arises when the car is on the verge of skidding sideways. Therefore, the maximum speed is given by

$$\frac{mv^2}{r} = 0.80mg \quad \text{or} \quad v = \sqrt{0.80gr} = \sqrt{(0.80)(9.81\ \text{m/s}^2)(25\ \text{m})} = 14\ \text{m/s}$$

9.11 [II] A spaceship orbits the Moon at a height of 20 000 m. Assuming it to be subject only to the gravitational pull of the Moon, find its speed and the time it takes for one orbit. For the Moon, $m_m = 7.34 \times 10^{22}$ kg and $r = 1.738 \times 10^6$ m.

The gravitational force of the Moon on the ship supplies the required centripetal force:

$$G\frac{m_s m_m}{R^2} = \frac{m_s v^2}{R}$$

where R is the radius of the orbit. Solving, we find that

$$v = \sqrt{\frac{Gm_m}{R}} = \sqrt{\frac{(6.67 \times 10^{-11}\ \mathrm{N \cdot m^2/kg^2})(7.34 \times 10^{22}\ \mathrm{kg})}{(1.738 + 0.0200) \times 10^6\ \mathrm{m}}} = 1.67\ \mathrm{km/s}$$

from which we find that

$$\text{Time for one orbit} = \frac{2\pi R}{v} = 6.62 \times 10^3\ \mathrm{s} = 110\ \mathrm{min}$$

9.12 [II] As shown in Fig. 9-4, a ball B is fastened to one end of a 24-cm string, and the other end is held fixed at point Q. The ball whirls in the horizontal circle shown. Find the speed of the ball in its circular path if the string makes an angle of 30° to the vertical.

The only forces acting on the ball are the ball's weight mg and the tension F_T in the cord. The tension must do two things: (1) balance the weight of the ball by means of its vertical component, $F_T \cos 30°$; (2) supply the required centripetal force by means of its horizontal component, $F_T \sin 30°$. Therefore we can write

$$F_T \cos 30° = mg \qquad \text{and} \qquad F_T \sin 30° = \frac{mv^2}{r}$$

Solving for F_T in the first equation and substituting it in the second gives

$$\frac{mg \sin 30°}{\cos 30°} = \frac{mv^2}{r} \qquad \text{or} \qquad v = \sqrt{rg(0.577)}$$

However, $r = \overline{BC} = (0.24\ \mathrm{m}) \sin 30° = 0.12\ \mathrm{m}$ and $g = 9.81\ \mathrm{m/s^2}$, from which $v = 0.82\ \mathrm{m/s}$.

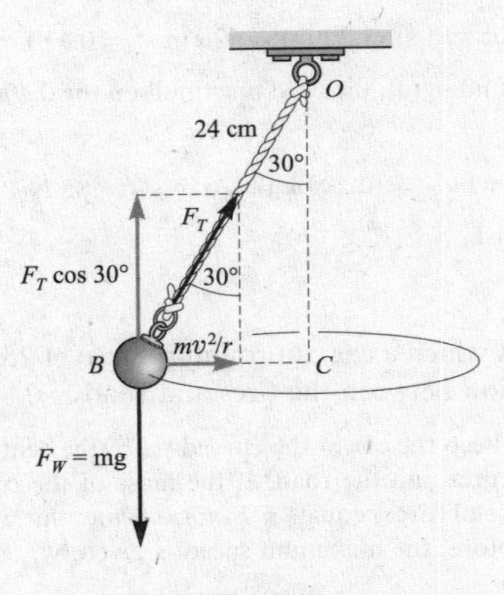

Fig. 9-4

9.13 [III] As shown in Fig. 9-5, a 20-g bead slides from rest at A along a frictionless wire. If h is 25 cm and R is 5.0 cm, how large a force must the wire exert on the bead when it is at (*a*) point B and (*b*) point D?

(*a*) As a general rule, remember to keep a few more numerical figures in the intermediate steps of the calculation than are to be found in the answer. This will avoid round-off errors. Let us first find the speed of the bead at point B. It has fallen through a distance $h - 2R$ and so its loss in $\mathrm{PE_G}$ is $mg(h - 2R)$. This must equal its KE at point B:

$$\tfrac{1}{2}mv^2 = mg(h - 2R)$$

where v is the speed of the bead at point B. Hence,

$$v = \sqrt{2g(h - 2R)} = \sqrt{2(9.81\ \text{m/s}^2)(0.15\ \text{m})} = 1.716\ \text{m/s}$$

As shown in Fig. 9-5(b), two forces act on the bead when it is at B: (1) the weight of the bead mg and (2) the (assumed downward) force F of the wire on the bead. Together, these two forces must supply the required centripetal force, mv^2/R, if the bead is to follow the circular path. We therefore write

$$mg + F = \frac{mv^2}{R}$$

or

$$F = \frac{mv^2}{R} - mg = (0.020\ \text{kg})\left[\left(\frac{1.716^2}{0.050} - 9.81\right)\ \text{m/s}^2\right] = 0.98\ \text{N}$$

The wire must exert a 0.98 N downward force on the bead to hold it in a circular path.

(b) The situation is similar at point D, but now the weight is perpendicular to the direction of the required centripetal force. Therefore the wire alone must furnish it. Proceeding as before, we have

$$v = \sqrt{2g(h - R)} = \sqrt{2(9.81\ \text{m/s}^2)(0.20\ \text{m})} = 1.98\ \text{m/s}$$

and

$$F = \frac{mv^2}{R} = \frac{(0.020\ \text{kg})(1.98\ \text{m/s})^2}{0.050\ \text{m}} = 1.6\ \text{N}$$

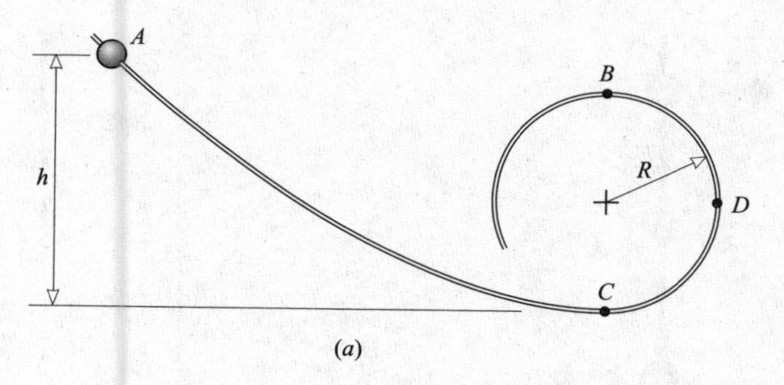

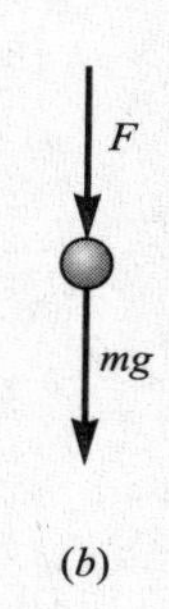

Fig. 9-5

9.14 [III] As shown in Fig. 9-6, a 0.90-kg body attached to a cord is whirled in a vertical circle of radius 2.50 m. (a) What minimum speed v_t must the body have at the top of the circle so as not to depart from the circular path? (b) Under condition (a), what speed v_b will the object have after it "falls" to the bottom of the circle? (c) Find the tension F_{Tb} in the cord when the body is at the bottom of the circle and moving with the critical speed v_b.

The object is moving at its slowest speed at the very top and increases its speed as it revolves downward because of gravity ($v_b > v_t$).

(a) As Fig. 9-6 shows, two radial forces act on the object at the top: (1) its weight mg and (2) the tension F_{Tt}. The resultant of these two forces must supply the required centripetal force.

$$\frac{mv_t^2}{r} = mg + F_{Tt}$$

For a given r, v will be smallest when $F_{Tt} = 0$. In that case,

$$\frac{mv_t^2}{r} = mg \qquad \text{or} \qquad v_t = \sqrt{rg}$$

Using $r = 2.50$ m and $g = 9.81$ m/s^2 gives $v_t = 4.95$ m/s as the speed at the top.

(*b*) In traveling from top to bottom, the body falls a distance $2r$. Therefore, with $v_t = 4.95$ m/s as the speed at the top and with v_b as the speed at the bottom, conservation of energy gives

$$\text{KE at bottom} = \text{KE at top} + \text{PE}_\text{G} \text{ at top}$$

$$\tfrac{1}{2}mv_b^2 = \tfrac{1}{2}mv_t^2 + mg(2r)$$

where we have chosen the bottom of the circle as the zero level for PE_G. Notice that m cancels. Using $v_t = 4.95$ m/s, $r = 2.50$ m, and $g = 9.81$ m/s^2 yields $v_b = 11.1$ m/s.

(*c*) When the object is at the bottom of its path, we see from Fig. 9-5 that the unbalanced upward radial force on it is $F_{Tb} - mg$. This force supplies the required centripetal force:

$$F_{Tb} - mg = \frac{mv_b^2}{r}$$

Using $m = 0.90$ kg, $g = 9.81$ m/s^2, $v_b = 11.1$ m/s, and $r = 2.50$ m gives

$$F_{Tb} = m\left(g + \frac{v_b^2}{r}\right) = 53 \text{ N}$$

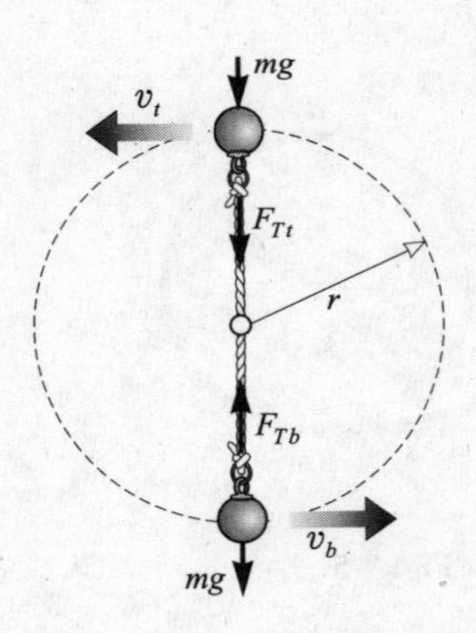

Fig. 9-6

9.15 [III] A curve of radius 30 m is to be banked so that a car may make the turn at a speed of 13 m/s without depending on friction. What must be the slope of the curve (the banking angle)?

The situation is shown in Fig. 9-7 if friction is absent. Only two forces act upon the car: (1) the weight mg of the car (which is straight downward) and (2) the normal force F_N (which is perpendicular to the road) exerted by the pavement on the car.

The force F_N must do two things: (1) its vertical component, $F_N \cos\theta$, must balance the car's weight; (2) its horizontal component, $F_N \sin\theta$, must supply the required centripetal force. In other words, the road pushes horizontally on the car keeping it moving in a circle. We can therefore write

$$F_N \cos\theta = mg \qquad \text{and} \qquad F_N \sin\theta = \frac{mv^2}{r}$$

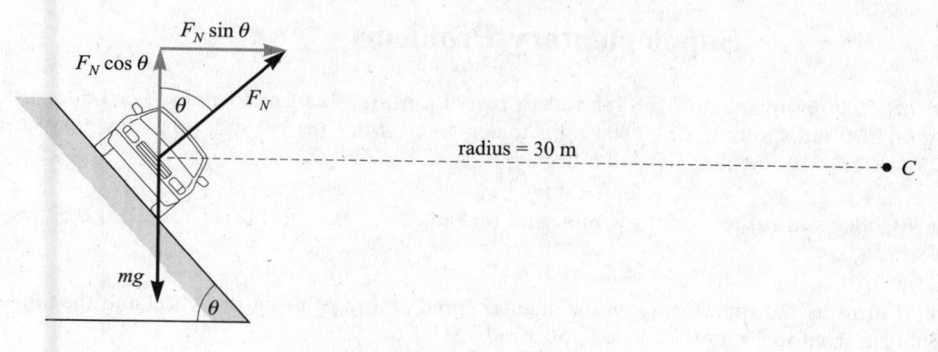

Fig. 9-7

Dividing the second equation by the first causes F_N and m to cancel and gives

$$\tan\theta = \frac{v^2}{gr} = \frac{(13\ \text{m/s})^2}{(9.81\ \text{m/s}^2)(30\ \text{m})} = 0.575$$

From this we find that θ, the banking angle, must be 30°.

9.16 [III] As shown in Fig. 9-8, a thin cylindrical shell of inner radius r rotates horizontally, about a vertical axis, at an angular speed of ω. A wooden block rests against the inner surface and rotates with it. If the coefficient of static friction between block and surface is μ_s, how fast must the shell be rotating if the block is not to slip and fall? Assume $r = 150$ cm and $\mu_s = 0.30$.

The surface holds the block in place by pushing on it with centripetal force $m\omega^2 r$. This force is perpendicular to the surface; it is the normal force that determines the friction on the block which in turn keeps it from sliding downward. Because $F_f = \mu_s F_N$ and $F_N = mr\omega^2$, we have

$$F_f = \mu_s F_N = \mu_s m r \omega^2$$

This friction force must balance the weight mg of the block if it is not to slip. Therefore,

$$mg = \mu_s m r \omega^2 \qquad \text{or} \qquad \omega = \sqrt{\frac{g}{\mu_s r}}$$

Inserting the given values, we find

$$\omega = \sqrt{\frac{9.81\ \text{m/s}^2}{(0.30)(1.50\ \text{m})}} = 4.7\ \text{rad/s} = 0.74\ \text{rev/s}$$

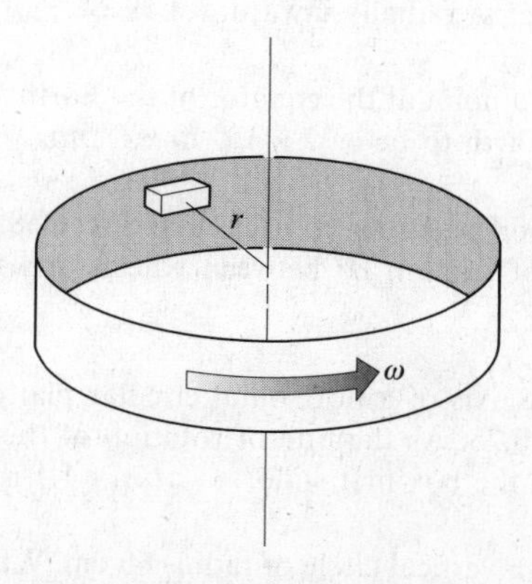

Fig. 9-8

Supplementary Problems

9.17 [I] Convert (*a*) 50.0 rev to radians, (*b*) 48π rad to revolutions, (*c*) 72.0 rps to rad/s, (*d*) 1.50×10^3 rpm to rad/s, (*e*) 22.0 rad/s to rpm, (*f*) 2.000 rad/s to deg/s. *Ans.* (*a*) 314 rad; (*b*) 24 rev; (*c*) 452 rad/s; (*d*) 157 rad/s; (*e*) 210 rev/min; (*f*) 114.6 deg/s

9.18 [I] Express 40.0 deg/s in (*a*) rev/s, (*b*) rev/min, and (*c*) rad/s. *Ans.* (*a*) 0.111 rev/s; (*b*) 6.67 rev/min; (*c*) 0.698 rad/s

9.19 [I] A flywheel turns at 480 rpm. Compute the angular speed at any point on the wheel and the tangential speed 30.0 cm from the center. *Ans.* 50.3 rad/s, 15.1 m/s

9.20 [I] It is desired that the outer edge of a grinding wheel 9.0 cm in radius move at a constant rate of 6.0 m/s. (*a*) Determine the angular speed of the wheel. (*b*) What length of thread could be wound on the rim of the wheel in 3.0 s when it is turning at this rate? *Ans.* (*a*) 67 rad/s; (*b*) 18 m

9.21 [I] Through how many radians does a point on the Earth's surface (off the poles) move in 6.00 h as a result of the Earth's rotation? What is the speed of a point on the equator? Take the radius of the Earth to be 6370 km. *Ans.* 1.57 rad, 463 m/s

9.22 [II] A wheel 25.0 cm in radius turning at 120 rpm uniformly increases its frequency to 660 rpm in 9.00 s. Find (*a*) the constant angular acceleration in rad/s^2, and (*b*) the tangential acceleration of a point on its rim. *Ans.* (*a*) 6.28 rad/s^2; (*b*) 157 cm/s^2

9.23 [II] The angular speed of a disk decreases uniformly from 12.00 to 4.00 rad/s in 16.0 s. Compute the angular acceleration and the number of revolutions made in this time. *Ans.* -0.500 rad/s^2, 20.4 rev

9.24 [II] A car wheel 30 cm in radius is turning at a rate of 8.0 rev/s when the car begins to slow uniformly to rest in a time of 14 s. Find the number of revolutions made by the wheel and the distance the car goes in the 14 s. *Ans.* 56 rev, 0.11 km

9.25 [II] A wheel revolving at 6.00 rev/s has an angular acceleration of 4.00 rad/s^2. Find the number of turns the wheel must make to reach 26.0 rev/s, and the time required. *Ans.* 502 rev, 31.4 s

9.26 [II] A string wound on the rim of a wheel 20 cm in diameter is pulled out at a rate of 75 cm/s. Through how many revolutions will the wheel have turned by the time that 9.0 m of string has been unwound? How long will it take? *Ans.* 14 rev, 12 s

9.27 [II] A mass of 1.5 kg out in space moves in a circle of radius 25 cm at a constant 2.0 rev/s. Calculate (*a*) the tangential speed, (*b*) the acceleration, and (*c*) the required centripetal force for the motion. *Ans.* (*a*) 3.1 m/s; (*b*) 39 m/s^2 radially inward; (*c*) 59 N

9.28 [II] (*a*) Compute the radial acceleration of a point at the equator of the Earth. (*b*) Repeat for the north pole of the Earth. Take the radius of the Earth to be 6.37×10^6 m. *Ans.* (*a*) 0.033 7 m/s^2; (*b*) zero

9.29 [II] A car moving at 5.0 m/s tries to round a corner in a circular arc of 8.0 m radius. The roadway is flat. How large must the coefficient of friction be between wheels and roadway if the car is not to skid? *Ans.* 0.32

9.30 [II] A box rests at a point 2.0 m from the axis of a horizontal circular platform. The coefficient of static friction between box and platform is 0.25. As the rate of rotation of the platform is slowly increased from zero, at what angular speed will the box first slide? *Ans.* 1.1 rad/s

9.31 [II] A stone rests in a pail that is moved in a vertical circle of radius 60 cm. What is the least speed the stone must have as it rounds the top of the circle if it is to remain in contact with the pail? *Ans.* 2.4 m/s

9.32 [II] A pendulum 80.0 cm long is pulled to the side, so that its bob is raised 20.0 cm from its lowest position, and is then released. As the 50.0 g bob moves through its lowest position, (*a*) what is its speed and (*b*) what is the tension in the pendulum cord? *Ans.* (*a*) 1.98 m/s; (*b*) 0.735 N

9.33 [II] Refer back to Fig. 9-4. How large must h be (in terms of R) if the frictionless wire is to exert no force on the bead as it passes through point B? Assume the bead is released from rest at A. *Ans.* $2.5R$

9.34 [II] If, in Fig. 9-4 and in Problem 9.33, $h = 2.5R$, how large a force will the 50-g bead exert on the wire as it passes through point C? *Ans.* 2.9 N

9.35 [II] A satellite orbits the Earth at a height of 200 km in a circle of radius 6570 km. Find the speed of the satellite and the time taken to complete one revolution. Assume the Earth's mass is 6.0×10^{24} kg. (*Hint*: The gravitational force provides the centripetal force.) *Ans.* 7.8 km/s, 88 min

9.36 [III] A roller coaster is just barely moving as it goes over the top of the hill. It rolls nearly without friction down the hill and then up over a lower hill that has a radius of curvature of 15 m. How much higher must the first hill be than the second if the passengers are to exert no force on the seat as they pass over the top of the lower hill? *Ans.* 7.5 m

9.37 [III] The human body can safely tolerate an acceleration 9.00 times that due to gravity. With what minimum radius of curvature may a pilot safely turn the plane upward at the end of a dive if the plane's speed is 770 km/h? *Ans.* 519 m

9.38 [III] A 60.0 kg glider pilot traveling in a glider at 40.0 m/s wishes to turn an inside vertical loop such that his body exerts a force of 350 N on the seat when the glider is at the top of the loop. What must be the radius of the loop under these conditions? (*Hint*: Gravity and the seat exert forces on the pilot.) *Ans.* 102 m

9.39 [III] Suppose the Earth is a perfect sphere with $R = 6370$ km. If a person weighs exactly 600.0 N at the north pole, how much will the person weigh at the equator? (*Hint*: The upward push of the scale on the person is what the scale will read and is what we are calling the weight in this case.) *Ans.* 597.9 N

9.40 [III] A mass m hangs at the end of a pendulum of length L which is released at an angle of 40.0° to the vertical. Find the tension in the pendulum cord when it makes an angle of 20.0° to the vertical. (*Hint*: Resolve the weight along and perpendicular to the cord.) *Ans.* $1.29mg$

Chapter 10

Rigid-Body Rotation

THE TORQUE (τ) due to a force about an axis was defined in Chapter 5. It's also sometimes called the moment of the force.

THE MOMENT OF INERTIA (I) of a body is a measure of the rotational inertia of the body. If an object that is free to rotate about an axis is difficult to set into rotation, its moment of inertia about that axis is large. An object with a small I has little rotational inertia.

If a body is considered to be made up of tiny masses $m_1, m_2, m_3 \ldots$, at respective distances $r_1, r_2, r_3, \ldots$, from an axis, its moment of inertia about the axis is

$$I = m_1 r_1^2 + m_2 r_2^2 + m_3 r_3^2 + \cdots = \sum m_i r_i^2$$

The units of I are $\text{kg} \cdot \text{m}^2$.

It is convenient to define a **radius of gyration** (k) for an object about an axis by the relation

$$I = Mk^2$$

where M is the total mass of the object. Hence k is the distance a point mass M must be from the axis if the point mass is to have the same I as the object.

TORQUE AND ANGULAR ACCELERATION: A torque τ, acting on a body having a ***moment of inertia*** I, produces in it an angular acceleration α given by

$$\tau = I\alpha$$

Here τ, I, and α are all computed with respect to the same axis. As for units, τ is in $\text{N} \cdot \text{m}$, I is in $\text{kg} \cdot \text{m}^2$, and α must be in rad/s^2. (Recall the translational equivalent, $F = ma$).

THE KINETIC ENERGY OF ROTATION (KE_r) of a mass whose moment of inertia about an axis is I, and which is rotating about that axis with an angular velocity ω, is

$$\text{KE}_r = \tfrac{1}{2} I\omega^2$$

where the energy is in joules and ω must be in rad/s. (Recall the translational equivalent, $\text{KE} = \frac{1}{2}mv^2$).

COMBINED ROTATION AND TRANSLATION: The KE of a rolling ball or other rolling object of mass M is the sum of (1) its rotational KE *about an axis through its center of mass* (i.e., c.m.) (Chapter 8) and (2) the translational KE of an equivalent point mass moving with the center of mass. In other words, putting it loosely, the total KE equals the KE around the *c.m.* plus the KE of the *c.m.* In symbols,

$$\text{KE}_{\text{total}} = \tfrac{1}{2} I\omega^2 + \tfrac{1}{2} Mv^2$$

Note that I is the moment of inertia of the object about an axis through its mass center.

THE WORK (W) done on a rotating body during an angular displacement θ by a constant torque τ is given by

$$W = \tau\theta$$

where W is in joules and θ must be in radians. (Recall the translational equivalent, $W = Fs$.)

THE POWER (P) transmitted to a body by a torque is given by

$$P = \tau\omega$$

where τ is the applied torque about the axis of rotation, and ω is the angular speed, about that same axis. Radian measure must be used for ω. (Recall the translational equivalent, $P = Fv$.)

ANGULAR MOMENTUM ($\vec{\mathbf{L}}$) is a vector quantity that has magnitude $I\omega$ and is directed along the axis of rotation. (Recall the translational equivalent $\vec{\mathbf{p}} = m\vec{\mathbf{v}}$.) If the net torque on a body is zero, its angular momentum will remain unchanged in both magnitude and direction. This is the **Law of Conservation of Angular Momentum.**

ANGULAR IMPULSE has magnitude τt, where t is the time during which the constant torque τ acts on the object. In analogy to the linear case, an angular impulse τt on a body causes a change in angular momentum of the body given by

$$\tau t = I\omega_f - I\omega_i$$

PARALLEL-AXIS THEOREM: The moment of inertia I of a body about an axis parallel to an axis through the center of mass is

$$I = I_{cm} + Mh^2$$

where I_{cm} = moment of inertia about an axis through the center of mass
M = total mass of the body
h = perpendicular distance between the two parallel axes

The moments of inertia (about an axis through the center of mass) of several uniform objects, each of mass M, are shown in Fig. 10-1.

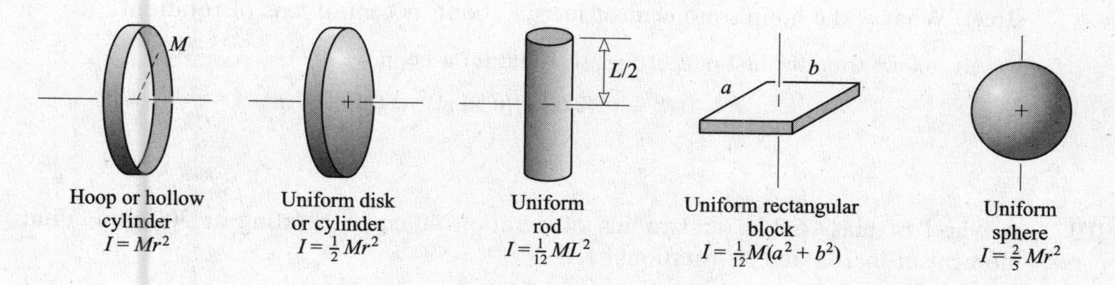

Fig. 10-1

ANALOGOUS LINEAR AND ANGULAR QUANTITIES:

Linear displacement	s	$\leftrightarrow$	angular displacement	θ
Linear speed	v	$\leftrightarrow$	angular speed	ω
Linear acceleration	a_T	$\leftrightarrow$	angular acceleration	α
Mass (inertia)	m	$\leftrightarrow$	moment of inertia	I
Force	F	$\leftrightarrow$	torque	τ
Linear momentum	mv	$\leftrightarrow$	angular momentum	$I\omega$
Linear impulse	Ft	$\leftrightarrow$	angular impulse	τt

If, in the equations for linear motion, we replace linear quantities by the corresponding angular quantities, we get the corresponding equations for angular motion. Thus, we have

Linear: $F = ma$ $\quad$ $\text{KE} = \frac{1}{2}mv^2$ $\quad$ $W = Fs$ $\quad$ $\text{P} = Fv$

Angular: $\tau = I\alpha$ $\quad$ $\text{KE}_r = \frac{1}{2}I\omega^2$ $\quad$ $W = \tau\theta$ $\quad$ $\text{P} = \tau\omega$

In these equations, θ, ω, and α must be expressed in radian measure.

Solved Problems

10.1 [I] A small sphere of mass 2.0 kg revolves at the end of a 1.2 m long string in a horizontal plane around a vertical axis. Determine its moment of inertia with respect to that axis.

A small sphere at the end of a long string resembles a point mass revolving about an axis at a radial distance r. Consequently its moment of inertia is given by

$$I_\bullet = m_\bullet r^2 = (2.0 \text{ kg})(1.2 \text{ m})^2 = 2.9 \text{ kg} \cdot \text{m}^2$$

10.2 [I] What is the moment of inertia of a homogeneous solid sphere of mass 10 kg and radius 20 cm, about an axis passing through its center?

It follows from the last part of Fig. 10-1 that for a sphere

$$I = \frac{2}{5}MR^2 = \frac{2}{5}(10 \text{ kg})(0.20 \text{ m})^2 = 0.16 \text{ kg} \cdot \text{m}^2$$

10.3 [I] A thin cylindrical hoop having a diameter of 1.0 m and a mass of 400 g, rolls down the street. What is the hoop's moment of inertia about its central axis of rotation?

It follows from the first part of Fig. 10-1 that for a hoop

$$I = MR^2 = (0.400 \text{ kg})(0.50 \text{ m})^2 = 0.10 \text{ kg} \cdot \text{m}^2$$

10.4 [II] A wheel of mass 6.0 kg and radius of gyration 40 cm is rotating at 300 rpm. Find its moment of inertia and its rotational KE.

$$I = Mk^2 = (6.0 \text{ kg})(0.40 \text{ m})^2 = 0.96 \text{ kg} \cdot \text{m}^2$$

The rotational KE is $\frac{1}{2}I\omega^2$, where ω must be in rad/s. We have

$$\omega = \left(300\frac{\text{rev}}{\text{min}}\right)\left(\frac{1\text{ min}}{60.0\text{ s}}\right)\left(\frac{2\pi\text{ rad}}{1\text{ rev}}\right) = 31.4\text{ rad/s}$$

so

$$\text{KE}_r = \tfrac{1}{2}I\omega^2 = \tfrac{1}{2}(0.96\text{ kg}\cdot\text{m}^2)(31.4\text{ rad/s})^2 = 0.47\text{ kJ}$$

10.5 [II] A 500-g uniform sphere of 7.0-cm radius spins at 30 rev/s on an axis through its center. Find its (*a*) KE_r, (*b*) angular momentum, and (*c*) radius of gyration.

We need the moment of inertia of a uniform sphere about an axis through its center. From Fig. 10-1,

$$I = \tfrac{2}{5}Mr^2 = (0.40)(0.50\text{ kg})(0.070\text{ m})^2 = 0.000\,98\text{ kg}\cdot\text{m}^2$$

(*a*) Knowing that $\omega = 30$ rev/s $= 188$ rad/s, we have

$$\text{KE}_r = \tfrac{1}{2}I\omega^2 = \tfrac{1}{2}(0.000\,98\text{ kg}\cdot\text{m}^2)(188\text{ rad/s})^2 = 0.017\text{ kJ}$$

Notice that ω must be in rad/s.

(*b*) Its angular momentum is

$$I\omega = (0.000\,98\text{ kg}\cdot\text{m}^2)(188\text{ rad/s}) = 0.18\text{ kg}\cdot\text{m}^2/\text{s}$$

(*c*) For any object, $I = Mk^2$, where k is the radius of gyration. Therefore,

$$k = \sqrt{\frac{I}{M}} = \sqrt{\frac{0.000\,98\text{ kg}\cdot\text{m}^2}{0.50\text{ kg}}} = 0.044\text{ m} = 4.4\text{ cm}$$

Notice that this is a reasonable value in view of the fact that the radius of the sphere is 7.0 cm.

10.6 [II] An airplane propeller has a mass of 70 kg and a radius of gyration of 75 cm. Find its moment of inertia. How large a torque is needed to give it an angular acceleration of 4.0 rev/s^2?

$$I = Mk^2 = (70\text{ kg})(0.75\text{ m})^2 = 39\text{ kg}\cdot\text{m}^2$$

To use $\tau = I\alpha$, we must have α in rad/s^2:

$$\alpha = \left(4.0\frac{\text{rev}}{\text{s}^2}\right)\left(2\pi\frac{\text{rad}}{\text{rev}}\right) = 8.0\pi\text{ rad/s}^2$$

Then

$$\tau = I\alpha = (39\text{ kg}\cdot\text{m}^2)(8.0\pi\text{ rad/s}^2) = 0.99\text{ kN}\cdot\text{m}$$

10.7 [III] As shown in Fig. 10-2, a constant force of 40 N is applied tangentially to the rim of a wheel with 20-cm radius. The wheel has a moment of inertia of 30 kg·m^2. Find (*a*) the angular acceleration, (*b*) the angular speed after 4.0 s from rest, and (*c*) the number of revolutions made in that 4.0 s. (*d*) Show that the work done on the wheel in the 4.0 s is equal to the KE_r of the wheel after 4.0 s.

(*a*) Using $\tau = I\alpha$, we have

$$(40\text{ N})(0.20\text{ m}) = (30\text{ kg}\cdot\text{m}^2)\alpha$$

from which $\alpha = 0.267$ rad/s^2 or 0.27 rad/s^2.

(*b*) We use $\omega_f = \omega_i + \alpha t$ to find the final angular speed,

$$\omega_f = 0 + (0.267\text{ rad/s}^2)(4.0\text{ s}) = 1.07\text{ rad/s} = 1.1\text{ rad/s}$$

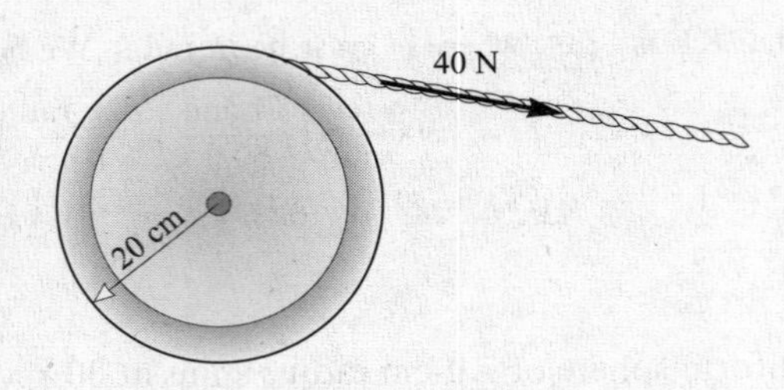

Fig. 10-2

(*c*) Because $\theta = \omega_{av}t = \frac{1}{2}(\omega_f + \omega_i)t$, we have

$$\theta = \tfrac{1}{2}(1.07 \text{ rad/s})(4.0 \text{ s}) = 2.14 \text{ rad}$$

which is equivalent to 0.34 rev.

(*d*) We know that work = torque $\times\ \theta$, and so

$$\text{Work} = (40 \text{ N} \times 0.20 \text{ m})(2.14 \text{ rad}) = 17 \text{ J}$$

Notice that radian measure must be used. The final KE_r is $\frac{1}{2}I\omega_f^2$ and so

$$\text{KE}_r = \tfrac{1}{2}(30 \text{ kg}\cdot\text{m}^2)(1.07 \text{ rad/s})^2 = 17 \text{ J}$$

The work done equals KE_r.

10.8 [II] The wheel on a grinder is a uniform 0.90-kg disk of 8.0-cm radius. It coasts uniformly to rest from 1400 rpm in a time of 35 s. How large a friction torque slows its motion?

Let's first find α from the change in ω; then we can use $\tau = I\alpha$ to find τ. We know that $f = 1400 \text{ rev/min} = 23.3 \text{ rev/s}$, and since $\omega = 2\pi f$, $\omega_i = 146$ rad/s and $\omega_f = 0$. Therefore,

$$\alpha = \frac{\omega_f - \omega_i}{t} = \frac{-146 \text{ rad/s}}{35 \text{ s}} = -4.2 \text{ rad/s}^2$$

We also need I. For a uniform disk,

$$I = \tfrac{1}{2}Mr^2 = \tfrac{1}{2}(0.90 \text{ kg})(0.080 \text{ m})^2 = 2.9 \times 10^{-3} \text{ kg}\cdot\text{m}^2$$

Then $$\tau = I\alpha = (0.0029 \text{ kg}\cdot\text{m}^2)(-4.2 \text{ rad/s}^2) = -1.2 \times 10^{-2} \text{ N}\cdot\text{m}$$

10.9 [II] Rework Problem 10.8 using the relation between work and energy.

The wheel originally had KE_r, but, as the wheel slowed, this energy was lost doing friction work. We therefore write

$$\text{Initial KE}_r = \text{work done against friction torque}$$

$$\tfrac{1}{2}I\omega_i^2 = \tau\theta$$

To find θ, we note that since α = constant,

$$\theta = \omega_{av}t = \tfrac{1}{2}(\omega_i + \omega_f)t = \tfrac{1}{2}(146 \text{ rad/s})(35 \text{ s}) = 2550 \text{ rad}$$

From Problem 10.5, $I = 0.0029 \text{ kg}\cdot\text{m}^2$ and so the work-energy equation is

$$\tfrac{1}{2}(0.0029 \text{ kg}\cdot\text{m}^2)(146 \text{ rad/s})^2 = \tau(2550 \text{ rad})$$

from which $\tau = 0.012 \text{ N}\cdot\text{m}$ or $1.2 \times 10^{-2} \text{ N}\cdot\text{m}$.

10.10 [II] A flywheel has a moment of inertia of 3.8 kg·m². What constant torque is required to increase its frequency from 2.0 rev/s to 5.0 rev/s in 6.0 revolutions?

Given

$$\theta = 12\pi \text{ rad} \qquad \omega_i = 4.0\pi \text{ rad/s} \qquad \omega_f = 10\pi \text{ rad/s}$$

we can write

$$\text{Work done on wheel} = \text{change in KE}_r \text{ of wheel}$$
$$\tau\theta = \tfrac{1}{2}I\omega_f^2 - \tfrac{1}{2}I\omega_i^2$$
$$(\tau)(12\pi \text{ rad}) = \tfrac{1}{2}(3.8 \text{ kg}\cdot\text{m}^2)[(100\pi^2 - 16\pi^2)\ (\text{rad/s})^2]$$

which gives $\tau = 42$ N·m. Notice in all of these problems that radians and seconds must be used.

10.11 [III] As shown in Fig. 10-3, a mass $m = 400$ g hangs from the rim of a wheel of radius $r = 15$ cm. When released from rest, the mass falls 2.0 m in 6.5 s. Find the moment of inertia of the wheel.

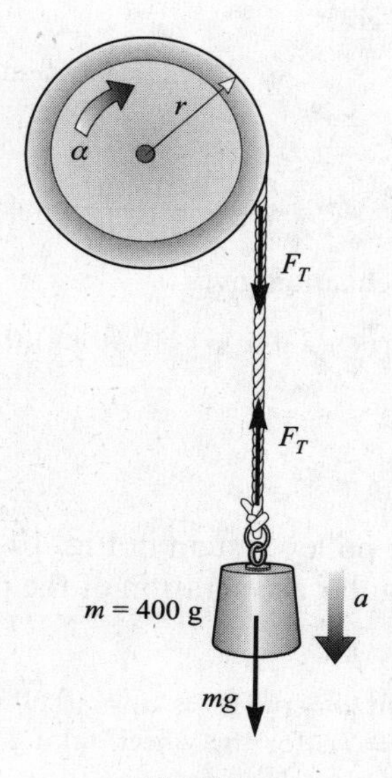

Fig. 10-3

We will write $\tau = I\alpha$ for the wheel and $F = ma$ for the mass. But first we find a using $y = v_i t + \frac{1}{2}at^2$:

$$2.0 \text{ m} = 0 + \tfrac{1}{2}a(6.5 \text{ s})^2$$

which gives $a = 0.095$ m/s². Then, from $a_T = \alpha r$,

$$\alpha = \frac{a_T}{r} = \frac{0.095 \text{ m/s}^2}{0.15 \text{ m}} = 0.63 \text{ rad/s}^2$$

The net force on the mass m is $mg - F_T$ and so $F = ma$ becomes

$$mg - F_T = ma_T$$
$$(0.40 \text{ kg})(9.81 \text{ m/s}^2) - F_T = (0.40 \text{ kg})(0.095 \text{ m/s}^2)$$

from which $F_T = 3.88$ N.

Now we write $\tau = I\alpha$ for the wheel:

$$(F_T)(r) = I\alpha \qquad \text{or} \qquad (3.88\ \text{N})(0.15\ \text{m}) = I(0.63\ \text{rad/s}^2)$$

from which $I = 0.92\ \text{kg}\cdot\text{m}^2$.

10.12 [III] Repeat Problem 10.11 using energy considerations.

Originally the mass m had $\text{PE}_\text{G} = mgh$, where $h = 2.0$ m. It loses all this PE_G, and an equal amount of KE results. Part of this KE is translational KE of the mass, and the rest is KE_r of the wheel:

$$\text{Original PE}_\text{G} = \text{final KE of } m + \text{final KE}_r \text{ of wheel}$$
$$mgh = \tfrac{1}{2}mv_f^2 + \tfrac{1}{2}I\omega_f^2$$

To find v_f, we note that $v_i = 0$, $y = 2$ m, and $t = 6.5$ s. (Observe that $a \neq g$ for the mass, because it does not fall freely.) Then

$$v_{av} = \frac{y}{t} = \frac{2.0\ \text{m}}{6.5\ \text{s}} = 0.308\ \text{m/s}$$

and $v_{av} = \frac{1}{2}(v_i + v_f)$ with $v_i = 0$ gives

$$v_f = 2v_{av} = 0.616\ \text{m/s}$$

Moreover, $v = \omega r$ gives

$$\omega_f = \frac{v_f}{r} = \frac{0.616\ \text{m/s}}{0.15\ \text{m}} = 4.1\ \text{rad/s}$$

Substitution in the energy equation gives

$$(0.40\ \text{kg})(9.81\ \text{m/s}^2)(2.0\ \text{m}) = \tfrac{1}{2}(0.40\ \text{kg})(0.62\ \text{m/s})^2 + \tfrac{1}{2}I(4.1\ \text{rad/s})^2$$

from which $I = 0.92\ \text{kg}\cdot\text{m}^2$.

10.13 [III] The moment of inertia of the pulley system in Fig. 10-4 is $I = 1.70\ \text{kg}\cdot\text{m}^2$, while $r_1 = 50$ cm and $r_2 = 20$ cm. Find the angular acceleration of the pulley system and the tensions F_{T1} and F_{T2}.

Note at the beginning that $a = \alpha r$ gives $a_1 = (0.50\ \text{m})\alpha$ and $a_2 = (0.20\ \text{m})\alpha$. We shall write $F = ma$ for both masses and $\tau = I\alpha$ for the wheel, taking the direction of motion to be the positive direction:

$$(2.0)(9.81)\ \text{N} - F_{T1} = 2a_1 \qquad \text{or} \qquad 19.6\ \text{N} - F_{T1} = (1.0\ \text{m})\alpha$$
$$F_{T2} - (1.8)(9.81)\ \text{N} = 1.8a_2 \qquad \text{or} \qquad F_{T2} - 17.6\ \text{N} = (0.36\ \text{m})\alpha$$
$$(F_{T1})(r_1) - (F_{T2})(r_2) = I\alpha \qquad \text{or} \qquad (0.50\ \text{m})F_{T1} - (0.20\ \text{m})F_{T2} = (1.70\ \text{kg}\cdot\text{m}^2)\alpha$$

These three equations have three unknowns. We solve for F_{T1} in the first equation and substitute it in the third to obtain

$$(9.81\ \text{N}\cdot\text{m}) - (0.50\ \text{m})\alpha - (0.20\ \text{m})F_{T2} = (1.70\ \text{kg}\cdot\text{m}^2)\alpha$$

We solve this equation for F_{T2} and substitute in the second equation to obtain

$$-11\alpha + 49 - 17.6 = 0.36\alpha$$

from which $\alpha = 2.8\ \text{rad/s}^2$.

We can now go back to the first equation to find $F_{T1} = 17$ N, and to the second to find $F_{T2} = 19$ N.

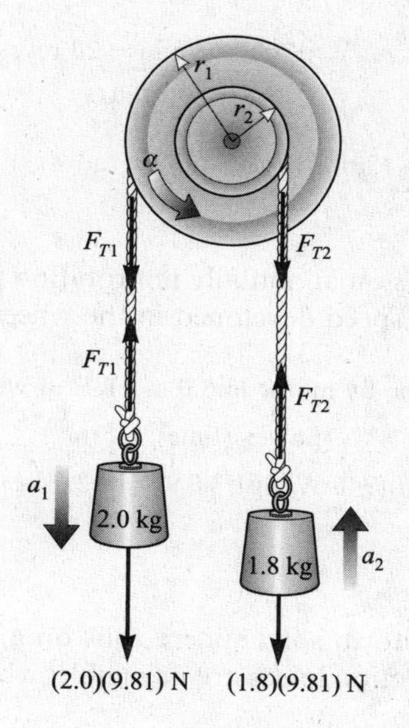

Fig. 10-4

10.14 [II] Use energy methods to find how fast the 2.0-kg mass in Fig. 10-4 is falling after it has fallen 1.5 m from rest. Use the same values for I, r_1, and r_2 as in Problem 10.10.

If the angular speed of the wheel is ω, then $v_1 = r_1\omega$ and $v_2 = r_2\omega$. As the wheel turns through an angle θ, the 2.0-kg mass falls through a distance s_1 and the 1.8-kg mass rises a distance s_2:

$$\theta = \frac{s_1}{r_1} = \frac{s_2}{r_2} \qquad \text{from which} \qquad s_2 = s_1\frac{r_2}{r_1}$$

From energy conservation, because PE_G is lost and KE is gained,

$$m_1gs_1 - m_2gs_2 = \tfrac{1}{2}m_1v_1^2 + \tfrac{1}{2}m_2v_2^2 + \tfrac{1}{2}I\omega^2$$

Since

$$s_2 = (20/50)(1.5\text{ m}) = 0.60\text{ m} \qquad v_1 = (0.50\text{ m})\,\omega \qquad v_2 = (0.20\text{ m})\,\omega$$

we can solve to find $\omega = 4.07$ rad/s. Then

$$v_1 = r_1\omega = (0.50\text{ m})(4.07\text{ rad/s}) = 2.0\text{ m/s}$$

10.15 [I] A motor runs at 20 rev/s and supplies a torque of 75 N·m. What horsepower is it delivering?

Using $\omega = 20$ rev/s $= 40\pi$ rad/s, we have

$$\text{P} = \tau\omega = (75\text{ N}\cdot\text{m})(40\pi\text{ rad/s}) = 9.4\text{ kW} = 13\text{ hp}$$

10.16 [I] The driving wheel of a belt drive attached to an electric motor has a diameter of 38 cm and operates at 1200 rpm. The tension in the belt is 130 N on the slack side, and 600 N on the tight side. Find the horsepower transmitted to the wheel by the belt.

We make use of $\text{P} = \tau\omega$. In this case two torques, due to the two parts of the belt, act on the wheel. We have

$$f = 1200 \text{ rev/min} = 20 \text{ rev/s}$$

and $$\omega = 40\pi \text{ rad/s}$$

therefore $$P = [(600 - 130)(0.19) \text{ N}\cdot\text{m}](40\pi \text{ rad/s}) = 11 \text{ kW} = 15 \text{ hp}$$

10.17 [I] A 0.75-hp motor acts for 8.0 s on an initially nonrotating wheel having a moment of inertia $2.0 \text{ kg}\cdot\text{m}^2$. Find the angular speed developed in the wheel, assuming no losses.

$$\text{Work done by motor in 8.0 s} = \text{KE of wheel after 8.0 s}$$
$$(\text{power})(\text{time}) = \tfrac{1}{2}I\omega^2$$
$$(0.75 \text{ hp})(746 \text{ W/hp})(8.0 \text{ s}) = \tfrac{1}{2}(2.0 \text{ kg}\cdot\text{m}^2)\omega^2$$

from which $\omega = 67$ rad/s.

10.18 [II] As shown in Fig. 10-5, a uniform solid sphere rolls on a horizontal surface at 20 m/s and then rolls up the incline. If friction losses are negligible, what will be the value of h where the ball stops?

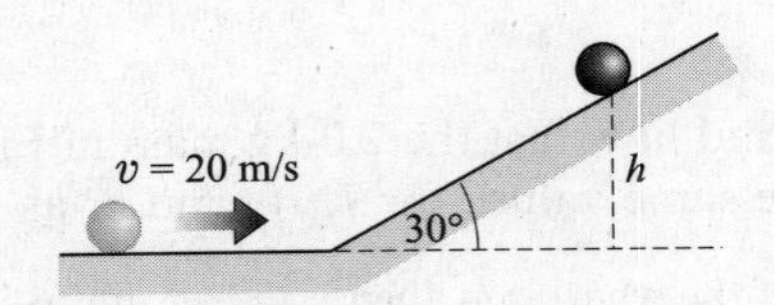

Fig. 10-5

The rotational and translational KE of the sphere at the bottom will be changed to PE_G when it stops. We therefore write

$$(\tfrac{1}{2}Mv^2 + \tfrac{1}{2}I\omega^2)_{\text{start}} = (Mgh)_{\text{end}}$$

But for a solid sphere, $I = \frac{2}{5}Mr^2$. Also, $\omega = v/r$. The above equation becomes

$$\frac{1}{2}Mv^2 + \frac{1}{2}\left(\frac{2}{5}\right)(Mr^2)\left(\frac{v}{r}\right)^2 = Mgh \qquad \text{or} \qquad \frac{1}{2}v^2 + \frac{1}{5}v^2 = (9.81 \text{ m/s}^2)h$$

Using $v = 20$ m/s gives $h = 29$ m. Notice that the answer does not depend upon the mass of the ball or the angle of the incline.

10.19 [II] Starting from rest, a hoop of 20-cm radius rolls down a hill to a point 5.0 m below its starting point. How fast is it rotating at that point?

$$PE_G \text{ at start } = (KE_r + KE_t) \text{ at end}$$
$$Mgh = \tfrac{1}{2}I\omega^2 + \tfrac{1}{2}Mv^2$$

But $I = Mr^2$ for a hoop and $v = \omega r$. The above equation becomes

$$Mgh = \tfrac{1}{2}M\omega^2 r^2 + \tfrac{1}{2}M\omega^2 r^2$$

from which $$\omega = \sqrt{\frac{gh}{r^2}} = \sqrt{\frac{(9.81 \text{ m/s}^2)(5.0 \text{ m})}{(0.20 \text{ m})^2}} = 35 \text{ rad/s}$$

10.20 [II] As a solid disk rolls over the top of a hill on a track, its speed is 80 cm/s. If friction losses are negligible, how fast is the disk moving when it is 18 cm below the top?

At the top, the disk has translational and rotational KE, plus its PE_G relative to the point 18 cm lower. At the final point, the PE_G has been transformed to more KE of rotation and translation. We therefore write, with $h = 18$ cm

$$(\mathrm{KE}_t + \mathrm{KE}_r)_{\mathrm{start}} + Mgh = (\mathrm{KE}_t + \mathrm{KE}_r)_{\mathrm{end}}$$

$$\tfrac{1}{2}Mv_i^2 + \tfrac{1}{2}I\omega_i^2 + Mgh = \tfrac{1}{2}Mv_f^2 + \tfrac{1}{2}I\omega_f^2$$

For a solid disk, $I = \frac{1}{2}Mr^2$. Also, $\omega = v/r$. Substituting these values and simplifying give

$$\tfrac{1}{2}v_i^2 + \tfrac{1}{4}v_i^2 + gh = \tfrac{1}{2}v_f^2 + \tfrac{1}{4}v_f^2$$

But $v_i = 0.80$ m/s and $h = 0.18$ m. Substitution gives $v_f = 1.7$ m/s.

10.21 [II] Find the moment of inertia of the four masses shown in Fig. 10-6 relative to an axis perpendicular to the page and extending (*a*) through point *A* and (*b*) through point *B*.

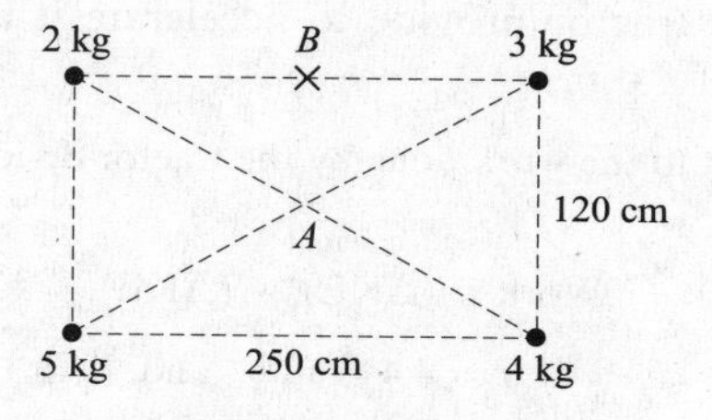

Fig. 10-6

(*a*) From the definition of moment of inertia,

$$I = m_1r_1^2 + m_2r_2^2 + \cdots + m_Nr_N^2 = (2.0\text{ kg} + 3.0\text{ kg} + 4.0\text{ kg} + 5.0\text{ kg})(r^2)$$

where r is half the length of the diagonal:

$$r = \tfrac{1}{2}\sqrt{(1.20\text{ m})^2 + (2.50\text{ m})^2} = 1.39\text{ m}$$

Thus, $I = 27\text{ kg}\cdot\text{m}^2$.

(*b*) We cannot use the parallel-axis theorem here because neither *A* nor *B* is at the center of mass. Hence we proceed as before. Because $r = 1.25$ m for the 2.0- and 3.0-kg masses, while $r = \sqrt{(1.20)^2 + (1.25)^2} = 1.733$ for the other two masses,

$$I_B = (2.0\text{ kg} + 3.0\text{ kg})(1.25\text{ m})^2 + (5.0\text{ kg} + 4.0\text{ kg})(1.733\text{ m})^2 = 33\text{ kg}\cdot\text{m}^2$$

10.22 [II] The uniform circular disk in Fig. 10-7 has mass 6.5 kg and diameter 80 cm. Compute its moment of inertia about an axis perpendicular to the page (*a*) through *G* and (*b*) through *A*.

(*a*) $$I_G = \tfrac{1}{2}Mr^2 = \tfrac{1}{2}(6.5\text{ kg})(0.40\text{ m})^2 = 0.52\text{ kg}\cdot\text{m}^2$$

(*b*) By the result of (*a*) and the parallel-axis theorem,

$$I_A = I_G + Mh^2 = 0.52\text{ kg}\cdot\text{m}^2 + (6.5\text{ kg})(0.22\text{ m})^2 = 0.83\text{ kg}\cdot\text{m}^2$$

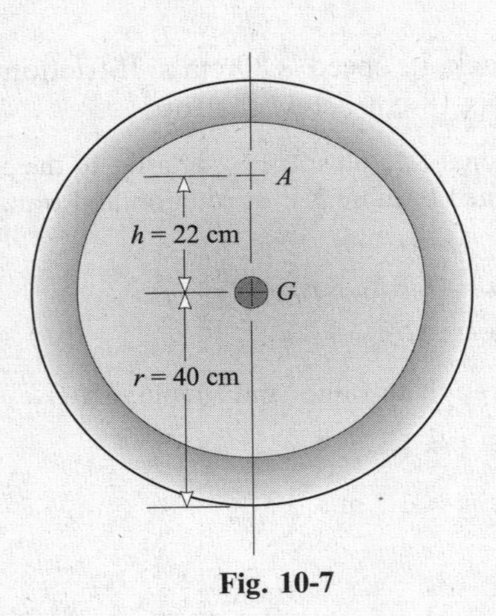

Fig. 10-7

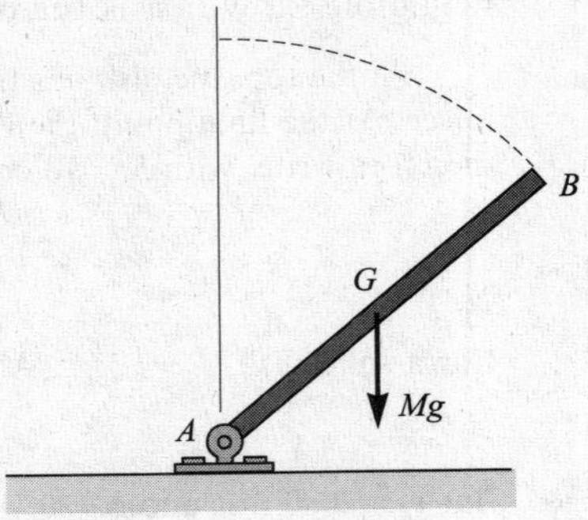

Fig. 10-8

10.23 [III] A large roller in the form of a uniform cylinder is pulled by a tractor to compact earth; it has a 1.80-m diameter and weighs 10 kN. If frictional losses can be ignored, what average horsepower must the tractor provide to accelerate it from rest to a speed of 4.0 m/s in a horizontal distance of 3.0 m?

The power is equal to the work done by the tractor divided by the time it takes. The tractor does the following work:

$$\text{Work} = (\Delta\text{KE})_r + (\Delta\text{KE})_t = \tfrac{1}{2}I\omega_f^2 + \tfrac{1}{2}mv_f^2$$

We have $v_f = 4.0$ m/s, $\omega_f = v_f/r = 4.44$ rad/s, and $m = 10\,000/9.81 = 1019$ kg. The moment of inertia of the cylinder is

$$I = \tfrac{1}{2}mr^2 = \tfrac{1}{2}(1019\ \text{kg})(0.90\ \text{m})^2 = 413\ \text{kg}\cdot\text{m}^2$$

Substituting these values, we find the work required to be 12.23 kJ.

We still need the time taken to do this work. Because the roller went 3.0 m with an average velocity $v_{av} = \tfrac{1}{2}(4+0) = 2.0$ m/s, we have

$$t = \frac{s}{v_{av}} = \frac{3.0\ \text{m}}{2.0\ \text{m/s}} = 1.5\ \text{s}$$

Then $$\text{Power} = \frac{\text{work}}{\text{time}} = \frac{12\,230\ \text{J}}{1.5\ \text{s}} = (8150\ \text{W})\left(\frac{1\ \text{hp}}{746\ \text{W}}\right) = 11\ \text{hp}$$

10.24 [III] As shown in Fig. 10-8, a thin uniform rod AB of mass M and length L is hinged at end A to the level floor. It originally stands vertically. If allowed to fall to the floor as shown, with what angular speed will it strike the floor?

The moment of inertia about a transverse axis through end A is

$$I_A = I_G + Mh^2 = \frac{1}{12}ML^2 + M\left(\frac{L}{2}\right)^2 = \frac{ML^2}{3}$$

As the rod falls to the floor, the center of mass G falls a distance $L/2$. We can write

$$\text{PE}_\text{G}\ \text{lost by rod} = \text{KE}_r\ \text{gained by rod}$$

$$Mg\left(\frac{L}{2}\right) = \frac{1}{2}\left(\frac{ML^2}{3}\right)\omega^2$$

from which $\omega = \sqrt{3g/L}$.

10.25 [I] A man stands on a freely rotating platform, as shown in Fig. 10-9. With his arms extended, his rotation frequency is 0.25 rev/s. But when he draws them in, his frequency is 0.80 rev/s. Find the ratio of his moment of inertia in the first case to that in the second.

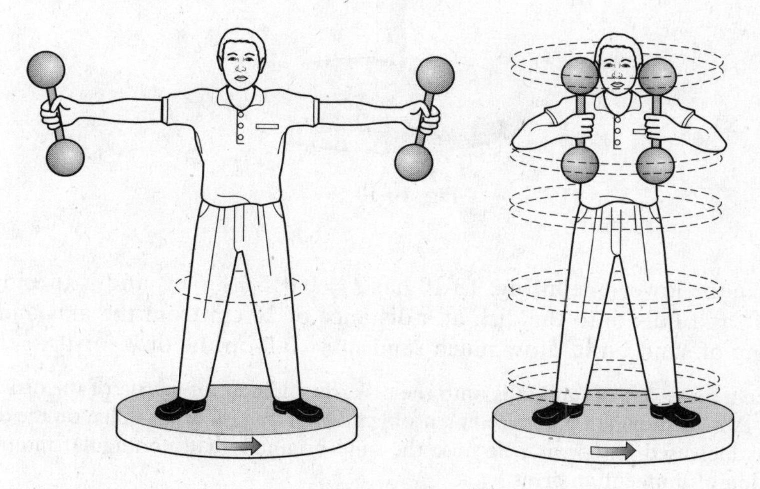

Fig. 10-9

Because there is no torque on the system (why?), the law of conservation of angular momentum tells us that

$$\text{Angular momentum before} = \text{angular momentum after}$$
$$I_i\omega_i = I_f\omega_f$$

Or, since we desire I_i/I_f,

$$\frac{I_i}{I_f} = \frac{\omega_f}{\omega_i} = \frac{0.80 \text{ rev/s}}{0.25 \text{ rev/s}} = 3.2$$

10.26 [II] A disk of moment of inertia I_1 is rotating freely with angular speed ω_1 when a second, nonrotating, disk with moment of inertia I_2 is dropped on it (Fig. 10-10). The two then rotate as a unit. Find the final angular speed.

From the law of conservation of angular momentum,

$$\text{Angular momentum before} = \text{angular momentum after}$$
$$I_1\omega_1 + I_2(0) = I_1\omega + I_2\omega$$

Solving gives
$$\omega = \frac{I_1\omega_1}{I_1 + I_2}$$

10.27 [II] A disk like the lower one in Fig. 10-10 has moment of inertia I_1 about the axis shown. What will be its new moment of inertia if a tiny mass M is set on it at a distance R from its center?

The definition of moment of inertia tells us that, for the disk plus added mass,

$$I = \sum_{\text{disk}} m_i r_i^2 + MR^2$$

where the sum extends over all the masses composing the original disk. Since the value of that sum is given as I_1, the new moment of inertia is $I = I_1 + MR^2$.

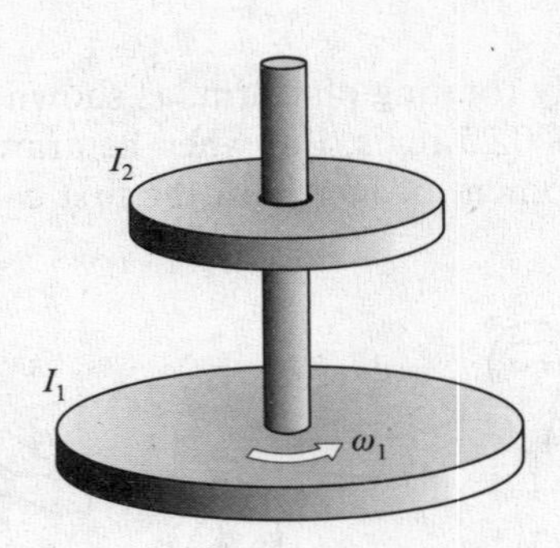

Fig. 10-10

10.28 [II] A disk like the lower one in Fig. 10-10 has $I = 0.0150 \text{ kg}\cdot\text{m}^2$ and is turning at 3.0 rev/s. A trickle of sand falls onto the disk at a distance of 20 cm from the axis and builds a 20-cm radius ring of sand on it. How much sand must fall on the disk for it to slow to 2.0 rev/s?

When a mass Δm of sand falls onto the disk, the moment of inertia of the disk is increased by an amount $r^2 \Delta m$, as shown in the preceding problem. After a mass m has fallen on the disk, its moment of inertia has increased to $I + mr^2$. Because the sand originally had no angular momentum, the law of conservation of momentum gives

$$(\text{momentum before}) = (\text{momentum after}) \qquad \text{or} \qquad I\omega_i = (I + mr^2)\omega_f$$

from which

$$m = \frac{I(\omega_i - \omega_f)}{r^2 \omega_f} = \frac{(0.0150 \text{ kg}\cdot\text{m}^2)(6.0\pi - 4.0\pi) \text{ rad/s}}{(0.040 \text{ m}^2)(4.0\pi \text{ rad/s})} = 0.19 \text{ kg}$$

Supplementary Problems

10.29 [I] A force of 200 N acts tangentially on the rim of a wheel 25 cm in radius. (*a*) Find the torque. (*b*) Repeat if the force makes an angle of 40° to a spoke of the wheel. *Ans.* (*a*) 50 N·m; (*b*) 32 N·m

10.30 [I] A certain 8.0-kg wheel has a radius of gyration of 25 cm. (*a*) What is its moment of inertia? (*b*) How large a torque is required to give it an angular acceleration of 3.0 rad/s²? *Ans.* (*a*) 0.50 kg·m²; (*b*) 1.5 N·m

10.31 [II] Determine the constant torque that must be applied to a 50-kg flywheel, with radius of gyration 40 cm, to give it a frequency of 300 rpm in 10 s if it's initially at rest. *Ans.* 25 N·m

10.32 [II] A 4.0-kg wheel of 20 cm radius of gyration is rotating at 360 rpm. The retarding frictional torque is 0.12 N·m. Compute the time it will take the wheel to coast to rest. *Ans.* 50 s

10.33 [II] Compute the rotational KE of a 25-kg wheel rotating at 6.0 rev/s if the radius of gyration of the wheel is 22 cm. *Ans.* 0.86 kJ

10.34 [II] A cord 3.0 m long is coiled around the axle of a wheel. The cord is pulled with a constant force of 40 N. When the cord leaves the axle, the wheel is rotating at 2.0 rev/s. Determine the moment of inertia of the wheel and axle. Neglect friction. (*Hint*: The easiest solution is by the energy method.) *Ans.* 1.5 kg·m²

10.35 [II] A 500-g wheel that has a moment of inertia of 0.015 $kg \cdot m^2$ is initially turning at 30 rev/s. It coasts to rest after 163 rev. How large is the torque that slowed it? *Ans.* 0.26 $N \cdot m$

10.36 [II] When 100 J of work is done upon a flywheel, its angular speed increases from 60 rev/min to 180 rev/min. What is its moment of inertia? *Ans.* 0.63 $kg \cdot m^2$

10.37 [II] A 5.0-kg wheel with radius of gyration 20 cm is to be given a frequency of 10 rev/s in 25 revolutions from rest. Find the constant unbalanced torque required. *Ans.* 2.5 $N \cdot m$

10.38 [II] An electric motor runs at 900 rpm and delivers 2.0 hp. How much torque does it deliver? *Ans.* 16 $N \cdot m$

10.39 [III] The driving side of a belt has a tension of 1600 N, and the slack side has 500 N tension. The belt turns a pulley 40 cm in radius at a rate of 300 rpm. This pulley drives a dynamo having 90 percent efficiency. How many kilowatts are being delivered by the dynamo? *Ans.* 12 kW

10.40 [III] A 25-kg wheel has a radius of 40 cm and turns freely on a horizontal axis. The radius of gyration of the wheel is 30 cm. A 1.2-kg mass hangs at the end of a cord that is wound around the rim of the wheel. This mass falls and causes the wheel to rotate. Find the acceleration of the falling mass and the tension in the cord. *Ans.* 0.77 m/s^2, 11 N

10.41 [III] A wheel and axle having a total moment of inertia of 0.0020 $kg \cdot m^2$ is caused to rotate about a horizontal axis by means of an 800-g mass attached to a cord wrapped around the axle. The radius of the axle is 2.0 cm. Starting from rest, how far must the mass fall to give the wheel a rotational rate of 3.0 rev/s? *Ans.* 5.3 cm

10.42 [II] A 20-kg solid disk ($I = \frac{1}{2}Mr^2$) rolls on a horizontal surface at the rate of 4.0 m/s. Compute its total KE. *Ans.* 0.24 kJ

10.43 [II] A 6.0-kg bowling ball ($I = 2Mr^2/5$) starts from rest and rolls down a gradual slope until it reaches a point 80 cm lower than its starting point. How fast is it then moving? Ignore friction losses. *Ans.* 3.3 m/s

10.44 [II] A tiny solid ball ($I = 2Mr^2/5$) rolls without slipping on the inside surface of a hemisphere as shown in Fig. 10-11. (The ball is much smaller than shown.) If the ball is released at A, how fast is it moving as it passes (a) point B, and (b) point C? *Ans.* (a) 2.65 m/s; (b) 2.32 m/s

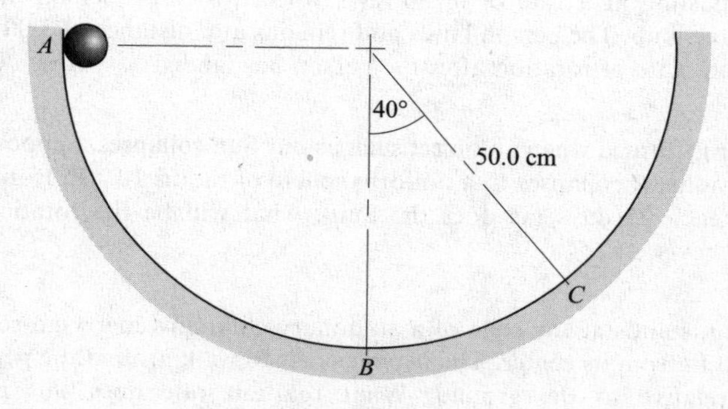

Fig. 10-11

10.45 [I] Compute the radius of gyration of a solid disk of diameter 24 cm about an axis through its center of mass and perpendicular to its face. *Ans.* 8.5 cm

10.46 [I] In Fig. 10-12 are shown four masses that are held at the corners of a square by a very light frame. What is the moment of inertia of the system about an axis perpendicular to the page (*a*) through *A* and (*b*) through *B*? *Ans.* (*a*) 1.4 kg·m^2; (*b*) 2.1 kg·m^2

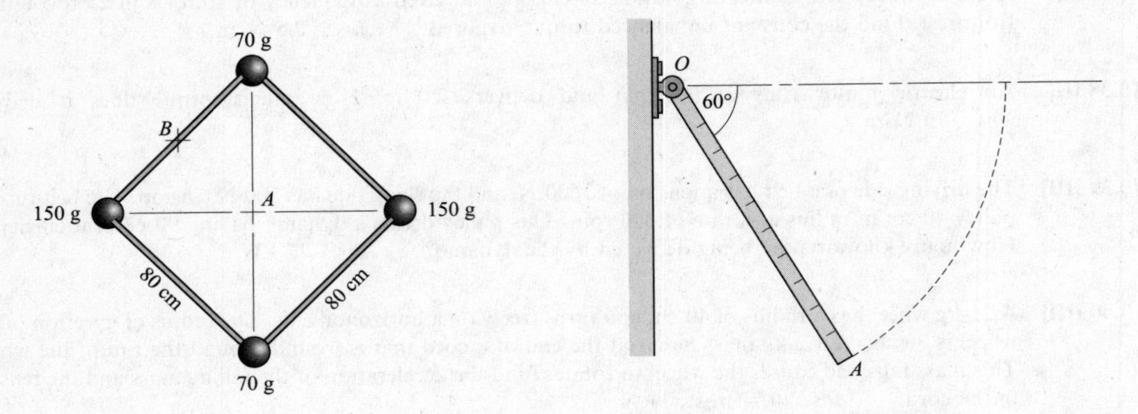

Fig. 10-12

Fig. 10-13

10.47 [I] Determine the moment of inertia (*a*) of a vertical thin hoop of mass 2 kg and radius 9 cm about a horizontal, parallel axis at its rim; (*b*) of a solid sphere of mass 2 kg and radius 5 cm about an axis tangent to the sphere. *Ans.* (*a*) $I = Mr^2 + Mr^2 = 0.03$ kg·m^2; (*b*) $I = \frac{2}{5}Mr^2 + Mr^2 = 7 \times 10^{-3}$ kg·m^2

10.48 [II] Rod *OA* in Fig. 10-13 is a meterstick. It is hinged at *O* so that it can turn in a vertical plane. It is held horizontally and then released. Compute the angular speed of the rod and the linear speed of its free end as it passes through the position shown in the figure. (*Hint*: Show that $I = mL^2/3$.) *Ans.* 5.0 rad/s, 5.0 m/s

10.49 [II] Suppose that a satellite orbits the Moon in an elliptical orbit. At its closest point to the Moon it has a speed v_c and a radius r_c from the center of the Moon. At its farthest point, it has a speed v_f and a radius r_f. Find the ratio v_c/v_f. (*Hint*: At the closest and farthest points, the relation $v = r\omega$ is valid.) *Ans.* r_f/r_c

10.50 [II] A large horizontal disk is rotating on a vertical axis through its center; for the disk, $I = 4000$ kg·m^2. The disk is coasting at a rate of 0.150 rev/s when a 90.0-kg person drops onto the disk from an overhanging tree limb. The person lands and remains at a distance of 3.00 m from the axis of rotation. What will be the rate of rotation after the person has landed? *Ans.* 0.125 rev/s

10.51 [II] A neutron star is formed when an object such as our Sun collapses. Suppose a uniform spherical star of mass *M* and radius *R* collapses to a uniform sphere of radius $10^{-5}R$. If the original star has a rotation rate of 1 rev each 25 days (as does the Sun), what will be the rotation rate of the neutron star? *Ans.* 5×10^3 rev/s

10.52 [II] A 90-kg person stands at the edge of a stationary children's merry-go-round (essentially a disk) at a distance of 5.0 m from its center. The person starts to walk around the perimeter of the disk at a speed of 0.80 m/s relative to the ground. What rotation rate does this motion give to the disk if $I_{\text{disk}} = 20\,000$ kg·m^2? (*Hint*: For the person, $I = mr^2$.) *Ans.* 0.018 rad/s

Chapter 11

Simple Harmonic Motion and Springs

THE PERIOD (T) of a cyclic motion of a system, one that is vibrating or rotating in a repetitive fashion, is the time required for the system to complete one full cycle. In the case of vibration it is the total time for the combined back and forth motion of the system. The **period** is *the number of seconds per cycle.*

THE FREQUENCY (f) is the number of vibrations made per unit time or *the number of cycles per second.* Because (T) is the time for one cycle, $f = 1/T$. The unit of frequency is the *hertz* where one cycle/s is one hertz (Hz).

THE GRAPH OF A VIBRATORY MOTION shown in Fig. 11-1 depicts the up-and-down oscillation of a mass at the end of a spring. One complete cycle is from a to b, or from c to d, or from e to f. The time taken for one cycle is T, the period.

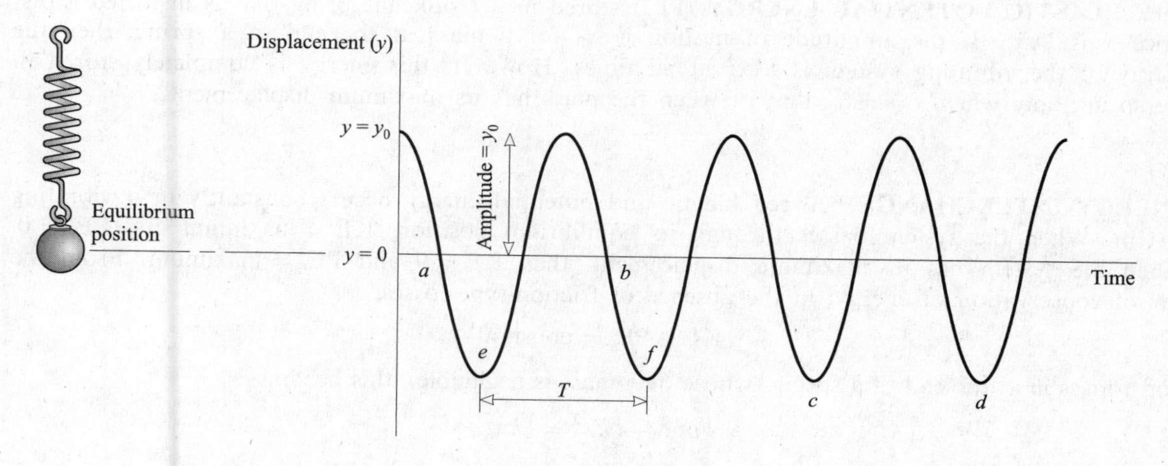

Fig. 11-1

THE DISPLACEMENT (x or y) is the distance of the vibrating object from its equilibrium position (normal rest position), i.e., from the center of its vibration path. The maximum displacement is called the **amplitude** (see Fig. 11-1).

A RESTORING FORCE is one that opposes the displacement of the system; it is necessary if vibration is to occur. In other words, a restoring force is always directed so as to push or pull the system back to its equilibrium (normal rest) position. For a mass at the end of a spring, the stretched spring pulls the mass back toward the equilibrium position, while the compressed spring pushes the mass back toward the equilibrium position.

A HOOKEAN SYSTEM (a spring, wire, rod, etc.) is one that returns to its original configuration after being distorted and then released. Moreover, when such a system is stretched a distance x (for compression, x is negative), the ***restoring force*** exerted by the spring is given by **Hooke's Law**

$$F = -kx$$

The minus sign indicates that the restoring force is always opposite in direction to the displacement. The ***spring (or elastic) constant*** k has units of N/m and is a measure of the stiffness of the spring. Most springs obey Hooke's Law for small distortions.

It is sometimes useful to express Hooke's Law in terms of F_{ext}, the external force needed to stretch the spring a given amount x. This force is the negative of the restoring force, and so

$$F_{\text{ext}} = kx$$

SIMPLE HARMONIC MOTION (SHM) is the vibratory motion which a system that obeys Hooke's Law undergoes. The motion illustrated in Fig. 11-1 is SHM. Because of the resemblance of its graph to a sine or cosine curve, SHM is frequently called *sinusoidal or* ***harmonic motion***. A central feature of SHM is that the system oscillates at a single constant frequency. That's what makes it "simple" harmonic.

THE ELASTIC POTENTIAL ENERGY (PE_e) stored in a Hookean spring that is distorted a distance x is $\frac{1}{2}kx^2$. If the amplitude of motion is x_0 for a mass at the end of a spring, then the energy of the vibrating system is $\frac{1}{2}kx_0^2$ at all times. However, this energy is completely stored in the spring only when $x = \pm x_0$, that is, when the mass has its maximum displacement.

ENERGY INTERCHANGE between kinetic and potential energy occurs constantly in a vibrating system. When the system passes through its equilibrium position, KE = maximum and $\text{PE}_e = 0$. When the system has its maximum displacement, then KE = 0 and PE_e = maximum. From the law of conservation of energy, in the absence of friction-type losses,

$$\text{KE} + \text{PE}_e = \text{constant}$$

For a mass m at the end of a spring (whose own mass is negligible), this becomes

$$\tfrac{1}{2}mv^2 + \tfrac{1}{2}kx^2 = \tfrac{1}{2}kx_0^2$$

where x_0 is the amplitude of the motion.

SPEED IN SHM is determined via the above energy equation as

$$|v| = \sqrt{(x_0^2 - x^2)\frac{k}{m}}$$

Remember that speed is always a positive quantity.

ACCELERATION IN SHM is determined via Hooke's Law, $F = -kx$, and $F = ma$; once displaced and released the restoring force drives the system. Equating these two expressions for F gives

$$a = -\frac{k}{m}x$$

The minus sign indicates that the direction of $\vec{a}$ (and $\vec{F}$) is always opposite to the direction of the displacement $\vec{x}$. Keep in mind that neither $\vec{F}$ nor $\vec{a}$ are constant.

REFERENCE CIRCLE: Suppose that a point P moves with constant speed $|v_0|$ around a circle, as shown in Fig. 11-2. This circle is called the *reference circle* for SHM. Point A is the projection of point P on the x-axis, which coincides with the horizontal diameter of the circle. The motion of point A back and forth about point O as center is SHM. The amplitude of the motion is x_0, the radius of the circle. The time taken for P to go around the circle once is the period T of the motion. The velocity, $\vec{v}_0$, of point A has a scalar x-component of

$$v_x = -|v_0| \sin\theta$$

When this quantity is positive $\vec{v}_x$ points in the positive x-direction, when it's negative $\vec{v}_x$ points in the negative x-direction.

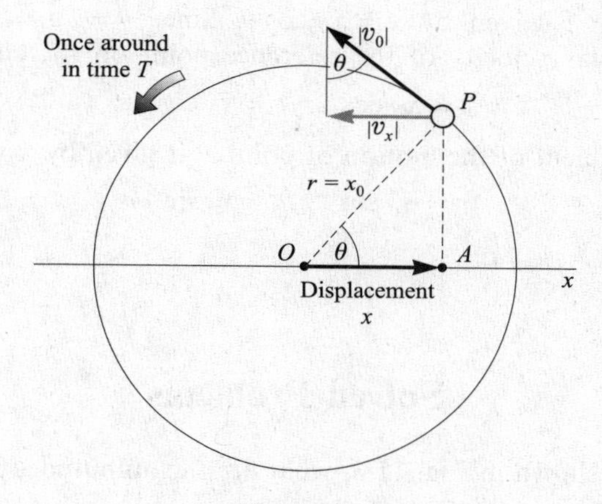

Fig. 11-2

PERIOD IN SHM: The period T of a SHM is the time taken for point P to go once around the reference circle in Fig. 11-2. Therefore,

$$T = \frac{2\pi r}{|v_0|} = \frac{2\pi x_0}{|v_0|}$$

But $|v_0|$ is the maximum speed of point A in Fig. 11-2, that is, $|v_0|$ is the value of $|v_x|$ in SHM when $x = 0$:

$$|v_x| = \sqrt{(x_0^2 - x^2)\frac{k}{m}} \qquad \text{gives} \qquad |v_0| = x_0\sqrt{\frac{k}{m}}$$

This then gives the period of SHM to be

$$T = 2\pi\sqrt{\frac{m}{k}}$$

for a Hookean spring system.

ACCELERATION IN TERMS OF T**:** Eliminating the quantity k/m between the two equations $a = -(k/m)x$ and $T = 2\pi\sqrt{m/k}$, we find

$$a = -\frac{4\pi^2}{T^2}x$$

THE SIMPLE PENDULUM very nearly undergoes SHM if its angle of swing is not too large. The period of vibration for a pendulum of length L at a location where the gravitational acceleration is g is given by

$$T = 2\pi\sqrt{\frac{L}{g}}$$

SHM can be expressed in analytic form by reference to Fig. 11-2 where we see that the horizontal displacement of point P is given by $x = x_0 \cos\theta$. Since $\theta = \omega t = 2\pi f t$, where the **angular frequency** $\omega = 2\pi f$ is the angular velocity of the reference point on the circle, we have

$$x = x_0 \cos 2\pi f t = x_0 \cos \omega t$$

Similarly, the vertical component of the motion of point P is given by

$$y = x_0 \sin 2\pi f t = x_0 \sin \omega t$$

Solved Problems

11.1 [I] For the motion shown in Fig. 11-3, what are the amplitude, period, and frequency?

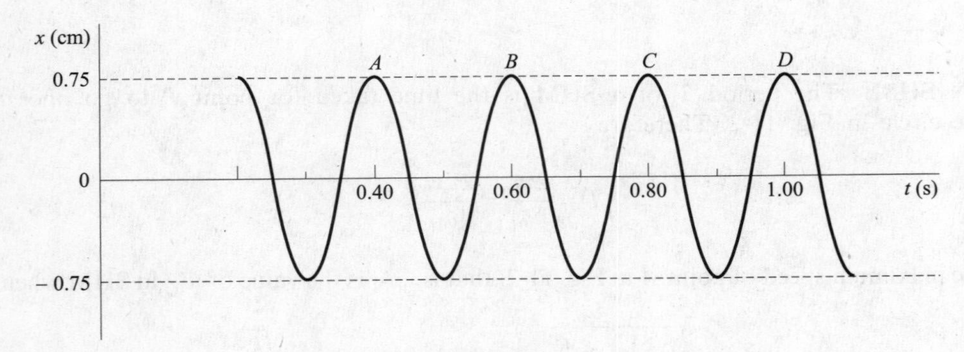

Fig. 11-3

The amplitude is the maximum displacement from the equilibrium position and so is 0.75 cm. The period is the time for one complete cycle, the time from A to B, for example. Therefore the period is 0.20 s. The frequency is

$$f = \frac{1}{T} = \frac{1}{0.20\ \text{s}} = 5.0\ \text{cycles/s} = 5.0\ \text{Hz}$$

11.2 [I] A spring makes 12 vibrations in 40 s. Find the period and frequency of the vibration.

$$T = \frac{\text{elapsed time}}{\text{vibrations made}} = \frac{40\ \text{s}}{12} = 3.3\ \text{s} \qquad f = \frac{\text{vibrations made}}{\text{elapsed time}} = \frac{12}{40\ \text{s}} = 0.30\ \text{Hz}$$

11.3 [I] When a 400-g mass is hung at the end of a vertical spring, the spring stetches 35 cm. What is the spring constant of the spring, and how much further will it stretch if an additional 400-g mass is hung from it?

We use $F_{\text{ext}} = ky$, where

$$F_{\text{ext}} = mg = (0.400\ \text{kg})(9.81\ \text{m/s}^2) = 3.92\ \text{N}$$

to get

$$k = \frac{F}{y} = \frac{3.92\ \text{N}}{0.35\ \text{m}} = 11\ \text{N/m}$$

With an additional 400-g load, the total force stretching the spring is 7.84 N. Then

$$y = \frac{F}{k} = \frac{7.84\ \text{N}}{11.2\ \text{N/m}} = 0.70\ \text{m} = 2 \times 35\ \text{cm}$$

Provided it's Hookean, each 400-g load stretches the spring by the same amount, whether or not the spring is already loaded.

11.4 [II] A 200-g mass vibrates horizontally without friction at the end of a horizontal spring for which $k = 7.0$ N/m. The mass is displaced 5.0 cm from equilibrium and released. Find (*a*) its maximum speed and (*b*) its speed when it is 3.0 cm from equilibrium. (*c*) What is its acceleration in each of these cases?

From the conservation of energy,

$$\tfrac{1}{2}kx_0^2 = \tfrac{1}{2}mv^2 + \tfrac{1}{2}kx^2$$

where $k = 7.0$ N/m, $x_0 = 0.050$ m, and $m = 0.200$ kg. Solving for $|v|$ gives

$$|v| = \sqrt{(x_0^2 - x^2)\frac{k}{m}}$$

(*a*) The speed is a maximum when $x = 0$; that is, when the mass is passing through the equilibrium position:

$$|v| = x_0\sqrt{\frac{k}{m}} = (0.050\ \text{m})\sqrt{\frac{7.0\ \text{N/m}}{0.200\ \text{kg}}} = 0.30\ \text{m/s}$$

(*b*) When $x = 0.030$ m,

$$|v| = \sqrt{\frac{7.0\ \text{N/m}}{0.200\ \text{kg}}[(0.050)^2 - (0.030)^2]\ \text{m}^2} = 0.24\ \text{m/s}$$

(*c*) By use of $F = ma$ and $F = kx$, we have

$$a = \frac{k}{m}x = (35\ \text{s}^{-2})(x)$$

which yields $a = 0$ when the mass is at $x = 0$ and $a = 1.1\ \text{m/s}^2$ when $x = 0.030$ m.

11.5 [II] A 50-g mass vibrates in SHM at the end of a spring. The amplitude of the motion is 12 cm, and the period is 1.70 s. Find: (*a*) the frequency, (*b*) the spring constant, (*c*) the maximum speed of the mass, (*d*) the maximum acceleration of the mass, (*e*) the speed when the displacement is 6.0 cm, and (*f*) the acceleration when $x = 6.0$ cm.

(*a*)
$$f = \frac{1}{T} = \frac{1}{1.70\ \text{s}} = 0.588\ \text{Hz}$$

(*b*) Since $T = 2\pi\sqrt{m/k}$,
$$k = \frac{4\pi^2 m}{T^2} = \frac{4\pi^2(0.050\ \text{kg})}{(1.70\ \text{s})^2} = 0.68\ \text{N/m}$$

(*c*)
$$|v|_0 = x_0\sqrt{\frac{k}{m}} = (0.12\ \text{m})\sqrt{\frac{0.68\ \text{N/m}}{0.050\ \text{kg}}} = 0.44\ \text{m/s}$$

(*d*) From $a = -(k/m)x$ it is seen that a has maximum magnitude when x has maximum magnitude, that is, at the endpoints $x = \pm x_0$. Thus,
$$a_0 = \frac{k}{m}x_0 = \frac{0.68\ \text{N/m}}{0.050\ \text{kg}}(0.12\ \text{m}) = 1.6\ \text{m/s}^2$$

(*e*) From $|v| = \sqrt{(x_0^2 - x^2)(k/m)}$,
$$|v| = \sqrt{\frac{[(0.12\ \text{m})^2 - (0.06\ \text{m})^2](0.68\ \text{N/m})}{(0.050\ \text{kg})}} = 0.38\ \text{m/s}$$

(*f*)
$$a = -\frac{k}{m}x = -\frac{0.68\ \text{N/m}}{0.050\ \text{kg}}(0.060\ \text{m}) = -0.82\ \text{m/s}^2$$

11.6 [II] A 50-g mass hangs at the end of a Hookean spring. When 20 g more is added to the end of the spring, it stretches 7.0 cm more. (*a*) Find the spring constant. (*b*) If the 20-g mass is now removed, what will be the period of the motion?

(*a*) Under the weight of the 50-g mass, $F_{\text{ext }1} = kx_1$, where x_1 is the original stretching of the spring. When 20 g more is added, the force becomes $F_{\text{ext }1} + F_{\text{ext }2} = k(x_1 + x_2)$, where $F_{\text{ext }2}$ is the weight of 20 g and x_2 is the stretching it causes. Subtracting the two force equations gives
$$F_{\text{ext }2} = kx_2$$
(Note that this is the same as $F_{\text{ext}} = kx$, where F_{ext} is the additional stretching force and x is the amount of stretch due to it. Hence we could have ignored the fact that the spring had the 50-g mass at its end to begin with.) Solving for k, we get
$$k = \frac{F_{\text{ext }2}}{x_2} = \frac{(0.020\ \text{kg})(9.81\ \text{m/s}^2)}{0.070\ \text{m}} = 2.8\ \text{N/m}$$

(*b*)
$$T = 2\pi\sqrt{\frac{m}{k}} = 2\pi\sqrt{\frac{0.050\ \text{kg}}{2.8\ \text{N/m}}} = 0.84\ \text{s}$$

11.7 [II] As shown in Fig. 11-4, a long, light piece of spring steel is clamped at its lower end and a 2.0 kg ball is fastened to its top end. A horizontal force of 8.0 N is required to displace the ball 20 cm to one side as shown. Assume the system to undergo SHM when released. Find (*a*) the force constant of the spring and (*b*) the period with which the ball will vibrate back and forth.

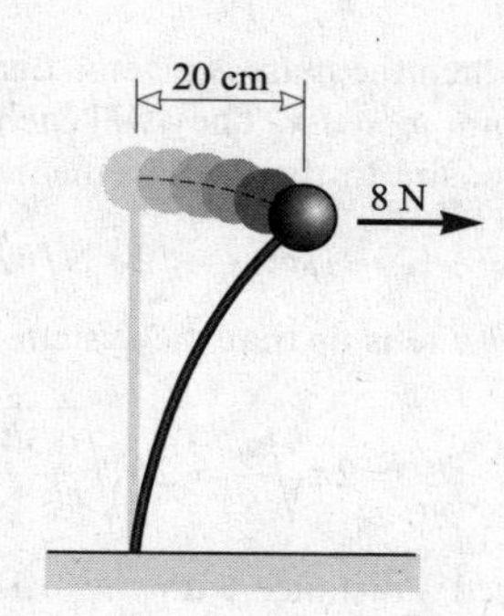

Fig. 11-4

(*a*)
$$k = \frac{\text{external force } F_{ext}}{\text{displacement } x} = \frac{8.0 \text{ N}}{0.20 \text{ m}} = 40 \text{ N/m}$$

(*b*)
$$T = 2\pi\sqrt{\frac{m}{k}} = 2\pi\sqrt{\frac{2.0 \text{ kg}}{40 \text{ N/m}}} = 1.4 \text{ s}$$

11.8 [II] When a mass m is hung on a spring, the spring stretches 6.0 cm. Determine its period of vibration if it is then pulled down a little and released.

Since

$$k = \frac{F_{ext}}{x} = \frac{mg}{0.060 \text{ m}}$$

we have
$$T = 2\pi\sqrt{\frac{m}{k}} = 2\pi\sqrt{\frac{0.060 \text{ m}}{g}} = 0.49 \text{ s}$$

Notice how the mass m cancels out of the equation.

11.9 [II] Two identical springs have $k = 20$ N/m. A 0.30-kg mass is connected to them as shown in Fig. 11-5(*a*) and (*b*). Find the period of motion for each system. Ignore friction forces.

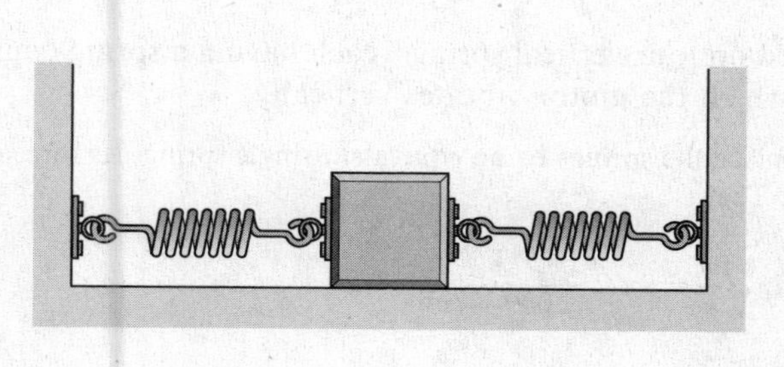

(*a*)

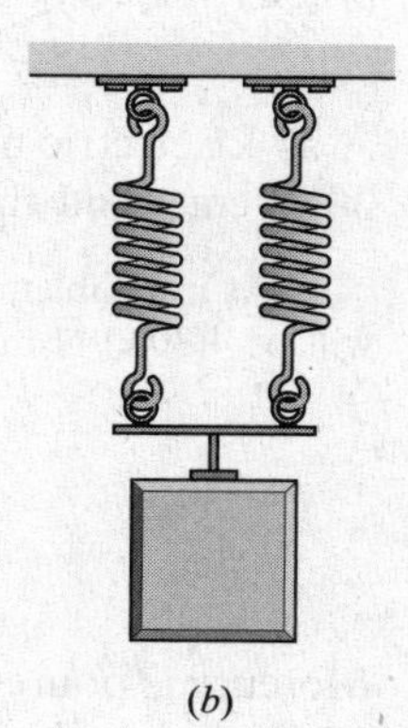

(*b*)

Fig. 11-5

(*a*) Consider what happens when the mass is given a displacement $x > 0$. One spring will be stretched x and the other will be compressed x. They will each exert a force of magnitude (20 N/m)x on the mass in the direction opposite to the displacement. Hence the total restoring force will be

$$F = -(20\ \text{N/m})x - (20\ \text{N/m})x = -(40\ \text{N/m})x$$

Comparison with $F = -kx$ tells us that the system has a spring constant of $k = 40$ N/m. Hence,

$$T = 2\pi\sqrt{\frac{m}{k}} = 2\pi\sqrt{\frac{0.30\ \text{kg}}{40\ \text{N/m}}} = 0.54\ \text{s}$$

(*b*) When the mass is displaced a distance y downward, each spring is stretched a distance y. The net restoring force on the mass is then

$$F = -(20\ \text{N/m})y - (20\ \text{N/m})y = -(40\ \text{N/m})y$$

Comparison with $F = -ky$ shows k to be 40 N/m, the same as in (*a*). Hence the period in this case is also 0.54 s.

11.10 [II] In a certain engine, a piston undergoes vertical SHM with amplitude 7.0 cm. A washer rests on top of the piston. As the motor speed is slowly increased, at what frequency will the washer no longer stay in contact with the piston?

The maximum downward acceleration of the washer will be that for free fall, g. If the piston accelerates downward faster than this, the washer will lose contact.

In SHM, the acceleration is given in terms of the displacement and the period as

$$a = -\frac{4\pi^2}{T^2}x$$

(To see this, notice that $a = -F/m = -kx/m$. But from $T = 2\pi\sqrt{m/k}$, we have $k = 4\pi^2 m/T^2$, which then gives the above expression for a.) With the upward direction chosen as positive, the largest downward (most negative) acceleration occurs for $x = +x_0 = 0.070$ m; it is

$$a_0 = \frac{4\pi^2}{T^2}(0.070\ \text{m})$$

The washer will separate from the piston when a_0 first becomes equal to g. Therefore, the critical period for the SHM, T_c, is given by

$$\frac{4\pi^2}{T_c^2}(0.070\ \text{m}) = g \qquad \text{or} \qquad T_c = 2\pi\sqrt{\frac{0.070\ \text{m}}{g}} = 0.53\ \text{s}$$

This corresponds to the frequency $f_c = 1/T_c = 1.9$ Hz. The washer will separate from the piston if the piston's frequency exceeds 1.9 cycles/s.

11.11 [II] A 20-kg electric motor is mounted on four vertical springs, each having a spring constant of 30 N/cm. Find the period with which the motor vibrates vertically.

As in Problem 11.9, we may replace the springs by an equivalent single spring. Its force constant will be 4(3000 N/m) or 12 000 N/m. Then

$$T = 2\pi\sqrt{\frac{m}{k}} = 2\pi\sqrt{\frac{20\ \text{kg}}{12\,000\ \text{N/m}}} = 0.26\ \text{s}$$

11.12 [II] Mercury is poured into a glass U-tube. Normally, the mercury stands at equal heights in the two columns, but, when disturbed, it oscillates back and forth from arm to arm. (See Fig. 11-6.) One centimeter of the mercury column has a mass of 15.0 g. Suppose the column is displaced as shown and released, and it vibrates back and forth without friction. Compute (*a*) the effective spring constant of the motion and (*b*) its period of oscillation.

(*a*) When the mercury is displaced x m from equilibrium as shown, the restoring force is the weight of the unbalanced column of length $2x$. The mercury has a mass of 1.50 kilograms per meter. The mass of the column is therefore $(2x)(1.50 \text{ kg})$, and so its weight is $mg = (29.4 \text{ kg}\cdot\text{m/s}^2)(x)$. Therefore, the restoring force is

$$F = (29.4 \text{ N/m})(x)$$

which is of the form $F = kx$ with $k = 29.4$ N/m. This is the effective spring constant for the motion.

(*b*) The period of motion is

$$T = 2\pi\sqrt{\frac{M}{k}} = 1.16\sqrt{M} \text{ s}$$

where M is the total mass of mercury in the U-tube, i.e., the total mass being moved by the restoring force.

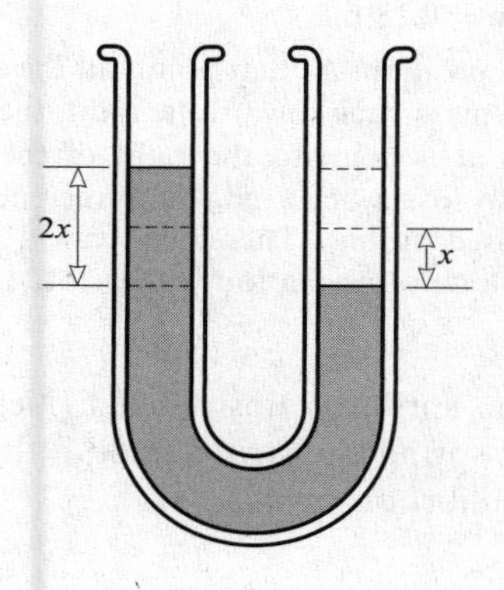

Fig. 11-6

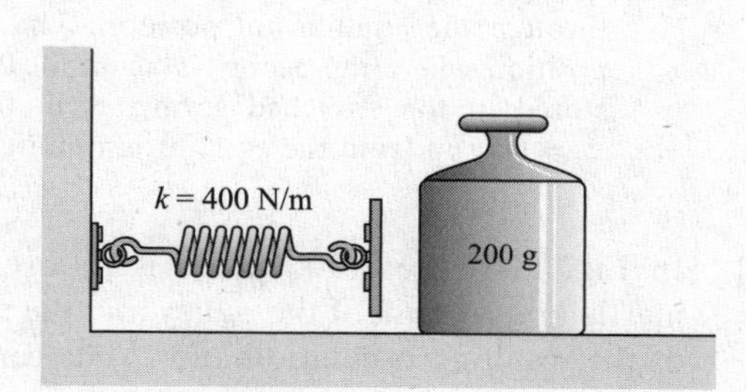

Fig. 11-7

11.13 [II] Compute the acceleration due to gravity at a place where a simple pendulum 150.3 cm long makes 100.0 cycles in 246.7 s.

We have $$T = \frac{246.7 \text{ s}}{100.0} = 2.467 \text{ s}$$

Squaring $T = 2\pi\sqrt{L/g}$ and solving for g gives us

$$g = \frac{4\pi^2}{T^2}L = 9.749 \text{ m/s}^2$$

11.14 [II] The 200-g mass shown in Fig. 11-7 is pushed to the left against the spring and compresses the spring 15 cm from its equilibrium position. The system is then released, and the mass shoots to the right. If friction can be ignored, how fast will the mass be moving as it shoots away? Assume the mass of the spring to be very small.

When the spring is compressed, energy is stored in it. That energy is $\frac{1}{2}kx_0^2$, where $x_0 = 0.15$ m. After release, this energy will be given to the mass as KE. When the spring passes through the equilibrium position, all the PE_e will be changed to KE. (Since the mass of the spring is small, its KE can be ignored.) Therefore,

$$\text{Original PE}_e = \text{final KE of mass}$$

$$\tfrac{1}{2}kx_0^2 = \tfrac{1}{2}mv^2$$

$$\tfrac{1}{2}(400\ \text{N/m})(0.15\ \text{m})^2 = \tfrac{1}{2}(0.200\ \text{kg})v^2$$

from which $v = 6.7$ m/s.

11.15 [II] Suppose that, in Fig. 11-7, the 200-g mass initially moves to the left at a speed of 8.0 m/s. It strikes the spring and becomes attached to it. (*a*) How far does it compress the spring? (*b*) If the system then oscillates back and forth, what is the amplitude of the oscillation? Ignore friction and the small mass of the spring.

(*a*) Because the spring can be considered massless, all the KE of the mass will go into compressing the spring. We can therefore write

$$\text{Original KE of mass} = \text{final PE}_e$$

$$\tfrac{1}{2}mv_0^2 = \tfrac{1}{2}kx_0^2$$

where $v_0 = 8.0$ m/s and x_0 is the maximum compression of the spring. For $m = 0.200$ kg and $k = 400$ N/m, the above relation gives $x_0 = 0.179\ \text{m} = 0.18\ \text{m}$.

(*b*) The spring compresses 0.179 m from its equilibrium position. At that point, all the energy of the spring–mass system is PE_e. As the spring pushes the mass back toward the right, the mass moves through the equilibrium position. The mass stops at a point to the right of the equilibrium position where the energy is again all PE_e. Since no losses occurred, the same energy must be stored in the stretched spring as in the compressed spring. Therefore, it will be stretched $x_0 = 0.18$ m from the equilibrium point. The amplitude of oscillation is therefore 0.18 m.

11.16 [II] In Fig. 11-8, the 2.0-kg mass is released when the spring is unstretched. Neglecting the inertia and friction of the pulley and the mass of the spring and string, find (*a*) the amplitude of the resulting oscillation and (*b*) its center or equilibrium point.

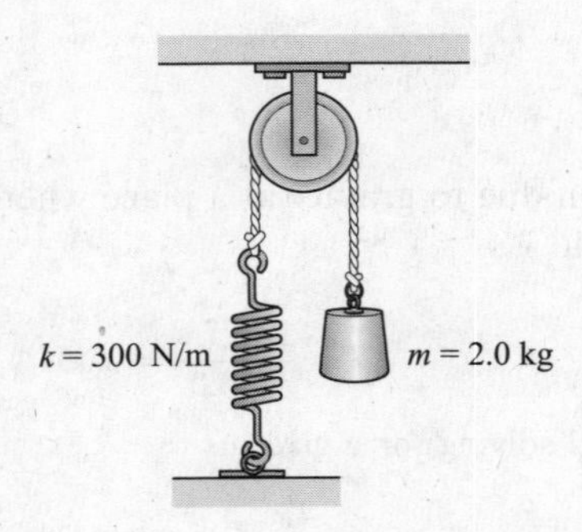

Fig. 11-8

(*a*) Suppose the mass falls a distance h before stopping. At that time, the PE_G it lost (mgh) will be stored in the spring, so that

$$mgh = \frac{1}{2}kh^2 \quad \text{or} \quad h = 2\frac{mg}{k} = 0.13\ \text{m}$$

The mass will stop in its upward motion when the energy of the system is all recovered as PE_G. Therefore it will rise 0.13 m above its lowest position. The amplitude is thus $0.13/2 = 0.065$ m.

(*b*) The center point of the motion is a distance of 0.065 m below the point from which the mass was released, i.e., a distance equal to half the total travel below the highest point.

11.17 [II] A 3.0-g particle at the end of a spring moves according to the equation $y = 0.75 \sin 63t$ where y is in centimeters and t is in seconds. Find the amplitude and frequency of its motion, its position at $t = 0.020$ s, and the spring constant.

The equation of motion is $y = y_0 \sin 2\pi ft$. By comparison, we see that the amplitude is $y_0 = 0.75$ cm. Also,

$$2\pi f = 63\ \mathrm{s}^{-1} \qquad \text{from which} \qquad f = 10\ \mathrm{Hz}$$

(Note that the argument of the sine must be dimensionless; because t is in seconds, $2\pi f$ must have the unit 1/s.)

When $t = 0.020$ s, we have

$$y = 0.75 \sin (1.26\ \mathrm{rad}) = (0.75)(0.952) = 0.71\ \mathrm{cm}$$

Notice that the argument of the sine is in radians, not degrees.

To find the spring constant, we use $f = (1/2\pi)\sqrt{k/m}$ to get

$$k = 4\pi^2 f^2 m = 11.9\ \mathrm{N/m} = 12\ \mathrm{N/m}$$

Supplementary Problems

11.18 [I] A small metal sphere weighing 10.0 N is hung from a vertical spring which comes to rest after stretching 2.0 cm. Determine the spring constant. *Ans.* 5.0×10^2 N/m

11.19 [I] How much energy is stored in a spring which has an elastic constant of 1000 N/m when it is compressed 10 cm? *Ans.* 5.0 J

11.20 [I] A pendulum is timed as it swings back and forth. The clock is started when the bob is at the left end of its swing. When the bob returns to the left end for the 90th return, the clock reads 60.0 s. What is the period of vibration? The frequency? *Ans.* 0.667 s, 1.50 Hz

11.21 [II] A 300-g mass at the end of a Hookean spring vibrates up and down in such a way that it is 2.0 cm above the tabletop at its lowest point and 16 cm above at its highest point. Its period is 4.0 s. Determine (*a*) the amplitude of vibration, (*b*) the spring constant, (*c*) the speed and acceleration of the mass when it is 9 cm above the table top, (*d*) the speed and acceleration of the mass when it is 12 cm above the tabletop. *Ans.* (*a*) 7.0 cm; (*b*) 0.74 N/m; (*c*) 0.11 m/s; zero; (*d*) 0.099 m/s, 0.074 m/s^2

11.22 [II] A coiled Hookean spring is stretched 10 cm when a 1.5-kg mass is hung from it. Suppose a 4.0-kg mass hangs from the spring and is set into vibration with an amplitude of 12 cm. Find (*a*) the force constant of the spring, (*b*) the maximum restoring force acting on the vibrating body, (*c*) the period of vibration, (*d*) the maximum speed and the maximum acceleration of the vibrating object, and (*e*) the speed and acceleration when the displacement is 9 cm. *Ans.* (*a*) 0.15 kN/m; (*b*) 18 N; (*c*) 1.0 s; (*d*) 0.73 m/s, 4.4 m/s^2; (*e*) 0.48 m/s, 3.3 m/s^2

11.23 [II] A 2.5-kg mass undergoes SHM and makes exactly 3 vibrations each second. Compute the acceleration and the restoring force acting on the body when its displacement from the equilibrium position is 5.0 cm. *Ans.* 18 m/s^2, 44 N

11.24 [II] A 300-g mass at the end of a spring oscillates with an amplitude of 7.0 cm and a frequency of 1.80 Hz. (*a*) Find its maximum speed and maximum acceleration. (*b*) What is its speed when it is 3.0 cm from its equilibrium position? *Ans.* (*a*) 0.79 m/s, 8.9 m/s^2; (*b*) 0.72 m/s

11.25 [II] A certain Hookean spring is stretched 20 cm when a given mass is hung from it. What is the frequency of vibration of the mass if pulled down a little and released? *Ans.* 1.1 Hz

11.26 [II] A 300-g mass at the end of a spring executes SHM with a period of 2.4 s. Find the period of oscillation when the 300-g mass is replaced by a 133-g mass on the same spring. *Ans.* 1.6 s

11.27 [II] With a 50-g mass at its end, a spring undergoes SHM with a frequency of 0.70 Hz. How much work is done in stretching the spring 15 cm from its unstretched length? How much energy is then stored in the spring? *Ans.* 0.011 J, 0.011 J

11.28 [II] In a situation similar to that shown in Fig. 11-7, a mass is pressed back against a light spring for which $k = 400$ N/m. The mass compresses the spring 8.0 cm and is then released. After sliding 55 cm along the flat table from the point of release, the mass comes to rest. How large a friction force opposed its motion? *Ans.* 2.3 N

11.29 [II] A 500-g mass is attached to the end of an initially unstretched vertical spring for which $k = 30$ N/m. The mass is then released, so that it falls and stretches the spring. How far will it fall before stopping? (*Hint*: The PE_G lost by the mass must appear as PE_e.) *Ans.* 33 cm

11.30 [II] A popgun uses a spring for which $k = 20$ N/cm. When cocked, the spring is compressed 3.0 cm. How high can the gun shoot a 5.0-g projectile? *Ans.* 18 m

11.31 [II] A cubical block vibrates horizontally in SHM with an amplitude of 8.0 cm and a frequency of 1.50 Hz. If a smaller block sitting on it is not to slide, what is the minimum value that the coefficient of static friction between the two blocks can have? *Ans.* 0.72

11.32 [II] Find the frequency of vibration on Mars for a simple pendulum that is 50 cm long. Objects weigh 0.40 as much on Mars as on the Earth. *Ans.* 0.45 Hz

11.33 [II] A "seconds pendulum" beats seconds; that is, it takes 1 s for half a cycle. (*a*) What is the length of a simple "seconds pendulum" at a place where $g = 9.80$ m/s^2? (*b*) What is the length there of a pendulum for which $T = 1.00$ s? *Ans.* (*a*) 99.3 cm; (*b*) 24.8 cm

11.34 [II] Show that the natural period of vertical oscillation of a mass hung on a Hookean spring is the same as the period of a simple pendulum whose length is equal to the elongation the mass causes when hung on the spring.

11.35 [II] A particle that is at the origin of coordinates at exactly $t = 0$ vibrates about the origin along the y-axis with a frequency of 20 Hz and an amplitude of 3.0 cm. Give its equation of motion in centimeters. *Ans.* $y = 3.0 \sin 125.6t$

11.36 [II] A particle vibrates according to the equation $x = 20 \cos 16t$, where x is in centimeters. Find its amplitude, frequency, and position at exactly $t = 0$ s. *Ans.* 20 cm, 2.6 Hz, $x = 20$ cm

11.37 [II] A particle oscillates according to the equation $y = 5.0 \cos 23t$, where y is in centimeters. Find its frequency of oscillation and its position at $t = 0.15$ s. *Ans.* 3.7 Hz, -4.8 cm

Chapter 12

Density; Elasticity

THE MASS DENSITY (ρ) of a material is its mass per unit volume:

$$\rho = \frac{\text{mass of body}}{\text{volume of body}} = \frac{m}{V}$$

The SI unit for mass density is kg/m^3, although g/cm^3 is also used: $1000\ \text{kg/m}^3 = 1\ \text{g/cm}^3$. The density of water is close to $1000\ \text{kg/m}^3$.

THE SPECIFIC GRAVITY (sp gr) of a substance is the ratio of the density of the substance to the density of some standard substance. The standard is usually water (at 4°C) for liquids and solids, while for gases, it is usually air.

$$\text{sp gr} = \frac{\rho}{\rho_{\text{standard}}}$$

Since sp gr is a dimensionless ratio, it has the same value for all systems of units.

ELASTICITY is the property by which a body returns to its original size and shape when the forces that deformed it are removed.

THE STRESS (σ) experienced within a solid is the magnitude of the force acting (F), divided by the area (A) over which it acts:

$$\text{Stress} = \frac{\text{force}}{\text{area of surface on which force acts}}$$

$$\sigma = \frac{F}{A}$$

The SI unit of stress is the **pascal** (Pa), where $1\ \text{Pa} = 1\ \text{N/m}^2$. Thus, if a cane supports a load the stress at any point within the cane is the load divided by the cross-sectional area at that point; the narrowest regions experience the greatest stress.

STRAIN (ε) is the fractional deformation resulting from a stress. It is measured as the ratio of the change in some dimension of a body to the original dimension in which the change occurred.

$$\text{Strain} = \frac{\text{change in dimension}}{\text{original dimension}}$$

Thus, the normal strain under an axial load is the change in length (ΔL) over the original length L_0:

$$\varepsilon = \frac{\Delta L}{L_0}$$

Strain has no units because it is a ratio of like quantities.

THE ELASTIC LIMIT of a body is the smallest stress that will produce a permanent distortion in the body. When a stress in excess of this limit is applied, the body will not return exactly to its original state after the stress is removed.

YOUNG'S MODULUS (Y) or the **modulus of elasticity**, is defined as

$$\text{Modulus of elasticity} = \frac{\text{stress}}{\text{strain}}$$

The modulus has the same units as stress which are N/m^2 or Pa. A large modulus means that a large stress is required to produce a given strain – the object is rigid.

Accordingly,
$$Y = \frac{F/A}{\Delta L/L_0} = \frac{FL_0}{A\,\Delta L}$$

Unlike the constant k in Hooke's Law, the value of Y depends only on the material of the wire or rod, and not on its dimensions or configuration. Consequently, Young's modulus is an important basic measure of the mechanical behavior of materials.

THE BULK MODULUS (B) describes the volume elasticity of a material. Suppose that a uniformly distributed compressive force acts on the surface of an object and is directed perpendicular to the surface at all points. Then if F is the force acting on and perpendicular to an area A, we define

$$\text{Pressure on } A = P = \frac{F}{A}$$

The SI unit for pressure is Pa.

Suppose that the pressure on an object of original volume V_0 is increased by an amount ΔP. The pressure increase causes a volume change ΔV, where ΔV will be negative. We then define

$$\text{Volume stress} = \Delta P \qquad \text{Volume strain} = -\frac{\Delta V}{V_0}$$

Then
$$\text{Bulk modulus} = \frac{\text{stress}}{\text{strain}}$$

$$B = -\frac{\Delta P}{\Delta V/V_0} = -\frac{V_0\,\Delta P}{\Delta V}$$

The minus sign is used so as to cancel the negative numerical value of ΔV and thereby make B a positive number. The bulk modulus has the units of pressure.

The reciprocal of the bulk modulus is called the *compressibility* K of the substance.

THE SHEAR MODULUS (S) describes the shape elasticity of a material. Suppose, as shown in Fig. 12-1, that equal and opposite tangential forces F act on a rectangular block. These *shearing forces* distort the block as indicated, but its volume remains unchanged. We define

$$\text{Shearing stress} = \frac{\text{tangential force acting}}{\text{area of surface being sheared}}$$

$$\sigma_s = \frac{F}{A}$$

$$\text{Shearing strain} = \frac{\text{distance sheared}}{\text{distance between surfaces}}$$

$$\varepsilon_s = \frac{\Delta L}{L_0}$$

Then

$$\text{Shear modulus} = \frac{\text{stress}}{\text{strain}}$$

$$S = \frac{F/A}{\Delta L/L_0} = \frac{FL_0}{A\,\Delta L}$$

Since ΔL is usually very small, the ratio $\Delta L/L_0$ is equal approximately to the shear angle γ in radians. In that case

$$S = \frac{F}{A\gamma}$$

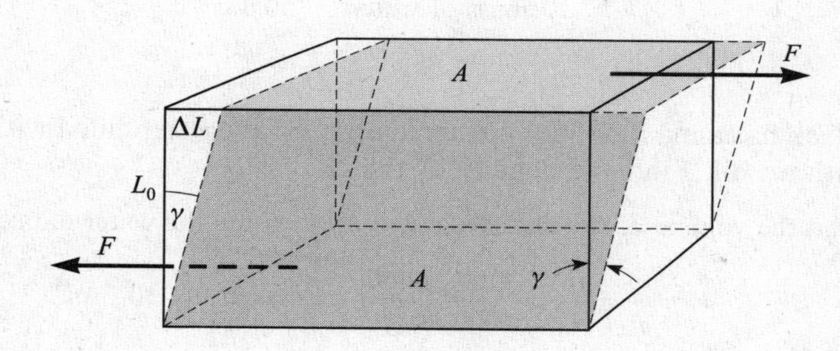

Fig. 12-1

Solved Problems

12.1 [I] Find the density and specific gravity of gasoline if 51 g occupies 75 cm³.

$$\text{Density} = \frac{\text{mass}}{\text{volume}} = \frac{0.051\ \text{kg}}{75 \times 10^{-6}\ \text{m}^3} = 6.8 \times 10^2\ \text{kg/m}^3$$

$$\text{sp gr} = \frac{\text{density of gasoline}}{\text{density of water}} = \frac{6.8 \times 10^2\ \text{kg/m}^3}{1000\ \text{kg/m}^3} = 0.68$$

or

$$\text{sp gr} = \frac{\text{mass of 75 cm}^3\ \text{gasoline}}{\text{mass of 75 cm}^3\ \text{water}} = \frac{51\ \text{g}}{75\ \text{g}} = 0.68$$

12.2 [I] What volume does 300 g of mercury occupy? The density of mercury is 13 600 kg/m³.

From $\rho = m/V$,

$$V = \frac{m}{\rho} = \frac{0.300 \text{ kg}}{13\,600 \text{ kg/m}^3} = 2.21 \times 10^{-5} \text{ m}^3 = 22.1 \text{ cm}^3$$

12.3 [I] The specific gravity of cast iron is 7.20. Find its density and the mass of 60.0 cm^3 of it.

We make use of

$$\text{sp gr} = \frac{\text{density of substance}}{\text{density of water}} \qquad \text{and} \qquad \rho = \frac{m}{V}$$

From the first equation,

$$\text{Density of iron} = (\text{sp gr})(\text{density of water}) = (7.20)(1000 \text{ kg/m}^3) = 7200 \text{ kg/m}^3$$

so

$$\text{Mass of } 60.0 \text{ cm}^3 = \rho V = (7200 \text{ kg/m}^3)(60.0 \times 10^{-6} \text{ m}^3) = 0.432 \text{ kg}$$

12.4 [I] The mass of a calibrated flask is 25.0 g when empty, 75.0 g when filled with water, and 88.0 g when filled with glycerin. Find the specific gravity of glycerin.

From the data, the mass of the glycerin in the flask is 63.0 g, while an equal volume of water has a mass of 50.0 g. Then

$$\text{sp gr} = \frac{\text{mass of glycerin}}{\text{mass of water}} = \frac{63.0 \text{ g}}{50.0 \text{ g}} = 1.26$$

12.5 [I] A calibrated flask has a mass of 30.0 g when empty, 81.0 g when filled with water, and 68.0 g when filled with an oil. Find the density of the oil.

We first find the volume of the flask from $\rho = m/V$, using the water data:

$$V = \frac{m}{\rho} = \frac{(81.0 - 30.0) \times 10^{-3} \text{ kg}}{1000 \text{ kg/m}^3} = 51.0 \times 10^{-6} \text{ m}^3$$

Then, for the oil,

$$\rho_{\text{oil}} = \frac{m_{\text{oil}}}{V} = \frac{(68.0 - 30.0) \times 10^{-3} \text{ kg}}{51.0 \times 10^{-6} \text{ m}^3} = 745 \text{ kg/m}^3$$

12.6 [I] A solid cube of aluminum is 2.00 cm on each edge. The density of aluminum is 2700 kg/m^3. Find the mass of the cube.

$$\text{Mass of cube} = \rho V = (2700 \text{ kg/m}^3)(0.0200 \text{ m})^3 = 0.0216 \text{ kg} = 21.6 \text{ g}$$

12.7 [I] What is the mass of one liter (1000 cm^3) of cottonseed oil of density 926 kg/m^3? How much does it weigh?

$$m = \rho V = (926 \text{ kg/m}^3)(1000 \times 10^{-6} \text{ m}^3) = 0.926 \text{ kg}$$

$$\text{Weight} = mg = (0.926 \text{ kg})(9.81 \text{ m/s}^2) = 9.08 \text{ N}$$

12.8 [I] An electrolytic tin-plating process gives a tin coating that is 7.50×10^{-5} cm thick. How large an area can be coated with 0.500 kg of tin? The density of tin is 7300 kg/m^3.

The volume of 0.500 kg of tin is given by $\rho = m/V$ to be

$$V = \frac{m}{\rho} = \frac{0.500 \text{ kg}}{7300 \text{ kg/m}^3} = 6.85 \times 10^{-5} \text{ m}^3$$

The volume of a film with area A and thickness t is $V = At$. Solving for A, we find

$$A = \frac{V}{t} = \frac{6.85 \times 10^{-5} \text{ m}^3}{7.50 \times 10^{-7} \text{ m}} = 91.3 \text{ m}^2$$

as the area that can be covered.

12.9 [I] A thin sheet of gold foil has an area of 3.12 cm^2 and a mass of 6.50 mg. How thick is the sheet? The density of gold is 19 300 kg/m^3.

One milligram is 10^{-6} kg, so the mass of the sheet is 6.50×10^{-6} kg. Its volume is

$$V = (\text{area}) \times (\text{thickness}) = (3.12 \times 10^{-4} \text{ m}^2)(\tau)$$

where τ is the thickness of the sheet. We equate this expression for the volume to m/ρ to get

$$(3.12 \times 10^{-4} \text{ m}^2)(\tau) = \frac{6.50 \times 10^{-6} \text{ kg}}{19\,300 \text{ kg/m}^3}$$

from which $\tau = 1.08 \times 10^{-6} \text{ m} = 1.08\ \mu\text{m}$.

12.10 [I] The mass of a liter of milk is 1.032 kg. The butterfat that it contains has a density of 865 kg/m^3 when pure, and it constitutes exactly 4 percent of the milk by volume. What is the density of the fat-free skimmed milk?

$$\text{Volume of fat in 1000 cm}^3 \text{ of milk} = 4\% \times 1000 \text{ cm}^3 = 40.0 \text{ cm}^3$$

$$\text{Mass of 40.0 cm}^3 \text{ fat} = V\rho = (40.0 \times 10^{-6} \text{ m}^3)(865 \text{ kg/m}^3) = 0.0346 \text{ kg}$$

$$\text{Density of skimmed milk} = \frac{\text{mass}}{\text{volume}} = \frac{(1.032 - 0.0346) \text{ kg}}{(1000 - 40.0) \times 10^{-6} \text{ m}^3} = 1.04 \times 10^3 \text{ kg/m}^3$$

12.11 [II] A metal wire 75.0 cm long and 0.130 cm in diameter stretches 0.0350 cm when a load of 8.00 kg is hung on its end. Find the stress, the strain, and the Young's modulus for the material of the wire.

$$\sigma = \frac{F}{A} = \frac{(8.00 \text{ kg})(9.81 \text{ m/s}^2)}{\pi(6.50 \times 10^{-4} \text{ m})^2} = 5.91 \times 10^7 \text{ N/m}^2 = 5.91 \times 10^7 \text{ Pa}$$

$$\varepsilon = \frac{\Delta L}{L_0} = \frac{0.0350 \text{ cm}}{75.0 \text{ cm}} = 4.67 \times 10^{-4}$$

$$Y = \frac{\sigma}{\varepsilon} = \frac{5.91 \times 10^7 \text{ Pa}}{4.67 \times 10^{-4}} = 1.27 \times 10^{11} \text{ Pa} = 127 \text{ GPa}$$

12.12 [II] A solid cylindrical steel column is 4.0 m long and 9.0 cm in diameter. What will be its decrease in length when carrying a load of 80 000 kg? $Y = 1.9 \times 10^{11}$ Pa.

We first find

$$\text{Cross-sectional area of column} = \pi r^2 = \pi(0.045 \text{ m})^2 = 6.36 \times 10^{-3} \text{ m}^2$$

Then, from $Y = (F/A)/(\Delta L/L_0)$ we have

$$\Delta L = \frac{FL_0}{AY} = \frac{[(8.00 \times 10^4)(9.81)\ \text{N}](4.0\ \text{m})}{(6.36 \times 10^{-3}\ \text{m}^2)(1.9 \times 10^{11}\ \text{Pa})} = 2.6 \times 10^{-3}\ \text{m} = 2.6\ \text{mm}$$

12.13 [I] Atmospheric pressure is about 1.01×10^5 Pa. How large a force does the atmosphere exert on a 2.0 cm^2 area on the top of your head?

Because $P = F/A$, where F is perpendicular to A, we have $F = PA$. Assuming that 2.0 cm^2 of your head is flat (nearly correct) and that the force due to the atmosphere is perpendicular to the surface (as it is), we have

$$F = PA = (1.01 \times 10^5\ \text{N/m}^2)(2.0 \times 10^{-4}\ \text{m}^2) = 20\ \text{N}$$

12.14 [I] A 60-kg woman stands on a light, cubical box that is 5.0 cm on each edge. The box sits on the floor. What pressure does the box exert on the floor?

$$P = \frac{F}{A} = \frac{(60)(9.81)\ \text{N}}{(5.0 \times 10^{-2}\ \text{m})^2} = 2.4 \times 10^5\ \text{N/m}^2$$

12.15 [I] The bulk modulus of water is 2.1 GPa. Compute the volume contraction of 100 mL of water when subjected to a pressure of 1.5 MPa.

From $B = -\Delta P/(\Delta V/V_0)$, we get

$$\Delta V = -\frac{V_0\,\Delta P}{B} = -\frac{(100\ \text{mL})(1.5 \times 10^6\ \text{Pa})}{2.1 \times 10^9\ \text{Pa}} = -0.071\ \text{mL}$$

12.16 [II] A box-shaped piece of gelatin dessert has a top area of 15 cm^2 and a height of 3.0 cm. When a shearing force of 0.50 N is applied to the upper surface, the upper surface displaces 4.0 mm relative to the bottom surface. What are the shearing stress, the shearing strain, and the shear modulus for the gelatin?

$$\sigma_s = \frac{\text{tangential force}}{\text{area of face}} = \frac{0.50\ \text{N}}{15 \times 10^{-4}\ \text{m}^2} = 0.33\ \text{kPa}$$

$$\varepsilon_s = \frac{\text{displacement}}{\text{height}} = \frac{0.40\ \text{cm}}{3.0\ \text{cm}} = 0.13$$

$$S = \frac{0.33\ \text{kPa}}{0.13} = 2.5\ \text{kPa}$$

12.17 [III] A 15-kg ball of radius 4.0 cm is suspended from a point 2.94 m above the floor by an iron wire of unstretched length 2.85 m. The diameter of the wire is 0.090 cm, and its Young's modulus is 180 GPa. If the ball is set swinging so that its center passes through the lowest point at 5.0 m/s, by how much does the bottom of the ball clear the floor? Discuss any approximations that you make.

Call the tension in the wire F_T when the ball is swinging through the lowest point. Since F_T must supply the centripetal force as well as balance the weight,

$$F_T = mg + \frac{mv^2}{r} = m\left(9.81 + \frac{25}{r}\right)$$

all in proper SI units. This is complicated, because r is the distance from the pivot to the center of the ball when the wire is stretched, and so it is $r_0 + \Delta r$, where r_0, the unstretched length of the pendulum, is

$$r_0 = 2.85 \text{ m} + 0.040 \text{ m} = 2.89 \text{ m}$$

and where Δr is as yet unknown. However, the unstretched distance from the pivot to the bottom of the ball is $2.85 \text{ m} + 0.080 \text{ m} = 2.93 \text{ m}$, and so the maximum possible value for Δr is

$$2.94 \text{ m} - 2.93 \text{ m} = 0.01 \text{ m}$$

We will therefore incur no more than a 1/3 percent error in r by using $r = r_0 = 2.89$ m. This gives $F_T = 277$ N. Under this tension, the wire stretches by

$$\Delta L = \frac{FL_0}{AY} = \frac{(277 \text{ N})(2.85 \text{ m})}{\pi(4.5 \times 10^{-4} \text{ m})^2(1.80 \times 10^{11} \text{ Pa})} = 6.9 \times 10^{-3} \text{ m}$$

Hence the ball misses by

$$2.94 \text{ m} - (2.85 + 0.0069 + 0.080) \text{ m} = 0.0031 \text{ m} = 3.1 \text{ mm}$$

To check the approximation we have made, we could use $r = 2.90$ m, its maximum possible value. Then we find that $\Delta L = 6.9$ mm, showing that the approximation has caused negligible error.

12.18 [III] A vertical wire 5.0 m long and of 0.008 8 cm^2 cross-sectional area has a modulus $Y = 200$ GPa. A 2.0-kg object is fastened to its end and stretches the wire elastically. If the object is now pulled down a little and released, the object undergoes vertical SHM. Find the period of its vibration.

The force constant of the wire acting as a vertical spring is given by $k = F/\Delta L$, where ΔL is the deformation produced by the force (weight) F. But, from $F/A = Y(\Delta L/L_0)$,

$$k = \frac{F}{\Delta L} = \frac{AY}{L_0} = \frac{(8.8 \times 10^{-7} \text{ m}^2)(2.00 \times 10^{11} \text{ Pa})}{5.0 \text{ m}} = 35 \text{ kN/m}$$

Then for the period we have

$$T = 2\pi\sqrt{\frac{m}{k}} = 2\pi\sqrt{\frac{2.0 \text{ kg}}{35 \times 10^3 \text{ N/m}}} = 0.047 \text{ s}$$

Supplementary Problems

12.19 [I] Find the density and specific gravity of ethyl alcohol if 63.3 g occupies 80.0 mL. *Ans.* 791 kg/m^3, 0.791

12.20 [I] Determine the volume of 200 g of carbon tetrachloride, for which sp gr = 1.60. *Ans.* 125 mL

12.21 [I] The density of aluminum is 2.70 g/cm^3. What volume does 2.00 kg occupy? *Ans.* 740 cm^3

12.22 [I] Determine the mass of an aluminum cube that is 5.00 cm on each edge. The density of aluminum is 2700 kg/m^2. *Ans.* 0.338 kg

12.23 [I] A drum holds 200 kg of water or 132 kg of gasoline. Determine for the gasoline (*a*) its sp gr and (*b*) ρ in kg/m^3. *Ans.* (*a*) 0.660; (*b*) 660 kg/m^3

12.24 [I] Air has a density of 1.29 kg/m^3 under standard conditions. What is the mass of air in a room with dimensions 10.0 m × 8.00 m × 3.00 m? *Ans.* 310 kg

12.25 [I] What is the density of the material in the nucleus of the hydrogen atom? The nucleus can be considered to be a sphere of radius 1.2×10^{-15} m, and its mass is 1.67×10^{-27} kg. The volume of a sphere is $(4/3)\pi r^3$. *Ans.* 2.3×10^{17} kg/m^3

12.26 [I] To determine the inner radius of a uniform capillary tube, the tube is filled with mercury. A column of mercury 2.375 cm long is found to have a mass of 0.24 g. What is the inner radius r of the tube? The density of mercury is 13 600 kg/m^3, and the volume of a right circular cylinder is $\pi r^2 h$. *Ans.* 0.49 mm

12.27 [I] Battery acid has sp gr = 1.285 and is 38.0 percent sulfuric acid by weight. What mass of sulfuric acid is contained in a liter of battery acid? *Ans.* 488 g

12.28 [II] A thin, semitransparent film of gold (ρ = 19 300 kg/m^3) has an area of 14.5 cm^2 and a mass of 1.93 mg. (*a*) What is the volume of 1.93 mg of gold? (*b*) What is the thickness of the film in angstroms, where 1 Å = 10^{-10} m? (*c*) Gold atoms have a diameter of about 5 Å. How many atoms thick is the film? *Ans.* (*a*) 1.00×10^{-10} m^3; (*b*) 690 Å; (*c*) 138 atoms thick

12.29 [II] In an unhealthy, dusty cement mill, there were 2.6×10^9 dust particles (sp gr = 3.0) per cubic meter of air. Assuming the particles to be spheres of 2.0 μm diameter, calculate the mass of dust (*a*) in a 20 m × 15 m × 8.0 m room and (*b*) inhaled in each average breath of 400-cm^3 volume. *Ans.* (*a*) 78 g; (*b*) 13 μg

12.30 [II] An iron rod 4.00 m long and 0.500 cm^2 in cross-section stretches 1.00 mm when a mass of 225 kg is hung from its lower end. Compute Young's modulus for the iron. *Ans.* 176 GPa

12.31 [II] A load of 50 kg is applied to the lower end of a steel rod 80 cm long and 0.60 cm in diameter. How much will the rod stretch? Y = 190 GPa for steel. *Ans.* 73 μm

12.32 [II] A platform is suspended by four wires at its corners. The wires are 3.0 m long and have a diameter of 2.0 mm. Young's modulus for the material of the wires is 180 GPa. How far will the platform drop (due to elongation of the wires) if a 50-kg load is placed at the center of the platform? *Ans.* 0.65 mm

12.33 [II] Determine the fractional change in volume as the pressure of the atmosphere (1×10^5 Pa) around a metal block is reduced to zero by placing the block in vacuum. The bulk modulus for the metal is 125 GPa. *Ans.* 8×10^{-7}

12.34 [II] Compute the volume change of a solid copper cube, 40 mm on each edge, when subjected to a pressure of 20 MPa. The bulk modulus for copper is 125 GPa. *Ans.* -10 mm^3

12.35 [II] The compressibility of water is 5.0×10^{-10} m^2/N. Find the decrease in volume of 100 mL of water when subjected to a pressure of 15 MPa. *Ans.* 0.75 mL

12.36 [II] Two parallel and opposite forces, each 4000 N, are applied tangentially to the upper and lower faces of a cubical metal block 25 cm on a side. Find the angle of shear and the displacement of the upper surface relative to the lower surface. The shear modulus for the metal is 80 GPa. *Ans.* 8.0×10^{-7} rad, 2.0×10^{-7} m

12.37 [II] A 60-kg motor sits on four cylindrical rubber blocks. Each cylinder has a height of 3.0 cm and a cross-sectional area of 15 cm^2. The shear modulus for this rubber is 2.0 MPa. (*a*) If a sideways force of 300 N is applied to the motor, how far will it move sideways? (*b*) With what frequency will the motor vibrate back and forth sideways if disturbed? *Ans.* (*a*) 0.075 cm; (*b*) 13 Hz

Chapter 13

Fluids at Rest

THE AVERAGE PRESSURE on a surface of area A is defined as force divided by area, where it is stipulated that the force must be perpendicular (normal) to the area:

$$\text{Average pressure} = \frac{\text{force acting normal to an area}}{\text{area over which the force is distributed}}$$

$$P = \frac{F}{A}$$

Recall that the SI unit for pressure is the ***pascal*** (Pa), and 1 Pa = 1 N/m^2.

STANDARD ATMOSPHERIC PRESSURE (P_A) is 1.01×10^5 Pa, and this is equivalent to 14.7 lb/in.^{2}. Other units of pressure are

$$1 \text{ atmosphere (atm)} = 1.013 \times 10^5 \text{ Pa}$$
$$1 \text{ torr} = 1 \text{ mm of mercury (mmHg)} = 133.32 \text{ Pa}$$
$$1 \text{ lb/in.}^2 = 6.895 \text{ kPa}$$

THE HYDROSTATIC PRESSURE (P) due to a column of fluid of height h and mass density ρ is

$$P = \rho g h$$

PASCAL'S PRINCIPLE: When the pressure on any part of a confined fluid (liquid or gas) is changed, the pressure on every other part of the fluid is also changed by the same amount.

ARCHIMEDES' PRINCIPLE: A body wholly or partly immersed in a fluid is buoyed up by a force equal to the weight of the fluid it displaces. The buoyant force can be considered to act vertically upward through the center of gravity of the displaced fluid.

$$F_B = \text{buoyant force} = \text{weight of displaced fluid}$$

The buoyant force on an object of volume V that is *totally* immersed in a fluid of density ρ_f is $\rho_f Vg$, and the weight of the object is $\rho_0 Vg$, where ρ_0 is the density of the object. Therefore, the net upward force on the submerged object is

$$F_{\text{net}}(\text{upward}) = Vg(\rho_f - \rho_0)$$

Solved Problems

13.1 [I] An 80-kg metal cylinder, 2.0 m long and with each end of area 25 cm^2, stands vertically on one end. What pressure does the cylinder exert on the floor?

$$P = \frac{\text{normal force}}{\text{area}} = \frac{(80\text{ kg})(9.81\text{ m/s}^2)}{25 \times 10^{-4}\text{ m}^2} = 3.1 \times 10^5\text{ Pa}$$

13.2 [I] Atmospheric pressure is about 1.0×10^5 Pa. How large a force does the still air in a room exert on the inside of a window pane that is 40 cm × 80 cm?

The atmosphere exerts a force normal to any surface placed in it. Consequently, the force on the window pane is perpendicular to the pane and is given by

$$F = PA = (1.0 \times 10^5\text{ N/m}^2)(0.40 \times 0.80\text{ m}^2) = 3.2 \times 10^4\text{ N}$$

Of course, a nearly equal force due to the atmosphere on the outside keeps the window from breaking.

13.3 [I] Find the pressure due to the fluid at a depth of 76 cm in still (*a*) water ($\rho_w = 1.00\text{ g/cm}^3$) and (*b*) mercury ($\rho = 13.6\text{ g/cm}^3$).

(*a*) $$P = \rho_w gh = (1000\text{ kg/m}^3)(9.81\text{ m/s}^2)(0.76\text{ m}) = 7450\text{ N/m}^2 = 7.5\text{ kPa}$$

(*b*) $$P = \rho gh = (13\,600\text{ kg/m}^3)(9.81\text{ m/s}^2)(0.76\text{ m}) = 1.01 \times 10^5\text{ N/m}^2 \approx 1.0\text{ atm}$$

13.4 [I] When a submarine dives to a depth of 120 m, to how large a total pressure is its exterior surface subjected? The density of seawater is about 1.03 g/cm^3.

$$\begin{aligned} P &= \text{atmospheric pressure} + \text{pressure of water} \\ &= 1.01 \times 10^5\text{ N/m}^2 + \rho gh = 1.01 \times 10^5\text{ N/m}^2 + (1030\text{ kg/m}^3)(9.81\text{ m/s}^2)(120\text{ m}) \\ &= 1.01 \times 10^5\text{ N/m}^2 + 12.1 \times 10^5\text{ N/m}^2 = 13.1 \times 10^5\text{ N/m}^2 = 1.31\text{ MPa} \end{aligned}$$

13.5 [I] How high would water rise in the pipes of a building if the water pressure gauge shows the pressure at the ground floor to be 270 kPa (about 40 lb/in.^{2})?

Water pressure gauges read the excess pressure due to the water, that is, the difference between the pressure in the water and the pressure of the atmosphere. The water pressure at the bottom of the highest column that can be supported is 270 kPa. Therefore, $P = \rho_w gh$ gives

$$h = \frac{P}{\rho_w g} = \frac{2.70 \times 10^5\text{ N/m}^2}{(1000\text{ kg/m}^3)(9.81\text{ m/s}^2)} = 27.5\text{ m}$$

13.6 [I] A reservoir dam holds an 8.00-km^2 lake behind it. Just behind the dam, the lake is 12.0 m deep. What is the water pressure (*a*) at the base of the dam and (*b*) at a point 3.0 m down from the lake's surface?

The area of the lake behind the dam has no effect on the pressure against the dam. At any point, $P = \rho_w gh$.

(*a*) $$P = (1000\text{ kg/m}^3)(9.81\text{ m/s}^2)(12.0\text{ m}) = 118\text{ kPa}$$

(*b*) $$P = (1000\text{ kg/m}^3)(9.81\text{ m/s}^2)(3.0\text{ m}) = 29\text{ kPa}$$

13.7 [II] A weighted piston confines a fluid of density ρ in a closed container, as shown in Fig. 13.1. The combined weight of piston and weight is 200 N, and the cross-sectional area of the piston is $A = 8.0\ \text{cm}^2$. Find the total pressure at point B if the fluid is mercury and $h = 25$ cm ($\rho_{\text{Hg}} = 13\,600\ \text{kg/m}^3$). What would an ordinary pressure gauge read at B?

Recall what Pascal's principle tells us about the pressure applied to the fluid by the piston and atmosphere: This added pressure is applied at all points within the fluid. Therefore the total pressure at B is composed of three parts:

$$\text{Pressure of the atmosphere} = 1.0 \times 10^5\ \text{Pa}$$

$$\text{Pressure due to the piston and weight} = \frac{F_W}{A} = \frac{200\ \text{N}}{8.0 \times 10^{-4}\ \text{m}^2} = 2.5 \times 10^5\ \text{Pa}$$

$$\text{Pressure due to the height } h \text{ of fluid} = h\rho g = 0.33 \times 10^5\ \text{Pa}$$

In this case, the pressure of the fluid itself is relatively small. We have

$$\text{Total pressure at } B = 3.8 \times 10^5\ \text{Pa}$$

The gauge pressure does not include atmospheric pressure. Therefore,

$$\text{Gauge pressure at } B = 2.8 \times 10^5\ \text{Pa}$$

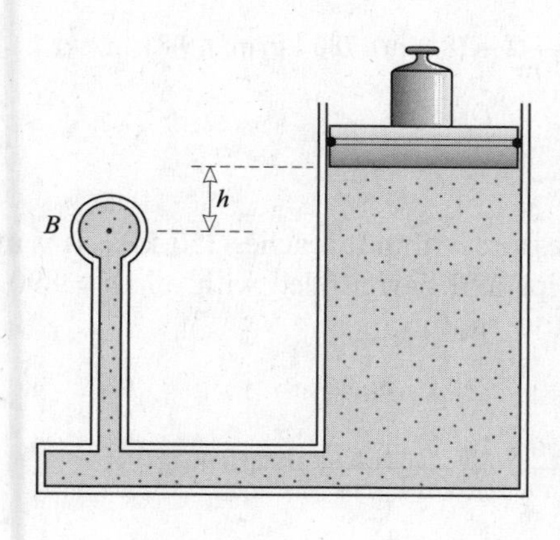

Fig. 13-1

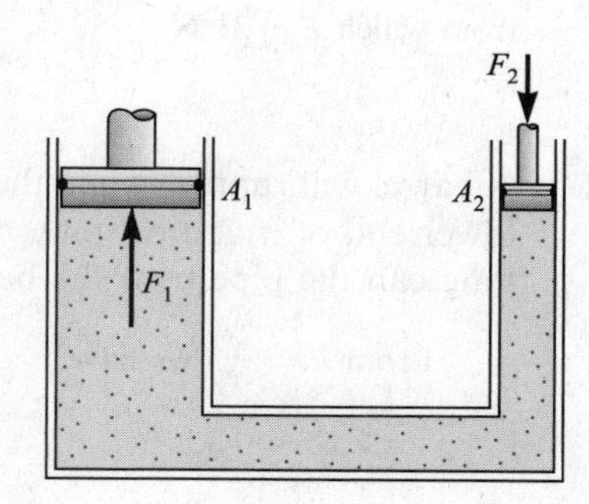

Fig. 13-2

13.8 [I] In a hydraulic press such as the one shown in Fig. 13-2, the large piston has cross-sectional area $A_1 = 200\ \text{cm}^2$ and the small piston has cross-sectional area $A_2 = 5.0\ \text{cm}^2$. If a force of 250 N is applied to the small piston, find the force F_1 on the large piston.

By Pascal's principle,

$$\text{Pressure under large piston} = \text{pressure under small piston} \qquad \text{or} \qquad \frac{F_1}{A_1} = \frac{F_2}{A_2}$$

so that

$$F_1 = \frac{A_1}{A_2} F_2 = \frac{200}{5.0} 250\ \text{N} = 10\ \text{kN}$$

13.9 [II] For the system shown in Fig. 13-3, the cylinder on the left, at L, has a mass of 600 kg and a cross-sectional area of 800 cm^2. The piston on the right, at S, has a cross-sectional area of 25 cm^2 and a negligible weight. If the apparatus is filled with oil ($\rho = 0.78\ \text{g/cm}^3$), find the force F required to hold the system in equilibrium as shown.

The pressures at points H_1 and H_2 are equal because they are at the same level in a single connected fluid. Therefore,

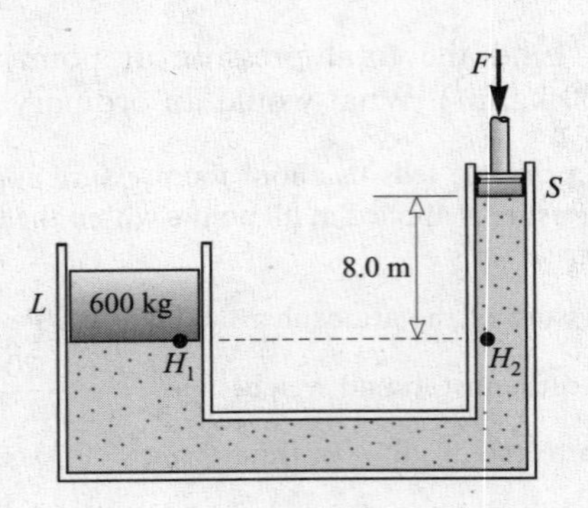

Fig. 13-3

$$\text{Pressure at } H_1 = \text{pressure at } H_2$$

$$\begin{pmatrix}\text{pressure due to}\\ \text{left piston}\end{pmatrix} = \begin{pmatrix}\text{pressure due to } F\\ \text{and right piston}\end{pmatrix} + (\text{pressure due to 8.0 m of oil})$$

$$\frac{(600)(9.81)\ \text{N}}{0.0800\ \text{m}^2} = \frac{F}{25 \times 10^{-4}\ \text{m}^2} + (8.0\ \text{m})(780\ \text{kg/m}^3)(9.81\ \text{m/s}^2)$$

from which $F = 31$ N.

13.10 [I] A barrel will rupture when the gauge pressure within it reaches 350 kPa. It is attached to the lower end of a vertical pipe, with the pipe and barrel filled with oil ($\rho = 890$ kg/m^3). How long can the pipe be if the barrel is not to rupture?

From $P = \rho g h$ we have

$$h = \frac{P}{\rho g} = \frac{350 \times 10^3\ \text{N/m}^2}{(9.81\ \text{m/s}^2)(890\ \text{kg/m}^3)} = 40.1\ \text{m}$$

13.11 [II] A vertical test tube has 2.0 cm of oil ($\rho = 0.80$ g/cm^3) floating on 8.0 cm of water. What is the pressure at the bottom of the tube due to the fluid in it?

$$P = \rho_1 g h_1 + \rho_2 g h_2 = (800\ \text{kg/m}^3)(9.81\ \text{m/s}^2)(0.020\ \text{m}) + (1000\ \text{kg/m}^3)(9.81\ \text{m/s}^2)(0.080\ \text{m})$$
$$= 0.94\ \text{kPa}$$

13.12 [II] As shown in Fig. 13-4, a column of water 40 cm high supports a 31-cm column of an unknown fluid. What is the density of the unknown fluid?

The pressures at point A due to the two fluids must be equal (or the one with the higher pressure would push the lower-pressure fluid away). Therefore,

$$\text{Pressure due to water} = \text{pressure due to unknown fluid}$$
$$\rho_1 g h_1 = \rho_2 g h_2$$

from which
$$\rho_2 = \frac{h_1}{h_2}\rho_1 = \frac{40}{31}(1000\ \text{kg/m}^3) = 1290\ \text{kg/m}^3 = 1.3 \times 10^3\ \text{kg/m}^3$$

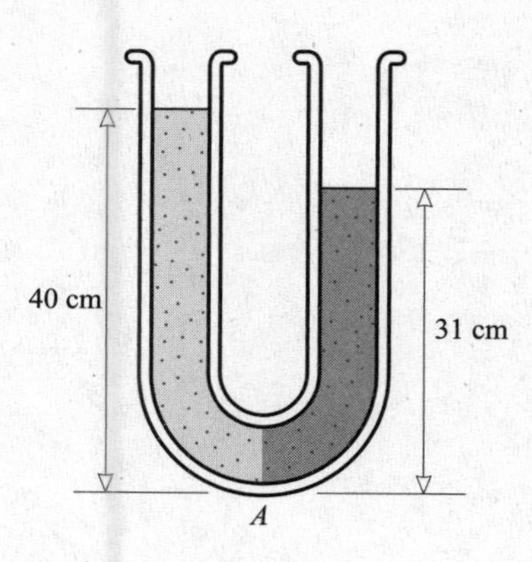

Fig. 13-4

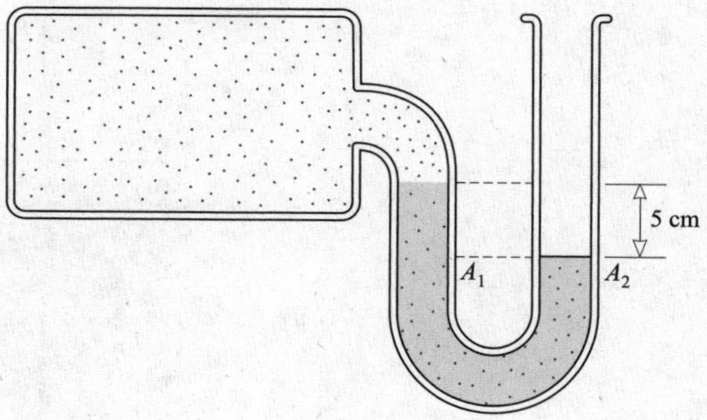

Fig. 13-5

13.13 [II] The U-tube device connected to the tank in Fig. 13-5 is called a *manometer*. As you can see, the mercury in the tube stands higher in one side than the other. What is the pressure in the tank if atmospheric pressure is 76 cm of mercury? The density of mercury is 13.6 g/cm^3.

$$\text{Pressure at } A_1 = \text{pressure at } A_2$$
$$(P \text{ in tank}) + (P \text{ due to 5 cm mercury}) = (P \text{ due to atmosphere})$$
$$P + (0.05\text{ m})(13\,600\text{ kg/m}^3)(9.81\text{ m/s}^2) = (0.76\text{ m})(13\,600\text{ kg/m}^3)(9.81\text{ m/s}^2)$$

from which $P = 95$ kPa.

Or, more simply perhaps, we could note that the pressure in the tank is 5.0 cm of mercury *lower* than atmospheric. So the pressure is 71 cm of mercury, which is 94.6 kPa.

13.14 [II] The mass of a block of aluminum is 25.0 g. (*a*) What is its volume? (*b*) What will be the tension in a string that suspends the block when the block is totally submerged in water? The density of aluminum is 2700 kg/m^3.

This problem is basically about buoyant force. (*a*) Because $\rho = m/V$, we have

$$V = \frac{m}{\rho} = \frac{0.0250\text{ kg}}{2700\text{ kg/m}^3} = 9.26 \times 10^{-6}\text{ m}^3 = 9.26\text{ cm}^3$$

(*b*) The block displaces 9.26×10^{-6} m^3 of water when submerged, so the buoyant force on it is

$$F_B = \text{weight of displaced water} = (\text{volume})(\rho \text{ of water})(g)$$
$$= (9.26 \times 10^{-6}\text{ m}^3)(1000\text{ kg/m}^3)(9.81\text{ m/s}^2) = 0.0908\text{ N}$$

The tension in the supporting cord plus the buoyant force must equal the weight of the block if it is to be in equilibrium (see Fig. 13-6). That is, $F_T + F_B = mg$, from which

$$F_T = mg - F_B = (0.0250\text{ kg})(9.81\text{ m/s}^2) - 0.0908\text{ N} = 0.154\text{ N}$$

13.15 [II] Using a scale, a piece of alloy has a measured mass of 86 g in air and 73 g when immersed in water. Find its volume and its density.

The apparent change in measured mass is due to the buoyant force of the water. Figure 13-6 shows the situation when the object is in water. From the figure, $F_B + F_T = mg$, so

$$F_B = (0.086)(9.81)\text{ N} - (0.073)(9.81)\text{ N} = (0.013)(9.81)\text{ N}$$

But F_B must be equal to the weight of the displaced water.

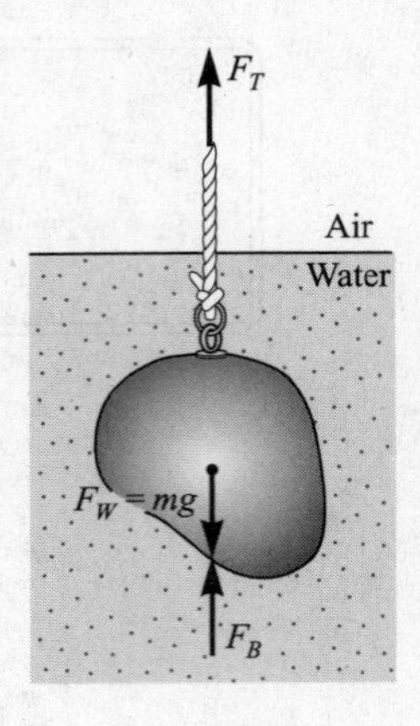

Fig. 13-6

$$F_B = \text{weight of water} = (\text{mass of water})(g)$$
$$= (\text{volume of water})(\text{density of water})(g)$$

or $$(0.013)(9.81)\text{ N} = V(1000\text{ kg/m}^3)(9.81\text{ m/s}^2)$$

from which $V = 1.3 \times 10^{-5}\text{ m}^3$. This is also the volume of the piece of alloy. Therefore,

$$\rho \text{ of alloy} = \frac{\text{mass}}{\text{volume}} = \frac{0.086\text{ kg}}{1.3 \times 10^{-5}\text{ m}^3} = 6.6 \times 10^3\text{ kg/m}^3$$

13.16 [II] A solid aluminum cylinder with $\rho = 2700\text{ kg/m}^3$ has a measured mass of 67 g in air and 45 g when immersed in turpentine. Determine the density of turpentine.

The F_B acting on the immersed cylinder is

$$F_B = (0.067 - 0.045)(9.81)\text{ N} = (0.022)(9.81)\text{ N}$$

This is also the weight of the displaced turpentine.

The volume of the cylinder is, from $\rho = m/V$,

$$V \text{ of cylinder} = \frac{m}{\rho} = \frac{0.067\text{ kg}}{2700\text{ kg/m}^3} = 2.5 \times 10^{-5}\text{ m}^3$$

This is also the volume of the displaced turpentine. We therefore have, for the turpentine,

$$\rho = \frac{\text{mass}}{\text{volume}} = \frac{(\text{weight})/g}{\text{volume}} = \frac{(0.022)(9.81)/(9.81)}{2.48 \times 10^{-5}}\ \frac{\text{kg}}{\text{m}^3} = 8.9 \times 10^2\text{ kg/m}^3$$

13.17 [II] A glass stopper has a mass of 2.50 g when measured in air, 1.50 g in water, and 0.70 g in sulfuric acid. What is the density of the acid? What is its specific gravity?

The F_B on the stopper in water is $(0.002\,50 - 0.001\,50)(9.81)$ N. This is the weight of the displaced water. Since $\rho = m/V$, or $\rho g = F_W/V$, we have

$$\text{Volume of stopper} = \text{volume of displaced water} = \frac{\text{weight}}{\rho g}$$

$$V = \frac{(0.001\,00)(9.81)\text{ N}}{(1000\text{ kg/m}^3)(9.81\text{ m/s}^2)} = 1.00 \times 10^{-6}\text{ m}^3$$

The buoyant force in acid is

$$[(2.50 - 0.70) \times 10^{-3}](9.81)\text{ N} = (0.001\,80)(9.81)\text{ N}$$

But this is equal to the weight of displaced acid, mg. Since $\rho = m/V$, and since $m = 0.001\,80$ kg and $V = 1.00 \times 10^{-6}$ m^3, we have

$$\rho \text{ of acid} = \frac{0.001\,80 \text{ kg}}{1.00 \times 10^{-6} \text{ m}^3} = 1.8 \times 10^3 \text{ kg/m}^3$$

Then, for the acid,

$$\text{sp gr} = \frac{\rho \text{ of acid}}{\rho \text{ of water}} = \frac{1800}{1000} = 1.8$$

Alternative Method

$$\text{Weight of displaced water} = [(2.50 - 1.50) \times 10^{-3}](9.81) \text{ N}$$
$$\text{Weight of displaced acid} = [(2.50 - 0.70) \times 10^{-3}](9.81) \text{ N}$$

so
$$\text{sp gr of acid} = \frac{\text{weight of displaced acid}}{\text{weight of equal volume of displaced water}} = \frac{1.80}{1.00} = 1.8$$

Then, since sp gr of acid = (ρ of acid)/(ρ of water), we have

$$\rho \text{ of acid} = (\text{sp gr of acid})(\rho \text{ of water}) = (1.8)(1000 \text{ kg/m}^3) = 1.8 \times 10^3 \text{ kg/m}^3$$

13.18 [II] The density of ice is 917 kg/m^3. What fraction of the volume of a piece of ice will be above water when floating in fresh water?

The piece of ice will float in the water, since its density is less than 1000 kg/m^3, the density of water. As it does,

$$F_B = \text{weight of displaced water} = \text{weight of piece of ice}$$

But the weight of the ice is $\rho_{ice} g V$, where V is the volume of the piece. In addition, the weight of the displaced water is $\rho_w g V'$, where V' is the volume of the displaced water. Substituting into the above equation gives

$$\rho_{ice} g V = \rho_w g V'$$
$$V' = \frac{\rho_{ice}}{\rho_w} V = \frac{917}{1000} V = 0.917V$$

The fraction of the volume that is above water is then

$$\frac{V - V'}{V} = \frac{V - 0.917V}{V} = 1 - 0.917 = 0.083$$

13.19 [II] A 60-kg rectangular box, open at the top, has base dimensions 1.0 m by 0.80 m and depth 0.50 m. (*a*) How deep will it sink in fresh water? (*b*) What weight F_{Wb} of ballast will cause it to sink to a depth of 30 cm?

(*a*) Assuming that the box floats, we have

$$F_B = \text{weight of displaced water} = \text{weight of box}$$
$$(1000 \text{ kg/m}^3)(9.81 \text{ m/s}^2)(1.0 \text{ m} \times 0.80 \text{ m} \times y) = (60 \text{ kg})(9.81 \text{ m/s}^2)$$

where y is the depth the box sinks. Solving gives $y = 0.075$ m. Because this is smaller than 0.50 m, our assumption is shown to be correct.

(*b*) $$F_B = \text{weight of box} + \text{weight of ballast}$$

But the F_B is equal to the weight of the displaced water. Therefore, the above equation becomes

$$(1000\ \text{kg/m}^3)(9.81\ \text{m/s}^2)(1.0\ \text{m} \times 0.80\ \text{m} \times 0.30\ \text{m}) = (60)(9.81)\ \text{N} + F_{Wb}$$

from which $F_{Wb} = 1760\ \text{N} = 1.8\ \text{kN}$. So the ballast must have a mass of (1760/9.81) kg = 180 kg.

13.20 [III] A foam plastic ($\rho_p = 0.58\ \text{g/cm}^3$) is to be used as a life preserver. What volume of plastic must be used if it is to keep 20 percent (by volume) of an 80-kg man above water in a lake? The average density of the man is 1.04 g/cm^3.

Keep in mind that a density of 1g/cm^3 equals 1000 kg/m^3. At equilibrium we have

$$F_B \text{ on man} + F_B \text{ on plastic} = \text{weight of man} + \text{weight of plastic}$$
$$(\rho_w)(0.80V_m)g + \rho_w V_p g = \rho_m V_m g + \rho_p V_p g$$

or

$$(\rho_w - \rho_p)V_p = (\rho_m - 0.80\rho_w)V_m$$

where subscripts m, w, and p refer to man, water, and plastic, respectively.

But $\rho_m V_m = 80$ kg and so $V_m = (80/1040)\ \text{m}^3$. Substitution gives

$$[(1000 - 580)\ \text{kg/m}^3]V_p = [(1040 - 800)\ \text{kg/m}^3][(80/1040)\ \text{m}^3]$$

from which $V_p = 0.044\ \text{m}^3$.

13.21 [III] A partly filled beaker of water sits on a scale, and its weight is 2.30 N. When a piece of metal suspended from a thread is totally immersed in the beaker (but not touching bottom), the scale reads 2.75 N. What is the volume of the metal?

The water exerts an upward buoyant force on the metal. According to Newton's third law of action and reaction, the metal exerts an equal downward force on the water. It is this force that increases the scale reading from 2.30 N to 2.75 N. Hence the buoyant force is 2.75 − 2.30 = 0.45 N. Then, because

$$F_B = \text{weight of displaced water} = \rho_w g V = (1000\ \text{kg/m}^3)(9.81\ \text{m/s}^2)(V)$$

we have the volume of the displaced water, and of the piece of metal, as

$$V = \frac{0.45\ \text{N}}{9810\ \text{kg/m}^2\cdot\text{s}^2} = 46 \times 10^{-6}\ \text{m}^3 = 46\ \text{cm}^3$$

13.22 [II] A piece of pure gold ($\rho = 19.3\ \text{g/cm}^3$) is suspected to have a hollow center. It has a mass of 38.25 g when measured in air and 36.22 g in water. What is the volume of the central hole in the gold?

Remember that you go from a density in g/cm^3 to kg/m^3 by multiplying by 1000. From $\rho = m/V$,

$$\text{Actual volume of 38.25 g of gold} = \frac{0.038\,25\ \text{kg}}{19\,300\ \text{kg/m}^3} = 1.982 \times 10^{-6}\ \text{m}^3$$

$$\text{Volume of displaced water} = \frac{(38.25 - 36.22) \times 10^{-3}\ \text{kg}}{1000\ \text{kg/m}^3} = 2.030 \times 10^{-6}\ \text{m}^3$$

$$\text{Volume of hole} = (2.030 - 1.982)\ \text{cm}^3 = 0.048\ \text{cm}^3$$

13.23 [III] A wooden cylinder has mass m and base area A. It floats in water with its axis vertical. Show that the cylinder undergoes SHM if given a small vertical displacement. Find the frequency of its motion.

When the cylinder is pushed down a distance y, it displaces an additional volume Ay of water. Because this additional displaced volume has mass $Ay\rho_w$, an additional buoyant force $Ay\rho_w g$ acts on the cylinder, where ρ_w is the density of water. This is an unbalanced force on the cylinder and is a restoring force. In addition, the force is proportional to the displacement and so is a Hooke's Law force. Therefore the cylinder will undergo SHM, as described in Chapter 11.

Comparing $F_B = A\rho_w gy$ with Hooke's Law in the form $F = ky$, we see that the spring constant for the motion is $k = A\rho_w g$. This, acting on the cylinder of mass m, causes it to have a vibrational frequency of

$$f = \frac{1}{2\pi}\sqrt{\frac{k}{m}} = \frac{1}{2\pi}\sqrt{\frac{A\rho_w g}{m}}$$

13.24 [II] What must be the volume V of a 5.0-kg balloon filled with helium ($\rho_{He} = 0.178\ \text{kg/m}^3$) if it is to lift a 30-kg load? Use $\rho_{air} = 1.29\ \text{kg/m}^3$.

The buoyant force, $V\rho_{air}g$, must lift the weight of the balloon, its load, and the helium within it:

$$V\rho_{air}g = (35\ \text{kg})(g) + V\rho_{He}g$$

which gives

$$V = \frac{35\ \text{kg}}{\rho_{air} - \rho_{He}} = \frac{35\ \text{kg}}{1.11\ \text{kg/m}^3} = 32\ \text{m}^3$$

13.25 [III] Find the density ρ of a fluid at a depth h in terms of its density ρ_0 at the surface.

If a mass m of fluid has volume V_0 at the surface, then it will have volume $V_0 - \Delta V$ at a depth h. The density at depth h is then

$$\rho = \frac{m}{V_0 - \Delta V} \qquad \text{while} \qquad \rho_0 = \frac{m}{V_0}$$

which gives

$$\frac{\rho}{\rho_0} = \frac{V_0}{V_0 - \Delta V} = \frac{1}{1 - (\Delta V/V_0)}$$

However, from Chapter 12, the bulk modulus is $B = P/(\Delta V/V_0)$ and so $\Delta V/V_0 = P/B$. Making this substitution, we obtain

$$\frac{\rho}{\rho_0} = \frac{1}{1 - P/B}$$

If we assume that ρ is close to ρ_0, then the pressure at depth h is approximately $\rho_0 gh$, and so

$$\frac{\rho}{\rho_0} = \frac{1}{1 - (\rho_0 gh/B)}$$

Supplementary Problems

13.26 [I] A 60-kg performer balances on a cane. The end of the cane in contact with the floor has an area of $0.92\ \text{cm}^2$. Find the pressure exerted on the floor by the cane. (Neglect the weight of the cane.) *Ans.* 6.4 MPa

13.27 [I] A certain town receives its water directly from a water tower. If the top of the water in the tower is 26.0 m above the water faucet in a house, what should be the water pressure at the faucet? (Neglect the effects of other water users.) *Ans.* 255 kPa

13.28 [II] At a height of 10 km (33 000 ft) above sea level, atmospheric pressure is about 210 mm of mercury. What is the net resultant normal force on a 600 cm^2 window of an airplane flying at this height? Assume the pressure inside the plane is 760 mm of mercury. The density of mercury is 13 600 kg/m^3. *Ans.* 4.4 kN

13.29 [II] A narrow tube is sealed onto a tank as shown in Fig. 13-7. The base of the tank has an area of 80 cm^2. (*a*) Remembering that pressure is determined by the height of the column of liquid, find the force on the bottom of the tank due to oil when the tank and capillary are filled with oil ($\rho = 0.72\ \text{g/cm}^3$) to the height h_1. (*b*) Repeat for an oil height of h_2. *Ans.* (*a*) 11 N downward; (*b*) 20 N downward

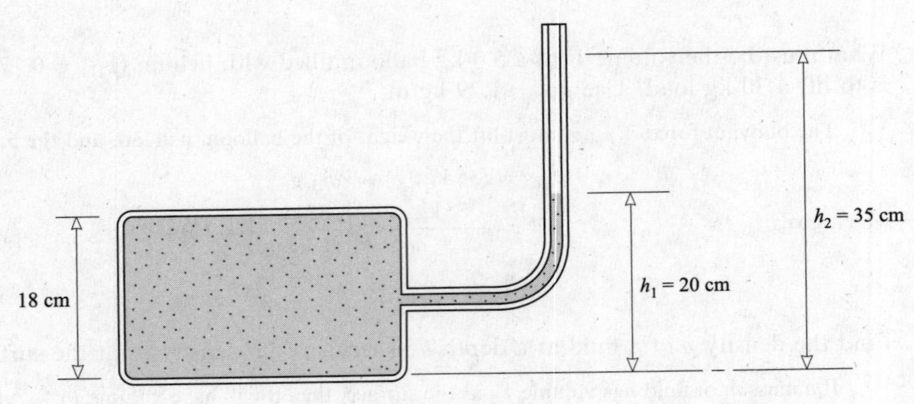

Fig. 13-7

13.30 [II] Repeat Problem 13.29, but now find the force on the top wall of the tank due to the oil. *Ans.* (*a*) 1.1 N upward; (*b*) 9.6 N upward

13.31 [II] Compute the pressure required for a water supply system that will raise water 50.0 m vertically. *Ans.* 490 kPa

13.32 [II] The area of a piston of a force pump is 8.0 cm^2. What force must be applied to the piston to raise oil ($\rho = 0.78\ \text{g/cm}^2$) to a height of 6.0 m? Assume the upper end of the oil is open to the atmosphere. *Ans.* 37 N

13.33 [II] The diameter of the large piston of a hydraulic press is 20 cm, and the area of the small piston is 0.50 cm^2. If a force of 400 N is applied to the small piston, (*a*) what is the resulting force exerted on the large piston? (*b*) What is the increase in pressure underneath the small piston? (*c*) Underneath the large piston? *Ans.* (*a*) 2.5×10^5 N; (*b*) 8.0 MPa; (*c*) 8.0 MPa

13.34 [II] A metal cube, 2.00 cm on each side, has a density of 6600 kg/m^3. Find its apparent mass when it is totally submerged in water. *Ans.* 44.8 g

13.35 [II] A solid wooden cube, 30.0 cm on each edge, can be totally submerged in water if it is pushed downward with a force of 54.0 N. What is the density of the wood? *Ans.* 800 kg/m^3

13.36 [II] A metal object "weighs" 26.0 g in air and 21.48 g when totally immersed in water. What is the volume of the object? Its mass density? *Ans.* 4.55 cm^3, $5.72 \times 10^3\ \text{kg/m}^3$

13.37 [II] A solid piece of aluminum ($\rho = 2.70$ g/cm^3) has a mass of 8.35 g when measured in air. If it is hung from a thread and submerged in a vat of oil ($\rho = 0.75$ g/cm^3), what will be the tension in the thread? *Ans.* 0.059 N

13.38 [II] A beaker contains oil of density 0.80 g/cm^3. By means of a thread, a 1.6-cm cube of aluminum ($\rho = 2.70$ g/cm^3) is submerged in the oil. Find the tension in the thread. *Ans.* 0.076 N

13.39 [II] A tank containing oil of sp gr = 0.80 rests on a scale and weighs 78.6 N. By means of a wire, a 6.0 cm cube of aluminum, sp gr = 2.70, is submerged in the oil. Find (*a*) the tension in the wire and (*b*) the scale reading if none of the oil overflows. *Ans.* (*a*) 4.0 N; (*b*) 80 N

13.40 [II] Downward forces of 45.0 N and 15.0 N, respectively, are required to keep a plastic block totally immersed in water and in oil. If the volume of the block is 8000 cm^3, find the density of the oil. *Ans.* 620 kg/m^3

13.41 [III] Determine the unbalanced force acting on an iron ball ($r = 1.5$ cm, $\rho = 7.8$ g/cm^3) when just released while totally immersed in (*a*) water and (*b*) mercury ($\rho = 13.6$ g/cm^3). What will be the initial acceleration of the ball in each case? *Ans.* (*a*) 0.94 N down, 8.6 m/s^2 down; (*b*) 0.80 N up, 7.3 m/s^2 up

13.42 [II] A 2.0 cm cube of metal is suspended by a thread attached to a scale. The cube appears to have a mass of 47.3 g when measured submerged in water. What will its mass appear to be when submerged in glycerin, sp gr = 1.26? (*Hint*: Find ρ too.) *Ans.* 45 g

13.43 [II] A balloon and its gondola have a total (empty) mass of 2.0×10^2 kg. When filled, the balloon contains 900 m^3 of helium at a density of 0.183 kg/m^3. Find the added load, in addition to its own weight, that the balloon can lift. The density of air is 1.29 kg/m^3. *Ans.* 7.8 kN

13.44 [I] A certain piece of metal has a measured mass of 5.00 g in air, 3.00 g in water, and 3.24 g in benzene. Determine the mass density of the metal and of the benzene. *Ans.* 2.50×10^3 kg/m^3, 880 kg/m^3

13.45 [II] A spring whose composition is not completely known might be either bronze (sp gr 8.8) or brass (sp gr 8.4). It has a mass of 1.26 g when measured in air and 1.11 g in water. Which is it made of? *Ans.* brass

13.46 [II] What fraction of the volume of a piece of quartz ($\rho = 2.65$ g/cm^3) will be submerged when it is floating in a container of mercury ($\rho = 13.6$ g/cm^3)? *Ans.* 0.195

13.47 [II] A cube of wood floating in water supports a 200-g mass resting on the center of its top face. When the mass is removed, the cube rises 2.00 cm. Determine the volume of the cube. *Ans.* 1.00×10^3 cm^3

13.48 [III] A cork has a measured mass of 5.0 g in air. A sinker has a measured mass of 86 g in water. The cork is attached to the sinker and both together have a measured mass of 71 g when under water. What is the density of the cork? *Ans.* 2.5×10^2 kg/m^3

13.49 [II] A glass of water has a 10-cm^3 ice cube floating in it. The glass is filled to the brim with cold water. By the time the ice cube has completely melted, how much water will have flowed out of the glass? The sp gr of ice is 0.92. *Ans.* none

13.50 [II] A glass tube is bent into the form of a U. A 50.0 cm height of olive oil in one arm is found to balance 46.0 cm of water in the other. What is the density of the olive oil? *Ans.* 920 kg/m^3

13.51 [II] On a day when the pressure of the atmosphere is 1.000×10^5 Pa, a chemist distills a liquid under slightly reduced pressure. The pressure within the distillation chamber is read by an oil-filled manometer (density of oil = 0.78 g/cm^3). The difference in heights on the two sides of the manometer is 27 cm. What is the pressure in the distillation chamber? *Ans.* 98 kPa

Chapter 14

Fluids in Motion

FLUID FLOW OR DISCHARGE (J): When a fluid that fills a pipe flows through the pipe with an average speed v, the *flow* or *discharge* J is

$$J = Av$$

where A is the cross-sectional area of the pipe. The units of J are m^3/s in the SI and ft^3/s in U.S. customary units. Sometimes J is called the *rate of flow* or the **discharge rate.**

EQUATION OF CONTINUITY: Suppose an *incompressible* (constant-density) fluid fills a pipe and flows through it. Suppose further that the cross-sectional area of the pipe is A_1 at one point and A_2 at another. Since the flow through A_1 must equal the flow through A_2, one has

$$J = A_1 v_1 = A_2 v_2 = \text{constant}$$

where v_1 and v_2 are the average fluid speeds over A_1 and A_2, respectively.

THE SHEAR RATE of a fluid is the rate at which the shear strain within the fluid is changing. Because strain has no units, the SI unit for shear rate is s^{-1}.

THE VISCOSITY (η) of a fluid is a measure of how large a shear stress is required to produce a shear rate of one. Its unit is that of stress per unit shear rate, or Pa·s in the SI. Another SI unit is the $N \cdot s/m^2$ (or kg/m·s), called the *poiseuille* (Pl): 1 Pl = 1 kg/m·s = 1 Pa·s. Other units used are the *poise* (P), where 1 P = 0.1 , and the *centipoise* (cP), where 1 cP = 10^{-3} Pl. A viscous fluid, such as tar, has a large η.

POISEUILLE'S LAW: The fluid flow through a cylindrical pipe of length L and cross-sectional radius R is given by

$$J = \frac{\pi R^4 (P_i - P_o)}{8 \eta L}$$

where $P_i - P_o$ is the pressure difference between the two ends of the pipe (input minus output).

THE WORK DONE BY A PISTON in forcing a volume V of fluid into a cylinder against an opposing pressure P is given by PV.

THE WORK DONE BY A PRESSURE P acting on a surface of area A as the surface moves through a distance Δx normal to the surface (thereby displacing a volume $A\,\Delta x = \Delta V$) is

$$\text{Work} = PA\,\Delta x = P\,\Delta V$$

BERNOULLI'S EQUATION for the steady flow of a continuous stream of fluid: Consider two different points along the stream path. Let point-1 be at a height h_1, and let v_1, ρ_1, and P_1 be the fluid-speed, density, and absolute pressure at that point. Similarly define h_2, v_2, ρ_2, and P_2 for point-2. Then, provided the fluid is incompressible and has negligible viscosity,

$$P_1 + \tfrac{1}{2}\rho v_1^2 + h_1 \rho g = P_2 + \tfrac{1}{2}\rho v_2^2 + h_2 \rho g$$

where $\rho_1 = \rho_2 = \rho$ and g is the acceleration due to gravity.

TORRICELLI'S THEOREM: Suppose that a tank contains liquid and is open to the atmosphere at its top. If an orifice (opening) exists in the tank at a distance h below the top of the liquid, then the speed of *outflow* from the orifice is $\sqrt{2gh}$, provided the liquid obeys Bernoulli's equation and the tank is big enough so that the top of the liquid may be regarded as motionless.

THE REYNOLDS NUMBER (N_R) is a dimensionless number that applies to a fluid of viscosity η and density ρ flowing with speed v through a pipe (or past an obstacle) with diameter D:

$$N_R = \frac{\rho v D}{\eta}$$

For systems of the same geometry, flows will usually be similar provided their Reynolds numbers are close. *Turbulent flow* occurs if N_R for the flow exceeds about 2000 for pipes or about 10 for obstacles.

Solved Problems

14.1 [I] Oil flows through a pipe 8.0 cm in diameter, at an average speed of 4.0 m/s. What is the flow J in m^3/s and m^3/h?

$$J = Av = \pi(0.040\ \text{m})^2(4.0\ \text{m/s}) = 0.020\ \text{m}^3/\text{s}$$
$$= (0.020\ \text{m}^3/\text{s})(3600\ \text{s/h}) = 72\ \text{m}^3/\text{h}$$

14.2 [I] Exactly 250 mL of fluid flows out of a tube whose inner diameter is 7.0 mm in a time of 41 s. What is the average speed of the fluid in the tube?

From $J = Av$, since $1\ \text{mL} = 10^{-6}\ \text{m}^3$,

$$v = \frac{J}{A} = \frac{(250 \times 10^{-6}\ \text{m}^3)/(41\ \text{s})}{\pi(0.003\,5\ \text{m})^2} = 0.16\ \text{m/s}$$

14.3 [I] A 14 cm inner diameter (i.d.) water main furnishes water (through intermediate pipes) to a 1.00 cm i.d. (i.e., inner diameter) faucet pipe. If the average speed in the faucet pipe is 3.0 cm/s, what will be the average speed it causes in the water main?

The two flows are equal. From the continuity equation, we have

$$J = A_1 v_1 = A_2 v_2$$

Letting 1 be the faucet and 2 be the water main, we have

$$v_2 = v_1 \frac{A_1}{A_2} = v_1 \frac{\pi r_1^2}{\pi r_2^2} = (3.0\ \text{cm/s})\left(\frac{1}{14}\right)^2 = 0.015\ \text{cm/s}$$

14.4 [II] How much water will flow in 30.0 s through 200 mm of capillary tube of 1.50 mm i.d., if the pressure differential across the tube is 5.00 cm of mercury? The viscosity of water is 0.801 cP and ρ for mercury is 13 600 kg/m^3.

We shall make use of Poiseuille's Law with

$$P_i - P_o = \rho g h = (13\,600\ \text{kg/m}^3)(9.81\ \text{m/s}^2)(0.0500\ \text{m}) = 6660\ \text{N/m}^2$$

and

$$\eta = (0.801\ \text{cP})\left(10^{-3}\frac{\text{kg/m}\cdot\text{s}}{\text{cP}}\right) = 8.01 \times 10^{-4}\ \text{kg/m}\cdot\text{s}$$

Thus, we have

$$J = \frac{\pi r^4 (P_i - P_o)}{8\eta L} = \frac{\pi(7.5 \times 10^{-4}\ \text{m})^4(6660\ \text{N/m}^2)}{8(8.01 \times 10^{-4}\ \text{kg/m}\cdot\text{s})(0.200\ \text{m})} = 5.2 \times 10^{-6}\ \text{m}^3/\text{s} = 5.2\ \text{mL/s}$$

In 30.0 s, the quantity that would flow out of the tube is (5.2 mL/s)(30 s) $= 1.6 \times 10^2$ mL.

14.5 [II] An artery in a certain person has been reduced to half its original inside diameter by deposits on the inner artery wall. By what factor will the blood flow through the artery be reduced if the pressure differential across the artery has remained unchanged?

From Poiseuille's Law, $J \propto r^4$. Therefore,

$$\frac{J_{\text{final}}}{J_{\text{original}}} = \left(\frac{r_{\text{final}}}{r_{\text{original}}}\right)^4 = \left(\frac{1}{2}\right)^4 = 0.0625$$

14.6 [II] Under the same pressure differential, compare the flow of water through a pipe to the flow of SAE No. 10 oil. η for water is 0.801 cP; η for the oil is 200 cP.

From Poiseuille's Law, $J \propto 1/\eta$. Therefore, since everything else cancels,

$$\frac{J_{\text{water}}}{J_{\text{oil}}} = \frac{200\ \text{cP}}{0.801\ \text{cP}} = 250$$

so the flow of water is 250 times as large as that of the oil under the same pressure differential.

14.7 [II] Calculate the power output of the heart if, in each heartbeat, it pumps 75 mL of blood at an average pressure of 100 mmHg. Assume 65 heartbeats per minute.

The work done by the heart is $P\Delta V$. In one minute, $\Delta V = (65)(75 \times 10^{-6}\ \text{m}^3)$. Also

$$P = (100\ \text{mmHg})\frac{1.01 \times 10^5\ \text{Pa}}{760\ \text{mmHg}} = 1.33 \times 10^4\ \text{Pa}$$

so

$$\text{Power} = \frac{\text{work}}{\text{time}} = \frac{(1.33 \times 10^4\ \text{Pa})[(65)(75 \times 10^{-6}\ \text{m}^3)]}{60\ \text{s}} = 1.1\ \text{W}$$

14.8 [II] What volume of water will escape per minute from an open-top tank through an opening 3.0 cm in diameter that is 5.0 m below the water level in the tank? (See Fig. 14-1.)

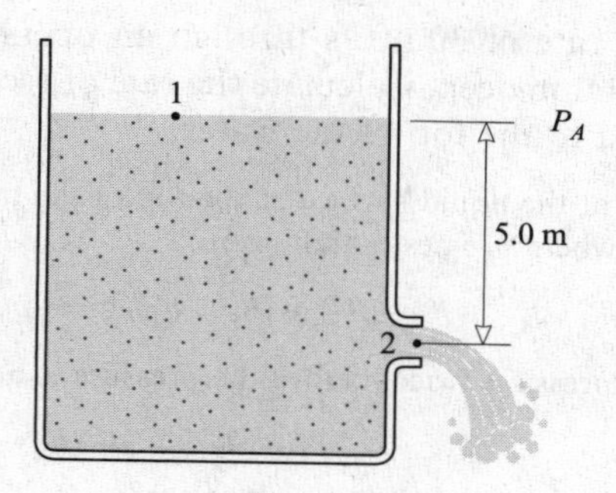

Fig. 14-1

We can use Bernoulli's equation, with 1 representing the top level and 2 the orifice. The pressure at the outlet inside the free jet is atmospheric. Then $P_1 = P_2$ and $h_1 = 5.0$ m, $h_2 = 0$.

$$P_1 + \tfrac{1}{2}\rho v_1^2 + h_1\rho g = P_2 + \tfrac{1}{2}\rho v_2^2 + h_2\rho g$$

$$\tfrac{1}{2}\rho v_1^2 + h_1\rho g = \tfrac{1}{2}\rho v_2^2 + h_2\rho g$$

If the tank is large, v_1 can be approximated as zero. Then, solving for v_2, we obtain Torricelli's equation:

$$v_2 = \sqrt{2g(h_1 - h_2)} = \sqrt{2(9.81 \text{ m/s}^2)(5.0 \text{ m})} = 9.9 \text{ m/s}$$

and the flow is given by

$$J = v_2 A_2 = (9.9 \text{ m/s})\pi(1.5 \times 10^{-2} \text{ m})^2 = 7.0 \times 10^{-3} \text{ m}^3/\text{s} = 0.42 \text{ m}^3/\text{min}$$

14.9 [II] An open water tank in air springs a leak at position-2 in Fig. 14-2, where the pressure due to the water at position-1 is 500 kPa. What is the velocity of escape of the water through the hole?

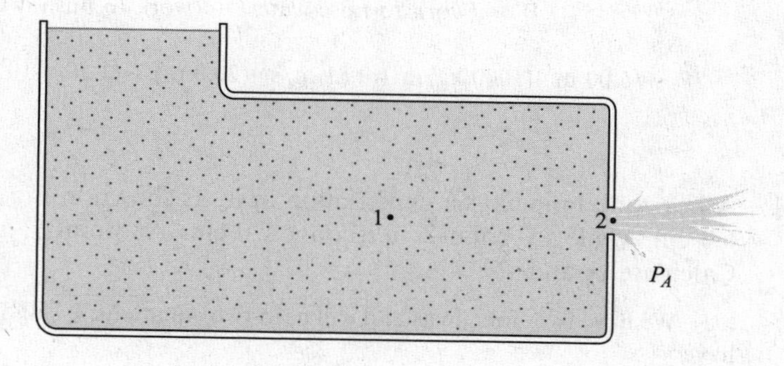

Fig. 14-2

The pressure at position-2 in the free jet is atmospheric. We use Bernoulli's equation with $P_1 - P_2 = 5.00 \times 10^5$ N/m^2, $h_1 = h_2$, and the approximation $v_1 = 0$. Then

$$(P_1 - P_2) + (h_1 - h_2)\rho g = \tfrac{1}{2}\rho v_2^2$$

whence
$$v_2 = \sqrt{\frac{2(P_1 - P_2)}{\rho}} = \sqrt{\frac{2(5.00 \times 10^5 \text{ N/m}^2)}{1000 \text{ kg/m}^3}} = 31.6 \text{ m/s}$$

14.10 [III] Water flows at the rate of 30 mL/s through an opening at the bottom of a large tank in which the water is 4.0 m deep. Calculate the rate of escape of the water if an added pressure of 50 kPa is applied to the top of the water.

Take position-1 at the liquid surface at the top of the tank, and position-2 at the opening. From Bernoulli's equation where v_1 is essentially zero,

$$(P_1 - P_2) + (h_1 - h_2)\rho g = \tfrac{1}{2}\rho v_2^2$$

We can apply this expression twice, before the pressure is added and after.

$$(P_1 - P_2)_{\text{before}} + (h_1 - h_2)\rho g = \tfrac{1}{2}\rho(v_2^2)_{\text{before}}$$

$$(P_1 - P_2)_{\text{before}} + 5 \times 10^4\ \text{N/m}^2 + (h_1 - h_2)\rho g = \tfrac{1}{2}\rho(v_2^2)_{\text{after}}$$

If the opening and the top of the tank are originally at atmospheric pressure, then

$$(P_1 - P_2)_{\text{before}} = 0$$

and division of the second equation by the first gives

$$\frac{(v_2^2)_{\text{after}}}{(v_2^2)_{\text{before}}} = \frac{5 \times 10^4\ \text{N/m}^2 + (h_1 - h_2)\rho g}{(h_1 - h_2)\rho g}$$

But $$(h_1 - h_2)\rho g = (4.0\ \text{m})(1000\ \text{kg/m}^3)(9.81\ \text{m/s}^2) = 3.9 \times 10^4\ \text{N/m}^2$$

Therefore, $$\frac{(v_2)_{\text{after}}}{(v_2)_{\text{before}}} = \sqrt{\frac{8.9 \times 10^4\ \text{N/m}^2}{3.9 \times 10^4\ \text{N/m}^2}} = 1.51$$

Since $J = Av$, this can be written as

$$\frac{J_{\text{after}}}{J_{\text{before}}} = 1.51 \qquad \text{or} \qquad J_{\text{after}} = (30\ \text{mL/s})(1.51) = 45\ \text{mL/s}$$

14.11 [II] How much work W is done by a pump in raising 5.00 m^3 of water 20.0 m and forcing it into a main at a gauge pressure of 150 kPa?

$$W = (\text{work to raise water}) + (\text{work to push it in}) = mgh + P\,\Delta V$$

$$W = (5.00\ \text{m}^3)(1000\ \text{kg/m}^3)(9.81\ \text{m/s}^2)(20.0\ \text{m}) + (1.50 \times 10^5\ \text{N/m}^2)(5.00\ \text{m}^3) = 1.73 \times 10^6\ \text{J}$$

14.12 [II] A horizontal pipe has a constriction in it, as shown in Fig. 14-3. At point-1 the diameter is 6.0 cm, while at point-2 it is only 2.0 cm. At point-1, $v_1 = 2.0$ m/s and $P_1 = 180$ kPa. Calculate v_2 and P_2.

We have two unknowns and will need two equations. Using Bernoulli's equation with $h_1 = h_2$, we have

$$P_1 + \tfrac{1}{2}\rho v_1^2 = P_2 + \tfrac{1}{2}\rho v_2^2 \qquad \text{or} \qquad P_1 + \tfrac{1}{2}\rho(v_1^2 - v_2^2) = P_2$$

Furthermore, $v_1 = 2.0$ m/s, and the equation of continuity tells us that

$$v_2 = v_1\frac{A_1}{A_2} = (2.0\ \text{m/s})\left(\frac{r_1}{r_2}\right)^2 = (2.0\ \text{m/s})(9.0) = 18\ \text{m/s}$$

Substituting then gives

$$1.80 \times 10^5\ \text{N/m}^2 + \tfrac{1}{2}(1000\ \text{kg/m}^3)[(2.0\ \text{m/s})^2 - (18\ \text{m/s})^2] = P_2$$

from which $P_2 = 0.20 \times 10^5\ \text{N/m}^2 = 20$ kPa.

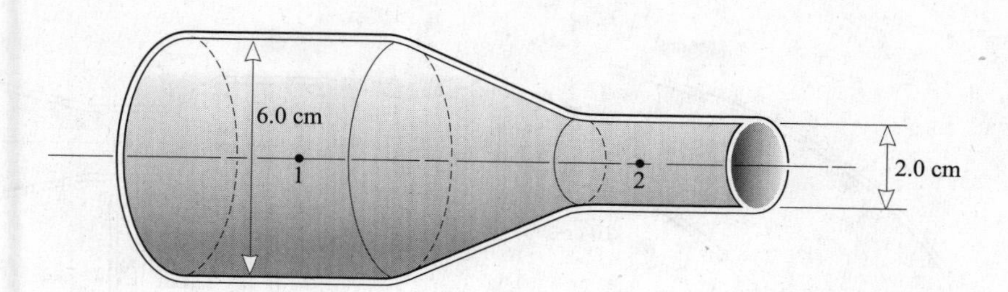

Fig. 14-3

14.13 [III] What must be the gauge pressure in a large-diameter hose if the nozzle is to shoot water straight upward to a height of 30.0 m?

To rise to a height h, a projectile must have an initial speed $\sqrt{2gh}$. (We obtain this by equating $\frac{1}{2}mv_0^2$ to mgh.) We can find this speed in terms of the difference between the pressures inside and outside the hose by writing Bernoulli's equation for points just inside and outside the nozzle in terms of absolute pressure:

$$P_{\text{in}} + \tfrac{1}{2}\rho v_{\text{in}}^2 + h_{\text{in}}\rho g = P_{\text{out}} + \tfrac{1}{2}\rho v_{\text{out}}^2 + h_{\text{out}}\rho g$$

Here $h_{\text{out}} \approx h_{\text{in}}$ and because the hose is large $v_{\text{in}} \approx 0$, therefore

$$P_{\text{in}} - P_{\text{out}} = \tfrac{1}{2}\rho v_{\text{out}}^2$$

Substitution of $\sqrt{2gh}$ for v_{out} gives

$$P_{\text{in}} - P_{\text{out}} = \rho g h = (1000\ \text{kg/m}^3)(9.81\ \text{m/s}^2)(30.0\ \text{m}) = 294\ \text{kPa}$$

Since $P_{\text{out}} = P_A$ this is the gauge pressure inside the hose. How could you obtain this latter equation directly from Torricelli's Theorem?

14.14 [III] At what rate does water flow from a 0.80 cm i.d. faucet if the water (or gauge) pressure is 200 kPa?

We apply Bernoulli's equation for points just inside and outside the faucet (using absolute pressure):

$$P_{\text{in}} + \tfrac{1}{2}\rho v_{\text{in}}^2 + h_{\text{in}}\rho g = P_{\text{out}} + \tfrac{1}{2}\rho v_{\text{out}}^2 + h_{\text{out}}\rho g$$

Note that the pressure inside due only to the water is 200 kPa and therefore $P_{\text{in}} = P_{\text{out}} = 200$ kPa since $P_{\text{out}} = P_A$. Taking $h_{\text{out}} = h_{\text{in}}$, we have

$$v_{\text{out}}^2 - v_{\text{in}}^2 = (200 \times 10^3\ \text{Pa})\frac{2}{\rho}$$

Assuming $v_{\text{in}}^2 \ll v_{\text{out}}^2$, we solve to obtain $v_{\text{out}} = 20$ m/s. The flow rate is then

$$J = vA = (20\ \text{m/s})(\pi r^2) = (20\ \text{m/s})(\pi)(0.16 \times 10^{-4}\ \text{m}^2) = 1.0 \times 10^{-3}\ \text{m}^3/\text{s}$$

14.15 [II] The pipe shown in Fig. 14-4 has a diameter of 16 cm at section-1 and 10 cm at section-2. At section-1 the pressure is 200 kPa. Point-2 is 6.0 m higher than point-1. When oil of density 800 kg/m^3 flows at a rate of 0.030 m^3/s, find the pressure at point-2 if viscous effects are negligible.

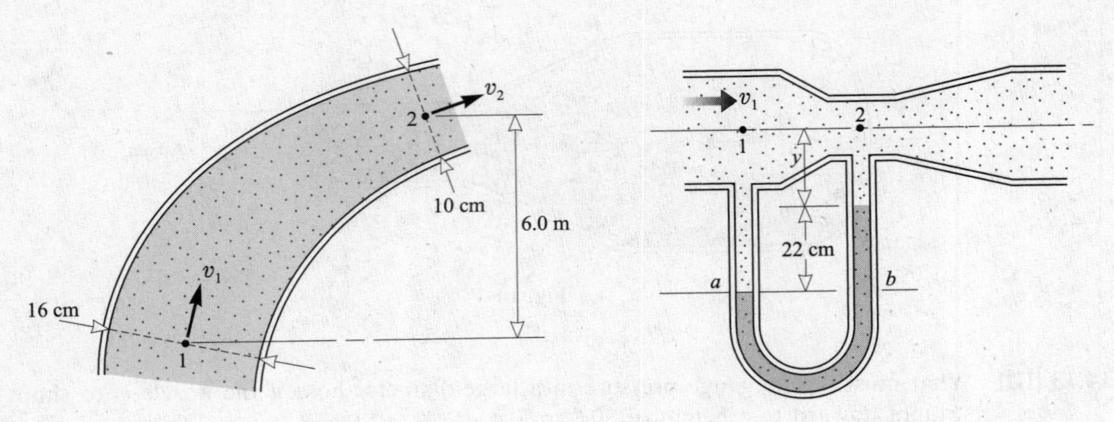

Fig. 14-4

Fig. 14-5

From $J = v_1 A_1 = v_2 A_2$ we have

$$v_1 = \frac{J}{A_1} = \frac{0.030 \text{ m}^3/\text{s}}{\pi(8.0 \times 10^{-2} \text{ m})^2} = 1.49 \text{ m/s}$$

$$v_2 = \frac{J}{A_2} = \frac{0.030 \text{ m}^3/\text{s}}{\pi(5.0 \times 10^{-2} \text{ m})^2} = 3.82 \text{ m/s}$$

We can now use Bernoulli's equation:

$$P_1 + \tfrac{1}{2}\rho v_1^2 + \rho g(h_1 - h_2) = P_2 + \tfrac{1}{2}\rho v_2^2$$

Setting $P_1 = 2.00 \times 10^5 \text{ N/m}^2, h_2 - h_1 = 6 \text{ m}$ and $\rho = 800 \text{ kg/m}^3$ gives

$$\begin{aligned} P_2 &= 2.00 \times 10^5 \text{ N/m}^3 + \tfrac{1}{2}(800 \text{ kg/m}^3)[(1.49 \text{ m/s})^2 - (3.82 \text{ m/s})^2] - (800 \text{ kg/m}^3)(9.81 \text{ m/s}^2)(6.0 \text{ m}) \\ &= 1.48 \times 10^5 \text{ N/m}^2 = 1.5 \times 10^5 \text{ kPa.} \end{aligned}$$

14.16 [III] A venturi meter equipped with a differential mercury manometer is shown in Fig. 14-5. At the inlet, point-1, the diameter is 12 cm, while at the throat, point-2, the diameter is 6.0 cm. What is the flow J of water through the meter if the mercury manometer reading is 22 cm? The density of mercury is 13.6 g/cm^3.

From the manometer reading (remembering that 1 g/cm^3 = 1000 kg/m^3) we obtain

$$P_1 - P_2 = \rho g h = (13\,600 \text{ kg/m}^3)(9.81 \text{ m/s}^2)(0.22 \text{ m}) = 2.93 \times 10^4 \text{ N/m}^2$$

Since $J = v_1 A_1 = v_2 A_2$, we have $v_1 = J/A_1$ and $v_2 = J/A_2$. Using Bernoulli's equation with $h_1 - h_2 = 0$ gives

$$(P_1 - P_2) + \tfrac{1}{2}\rho(v_1^2 - v_2^2) = 0$$

$$2.93 \times 10^4 \text{ N/m}^2 + \tfrac{1}{2}(1000 \text{ kg/m}^3)\left(\frac{1}{A_1^2} - \frac{1}{A_2^2}\right)J^2 = 0$$

where

$$A_1 = \pi r_1^2 = \pi(0.060)^2 \text{ m}^2 = 0.011\,31 \text{ m}^2 \quad \text{and} \quad A_2 = \pi r_2^2 = \pi(0.030)^2 \text{ m}^2 = 0.002\,8 \text{ m}^2$$

Substitution then gives $J = 0.022 \text{ m}^3/\text{s}$.

14.17 [III] A wind tunnel is to be used with a 20 cm high model car to approximately reproduce the situation in which a 550 cm high car is moving at 15 m/s. What should be the wind speed in the tunnel? Is the flow likely to be turbulent?

We want the Reynolds number N_R to be the same in both cases, so that the situations will be similar. That is, we want

$$N_R = \left(\frac{\rho v D}{\eta}\right)_{\text{tunnel}} = \left(\frac{\rho v D}{\eta}\right)_{\text{air}}$$

Both ρ and η are the same in the two cases, so we have

$$v_t D_t = v_a D_a \qquad \text{from which} \qquad v_t = v_a \frac{D_a}{D_t} = (15\ \text{m/s})(550/20) = 0.41\ \text{km/s}$$

To investigate turbulence, we evaluate N_R using $\rho = 1.29\ \text{kg/m}^3$ and $\eta = 1.8 \times 10^{-5}\ \text{Pa}\cdot\text{s}$ for air. We find that $N_R = 5.9 \times 10^6$, a value far in excess of that required for turbulent flow. The flow will certainly be turbulent.

Supplementary Problems

14.18 [I] Oil flows through a 4.0 cm i.d. pipe at an average speed of 2.5 m/s. Find the flow in m^3/s and cm^3/s. *Ans.* $3.1 \times 10^{-3}\ \text{m}^3/\text{s} = 3.1 \times 10^3\ \text{cm}^3/\text{s}$

14.19 [I] Compute the average speed of water in a pipe having an i.d. of 5.0 cm and delivering 2.5 m^3 of water per hour. *Ans.* 0.35 m/s

14.20 [II] The speed of glycerin flowing in a 5.0 cm i.d. pipe is 0.54 m/s. Find the fluid's speed in a 3.0 cm i.d. pipe that connects with it, both pipes flowing full. *Ans.* 1.5 m/s

14.21 [II] How long will it take for 500 mL of water to flow through a 15 cm long, 3.0 mm i.d. pipe, if the pressure differential across the pipe is 4.0 kPa? The viscosity of water is 0.80 cP. *Ans.* 7.5 s

14.22 [II] A molten plastic flows out of a tube that is 8.0 cm long at a rate of 13 cm^3/min when the pressure differential between the two ends of the tube is 18 cm of mercury. Find the viscosity of the plastic. The i.d. of the tube is 1.30 mm. The density of mercury is 13.6 g/cm^3. *Ans.* $0.097\ \text{kg/m}\cdot\text{s} = 97\ \text{cP}$

14.23 [II] In a horizontal pipe system, a pipe (i.d. 4.0 mm) that is 20 cm long connects in line to a pipe (i.d. 5.0 mm) that is 30 cm long. When a viscous fluid is being pushed through the pipes at a steady rate, what is the ratio of the pressure difference across the 20-cm pipe to that across the 30-cm pipe? *Ans.* 1.6

14.24 [II] A hypodermic needle of length 3.0 cm and i.d. 0.45 mm is used to draw blood ($\eta = 4.0$ mPl). Assuming the pressure differential across the needle is 80 cmHg, how long does it take to draw 15 mL? *Ans.* 17 s

14.25 [II] In a blood transfusion, blood flows from a bottle at atmospheric pressure into a patient's vein in which the pressure is 20 mmHg higher than atmospheric. The bottle is 95 cm higher than the vein, and the needle into the vein has a length of 3.0 cm and an i.d. of 0.45 mm. How much blood flows into the vein each minute? For blood, $\eta = 0.0040\ \text{Pa}\cdot\text{s}$ and $\rho = 1005\ \text{kg/m}^3$. *Ans.* 3.4 cm^3

14.26 [I] How much work does the piston in a hydraulic system do during one 2.0-cm stroke if the end area of the piston is 0.75 cm^2 and the pressure in the hydraulic fluid is 50 kPa? *Ans.* 75 mJ

14.27 [II] A large tank of nonviscous liquid, which is open to the surrounding air, springs a leak 4.5 m below the top of the liquid. What is the theoretical velocity of outflow from the hole? If the area of the hole is 0.25 cm^2, how much liquid would escape in exactly 1 minute? *Ans.* 9.4 m/s, 0.014 1 m^3

14.28 [II] Find the flow in liters/s of a nonviscous liquid through an opening 0.50 cm^2 in area and 2.5 m below the level of the liquid in an open tank surrounded by air. *Ans.* 0.35 liter/s

14.29 [II] Calculate the theoretical velocity of efflux of water, into the surrounding air, from an aperture that is 8.0 m below the surface of water in a large tank, if an added pressure of 140 kPa is applied to the surface of the water. *Ans.* 21 m/s

14.30 [II] What horsepower is required to force 8.0 m^3 of water per minute into a water main at a pressure of 220 kPa? *Ans.* 39 hp

14.31 [II] A pump lifts water at the rate of 9.0 liters/s from a lake through a 5.0 cm i.d. pipe and discharges it into the air at a point 16 m above the level of the water in the lake. What are the theoretical (*a*) velocity of the water at the point of discharge and (*b*) power delivered by the pump. *Ans.* (*a*) 4.6 m/s; (*b*) 2.0 hp

14.32 [II] Water flows steadily through a horizontal pipe of varying cross-section. At one place the pressure is 130 kPa and the speed is 0.60 m/s. Determine the pressure at another place in the same pipe where the speed is 9.0 m/s. *Ans.* 90 kPa.

14.33 [II] A pipe of varying inner diameter carries water. At point-1 the diameter is 20 cm and the pressure is 130 kPa. At point-2, which is 4.0 m higher than point-1, the diameter is 30 cm. If the flow is 0.080 m^3/s, what is the pressure at the second point? *Ans.* 93 kPa

14.34 [II] Fuel oil of density 820 kg/m^3 flows through a venturi meter having a throat diameter of 4.0 cm and an entrance diameter of 8.0 cm. The pressure drop between entrance and throat is 16 cm of mercury. Find the flow. The density of mercury is 13 600 kg/m^3. *Ans.* 9.3×10^{-3} m^3/s

14.35 [II] Find the maximum amount of water that can flow through a 3.0 cm i.d. pipe per minute without turbulence. Take the maximum Reynolds number for nonturbulent flow to be 2000. For water at 20°C, $\eta = 1.0 \times 10^{-3}$ Pa·s. *Ans.* 0.002 8 m^3

14.36 [I] How fast can a raindrop ($r = 1.5$ mm) fall through air if the flow around it is to be close to turbulent, i.e., for N_R close to 10? For air, $\eta = 1.8 \times 10^{-5}$ Pa·s and $\rho = 1.29$ kg/m^3. *Ans.* 4.6 cm/s

Chapter 15

Thermal Expansion

TEMPERATURE (T) may be measured on the *Celsius* scale, on which the freezing point of water is at 0 °C, and the boiling point (under standard conditions) is at 100 °C. The *Kelvin* (or *absolute*) scale is displaced 273.15 Celsius-size degrees from the Celsius scale, so that the freezing point of water is 273.15 K and the boiling point is 373.15 K. Absolute zero, a temperature discussed further in Chapter 16, is at 0 K (−273.15 °C). The still-used *Fahrenheit* scale is related to the Celsius scale by

$$\text{Fahrenheit temperature} = \tfrac{9}{5}\,(\text{Celsius temperature}) + 32$$

LINEAR EXPANSION OF SOLIDS: When a solid is subjected to a rise in temperature ΔT, its increase in length ΔL is very nearly proportional to its initial length L_0 multiplied by ΔT. That is,

$$\Delta L = \alpha L_0\,\Delta T$$

where the proportionality constant α is called the **coefficient of linear expansion**. The value of α depends on the nature of the substance. For our purposes we can take α to be constant independent of T, although that's rarely, if ever, exactly true.

From the above equation, α is the change in length per unit initial length per degree change in temperature. For example, if a 1.000 000 cm length of brass becomes 1.000 019 cm long when the temperature is raised 1.0 °C, the linear expansion coefficient for brass is

$$\alpha = \frac{\Delta L}{L_0\,\Delta T} = \frac{0.000\,019\ \text{cm}}{(1.0\ \text{cm})(1.0\ {}^\circ\text{C})} = 1.9 \times 10^{-5}\ {}^\circ\text{C}^{-1}$$

AREA EXPANSION: If an area A_0 expands to $A_0 + \Delta A$ when subjected to a temperature rise ΔT, then

$$\Delta A = \gamma A_0\,\Delta T$$

where γ is the **coefficient of area expansion.** For *isotropic* solids (those that expand in the same way in all directions), $\gamma \approx 2\alpha$.

VOLUME EXPANSION: If a volume V_0 changes by an amount ΔV when subjected to a temperature change of ΔT, then

$$\Delta V = \beta V_0\,\Delta T$$

where β is the **coefficient of volume expansion**. This can be either an increase or decrease in volume. For isotropic solids, $\beta \approx 3\alpha$.

Solved Problems

15.1 [I] A copper bar is 80 cm long at 15 °C. What is the increase in length when it is heated to 35 °C? The linear expansion coefficient for copper is $1.7 \times 10^{-5}\,°\text{C}^{-1}$.

$$\Delta L = \alpha L_0\,\Delta T = (1.7 \times 10^{-5}\,°\text{C}^{-1})(0.80\,\text{m})[(35 - 15)\,°\text{C}] = 2.7 \times 10^{-4}\ \text{m}$$

15.2 [II] A cylinder of diameter 1.000 00 cm at 30 °C is to be slid into a hole in a steel plate. The hole has a diameter of 0.999 70 cm at 30 °C. To what temperature must the plate be heated? For steel, $\alpha = 1.1 \times 10^{-5}\,°\text{C}^{-1}$.

The plate will expand in the same way whether or not there is a hole in it. Hence the hole expands in the same way a circle of steel filling it would expand. We want the diameter of the hole to change by

$$\Delta L = (1.000\,00 - 0.999\,70)\ \text{cm} = 0.000\,30\ \text{cm}$$

Using $\Delta L = \alpha L_0\,\Delta T$, we find

$$\Delta T = \frac{\Delta L}{\alpha L_0} = \frac{0.000\,30\ \text{cm}}{(1.1 \times 10^{-5}\,°\text{C}^{-1})(0.999\,70\ \text{cm})} = 27\,°\text{C}$$

The temperature of the plate must be $30 + 27 = 57\,°\text{C}$

15.3 [I] A steel tape is calibrated at 20 °C. On a cold day when the temperature is −15 °C, what will be the percent error in the tape? $\alpha_{\text{steel}} = 1.1 \times 10^{-5}\,°\text{C}^{-1}$.

For a temperature change from 20 °C to −15 °C, we have $\Delta T = -35\,°\text{C}$. Then,

$$\frac{\Delta L}{L_0} = \alpha\,\Delta T = (1.1 \times 10^{-5}\,°\text{C}^{-1})(-35\,°\text{C}) = -3.9 \times 10^{-4} = -0.039\%$$

15.4 [II] A copper rod $(\alpha = 1.70 \times 10^{-5}\,°\text{C}^{-1})$ is 20 cm longer than an aluminum rod $(\alpha = 2.20 \times 10^{-5}\,°\text{C}^{-1})$. How long should the copper rod be if the difference in their lengths is to be independent of temperature?

For their difference in lengths not to change with temperature, ΔL must be the same for both rods under the same temperature change. That is,

$$(\alpha L_0\,\Delta T)_{\text{copper}} = (\alpha L_0\,\Delta T)_{\text{aluminum}}$$

or

$$(1.70 \times 10^{-5}\,°\text{C}^{-1})L_0\,\Delta T = (2.20 \times 10^{-5}\,°\text{C}^{-1})(L_0 - 0.20\ \text{m})\,\Delta T$$

where L_0 is the length of the copper rod, and ΔT is the same for both rods. Solving, we find that $L_0 = 0.88$ m.

15.5 [II] At 20.0 °C a steel ball $(\alpha = 1.10 \times 10^{-5}\,°\text{C}^{-1})$ has a diameter of 0.900 0 cm, while the diameter of a hole in an aluminum plate $(\alpha = 2.20 \times 10^{-5}\,°\text{C}^{-1})$ is 0.899 0 cm. At what temperature (the same for both) will the ball just pass through the hole?

At a temperature ΔT higher than 20.0 °C, we wish the diameters of the hole and of the ball to be equal:

$$0.900\,0\ \text{cm} + (0.900\,0\ \text{cm})(1.10 \times 10^{-5}\,°\text{C}^{-1})\,\Delta T = 0.899\,0\ \text{cm} + (0.899\,0\ \text{cm})(2.20 \times 10^{-5}\,°\text{C}^{-1})\,\Delta T$$

Solving for ΔT, we find $\Delta T = 101\,°\text{C}$. Because the original temperature was 20.0 °C, the final temperature must be 121 °C.

15.6 [II] A steel tape measures the length of a copper rod as 90.00 cm when both are at 10 °C, the calibration temperature for the tape. What would the tape read for the length of the rod when both are at 30 °C? $\alpha_{\text{steel}} = 1.1 \times 10^{-5}\ {}^\circ\text{C}^{-1}$; $\alpha_{\text{copper}} = 1.7 \times 10^{-5}\ {}^\circ\text{C}^{-1}$.

At 30 °C, the copper rod will be of length

$$L_0(1 + \alpha_c\,\Delta T)$$

while adjacent "centimeter" marks on the steel tape will be separated by a distance of

$$(1.000\ \text{cm})(1 + \alpha_s\,\Delta T)$$

Therefore, the number of "centimeters" read on the tape will be

$$\frac{L_0(1 + \alpha_c\,\Delta T)}{(1\ \text{cm})(1 + \alpha_s\,\Delta T)} = \frac{(90.00\ \text{cm})[1 + (1.7 \times 10^{-5}\ {}^\circ\text{C}^{-1})(20\ {}^\circ\text{C})]}{(1.000\ \text{cm})[1 + (1.1 \times 10^{-5}\ {}^\circ\text{C}^{-1})(20\ {}^\circ\text{C})]} = 90.00\frac{1 + 3.4 \times 10^{-4}}{1 + 2.2 \times 10^{-4}}$$

Using the approximation

$$\frac{1}{1+x} \approx 1 - x$$

for x small compared to 1, we have

$$90.00\frac{1 + 3.4 \times 10^{-4}}{1 + 2.2 \times 10^{-4}} \approx 90.00(1 + 3.4 \times 10^{-4})(1 - 2.2 \times 10^{-4}) \approx 90.00(1 + 3.4 \times 10^{-4} - 2.2 \times 10^{-4})$$

$$= 90.00 + 0.010\,8$$

The tape will read 90.01 cm.

15.7 [II] A glass flask is filled "to the mark" with 50.00 cm^3 of mercury at 18 °C. If the flask and its contents are heated to 38 °C, how much mercury will be above the mark? $\alpha_{\text{glass}} = 9.0 \times 10^{-6}\ {}^\circ\text{C}^{-1}$ and $\beta_{\text{mercury}} = 182 \times 10^{-6}\ {}^\circ\text{C}^{-1}$.

We shall take $\beta_{\text{glass}} = 3\alpha_{\text{glass}}$ as a good approximation. The flask interior will expand just as though it were a solid piece of glass. Thus,

$$\begin{aligned}\text{Volume of mercury above mark} &= (\Delta V \text{ for mercury}) - (\Delta V \text{ for glass})\\ &= \beta_m V_0\,\Delta T - \beta_g V_0\,\Delta T = (\beta_m - \beta_g)V_0\,\Delta T\\ &= [(182 - 27) \times 10^{-6}\ {}^\circ\text{C}^{-1}](50.00\ \text{cm}^3)[(38 - 18)\ {}^\circ\text{C}]\\ &= 0.15\ \text{cm}^3\end{aligned}$$

15.8 [II] The density of mercury at exactly 0 °C is 13 600 kg/m^3, and its volume expansion coefficient is $1.82 \times 10^{-4}\ {}^\circ\text{C}^{-1}$. Calculate the density of mercury at 50.0 °C.

Let

$$\begin{aligned}\rho_0 &= \text{density of mercury at } 0\ {}^\circ\text{C}\\ \rho_1 &= \text{density of mercury at } 50\ {}^\circ\text{C}\\ V_0 &= \text{volume of } m \text{ kg of mercury at } 0\ {}^\circ\text{C}\\ V_1 &= \text{volume of } m \text{ kg of mercury at } 50\ {}^\circ\text{C}\end{aligned}$$

Since the mass does not change, $m = \rho_0 V_0 = \rho_1 V_1$, from which it follows that

$$\rho_1 = \rho_0\frac{V_0}{V_1} = \rho_0\frac{V_0}{V_0 + \Delta V} = \rho_0\frac{1}{1 + (\Delta V/V_0)}$$

But
$$\frac{\Delta V}{V_0} = \beta\,\Delta T = (1.82 \times 10^{-4}\ {}^\circ\text{C}^{-1})(50.0\ {}^\circ\text{C}) = 0.009\,10$$

Substitution into the first equation then gives

$$\rho_1 = (13\,600 \text{ kg/m}^3)\frac{1}{1 + 0.009\,10} = 13.5 \times 10^3 \text{ kg/m}^3$$

15.9 [II] Show that the density of a liquid or solid changes in the following way with temperature: $\Delta\rho = -\rho\beta\,\Delta T \approx -\rho_0\beta\Delta T$.

Consider a mass m of liquid having a volume V_0, for which $\rho_0 = m/V_0$. After a temperature change ΔT, the volume will be

$$V = V_0 + V_0\beta\,\Delta T$$

and the density will be

$$\rho = \frac{m}{V} = \frac{m}{V_0(1 + \beta\,\Delta T)}$$

But $m/V_0 = \rho_0$, and so this can be written as

$$\rho(1 + \beta\,\Delta T) = \rho_0$$

Thus we find that

$$\Delta\rho = \rho - \rho_0 = -\rho\beta\,\Delta T$$

In practice, ρ is close enough to ρ_0 so that we can say $\Delta\rho \approx -\rho_0\beta\,\Delta T$.

15.10 [II] Solve Problem 15.8 using the result of Problem 15.9.

We have

$$\Delta\rho \approx -\rho_0\beta\Delta T$$

Hence $$\Delta\rho \approx -(13\,600 \text{ kg/m}^3)(182 \times 10^{-6}\,{}^\circ\text{C}^{-1})(50.0\,{}^\circ\text{C}) = -124 \text{ kg/m}^3$$

so $$\rho_{50\,{}^\circ\text{C}} = \rho_{0\,{}^\circ\text{C}} - 124 \text{ kg/m}^3 = 13.5 \times 10^3 \text{ kg/m}^3$$

15.11 [III] A steel wire of 2.0 mm^2 cross-section at 30 °C is held straight (but under no tension) by attaching its ends firmly to two points a distance 1.50 m apart. (Of course this will have to be done out in space so the wire is weightless, but don't worry about that.) If the temperature now decreases to −10 °C, and if the two tie points remain fixed, what will be the tension in the wire? For steel, $\alpha = 1.1 \times 10^{-5}\,{}^\circ\text{C}^{-1}$ and $Y = 2.0 \times 10^{11}$ N/m^2.

If it were free to do so, the wire would contract a distance ΔL as it cooled, where

$$\Delta L = \alpha L_0\,\Delta T = (1.1 \times 10^{-5}\,{}^\circ\text{C}^{-1})(1.5 \text{ m})(40\,{}^\circ\text{C}) = 6.6 \times 10^{-4} \text{ m}$$

But the ends are fixed. As a result, forces at the ends must, in effect, stretch the wire this same length ΔL. Therefore, from $Y = (F/A)(\Delta L/L_0)$, we have

$$\text{Tension} = F = \frac{YA\,\Delta L}{L_0} = \frac{(2.0 \times 10^{11} \text{ N/m}^2)(2.0 \times 10^{-6} \text{ m}^2)(6.6 \times 10^{-4} \text{ m})}{1.50 \text{ m}} = 176 \text{ N} = 0.18 \text{ kN}$$

Strictly, we should have substituted $(1.5 - 6.6 \times 10^{-4})$ m for L in the expression for the tension. However, the error incurred in not doing so is negligible.

15.12 [III] When a building is constructed at −10 °C, a steel beam (cross-sectional area 45 cm^2) is put in place with its ends cemented in pillars. If the sealed ends cannot move, what will be the compressional force on the beam when the temperature is 25 °C? For this kind of steel, $\alpha = 1.1 \times 10^{-5}\,{}^\circ\text{C}^{-1}$ and $Y = 2.0 \times 10^{11}$ N/m^2.

We proceed much as in Problem 15.11:

$$\frac{\Delta L}{L_0} = \alpha\,\Delta T = (1.1 \times 10^{-5}\,°\text{C}^{-1})(35\,°\text{C}) = 3.85 \times 10^{-4}$$

so

$$F = YA\frac{\Delta L}{L_0} = (2.0 \times 10^{11}\ \text{N/m}^2)(45 \times 10^{-4}\ \text{m}^2)(3.85 \times 10^{-4}) = 3.5 \times 10^5\ \text{N}$$

Supplementary Problems

15.13 [I] Compute the increase in length of 50 m of copper wire when its temperature changes from 12 °C to 32 °C. For copper, $\alpha = 1.7 \times 10^{-5}\,°\text{C}^{-1}$. *Ans.* 1.7 cm

15.14 [I] A rod 3.0 m long is found to have expanded 0.091 cm in length after a temperature rise of 60 °C. What is α for the material of the rod? *Ans.* $5.1 \times 10^{-6}\,°\text{C}^{-1}$

15.15 [I] At 15.0 °C, a bare wheel has a diameter of 30.000 cm, and the inside diameter of its steel rim is 29.930 cm. To what temperature must the rim be heated so as to slip over the wheel? For this type of steel, $\alpha = 1.10 \times 10^{-5}\,°\text{C}^{-1}$. *Ans.* 227 °C

15.16 [II] An iron ball has a diameter of 6 cm and is 0.010 mm too large to pass through a hole in a brass plate when the ball and plate are at a temperature of 30 °C. At what temperature (the same for ball and plate) will the ball just pass through the hole? $\alpha = 1.2 \times 10^{-5}\,°\text{C}^{-1}$ and $1.9 \times 10^{-5}\,°\text{C}^{-1}$ for iron and brass, respectively. *Ans.* 54 °C

15.17 [II] (*a*) An aluminum measuring rod, which is correct at 5.0 °C, measures a certain distance as 88.42 cm at 35.0 °C. Determine the error in measuring the distance due to the expansion of the rod. (*b*) If this aluminum rod measures a length of steel as 88.42 cm at 35.0 °C, what is the correct length of the steel at 35 °C? The coefficient of linear expansion of aluminum is $22 \times 10^{-6}\,°\text{C}^{-1}$. *Ans.* (*a*) 0.058 cm; (*b*) 88 cm

15.18 [II] A solid sphere of mass m and radius b is spinning freely on its axis with angular velocity ω_0. When heated by an amount ΔT, its angular velocity changes to ω. Find ω_0/ω if the linear expansion coefficient for the material of the sphere is α. *Ans.* $1 + 2\alpha\,\Delta T + (\alpha\,\Delta T)^2$

15.19 [I] Calculate the increase in volume of 100 cm^3 of mercury when its temperature changes from 10 °C to 35 °C. The volume coefficient of expansion of mercury is 0.000 18 °C^{-1}. *Ans.* 0.45 cm^3

15.20 [II] The coefficient of linear expansion of glass is $9.0 \times 10^{-6}\,°\text{C}^{-1}$. If a specific gravity bottle holds 50.000 mL at 15 °C, find its capacity at 25 °C. *Ans.* 50.014 mL

15.21 [II] Determine the change in volume of a block of cast iron 5.0 cm × 10 cm × 6.0 cm, when the temperature changes from 15 °C to 47 °C. The coefficient of linear expansion of cast iron is 0.000 010 °C^{-1}. *Ans.* 0.29 cm^3

15.22 [II] A glass vessel is filled with exactly 1 liter of turpentine at 20 °C. What volume of the liquid will overflow if the temperature is raised to 86 °C? The coefficient of linear expansion of the glass is $9.0 \times 10^{-6}\,°\text{C}^{-1}$; the coefficient of volume expansion of turpentine is $97 \times 10^{-5}\,°\text{C}^{-1}$. *Ans.* 62 mL

15.23 [II] The density of gold is 19.30 g/cm^3 at 20.0 °C, and the coefficient of linear expansion is $14.3 \times 10^{-6}\,°\text{C}^{-1}$. Compute the density of gold at 90.0°C. Take a look at Problem 15.9. *Ans.* 19.2 g/cm^3

Chapter 16

Ideal Gases

AN IDEAL (OR PERFECT) GAS is composed of tiny, moving, noninteracting patrticles and it obeys the ***Ideal Gas Law***, given below. At low to moderate pressures, and at temperatures not too low, the following common gases can be considered ideal: air, nitrogen, oxygen, helium, hydrogen, and neon. Almost any chemically stable gas behaves ideally if it is far removed from conditions under which it will liquefy or solidify. In other words, a real gas behaves like an ideal gas when its atoms or molecules are so far apart that they do not appreciably interact with one another.

ONE MOLE OF A SUBSTANCE is the amount of the substance that contains as many particles as there are atoms in exactly 12 grams (0.012 kg) of the isotope carbon-12. It follows that one **kilomole** (kmol) of a substance is the mass (in kg) that is numerically equal to the molecular (or atomic) mass of the substance. For example, the molecular mass of hydrogen gas, H_2 is 2 kg/kmol; hence there are 2 kg in 1 kmol of H_2. Similarly, there are 32 kg in 1 kmol of O_2, and 28 kg in 1 kmol of N_2. We shall always use *kilo*moles and *kilo*grams in our calculations. Sometimes the term molecular (or atomic) *weight* is used, rather than molecular *mass*, but the latter is correct.

IDEAL GAS LAW: The ***absolute pressure*** P of n kilomoles of gas contained in a volume V is related to the ***absolute temperature*** T by

$$PV = nRT$$

where $R = 8314$ J/kmol·K is called the **universal gas constant**. If the volume contains m kilograms of gas that has a molecular (or atomic) mass M, then $n = m/M$.

SPECIAL CASES of the Ideal Gas Law, obtained by holding all but two of its parameters constant, are

Boyle's Law (n, T constant) : $PV = \text{constant}$

Charles' Law (n, P constant) : $\dfrac{V}{T} = \text{constant}$

Gay-Lussac's Law (n, V constant) : $\dfrac{P}{T} = \text{constant}$

ABSOLUTE ZERO: With n and P constant (Charles' Law), the volume of an ideal gas decreases linearly with T and (if the gas remained ideal) would reach zero at $T = 0\,\text{K}$. Similarly, with n and V constant (Gay-Lussac's Law), the pressure would decrease to zero with the temperature. This unique temperature, at which P and V would reach zero, is called **absolute zero**.

STANDARD CONDITIONS OR STANDARD TEMPERATURE AND PRESSURE (S.T.P.) are defined to be

$$T = 273.15\text{ K} = 0\,^\circ\text{C} \qquad P = 1.013 \times 10^5\text{ Pa} = 1\text{ atm}$$

Under standard conditions, 1 kmol of *ideal gas* occupies a volume of 22.4 m^3. Therefore, at S.T.P., 2 kg of H_2 occupies the same volume as 32 kg of O_2 or 28 kg of N_2, namely 22.4 m^3.

DALTON'S LAW OF PARTIAL PRESSURES: Define the **partial pressure** of one component of a gas mixture to be the pressure the component gas would exert if it alone occupied the entire volume. Then, the total pressure of a mixture of ideal, nonreactive gases is the sum of the partial pressures of the component gases. Which makes sense since each gas is effectively "unaware" of the presence of any of the other gases.

GAS-LAW PROBLEMS involving a change of conditions from (P_1, V_1, T_1) to (P_2, V_2, T_2) are usually easily solved by writing the gas law as

$$\frac{P_1 V_1}{T_1} = \frac{P_2 V_2}{T_2} \qquad \text{(at constant } n\text{)}$$

Remember that's absolute temperature and absolute pressure. Notice that pressure, because it appears on both sides of the equation, can be expressed in any units you like.

Solved Problems

16.1 [II] A mass of oxygen occupies 0.0200 m^3 at atmospheric pressure, 101 kPa, and 5.0 °C. Determine its volume if its pressure is increased to 108 kPa while its temperature is changed to 30 °C.

From

$$\frac{P_1 V_1}{T_1} = \frac{P_2 V_2}{T_2} \qquad \text{we have} \qquad V_2 = V_1 \left(\frac{P_1}{P_2}\right)\left(\frac{T_2}{T_1}\right)$$

But $T_1 = 5 + 273 = 278$ K and $T_2 = 30 + 273 = 303$ K, so

$$V_2 = (0.0200\ \text{m}^3)\left(\frac{101}{108}\right)\left(\frac{303}{278}\right) = 0.0204\ \text{m}^3$$

16.2 [II] On a day when atmospheric pressure is 76 cmHg, the pressure gauge on a tank reads the pressure inside to be 400 cmHg. The gas in the tank has a temperature of 9 °C. If the tank is heated to 31 °C by the Sun, and if no gas exits from it, what will the pressure gauge read?

$$\frac{P_1 V_1}{T_1} = \frac{P_2 V_2}{T_2} \qquad \text{so} \qquad P_2 = P_1 \left(\frac{T_2}{T_1}\right)\left(\frac{V_1}{V_2}\right)$$

But gauges on tanks usually read the difference in pressure between inside and outside; this is called the *gauge pressure*. Therefore,

$$P_1 = 76\ \text{cmHg} + 400\ \text{cmHg} = 476\ \text{cmHg}$$

Also, $V_1 = V_2$. We then have

$$P_2 = (476\ \text{cmHg})\left(\frac{273 + 31}{273 + 9}\right)(1.00) = 513\ \text{cmHg}$$

The gauge will read 513 cmHg − 76 cmHg = 437 cmHg.

16.3 [II] The gauge pressure in a car tire is 305 kPa when its temperature is 15 °C. After running at high speed, the tire has heated up and its pressure is 360 kPa. What is then the temperature of the gas in the tire? Assume atmospheric pressure to be 101 kPa.

Being careful to use only absolute temperature and absolute pressures:

$$\frac{P_1 V_1}{T_1} = \frac{P_2 V_2}{T_2} \quad \text{or} \quad T_2 = T_1\left(\frac{P_2}{P_1}\right)\left(\frac{V_2}{V_1}\right)$$

with $P_1 = 305\ \text{kPa} + 101\ \text{kPa} = 406\ \text{kPa}$ and $P_2 = 360\ \text{kPa} + 101\ \text{kPa} = 461\ \text{kPa}$

Then
$$T_2 = (273 + 15)\left(\frac{461}{406}\right)(1.00) = 327\ \text{K}$$

So the final temperature of the tire is $327 - 273 = 54\ °\text{C}$.

16.4 [II] Gas at room temperature and pressure is confined to a cylinder by a piston. The piston is now pushed in so as to reduce the volume to one-eighth of its original value. After the gas temperature has returned to room temperature, what is the gauge pressure of the gas in kPa? Local atmospheric pressure is 740 mm of mercury.

$$\frac{P_1 V_1}{T_1} = \frac{P_2 V_2}{T_2} \quad \text{or} \quad P_2 = P_1\left(\frac{V_1}{V_2}\right)\left(\frac{T_2}{T_1}\right)$$

Remember that you can work in any pressure units you like. Here $T_1 = T_2$, $P_1 = 740$ mmHg, and $V_2 = V_1/8$. Substitution gives

$$P_2 = (740\ \text{mmHg})(8)(1) = 5920\ \text{mmHg}$$

Gauge pressure is the difference between actual and atmospheric pressure. Therefore,

$$\text{Gauge pressure} = 5920\ \text{mmHg} - 740\ \text{mmHg} = 5180\ \text{mmHg}$$

Since 760 mmHg = 101 kPa, the gauge reading in kPa is

$$(5180\ \text{mmHg})\left(\frac{101\ \text{kPa}}{760\ \text{mmHg}}\right) = 690\ \text{kPa}$$

16.5 [II] An ideal gas has a volume of exactly 1 liter at 1.00 atm and −20 °C. To how many atmospheres of pressure must it be subjected to be compressed to 0.500 liter when the temperature is 40 °C?

$$\frac{P_1 V_1}{T_1} = \frac{P_2 V_2}{T_2} \quad \text{or} \quad P_2 = P_1\left(\frac{V_1}{V_2}\right)\left(\frac{T_2}{T_1}\right)$$

from which
$$P_2 = (1.00\ \text{atm})\left(\frac{1.00\ \text{L}}{0.500\ \text{L}}\right)\left(\frac{273\ \text{K} + 40\ \text{K}}{273\ \text{K} - 20\ \text{K}}\right) = 2.47\ \text{atm}$$

16.6 [II] A certain mass of hydrogen gas occupies 370 mL at 16 °C and 150 kPa. Find its volume at −21 °C and 420 kPa.

$$\frac{P_1 V_1}{T_1} = \frac{P_2 V_2}{T_2} \quad \text{gives} \quad V_2 = V_1\left(\frac{P_1}{P_2}\right)\left(\frac{T_2}{T_1}\right)$$

$$V_2 = (370\ \text{mL})\left(\frac{150\ \text{kPa}}{420\ \text{kPa}}\right)\left(\frac{273\ \text{K} - 21\ \text{K}}{273\ \text{K} + 16\ \text{K}}\right) = 115\ \text{mL}$$

16.7 [II] The density of nitrogen is 1.25 kg/m^3 at S.T.P. Determine the density of nitrogen at 42 °C and 730 mm of mercury.

Since $\rho = m/V$, we have $V_1 = m/\rho_1$ and $V_2 = m/\rho_2$ for a given mass of gas under two sets of conditions. Then

$$\frac{P_1 V_1}{T_1} = \frac{P_2 V_2}{T_2} \qquad \text{gives} \qquad \frac{P_1}{\rho_1 T_1} = \frac{P_2}{\rho_2 T_2}$$

Since S.T.P. are 760 mmHg and 273 K,

$$\rho_2 = \rho_1 \left(\frac{P_2}{P_1}\right)\left(\frac{T_1}{T_2}\right) = (1.25\ \text{kg/m}^3)\left(\frac{730\ \text{mmHg}}{760\ \text{mmHg}}\right)\left(\frac{273\ \text{K}}{273\ \text{K} + 42\ \text{K}}\right) = 1.04\ \text{kg/m}^3$$

Notice that pressures in mmHg can be used here because the units cancel in the ratio P_2/P_1.

16.8 [II] A 3.0-liter tank contains oxygen gas at 20 °C and a gauge pressure of 25×10^5 Pa. What mass of oxygen is in the tank? The molecular mass of oxgyen gas is 32 kg/kmol. Assume atmospheric pressure to be 1×10^5 Pa.

The absolute pressure of the gas is

$$P = (\text{gauge pressure}) + (\text{atmospheric pressure}) = (25 + 1) \times 10^5\ \text{N/m}^2 = 26 \times 10^5\ \text{N/m}^2$$

From the gas law, with $M = 32$ kg/kmol,

$$PV = \left(\frac{m}{M}\right) RT$$

$$(26 \times 10^5\ \text{N/m}^2)(3.0 \times 10^{-3}\ \text{m}^3) = \left(\frac{m}{32\ \text{kg/kmol}}\right)\left(8314 \frac{\text{J}}{\text{kmol} \cdot \text{K}}\right)(293\ \text{K})$$

Solving gives m, the mass of gas in the tank, as 0.10 kg.

16.9 [II] Determine the volume occupied by 4.0 g of oxygen ($M = 32$ kg/kmol) at S.T.P.

Method 1

Use the gas law directly:

$$PV = \left(\frac{m}{M}\right) RT$$

$$V = \left(\frac{1}{P}\right)\left(\frac{m}{M}\right) RT = \frac{(4.0 \times 10^{-3}\ \text{kg})(8314\ \text{J/kmol} \cdot \text{K})(273\ \text{K})}{(1.01 \times 10^5\ \text{N/m}^2)(32\ \text{kg/kmol})} = 2.8 \times 10^{-3}\ \text{m}^3$$

Method 2

Under S.T.P., 1 kmol occupies 22.4 m^3. Therefore, 32 kg occupies 22.4 m^3, and so 4 g occupies

$$\left(\frac{4.0\ \text{g}}{32\,000\ \text{g}}\right)(22.4\ \text{m}^3) = 2.8 \times 10^{-3}\ \text{m}^3$$

16.10 [II] A 2.0-mg droplet of liquid nitrogen is present in a 30 mL tube as it is sealed off at very low temperature. What will be the nitrogen pressure in the tube when it is warmed to 20 °C? Express your answer in atmospheres. (M for nitrogen is 28 kg/kmol.)

We use $PV = (m/M)RT$ to find

$$P = \frac{mRT}{MV} = \frac{(2.0 \times 10^{-6}\ \text{kg})(8314\ \text{J/kmol} \cdot \text{K})(293\ \text{K})}{(28\ \text{kg/kmol})(30 \times 10^{-6}\ \text{m}^3)} = 5800\ \text{N/m}^2$$

$$= (5800\ \text{N/m}^2)\left(\frac{1.0\ \text{atm}}{1.01 \times 10^5\ \text{N/m}^2}\right) = 0.057\ \text{atm}$$

16.11 [II] A tank of volume 590 liters contains oxygen at 20 °C and 5.0 atm pressure. Calculate the mass of oxygen in the tank. $M = 32$ kg/kmol for oxygen.

We use $PV = (m/M)RT$ to get

$$m = \frac{PVM}{RT} = \frac{(5 \times 1.01 \times 10^5 \text{ N/m}^2)(0.59 \text{ m}^3)(32 \text{ kg/kmol})}{(8314 \text{ J/kmol}\cdot\text{K})(293 \text{ K})} = 3.9 \text{ kg}$$

16.12 [II] At 18 °C and 765 mmHg, 1.29 liters of an ideal gas has a mass of 2.71 g. Compute the molecular mass of the gas.

We use $PV = (m/M)RT$ and the fact that 760 mmHg = 1.00 atm to obtain

$$M = \frac{mRT}{PV} = \frac{(0.00271 \text{ kg})(8314 \text{ J/kmol}\cdot\text{K})(291 \text{ K})}{[(765/760)(1.01 \times 10^5 \text{ N/m}^2)](0.001\,29 \text{ m}^3)} = 50.0 \text{ kg/kmol}$$

16.13 [II] Compute the volume of 8.0 g of helium ($M = 4.0$ kg/kmol) at 15 °C and 480 mmHg.

We use $PV = (m/M)RT$ to obtain

$$V = \frac{mRT}{MP} = \frac{(0.0080 \text{ kg})(8314 \text{ J/kmol}\cdot\text{K})(288 \text{ K})}{(4.0 \text{ kg/kmol})[(480/760)(1.01 \times 10^5 \text{ N/m}^2)]} = 0.075 \text{ m}^3 = 75 \text{ liters}$$

16.14 [II] Find the density of methane ($M = 16$ kg/kmol) at 20 °C and 5.0 atm.

We use $PV = (m/M)RT$ and $\rho = m/V$ to get

$$\rho = \frac{PM}{RT} = \frac{(5.0 \times 1.01 \times 10^5 \text{ N/m}^2)(16 \text{ kg/kmol})}{(8314 \text{ J/kmol}\cdot\text{K})(293 \text{ K})} = 3.3 \text{ kg/m}^3$$

16.15 [II] A fish emits a 2.0 mm³ bubble at a depth of 15 m in a lake. Find the volume of the bubble as it reaches the surface. Assume its temperature does not change.

The absolute pressure in the bubble at depth h is

$$P = \rho gh + \text{atmospheric pressure}$$

where $\rho = 1000$ kg/m³ and atmospheric pressure is about 100 kPa. At 15 m,

$$P_1 = (1000 \text{ kg/m}^3)(9.8 \text{ m/s}^2)(15 \text{ m}) + 100 \text{ kPa} = 247 \text{ kPa}$$

and at the surface, $P_2 = 100$ kPa. Following the usual procedure, we get

$$V_2 = V_1\left(\frac{P_1}{P_2}\right)\left(\frac{T_2}{T_1}\right) = (2.0 \text{ mm}^3)\left(\frac{247}{100}\right)(1.0) = 4.9 \text{ mm}^3$$

16.16 [II] A 15 cm long test tube of uniform bore is lowered, open end down, into a fresh-water lake. How far below the surface of the lake must the water level be in the tube if one-third of the tube is to be filled with water?

Let h be the depth of the water in the tube below the lake's surface. The air pressure P_2 in the tube at depth h must equal atmospheric pressure P_a plus the pressure of water at that depth:

$$P_2 = P_a + \rho gh$$

The gas law gives us the value of P_2 as

$$P_2 = (P_1)\left(\frac{V_1}{V_2}\right)\left(\frac{T_2}{T_1}\right) = (1.01 \times 10^5 \text{ Pa})\left(\frac{3}{2}\right)(1.00) = 1.50 \times 10^5 \text{ Pa}$$

Then, from the relation between P_2 and h,

$$h = \frac{P_2 - P_a}{\rho g} = \frac{0.50 \times 10^5 \text{ Pa}}{(1000 \text{ kg/m}^3)(9.81 \text{ m/s}^2)} = 5.1 \text{ m}$$

where atmospheric pressure has been taken as 100 kPa.

16.17 [II] A tank contains 18 kg of N_2 gas ($M = 28$ kg/kmol) at a pressure of 4.50 atm. How much H_2 gas ($M = 2.0$ kg/kmol) at 3.50 atm would the same tank contain?

We write the gas law twice, once for each gas:

$$P_N V = n_N RT \qquad \text{and} \qquad P_H V = n_H RT$$

Division of one equation by the other eliminates V, R, and T:

$$\frac{n_H}{n_N} = \frac{P_H}{P_N} = \frac{3.50 \text{ atm}}{4.50 \text{ atm}} = 0.778$$

But

$$n_N = \frac{m}{M} = \frac{18 \text{ kg}}{28 \text{ kg/kmol}} = 0.643 \text{ kmol}$$

so

$$n_H = (n_N)(0.778) = (0.643 \text{ kmol})(0.778) = 0.500 \text{ kmol}$$

Then, from $n = m/M$, we have

$$m_H = (0.500 \text{ kmol})(2.0 \text{ kg/kmol}) = 1.0 \text{ kg}$$

16.18 [II] In a gaseous mixture at 20 °C the partial pressures of the components are as follows: hydrogen, 200 mmHg; carbon dioxide, 150 mmHg; methane, 320 mmHg; ethylene, 105 mmHg. What are (*a*) the total pressure of the mixture and (*b*) the mass fraction of hydrogen? ($M_H = 2.0$ kg/kmol, $M_{CO_2} = 44$ kg/kmol, $M_{methane} = 16$ kg/kmol, $M_{ethylene} = 30$ kg/kmol.)

(*a*) According to Dalton's Law,

Total pressure = sum of partial pressures = 200 mmHg + 150 mmHg + 320 mmHg + 105 mmHg = 775 mmHg

(*b*) From the Gas Law, $m = M(PV/RT)$. The mass of hydrogen gas present is

$$m_H = M_H P_H \left(\frac{V}{RT}\right)$$

The total mass of gas present, m_t, is the sum of similar terms:

$$m_t = (M_H P_H + M_{CO_2} P_{CO_2} + M_{methane} P_{methane} + M_{ethylene} P_{ethylene})\left(\frac{V}{RT}\right)$$

The required fraction is then

$$\frac{m_H}{m_t} = \frac{M_H P_H}{M_H P_H + M_{CO_2} P_{CO_2} + M_{methane} P_{methane} + M_{ethylene} P_{ethylene}}$$

$$\frac{m_H}{m_t} = \frac{(2.0 \text{ kg/kmol})(200 \text{ mmHg})}{(2.0 \text{ kg/kmol})(200 \text{ mmHg}) + (44 \text{ kg/kmol})(150 \text{ mmHg}) + (16 \text{ kg/kmol})(320 \text{ mmHg}) + (30 \text{ kg/kmol})(105 \text{ mmHg})} = 0.026$$

Supplementary Problems

16.19 [I] A certain mass of an ideal gas occupies a volume of 4.00 m^3 at 758 mmHg. Compute its volume at 635 mmHg if the temperature remains unchanged. *Ans.* 4.77 m^3

16.20 [I] A given mass of ideal gas occupies 38 mL at 20 °C. If its pressure is held constant, what volume does it occupy at a temperature of 45 °C? *Ans.* 41 mL

16.21 [I] On a day when atmospheric pressure is 75.83 cmHg, a pressure gauge on a tank of gas reads a pressure of 258.5 cmHg. What is the absolute pressure (in atmospheres and kPa) of the gas in the tank? *Ans.* 334.3 cmHg = 4.398 atm = 445.6 kPa

16.22 [II] A tank of ideal gas is sealed off at 20 °C and 1.00 atm pressure. What will be the pressure (in kPa and mmHg) in the tank if the gas temperature is decreased to −35 °C? *Ans.* 82 kPa = 6.2×10^2 mmHg

16.23 [II] Given 1000 mL of helium at 15 °C and 763 mmHg, determine its volume at −6 °C and 420 mmHg. *Ans.* 1.68×10^3 mL

16.24 [II] One kilomole of ideal gas occupies 22.4 m^3 at 0 °C and 1 atm. (*a*) What pressure is required to compress 1.00 kmol into a 5.00 m^3 container at 100 °C? (*b*) If 1.00 kmol was to be sealed in a 5.00 m^3 tank that could withstand a gauge pressure of only 3.00 atm, what would be the maximum temperature of the gas if the tank was not to burst? *Ans.* (*a*) 6.12 atm; (*b*) −30 °C

16.25 [II] Air is trapped in the sealed lower end of a capillary tube by a mercury column as shown in Fig. 16-1. The top of the tube is open. The temperature is 14 °C, and atmospheric pressure is 740 mmHg. What length would the trapped air column have if the temperature were 30 °C and atmospheric pressure were 760 mmHg? *Ans.* 12.4 cm

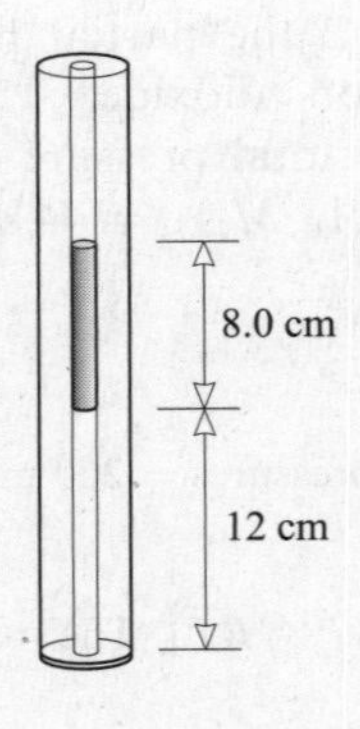

Fig. 16-1

16.26 [II] Air is trapped in the sealed lower part of the vertical capillary tube shown in Fig. 16-1 by an 8.0 cm long mercury column. The top is open, and the system is at equilibrium. What will be the length of the trapped air column if the tube is now tilted so it makes an angle of 65° to the vertical? Take $P_a = 76$ cmHg. *Ans.* 0.13 m

16.27 [II] On a day when the barometer reads 75.23 cm, a reaction vessel holds 250 mL of ideal gas at 20.0 °C. An oil manometer ($\rho = 810$ kg/m^3) reads the pressure in the vessel to be 41.0 cm of oil and below atmospheric pressure. What volume will the gas occupy under S.T.P.? *Ans.* 233 mL

16.28 [II] A 5000-cm^3 tank contains an ideal gas ($M = 40$ kg/kmol) at a gauge pressure of 530 kPa and a temperature of 25 °C. Assuming atmospheric pressure to be 100 kPa, what mass of gas is in the tank? *Ans.* 0.051 kg

16.29 [II] The pressure of air in a reasonably good vacuum might be 2.0×10^{-5} mmHg. What mass of air exists in a 250 mL volume at this pressure and 25 °C? Take $M = 28$ kg/kmol for air. *Ans.* 7.5×10^{-12} kg

16.30 [II] What volume will 1.216 g of SO_2 gas ($M = 64.1$ kg/kmol) occupy at 18.0 °C and 755 mmHg if it acts like an ideal gas? *Ans.* 457 mL

16.31 [II] Compute the density of H_2S gas ($M = 34.1$ kg/kmol) at 27 °C and 2.00 atm, assuming it to be ideal. *Ans.* 2.76 kg/m^3

16.32 [II] A 30-mL tube contains 0.25 g of water vapor ($M = 18$ kg/kmol) at a temperature of 340 °C. Assuming the gas to be ideal, what is its pressure? *Ans.* 2.4 MPa

16.33 [II] One method for estimating the temperature at the center of the Sun is based on the Ideal Gas Law. If the center is assumed to consist of gases whose average M is 0.70 kg/kmol, and if the density and pressure are 90×10^3 kg/m^3 and 1.4×10^{11} atm, respectively, calculate the temperature *Ans.* 1.3×10^7 K

16.34 [II] A 500-mL sealed flask contains nitrogen at a pressure of 76.00 cmHg. A tiny glass tube lies at the bottom of the flask. Its volume is 0.50 mL and it contains hydrogen gas at a pressure of 4.5 atm. Suppose the glass tube is now broken so that the hydrogen fills the flask. What is the new pressure in the flask? *Ans.* 76.34 cmHg

16.35 [II] As shown in Fig. 16-2, two flasks are connected by an initially closed stopcock. One flask contains krypton gas at 500 mmHg, while the other contains helium at 950 mmHg. The stopcock is now opened so that the gases mix. What is the final pressure in the system? Assume constant temperature. *Ans.* 789 mmHg

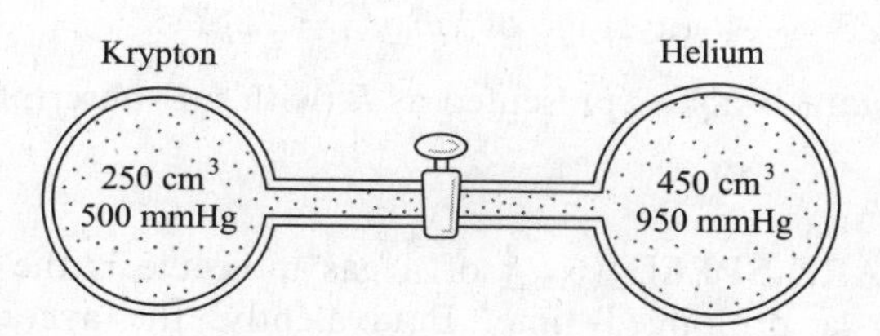

Fig. 16-2

16.36 [II] An air bubble of volume V_0 is released near the bottom of a lake at a depth of 11.0 m. What will be its new volume at the surface? Assume its temperature to be 4.0 °C at the release point and 12 °C at the surface. The water has a density of 1000 kg/m^3, and atmospheric pressure is 75 cmHg. *Ans.* $2.1 V_0$

16.37 [II] A cylindrical diving bell (a vertical cylinder with open bottom end and closed top end) 12.0 m high is lowered in a lake until water within the bell rises 8.0 m from the bottom end. Determine the distance from the top of the bell to the surface of the lake. (Atmospheric pressure = 1.00 atm.) *Ans.* 20.6 m − 4.0 m = 16.6 m

Chapter 17

Kinetic Theory

THE KINETIC THEORY considers matter to be composed of discrete particles (atoms and/or molecules) in continual motion. In a gas, the molecules are in random motion with a wide distribution of speeds ranging from zero to very large values.

AVOGADRO'S NUMBER (N_A) is the number of particles (molecules or atoms) in 1 kmol of substance. For all substances,

$$N_A = 6.022 \times 10^{26} \text{ particles/kmol} = 6.022 \times 10^{23} \text{ particles/mol}$$

As examples, $M = 2$ kg/kmol for H_2 and $M = 32$ kg/kmol for O_2. Therefore, 2 kg of H_2 and 32 kg of O_2 each contain 6.02×10^{26} molecules.

THE MASS OF A MOLECULE (or atom) can be found from the molecular (or atomic) mass M of the substance and Avogadro's number N_A. Since M kg of substance contains N_A particles, the mass m_0 of one particle is given by

$$m_0 = \frac{M}{N_A}$$

THE AVERAGE TRANSLATIONAL KINETIC ENERGY of a gas molecule is $3k_BT/2$, where T is the absolute temperature of the gas and $k_B = R/N_A = 1.381 \times 10^{-23}$ J/K is **Boltzmann's constant**. In other words, for a molecule of mass m_0,

$$(\text{average of } \tfrac{1}{2}m_0v^2) = \tfrac{3}{2}k_BT$$

Note that Boltzmann's constant is also represented as k (with no subscript) in the literature.

THE ROOT MEAN SQUARE SPEED (v_{rms}) of a gas molecule is the square root of the average of v^2 for a molecule over a prolonged time. Equivalently, the average may be taken over all molecules of the gas at a given instant. From the expression for the average kinetic energy, the rms speed is

$$v_{\text{rms}} = \sqrt{\frac{3k_BT}{m_0}}$$

THE ABSOLUTE TEMPERATURE (T) of an ideal gas has a meaning that is found by solving $\frac{1}{2}m_0v_{\text{rms}}^2 = \frac{3}{2}k_BT$. It leads to

$$T = \left(\frac{2}{3k_B}\right)\left(\frac{1}{2}m_0v_{\text{rms}}^2\right)$$

The absolute temperature of an ideal gas is a measure of its average translational kinetic energy (KE) per molecule.

THE PRESSURE (P) of an ideal gas was given in Chapter 16 in the form $PV = (m/M)RT$. Noticing that $m = Nm_0$, where N is the number of molecules in the volume V, and replacing T by the value determined above, we have

$$PV = \tfrac{1}{3} N m_0 v_{\text{rms}}^2$$

Further, since $Nm_0/V = \rho$, the density of the gas,

$$P = \tfrac{1}{3} \rho v_{\text{rms}}^2$$

THE MEAN FREE PATH (m.f.p.) of a gas molecule is the average distance such a molecule moves between collisions. For an ideal gas of spherical molecules with radius b,

$$\text{m.f.p.} = \frac{1}{4\pi\sqrt{2}b^2(N/V)}$$

where N/V is the number of molecules per unit volume.

Solved Problems

17.1 [I] Ordinary nitrogen gas consists of molecules of N_2. Find the mass of one such molecule. The molecular mass is 28 kg/kmol.

$$m_0 = \frac{M}{N_A} = \frac{28\text{ kg/kmol}}{6.02 \times 10^{26}\text{ kmol}^{-1}} = 4.7 \times 10^{-26}\text{ kg}$$

17.2 [II] Helium gas consists of separate He atoms rather than molecules. How many helium atoms, He, are there in 2.0 g of helium? $M = 4.0$ kg/kmol for He.

Method 1

One kilomole of He is 4.0 kg, and it contains N_A atoms. But 2.0 g is equivalent to

$$\frac{0.002\,0\text{ kg}}{4.0\text{ kg/kmol}} = 0.000\,50\text{ kmol}$$

of helium. Therefore,

$$\text{Number of atoms in 2.0 g} = (0.000\,50\text{ kmol})N_A$$
$$= (0.000\,50\text{ kmol})(6.02 \times 10^{26}\text{ kmol}^{-1}) = 3.0 \times 10^{23}$$

Method 2

The mass of a helium atom is

$$m_0 = \frac{M}{N_A} = \frac{4.0\text{ kg/kmol}}{6.02 \times 10^{26}\text{ kmol}^{-1}} = 6.64 \times 10^{-27}\text{ kg}$$

hence $$\text{Number in 2.0 g} = \frac{0.002\,0\text{ kg}}{6.64 \times 10^{-27}\text{ kg}} = 3.0 \times 10^{23}$$

17.3 [II] A droplet of mercury has a radius of 0.50 mm. How many mercury atoms are in the droplet? For Hg, $M = 202$ kg/kmol and $\rho = 13\,600$ kg/m^3.

The volume of the droplet is

$$V = \frac{4\pi r^3}{3} = \left(\frac{4\pi}{3}\right)(5.0 \times 10^{-4}\ \text{m})^3 = 5.24 \times 10^{-10}\ \text{m}^3$$

The mass of the droplet is

$$m = \rho V = (13\,600\ \text{kg/m}^3)(5.24 \times 10^{-10}\ \text{m}^3) = 7.1 \times 10^{-6}\ \text{kg}$$

The mass of a mercury atom is

$$m_0 = \frac{M}{N_A} = \frac{202\ \text{kg/kmol}}{6.02 \times 10^{26}\ \text{kmol}^{-1}} = 3.36 \times 10^{-25}\ \text{kg}$$

The number of atoms in the droplet is then

$$\text{Number} = \frac{m}{m_0} = \frac{7.1 \times 10^{-6}\ \text{kg}}{3.36 \times 10^{-25}\ \text{kg}} = 2.1 \times 10^{19}$$

17.4 [II] How many molecules are there in 70 mL of benzene? For benzene, $\rho = 0.88\ \text{g/cm}^3$ and $M = 78\ \text{kg/kmol}$.

Remember that $1\ \text{g/cm}^3 = 1000\ \text{kg/m}^3$ and so here $\rho = 880\ \text{kg/m}^3$.

$$\text{Mass of } 70\ \text{cm}^3 = m = \rho V = (880\ \text{kg/m}^3)(70 \times 10^{-6}\ \text{m}^3) = 0.061\,6\ \text{kg}$$

$$m_0 = \frac{M}{N_A} = \frac{78\ \text{kg/kmol}}{6.02 \times 10^{26}\ \text{kmol}^{-1}} = 1.30 \times 10^{-25}\ \text{kg}$$

$$\text{Number in } 70\ \text{cm}^3 = \frac{m}{m_0} = \frac{0.0616\ \text{kg}}{1.30 \times 10^{-25}\ \text{kg}} = 4.8 \times 10^{23}$$

17.5 [I] Find the rms speed of a nitrogen molecule ($M = 28$ kg/kmol) in air at 0 °C.

We know that $\frac{1}{2} m_0 v_{\text{rms}}^2 = \frac{3}{2} k_B T$ and so

$$v_{\text{rms}} = \sqrt{\frac{3k_B T}{m_0}}$$

But

$$m_0 = \frac{M}{N_A} = \frac{28\ \text{kg/kmol}}{6.02 \times 10^{26}\ \text{kmol}^{-1}} = 4.65 \times 10^{-26}\ \text{kg}$$

Therefore

$$v_{\text{rms}} = \sqrt{\frac{3(1.38 \times 10^{-23}\ \text{J/K})(273\ \text{K})}{4.65 \times 10^{-26}\ \text{kg}}} = 0.49\ \text{km/s}$$

17.6 [II] Suppose a particular gas molecule at the surface of the Earth happens to have the rms speed for that gas at exactly 0 °C. If it were to go straight up without colliding with other molecules, how high would it rise? Assume g is constant over the trajectory.

The molecule's KE is initially

$$\text{KE} = \tfrac{1}{2} m_0 v_{\text{rms}}^2 = \tfrac{3}{2} k_B T$$

The molecule will rise until its KE has been changed to PE_G. Therefore, calling the height to which it rises h, we have

$$\tfrac{3}{2} k_B T = m_0 g h$$

Solving for h yields

$$h = \left(\frac{1}{m_0}\right)\left(\frac{3k_BT}{2g}\right) = \left(\frac{1}{m_0}\right)\left[\frac{(3)(1.38 \times 10^{-23}\ \text{J/K})(273\ \text{K})}{2(9.81\ \text{m/s}^2)}\right]$$

$$= \frac{5.76 \times 10^{-22}\ \text{kg}\cdot\text{m}}{m_0}$$

where m_0 is in kg. The height varies inversely with the mass of the molecule. For an N_2 molecule, $m_0 = 4.65 \times 10^{-26}$ kg (Problem 17.5), and in this case h turns out to be 12.4 km.

17.7 [I] Air at room temperature has a density of about 1.29 kg/m^3. Assuming it to be entirely one gas, find v_{rms} for its molecules.

Because $P = \frac{1}{3}\rho v_{rms}^2$, we have

$$v_{rms} = \sqrt{\frac{3P}{\rho}} = \sqrt{\frac{3(100 \times 10^3\ \text{Pa})}{1.29\ \text{kg/m}^3}} \approx 480\ \text{m/s}$$

where we assumed atmospheric pressure to be 100 kPa.

17.8 [I] Find the translational kinetic energy of one gram mole of any ideal gas at 0 °C.

For an ideal gas, $\frac{3}{2}k_BT = \frac{1}{2}m_0v_{rms}^2$, which is the KE of each molecule. One gram mole contains $N_A \times 10^{-3}$ molecules. Hence the total KE per mole is

$$\text{KE}_{total} = (N_A \times 10^{-3})(\tfrac{3}{2}k_BT) = 3 \times 10^{-3}\frac{RT}{2} = 3.4\ \text{kJ}$$

where T was taken as 273 K, and use was made of the fact that $k_BN_A = R$.

17.9 [II] There is about one hydrogen atom per cm^3 in outer space, where the temperature (in the shade) is about 3.5 K. Find the rms speed of these atoms and the pressure they exert.

Keeping in mind that $k_BN_A = R$ and that $m_0 = M/N_A$,

$$v_{rms} = \sqrt{\frac{3k_BT}{m_0}} = \sqrt{\frac{3k_BT}{M/N_A}} = \sqrt{\frac{3RT}{M}} \approx 295\ \text{m/s or } 0.30\ \text{km/s}$$

where M for hydrogen is 1.0 kg/kmol and $T = 3.5$ K. We can now use $P = \rho v_{rms}^2/3$ to find the pressure. The mass m_0 of a hydrogen atom is (1.0 kg/kmol)/N_A. Since 1 m^3 = 10^6 cm^3 there are $N = 10^6$ atoms/m^3 and we have

$$\rho = \frac{Nm_0}{V} = \left(\frac{N}{V}\right)m_0 = 10^6\left(\frac{1}{N_A}\right)\ \text{kg/m}^3$$

and

$$P = \tfrac{1}{3}\rho v_{rms}^2 = \frac{1}{3}\left(\frac{10^6}{6.02 \times 10^{26}}\right)(295)^2 = 5 \times 10^{-17}\ \text{Pa}$$

17.10 [I] Find the following ratios for hydrogen ($M = 2.0$ kg/kmol) and nitrogen ($M = 28$ kg/kmol) gases at the same temperature: (*a*) $(\text{KE})_H/(\text{KE})_N$ and (*b*) (rms speed)$_H$/(rms speed)$_N$.

(*a*) The average translational KE of a molecule, $\frac{3}{2}k_BT$, depends only on temperature. Therefore the ratio $(\text{KE})_H/(\text{KE})_N = 1$.

(*b*)

$$\frac{(v_{rms})_H}{(v_{rms})_N} = \sqrt{\frac{3k_BT/m_{0H}}{3k_BT/m_{0N}}} = \sqrt{\frac{m_{0N}}{m_{0H}}}$$

But $m_0 = M/N_A$, so

$$\frac{(v_{\text{rms}})_{\text{H}}}{(v_{\text{rms}})_{\text{N}}} = \sqrt{\frac{M_{\text{N}}}{M_{\text{H}}}} = \sqrt{\frac{28}{2.0}} = 3.7$$

17.11 [II] Certain ideal gas molecules behave like spheres of radius 3.0×10^{-10} m. Find the mean free path of these molecules under S.T.P.

Method 1

We know that at S.T.P. 1.00 kmol of substance occupies 22.4 m^3. The number of molecules per unit volume, N/V, can be found from the fact that in 22.4 m^3 there are $N_A = 6.02 \times 10^{26}$ molecules. The mean free path is given by

$$\text{m.f.p.} = \frac{1}{4\pi\sqrt{2}b^2(N/V)} = \frac{1}{4\pi\sqrt{2}(3.0 \times 10^{-10}\ \text{m})^2}\left(\frac{22.4\ \text{m}^3}{6.02 \times 10^{26}}\right) = 2.4 \times 10^{-8}\ \text{m}$$

Method 2

Because $M = m_0 N_A = m_0(R/k_B)$ and $m = Nm_0$,

$$PV = \left(\frac{m}{M}\right)RT \qquad \text{becomes} \qquad PV = Nk_BT$$

and so

$$\frac{N}{V} = \frac{P}{k_BT} = \frac{1.01 \times 10^5\ \text{N/m}^2}{(1.38 \times 10^{-23}\ \text{J/K})(273\ \text{K})} = 2.68 \times 10^{25}\ \text{m}^{-3}$$

We then use the mean free path equation as in method 1.

17.12 [II] At what pressure will the mean free path be 50 cm for spherical molecules of radius 3.0×10^{-10} m? Assume an ideal gas at 20 °C.

From the expression for the mean free path, we obtain

$$\frac{N}{V} = \frac{1}{4\pi\sqrt{2}b^2(\text{m.f.p.})}$$

Combining this with the Ideal Gas Law in the form $PV = Nk_BT$ (see Problem 17.11) gives

$$P = \frac{k_BT}{4\pi\sqrt{2}b^2(\text{m.f.p.})} = \frac{(1.38 \times 10^{-23}\ \text{J/K})(293\ \text{K})}{4\pi\sqrt{2}(3.0 \times 10^{-10}\ \text{m})^2(0.50\ \text{m})} = 5.1\ \text{mPa}$$

Supplementary Problems

17.13 [I] Find the mass of a neon atom. The atomic mass of neon is 20.2 kg/kmol. *Ans.* 3.36×10^{-26} kg

17.14 [II] A typical polymer molecule in polyethylene might have a molecular mass of 15×10^3. (*a*) What is the mass in kilograms of such a molecule? (*b*) How many such molecules would make up 2 g of polymer? *Ans.* (*a*) 2.5×10^{-23} kg; (*b*) 8×10^{19}

17.15 [II] A certain strain of tobacco mosaic virus has $M = 4.0 \times 10^7$ kg/kmol. How many molecules of the virus are present in 1.0 mL of a solution that contains 0.10 mg of virus per mL? *Ans.* 1.5×10^{12}

17.16 [II] An electronic vacuum tube was sealed off during manufacture at a pressure of 1.2×10^{-7} mmHg at 27 °C. Its volume is 100 cm^3. (*a*) What is the pressure in the tube (in Pa)? (*b*) How many gas molecules remain in the tube? *Ans.* (*a*) 1.6×10^{-5} Pa; (*b*) 3.8×10^{11}

17.17 [II] The pressure of helium gas in a tube is 0.200 mmHg. If the temperature of the gas is 20 °C, what is the density of the gas? (Use $M_{He} = 4.0$ kg/kmol.) *Ans.* 4.4×10^{-5} kg/m^3

17.18 [II] At what temperature will the molecules of an ideal gas have twice the rms speed they have at 20 °C? *Ans.* 1170 K $\approx$ 900 °C

17.19 [II] An object must have a speed of at least 11.2 km/s to escape from the Earth's gravitational field. At what temperature will v_{rms} for H_2 molecules equal the escape speed? Repeat for N_2 molecules. ($M_{H_2} = 2.0$ kg/kmol and $M_{N_2} = 28$ kg/kmol.) *Ans.* 1.0×10^4 K; 1.4×10^5 K

17.20 [II] In a certain region of outer space there are an average of only five molecules per cm^3. The temperature there is about 3 K. What is the average pressure of this very dilute gas? *Ans.* 2×10^{-16} Pa

17.21 [II] A cube of aluminum has a volume of 1.0 cm^3 and a mass of 2.7 g. (*a*) How many aluminum atoms are there in the cube? (*b*) How large a volume is associated with each atom? (*c*) If each atom were a cube, what would be its edge length? $M = 108$ kg/kmol for aluminum. *Ans.* (*a*) 1.5×10^{22}; (*b*) 6.6×10^{-29} m^3; (*c*) 4.0×10^{-10} m

17.22 [II] The rms speed of nitrogen molecules in the air at S.T.P. is about 490 m/s. Find their mean free path and the average time between collisions. The radius of a nitrogen molecule can be taken to be 2.0×10^{-10} m. *Ans.* 5.2×10^{-8} m, 1.1×10^{-10} s

17.23 [II] What is the mean free path of a gas molecule (radius 2.5×10^{-10} m) in an ideal gas at 500 °C when the pressure is 7.0×10^{-6} mmHg? *Ans.* 10 m

Chapter 18

Heat Quantities

THERMAL ENERGY is the random kinetic energy of the particles (usually electrons, ions, atoms, and molecules) composing a system.

HEAT (Q) is thermal energy in transit from a system (or aggregate of electrons, ions, and atoms) at one temperature to a system that is in contact with it, but is at a lower temperature. Its SI unit is the joule. Other units used for heat are the ***calorie*** (1 cal = 4.184 J) and the British thermal unit (1 Btu = 1054 J). The "Calorie" used by nutritionists is called the "large calorie" and is actually a kilocalorie (1 Cal = 1 kcal = 10^3 cal).

THE SPECIFIC HEAT (or ***specific heat capacity***, c) of a substance is the quantity of heat required to change the temperature of a unit mass of the substance by one degree Celsius or equivalently by one kelvin.

If a quantity of heat ΔQ is required to produce a temperature change ΔT in a mass m of substance, then the specific heat is

$$c = \frac{\Delta Q}{m\,\Delta T} \qquad \text{or} \qquad \Delta Q = cm\,\Delta T$$

In the SI, c has the unit J/kg·K, which is equivalent to J/kg·°C. Also widely used is the unit cal/g·°C, where 1 cal/g·°C = 4184 J/kg·°C.

Each substance has a characteristic value of specific heat, which varies slightly with temperature. For water, c = 4180 J/kg·°C = 1.00 cal/g·°C.

THE HEAT GAINED (OR LOST) by a body (whose phase does not change) as it undergoes a temperature change ΔT, is given by

$$\Delta Q = mc\,\Delta T$$

THE HEAT OF FUSION (L_f) of a crystalline solid is the quantity of heat required to melt a unit mass of the solid at constant temperature. It is also equal to the quantity of heat given off by a unit mass of the molten solid as it crystallizes at this same temperature. The heat of fusion of water at 0 °C is about 335 kJ/kg or 80 cal/g.

THE HEAT OF VAPORIZATION (L_v) of a liquid is the quantity of heat required to vaporize a unit mass of the liquid at constant temperature. For water at 100 °C, L_v is about 2.26 MJ/kg or 540 cal/g.

THE HEAT OF SUBLIMATION of a solid substance is the quantity of heat required to convert a unit mass of the substance from the solid to the gaseous state at constant temperature.

CALORIMETRY PROBLEMS involve the sharing of thermal energy among initially hot objects and cold objects. Since energy must be conserved, one can write the following equation:

$$\text{Sum of heat changes for all objects} = 0$$

Here the heat flowing out of the high temperature system ($\Delta Q_{out} < 0$) numerically equals the heat flowing into the low temperature system ($\Delta Q_{in} > 0$) and so the sum is zero. This, of course, assumes that no thermal energy is otherwise lost from the system.

ABSOLUTE HUMIDITY is the mass of water vapor present per unit volume of gas (usually the atmosphere). Typical units are kg/m^3 and g/cm^3.

RELATIVE HUMIDITY (R.H.) is the ratio obtained by dividing the mass of water vapor per unit volume *present in the air* by the mass of water vapor per unit volume *present in saturated air at the same temperature*. When it is expressed in percent, the ratio is multiplied by 100.

DEW POINT: Cooler air at saturation contains less water than warmer air does at saturation. When air is cooled, it eventually reaches a temperature at which it is saturated. This temperature is called the *dew point*. At temperatures lower than this, water condenses out of the air.

Solved Problems

18.1 [I] (*a*) How much heat is required to raise the temperature of 250 mL of water from 20.0 °C to 35.0 °C? (*b*) How much heat is lost by the water as it cools back down to 20.0 °C?

Since 250 mL of water has a mass of 250 g, and since $c = 1.00\ \text{cal/g}\cdot{}^\circ\text{C}$ for water, we have

(*a*) $$\Delta Q = mc\,\Delta T = (250\ \text{g})(1.00\ \text{cal/g}\cdot{}^\circ\text{C})(15.0\ {}^\circ\text{C}) = 3.75 \times 10^3\ \text{cal} = 15.7\ \text{kJ}$$

(*b*) $$\Delta Q = mc\,\Delta T = (250\ \text{g})(1.00\ \text{cal/g}\cdot{}^\circ\text{C})(-15.0\ {}^\circ\text{C}) = -3.75 \times 10^3\ \text{cal} = -15.7\ \text{kJ}$$

Notice that heat-in (i.e., the heat that enters an object) is positive, whereas heat-out (i.e., the heat that leaves an object) is negative.

18.2 [I] How much heat does 25 g of aluminum give off as it cools from 100 °C to 20 °C? For aluminum, $c = 880\ \text{J/kg}\cdot{}^\circ\text{C}$.

$$\Delta Q = mc\,\Delta T = (0.025\ \text{kg})(880\ \text{J/kg}\cdot{}^\circ\text{C})(-80\ {}^\circ\text{C}) = -1.8\ \text{kJ} = -0.42\ \text{kcal}$$

18.3 [I] A certain amount of heat is added to a mass of aluminum ($c = 0.21\ \text{cal/g}\cdot{}^\circ\text{C}$), and its temperature is raised 57 °C. Suppose that the same amount of heat is added to the same mass of copper ($c = 0.093\ \text{cal/g}\cdot{}^\circ\text{C}$). How much does the temperature of the copper rise?

Because ΔQ is the same for both, we have

$$mc_{\text{Al}}\Delta T_{\text{Al}} = mc_{\text{Cu}}\,\Delta T_{\text{Cu}}$$

or

$$\Delta T_{\text{Cu}} = \left(\frac{c_{\text{Al}}}{c_{\text{Cu}}}\right)(\Delta T_{\text{Al}}) = \left(\frac{0.21}{0.093}\right)(57\,^\circ\text{C}) = 1.3 \times 10^2\,^\circ\text{C}$$

18.4 [I] Two identical metal plates (mass $= m$, specific heat $= c$) have different temperatures; one is at 20 °C, and the other is at 90 °C. They are placed in good thermal contact. What is their final temperature?

Because the plates are identical, we would guess the final temperature to be midway between 20 °C and 90 °C, namely 55 °C. This is correct, but let us show it mathematically. From the law of conservation of energy, the *heat lost by one plate must equal the heat gained by the other*. Thus *the total heat change of the system is zero*. In equation form,

$$\text{(heat change of hot plate)} + \text{(heat change of cold plate)} = 0$$
$$mc(\Delta T)_{\text{hot}} + mc(\Delta T)_{\text{cold}} = 0$$

which is short-hand for $m_{\text{hot}}c_{\text{hot}}\Delta T_{\text{hot}} + m_{\text{cold}}c_{\text{cold}}\Delta T_{\text{cold}} = 0$

Be careful about ΔT: It is the final temperature (which we denote by T_f in this case) minus the initial temperature. The above equation thus becomes

$$mc(T_f - 90\,^\circ\text{C}) + mc(T_f - 20\,^\circ\text{C}) = 0$$

After canceling mc from each term, we solve and find $T_f = 55\,^\circ\text{C}$, the expected answer.

18.5 [II] A thermos bottle contains 250 g of coffee at 90 °C. To this is added 20 g of milk at 5 °C. After equilibrium is established, what is the temperature of the liquid? Assume no heat loss to the thermos bottle.

Water, coffee, and milk all have the same value of c, 1.00 cal/g·°C. The law of energy conservation allows us to write

$$\text{(heat change of coffee)} + \text{(heat change of milk)} = 0$$
$$(cm\,\Delta T)_{\text{coffee}} + (cm\,\Delta T)_{\text{milk}} = 0$$

In other words, the heat lost by the coffee equals the heat gained by the milk. If the final temperature of the liquid is T_f, then

$$\Delta T_{\text{coffee}} = T_f - 90\,^\circ\text{C} \qquad \Delta T_{\text{milk}} = T_f - 5\,^\circ\text{C}$$

Substituting and canceling c yields

$$(250\text{ g})(T_f - 90\,^\circ\text{C}) + (20\text{ g})(T_f - 5\,^\circ\text{C}) = 0$$

Solving gives $T_f = 84\,^\circ\text{C}$.

18.6 [II] A thermos bottle contains 150 g of water at 4 °C. Into this is placed 90 g of metal at 100 °C. After equilibrium is established, the temperature of the water and metal is 21 °C. What is the specific heat of the metal? Assume no heat loss to the thermos bottle.

$$\text{(heat change of metal)} + \text{(heat change of water)} = 0$$
$$(cm\,\Delta T)_{\text{metal}} + (cm\,\Delta T)_{\text{water}} = 0$$
$$c_{\text{metal}}(90\text{ g})(-79\,^\circ\text{C}) + (1.00\text{ cal/g}\cdot{}^\circ\text{C})(150\text{ g})(17\,^\circ\text{C}) = 0$$

Solving yields $c_{\text{metal}} = 0.36\text{ cal/g}\cdot{}^\circ\text{C}$. Notice that $\Delta T_{\text{metal}} = 21 - 90 = -79\,^\circ\text{C}$.

18.7 [II] A 200-g copper calorimeter can contains 150 g of oil at 20 °C. To the oil is added 80 g of aluminum at 300 °C. What will be the temperature of the system after equilibrium is established? $c_{\text{Cu}} = 0.093\ \text{cal/g}\cdot{}^\circ\text{C}$, $c_{\text{Al}} = 0.21\ \text{cal/g}\cdot{}^\circ\text{C}$, $c_{\text{oil}} = 0.37\ \text{cal/g}\cdot{}^\circ\text{C}$.

$$\text{(heat change of aluminum)} + \text{(heat change of can and oil)} = 0$$

$$(cm\,\Delta T)_{\text{Al}} + (cm\,\Delta T)_{\text{Cu}} + (cm\,\Delta T)_{\text{oil}} = 0$$

With given values substituted, this becomes

$$\left(0.21\frac{\text{cal}}{\text{g}\cdot{}^\circ\text{C}}\right)(80\ \text{g})(T_f - 300\,{}^\circ\text{C}) + \left(0.093\frac{\text{cal}}{\text{g}\cdot{}^\circ\text{C}}\right)(200\ \text{g})(T_f - 20\,{}^\circ\text{C})$$
$$+ \left(0.37\frac{\text{cal}}{\text{g}\cdot{}^\circ\text{C}}\right)(150\ \text{g})(T_f - 20\,{}^\circ\text{C}) = 0$$

Solving gives T_f as 72 °C.

18.8 [II] Exactly 3.0 g of carbon was burned to CO_2 in a copper calorimeter. The mass of the calorimeter is 1500 g, and there is 2000 g of water in the calorimeter. The initial temperature was 20 °C, and the final temperature is 31 °C. Calculate the heat given off per gram of carbon. $c_{\text{Cu}} = 0.093\ \text{cal/g}\cdot{}^\circ\text{C}$. Neglect the small heat capacity of the carbon and carbon dioxide.

The law of energy conservation tells us that

$$\text{(heat change of carbon)} + \text{(heat change of calorimeter)} + \text{(heat change of water)} = 0$$

$$\text{(heat change of carbon)} + (0.093\ \text{cal/g}\cdot{}^\circ\text{C})(1500\ \text{g})(11\,{}^\circ\text{C}) + (1\ \text{cal/g}\cdot{}^\circ\text{C})(2000\ \text{g})(11\,{}^\circ\text{C}) = 0$$

$$\text{(heat change of carbon)} = -23\,500\ \text{cal}$$

Therefore, the heat given off by one gram of carbon as it burns is

$$\frac{23\,500\ \text{cal}}{3.0\ \text{g}} = 7.8\ \text{kcal/g}$$

18.9 [II] Determine the temperature T_f that results when 150 g of ice at 0 °C is mixed with 300 g of water at 50 °C.

From energy conservation,

$$\text{(heat change of ice)} + \text{(heat change of water)} = 0$$

$$\text{(heat to melt ice)} + \text{(heat to warm ice water)} + \text{(heat change of water)} = 0$$

$$(mL_f)_{\text{ice}} + (cm\,\Delta T)_{\text{ice water}} + (cm\,\Delta T)_{\text{water}} = 0$$

$$(150\ \text{g})(80\ \text{cal/g}) + (1.00\ \text{cal/g}\cdot{}^\circ\text{C})(150\ \text{g})(T_f - 0\,{}^\circ\text{C}) + (1.00\ \text{cal/g}\cdot{}^\circ\text{C})(300\ \text{g})(T_f - 50\,{}^\circ\text{C}) = 0$$

from which $T_f = 6.7\,{}^\circ\text{C}$.

18.10 [II] How much heat is given up when 20 g of steam at 100 °C is condensed and cooled to 20 °C?

$$\begin{aligned}\text{Heat change} &= \text{(condensation heat change)} + \text{(heat change of water during cooling)}\\ &= mL_v + cm\,\Delta T\\ &= (20\ \text{g})(-540\ \text{cal/g}) + (1.00\ \text{cal/g}\cdot{}^\circ\text{C})(20\ \text{g})(20\,{}^\circ\text{C} - 100\,{}^\circ\text{C})\\ &= -12\,400\ \text{cal} = -12\ \text{kcal}\end{aligned}$$

18.11 [II] A 20-g piece of aluminum ($c = 0.21\ \text{cal/g}\cdot{}^\circ\text{C}$) at 90 °C is dropped into a cavity in a large block of ice at 0 °C. How much ice does the aluminum melt?

$$\text{(heat change of Al as it cools to 0 °C)} + \text{(heat change of mass } m \text{ of ice melted)} = 0$$

$$(mc\,\Delta T)_{\text{Al}} + (L_f m)_{\text{ice}} = 0$$

$$(20\text{ g})(0.21\text{ cal/g}\cdot{}^\circ\text{C})(0\,{}^\circ\text{C} - 90\,{}^\circ\text{C}) + (80\text{ cal/g})m = 0$$

from which $m = 4.7$ g is the quantity of ice melted.

18.12 [II] In a calorimeter can (which behaves thermally as if it were equivalent to 40 g of water) are 200 g of water and 50 g of ice, all at exactly 0 °C. Into this is poured 30 g of water at 90 °C. What will be the final condition of the system?

Let us start by assuming (perhaps incorrectly) that the final temperature is $T_f > 0\,{}^\circ\text{C}$. Then

$$\begin{pmatrix}\text{heat change of}\\ \text{hot water}\end{pmatrix} + \begin{pmatrix}\text{heat to}\\ \text{melt ice}\end{pmatrix} + \begin{pmatrix}\text{heat to warm}\\ \text{250 g of water}\end{pmatrix} + \begin{pmatrix}\text{heat to warm}\\ \text{calorimeter}\end{pmatrix} = 0$$

$$(30\text{ g})(1.00\text{ cal/g}\cdot{}^\circ\text{C})(T_f - 90\,{}^\circ\text{C}) + (50\text{ g})(80\text{ cal/g}) + (250\text{ g})(1\text{ cal/g}\cdot{}^\circ\text{C})(T_f - 0\,{}^\circ\text{C})$$
$$+(40\text{ g})(1.00\text{ cal/g}\cdot{}^\circ\text{C})(T_f - 0\,{}^\circ\text{C}) = 0$$

Solving gives $T_f = -4.1\,{}^\circ\text{C}$, contrary to our assumption that the final temperature is above 0 °C. Apparently, not all the ice melts. Therefore, $T_f = 0\,{}^\circ\text{C}$.

To find how much ice melts, we write

$$\text{Heat lost by hot water} = \text{heat gained by melting ice}$$

$$(30\text{ g})(1.00\text{ cal/g}\cdot{}^\circ\text{C})(90\,{}^\circ\text{C}) = (80\text{ cal/g})m$$

where m is the mass of ice that melts. Solving gives $m = 34$ g. The final system has 50 g − 34 g = 16 g of ice not melted.

18.13 [I] An electric heater that produces 900 W of power is used to vaporize water. How much water at 100 °C can be changed to steam at 100 °C in 3.00 min by the heater? (For water at 100 °C, $L_v = 2.26 \times 10^6$ J/kg.)

The heater produces 900 J of heat energy per second. So the heat produced in 3.00 min is

$$\Delta Q = (900\text{ J/s})(180\text{ s}) = 162\text{ kJ}$$

The heat required to vaporize a mass m of water is

$$\Delta Q = mL_v = m(2.26 \times 10^6\text{ J/kg})$$

Equating these two expressions for ΔQ and solving for m gives $m = 0.0717$ kg = 71.7 g as the mass of water vaporized.

18.14 [I] A 3.00-g bullet ($c = 0.0305$ cal/g·°C = 128 J/kg·°C) moving at 180 m/s enters a bag of sand and stops. By what amount does the temperature of the bullet change if all its KE becomes thermal energy that is added to the bullet?

The bullet loses KE in the amount

$$\text{KE} = \tfrac{1}{2}mv^2 = \tfrac{1}{2}(3.00 \times 10^{-3}\text{ kg})(180\text{ m/s})^2 = 48.6\text{ J}$$

This results in the addition of $\Delta Q = 48.6$ J of thermal energy to the bullet. Then, since $\Delta Q = mc\,\Delta T$, we can find ΔT for the bullet:

$$\Delta T = \frac{\Delta Q}{mc} = \frac{48.6\text{ J}}{(3.00 \times 10^{-3}\text{ kg})(128\text{ J/kg}\cdot{}^\circ\text{C})} = 127\,{}^\circ\text{C}$$

Notice that we had to use c in J/kg·°C, and not in cal/g·°C.

18.15 [I] Suppose a 60-kg person consumes 2500 Cal of food in one day. If the entire heat equivalent of this food were retained by the person's body, how large a temperature change would it cause? (For the body, $c = 0.83\ \text{cal/g}\cdot{}^\circ\text{C}$.) Remember that 1 Cal = 1 kcal = 1000 cal.

The equivalent amount of heat added to the body in one day is

$$\Delta Q = (2500\ \text{Cal})(1000\ \text{cal/Cal}) = 2.5 \times 10^6\ \text{cal}$$

Then, by use of $\Delta Q = mc\,\Delta T$,

$$\Delta T = \frac{\Delta Q}{mc} = \frac{2.5 \times 10^6\ \text{cal}}{(60 \times 10^3\ \text{g})(0.83\ \text{cal/g}\cdot{}^\circ\text{C})} = 50\ {}^\circ\text{C}$$

18.16 [II] A thermometer in a 10 m × 8.0 m × 4.0 m room reads 22 °C and a humidistat reads the R.H. to be 35 percent. What mass of water vapor is in the room? Saturated air at 22 °C contains 19.33 g H_2O/m^3.

$$\%\text{R.H.} = \frac{\text{mass of water/m}^3}{\text{mass of water/m}^3\ \text{of saturated air}} \times 100$$

$$35 = \frac{\text{mass/m}^3}{0.019\ 33\ \text{kg/m}^3} \times 100$$

from which mass/m^3 = 6.77×10^{-3} kg/m^3. But the room in question has a volume of 10 m × 8.0 m × 4.0 m = 320 m^3. Therefore, the total mass of water in it is

$$(320\ \text{m}^3)(6.77 \times 10^{-3}\ \text{kg/m}^3) = 2.2\ \text{kg}$$

18.17 [II] On a certain day when the temperature is 28 °C, moisture forms on the outside of a glass of cold drink if the glass is at a temperature of 16 °C or lower. What is the R.H. on that day? Saturated air at 28 °C contains 26.93 g/m^3 of water, while, at 16 °C, it contains 13.50 g/m^3.

Dew forms at a temperature of 16 °C or lower, so the dew point is 16 °C. The air is saturated at that temperature and therefore contains 13.50 g/m^3. Then

$$\text{R.H.} = \frac{\text{mass present/m}^3}{\text{mass/m}^3\ \text{in saturated air}} = \frac{13.50}{26.93} = 0.50 = 50\%$$

18.18 [II] Outside air at 5 °C and 20 percent relative humidity is introduced into a heating and air conditioning plant where it is heated to 20 °C and its relative humidity is increased to a comfortable 50 percent. How many grams of water must be evaporated into a cubic meter of outside air to accomplish this? Saturated air at 5 °C contains 6.8 g/m^3 of water, and at 20 °C it contains 17.3 g/m^3.

$$\text{Mass/m}^3\ \text{of water vapor in air at } 5\ {}^\circ\text{C} = 0.20 \times 6.8\ \text{g/m}^3 = 1.36\ \text{g/m}^3$$

$$\text{Comfortable mass/m}^3\ \text{at } 20\ {}^\circ\text{C} = 0.50 \times 17.3\ \text{g/m}^3 = 8.65\ \text{g/m}^3$$

$$1\ \text{m}^3\ \text{of air at } 5\ {}^\circ\text{C expands to } (293/278)\ \text{m}^3 = 1.054\ \text{m}^3\ \text{at } 20\ {}^\circ\text{C}$$

$$\text{Mass of water vapor in } 1.054\ \text{m}^3\ \text{at } 20\ {}^\circ\text{C} = 1.054\ \text{m}^3 \times 8.65\ \text{g/m}^3 = 9.12\ \text{g}$$

$$\text{Mass of water to be added to each m}^3\ \text{of air at } 5\ {}^\circ\text{C} = (9.12 - 1.36)\ \text{g} = 7.8\ \text{g}$$

Supplementary Problems

18.19 [I] How many calories are required to heat each of the following from 15 °C to 65 °C? (*a*) 3.0 g of aluminum, (*b*) 5.0 g of pyrex glass, (*c*) 20 g of platinum. The specific heats, in cal/g · °C, for aluminum, pyrex, and platinum are 0.21, 0.20, and 0.032, respectively. *Ans.* (*a*) 32 cal; (*b*) 50 cal; (*c*) 32 cal

18.20 [I] When 5.0 g of a certain type of coal is burned, it raises the temperature of 1000 mL of water from 10 °C to 47 °C. Calculate the thermal energy produced per gram of coal. Neglect the small heat capacity of the coal. *Ans.* 7.4 kcal/g

18.21 [II] Furnace oil has a heat of combustion of 44 MJ/kg. Assuming that 70 percent of the heat is useful, how many kilograms of oil are required to raise the temperature of 2000 kg of water from 20 °C to 99 °C? *Ans.* 22 kg

18.22 [II] What will be the final temperature if 50 g of water at exactly 0 °C is added to 250 g of water at 90 °C? *Ans.* 75 °C

18.23 [II] A 50-g piece of metal at 95 °C is dropped into 250 g of water at 17.0 °C and warms it to 19.4 °C. What is the specific heat of the metal? *Ans.* 0.16 cal/g·°C

18.24 [II] How long does it take a 2.50-W heater to boil away 400 g of liquid helium at the temperature of its boiling point (4.2 K)? For helium, $L_v = 5.0$ cal/g. *Ans.* 56 min

18.25 [II] A 55-g copper calorimeter ($c = 0.093$ cal/g·°C) contains 250 g of water at 18.0 °C. When 75 g of an alloy at 100 °C is dropped into the calorimeter, the final resulting temperature is 20.4 °C. What is the specific heat of the alloy? *Ans.* 0.10 cal/g·°C

18.26 [II] Determine the temperature that results when 1.0 kg of ice at exactly 0 °C is mixed with 9.0 kg of water at 50 °C and no heat is lost. *Ans* 37 °C

18.27 [II] How much heat is required to change 10 g of ice at exactly 0 °C to steam at 100 °C? *Ans.* 7.2 kcal

18.28 [II] Ten kilograms of steam at 100 °C is condensed by passing it into 500 kg of water at 40.0 °C. What is the resulting temperature? *Ans.* 51.8 °C

18.29 [II] The heat of combustion of ethane gas is 373 kcal/mole. Assuming that 60.0 percent of the heat is useful, how many liters of ethane, measured at standard temperature and pressure, must be burned to convert 50.0 kg of water at 10.0 ° C to steam at 100.0 °C? One mole of a gas occupies 22.4 liters at precisely 0 °C and 1 atm. *Ans.* 3.15×10^3 liters

18.30 [II] Calculate the heat of fusion of ice from the following data for ice at 0 °C added to water:

Mass of calorimeter	60 g
Mass of calorimeter plus water	460 g
Mass of calorimeter plus water and ice	618 g
Initial temperature of water	38.0 °C
Final temperature of mixture	5.0 °C
Specific heat of calorimeter	0.10 cal/g·°C

Ans. 80 cal/g

18.31 [II] Determine the result when 100 g of steam at 100 °C is passed into a mixture of 200 g of water and 20 g of ice at exactly 0 °C in a calorimeter which behaves thermally as if it were equivalent to 30 g of water. *Ans.* 49 g of steam condensed, final temperature 100 °C

18.32 [II] Determine the result when 10 g of steam at 100 °C is passed into a mixture of 400 g of water and 100 g of ice at exactly 0 °C in a calorimeter which behaves thermally as if it were equivalent to 50 g of water. *Ans.* 80 g of ice melted, final temperature 0 °C

18.33 [II] Suppose a person who eats 2500 Cal of food each day loses the heat equivalent of the food through evaporation of water from the body. How much water must evaporate each day? At body temperature, L_v for water is about 600 cal/g. *Ans.* 4.17 kg

18.34 [II] How long will it take a 500-W heater to raise the temperature of 400 g of water from 15.0 °C to 98.0 °C. *Ans.* 278 s

18.35 [II] A 0.250-hp drill causes a dull 50.0-g steel bit to heat up rather than to deepen a hole in a block of hard wood. Assuming that 75.0 percent of the friction-loss energy causes heating of the bit, by what amount will its temperature change in 20.0 s? For steel, $c = 450$ J/kg·°C. *Ans.* 124 °C

18.36 [II] On a certain day the temperature is 20 °C and the dew point is 5.0 °C. What is the relative humidity? Saturated air at 20 °C and 5.0 °C contains 17.12 and 6.80 g/m^3 of water, respectively. *Ans.* 40%

18.37 [II] How much water vapor exists in a 105-m^3 room on a day when the relative humidity in the room is 32 percent and the room temperature is 20 °C? Saturated air at 20 °C contains 17.12 g/m^3 of water. *Ans.* 0.58 kg

18.38 [II] Air at 30 °C and 90 percent relative humidity is drawn into an air conditioning unit and cooled to 20 °C. The relative humidity is simultaneously reduced to 50 percent. How many grams of water are removed from a cubic meter of air at 30 °C by the air conditioner? Saturated air contains 30.4 g/m^3 and 17.1 g/m^3 of water at 30 °C and 20 °C, respectively. *Ans.* 19 g

Chapter 19

Transfer of Thermal Energy

THERMAL ENERGY CAN BE TRANSFERRED into or out of a system via the mechanisms of **conduction**, **convection**, and **radiation**. Remember that heat is the energy transferred from a system at a higher temperature to a system at a lower temperature (with which it is in contact) via the collisions of their constituent particles.

CONDUCTION occurs when thermal energy moves through a material as a result of collisions between the free electrons, ions, atoms, and molecules of the material. The hotter a substance, the higher the average KE of its atoms. When a temperature difference exists between materials in contact, the higher-energy atoms in the warmer substance transfer energy to the lower-energy atoms in the cooler substance when atomic collisions occur between the two. Heat thus flows from hot to cold.

Consider the slab of material shown in Fig. 19-1. Its thickness is L, and its cross-sectional area is A. The temperatures of its two faces are T_1 and T_2, so the temperature difference across the slab is $\Delta T = T_1 - T_2$. The quantity $\Delta T/L$ is called the **temperature gradient**. It is the rate-of-change of temperature with distance.

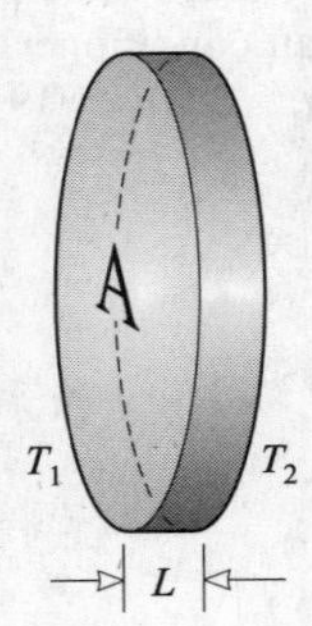

Fig. 19-1

The quantity of heat ΔQ transmitted from face 1 to face 2 in time Δt is given by

$$\frac{\Delta Q}{\Delta t} = k_T A \frac{\Delta T}{L}$$

where k_T depends on the material of the slab and is called the **thermal conductivity** of the material. In the SI, k_T has the unit W/m·K, and $\Delta Q/\Delta t$ is in J/s (i.e., W). Other units sometimes used to express k_T are related to W/m·K as follows:

$$1\ \text{cal/s}\cdot\text{cm}\cdot{}^\circ\text{C} = 418.4\ \text{W/m}\cdot\text{K} \qquad \text{and} \qquad 1\ \text{Btu}\cdot\text{in./h}\cdot\text{ft}^2\cdot{}^\circ\text{F} = 0.144\ \text{W/m}\cdot\text{K}$$

THE THERMAL RESISTANCE (or *R value*) of a slab is defined by the heat-flow equation in the form

$$\frac{\Delta Q}{\Delta t} = \frac{A\,\Delta T}{R} \qquad \text{where} \qquad R = \frac{L}{k_T}$$

Its SI unit is $m^2 \cdot K/W$. Its customary unit is $ft^2 \cdot h \cdot {}^\circ F/Btu$, where $1\ ft^2 \cdot h \cdot {}^\circ F/Btu = 0.176\ m^2 \cdot K/W$. (It is unlikely that you will have occasion to confuse this symbol R with the symbol for the universal gas constant.)

For several slabs of the same surface area in series, the combined R value is

$$R = R_1 + R_2 + \cdots + R_N$$

where $R_1, \ldots,$ are the R values of the individual slabs.

CONVECTION of thermal energy occurs in a fluid when warm material flows so as to displace cooler material. Typical examples are the flow of warm air from a register in a heating system and the flow of warm water in the Gulf Stream.

RADIATION is the mode of transport of radiant electromagnetic energy through vacuum and the empty space between atoms. Radiant energy is distinct from heat, though both correspond to energy in transit. Heat is heat; electromagnetic radiation is electromagnetic radiation – don't confuse the two.

A **blackbody** is a body that absorbs all the radiant energy falling on it. At thermal equilibrium, a body emits as much energy as it absorbs. Hence, a good absorber of radiation is also a good emitter of radiation.

Suppose a surface of area A has absolute temperature T and radiates only a fraction ϵ as much energy as would a blackbody surface. Then ϵ is called the **emissivity** of the surface, and the energy per second (i.e., the power) radiated by the surface is given by the **Stefan–Boltzmann Law**:

$$\mathrm{P} = \epsilon A \sigma T^4$$

where $\sigma = 5.67 \times 10^{-8}\ W/m^2 \cdot K^4$ is the *Stefan–Boltzmann constant*, and T is the absolute temperature. The emissivity of a blackbody is unity.

All objects whose temperature is above absolute zero radiate energy. When an object at absolute temperature T is in an environment where the temperature is T_e, the net energy radiated per second by the object is

$$\mathrm{P} = \epsilon A \sigma (T^4 - T_e^4)$$

Solved Problems

19.1 [I] An iron plate 2 cm thick has a cross-sectional area of 5000 cm^2. One face is at 150 °C, and the other is at 140 °C. How much heat passes through the plate each second? For iron, $k_T = 80\ W/m \cdot K$.

$$\frac{\Delta Q}{\Delta t} = k_T A \frac{\Delta T}{L} = (80\ W/m \cdot K)(0.50\ m^2)\left(\frac{10\ {}^\circ C}{0.02\ m}\right) = 20\ kJ/s$$

19.2 [I] A metal plate 4.00 mm thick has a temperature difference of 32.0 °C between its faces. It transmits 200 kcal/h through an area of 5.00 cm^2. Calculate the thermal conductivity of this metal in $W/m \cdot K$.

$$k_T = \frac{\Delta Q}{\Delta t}\frac{L}{A(T_1 - T_2)} = \frac{(2.00 \times 10^5\ \text{cal})(4.184\ \text{J/cal})}{(1.00\ \text{h})(3600\ \text{s/h})}\frac{4.00 \times 10^{-3}\ \text{m}}{(5.00 \times 10^{-4}\ \text{m}^2)(32.0\ \text{K})}$$
$$= 58.5\ \text{W/m}\cdot\text{K}$$

19.3 [II] Two metal plates are soldered together as shown in Fig. 19-2. It is known that $A = 80\ \text{cm}^2$, $L_1 = L_2 = 3.0\ \text{mm}$, $T_1 = 100\,°\text{C}$, $T_2 = 0\,°\text{C}$. For the plate on the left, $k_{T1} = 48.1\ \text{W/m}\cdot\text{K}$; for the plate on the right $k_{T2} = 68.2\ \text{W/m}\cdot\text{K}$. Find the heat flow rate through the plates and the temperature T of the soldered junction.

We assume equilibrium conditions so that the heat flowing through plate-1 equals that through plate-2. Then

$$k_{T1}A\frac{T_1 - T}{L_1} = k_{T2}A\frac{T - T_2}{L_2}$$

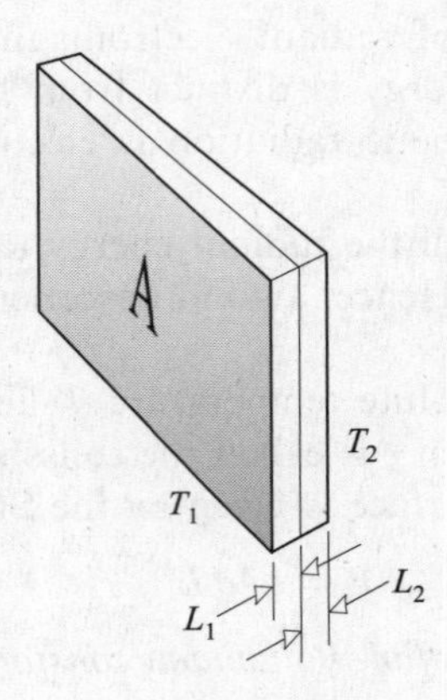

Fig. 19-2

But $L_1 = L_2$, so this becomes

$$k_{T1}(100\,°\text{C} - T) = k_{T2}(T - 0\,°\text{C})$$

from which
$$T = (100\,°\text{C})\left(\frac{k_{T1}}{k_{T1} + k_{T2}}\right) = (100\,°\text{C})\left(\frac{48.1}{48.1 + 68.2}\right) = 41.4\,°\text{C}$$

The heat flow rate is then

$$\frac{\Delta Q}{\Delta t} = k_{T1}A\frac{T_1 - T}{L_1} = \left(48.1\frac{\text{W}}{\text{m}\cdot\text{K}}\right)(0.008\,0\ \text{m}^2)\frac{(100 - 41.4)\text{K}}{0.003\,0\ \text{m}} = 7.5\ \text{kJ/s}$$

19.4 [II] A beverage cooler is in the shape of a cube, 42 cm on each inside edge. Its 3.0-cm thick walls are made of plastic ($k_T = 0.050\ \text{W/m}\cdot\text{K}$). When the outside temperature is $20\,°\text{C}$, how much ice will melt inside the cooler each hour?

We have to determine the amount of heat conducted into the box. The cubical box has six sides, each with an area of about $(0.42\ \text{m})^2$. Then, from $\Delta Q/\Delta t = k_T A\,\Delta T/L$, we have, with the ice inside at $0\,°\text{C}$

$$\frac{\Delta Q}{\Delta t} = (0.050\ \text{W/m}\cdot\text{k})(0.42\ \text{m})^2(6)\left(\frac{20\,°\text{C}}{0.030\ \text{m}}\right) = 35.3\ \text{J/s} = 8.43\ \text{cal/s}$$

In one hour, $\Delta Q = (60)^2(8.43) = 30\,350$ cal. To melt 1.0 g of ice requires 80 cal, so the mass of ice melted in one hour is

$$m = \frac{30\,350 \text{ cal}}{80 \text{ cal/g}} = 0.38 \text{ kg}$$

19.5 [III] A copper tube (length, 3.0 m; inner diameter, 1.500 cm; outer diameter, 1.700 cm) extends across a 3.0-m long vat of rapidly circulating water maintained at 20 °C. Live steam at 100 °C passes through the tube. (*a*) What is the heat flow rate from the steam into the vat? (*b*) How much steam is condensed each minute? For copper, $k_T = 1.0 \text{ cal/s}\cdot\text{cm}\cdot{}^\circ\text{C}$.

To determine the rate at which heat flows through the tube wall, approximate it as a flat sheet. Because the thickness of the tube is much smaller than its radius, the inner surface area of the tube,

$$2\pi r_i L = 2\pi(0.750 \text{ cm})(300 \text{ cm}) = 1410 \text{ cm}^2$$

nearly equals its outer surface area,

$$2\pi r_o L = 2\pi(0.850 \text{ cm})(300 \text{ cm}) = 1600 \text{ cm}^2$$

As an approximation, we can consider the tube to be a plate of thickness 0.100 cm and area given by

$$A = \tfrac{1}{2}(1410 \text{ cm}^2 + 1600 \text{ cm}^2) = 1500 \text{ cm}^2$$

(*a*)
$$\frac{\Delta Q}{\Delta t} = k_T A \frac{\Delta T}{L} = \left(1.0 \frac{\text{cal}}{\text{s}\cdot\text{cm}\cdot{}^\circ\text{C}}\right)\frac{(1500 \text{ cm}^2)(80\,{}^\circ\text{C})}{(0.100 \text{ cm})} = 1.2 \times 10^6 \text{ cals/s}$$

(*b*) In one minute, the heat conducted from the tube is

$$\Delta Q = (1.2 \times 10^6 \text{ cal/s})(60 \text{ s}) = 72 \times 10^6 \text{ cal}$$

It takes 540 cal to condense 1.0 g of steam at 100 °C. Therefore

$$\text{Steam condensed per min} = \frac{72 \times 10^6 \text{ cal}}{540 \text{ cal/g}} = 13.3 \times 10^4 \text{ g} = 1.3 \times 10^2 \text{ kg}$$

In practice, various factors would greatly reduce this theoretical value.

19.6 [I] (*a*) Calculate the R value for a wall consisting of the following layers: concrete block ($R = 1.93$), 1.0 inch of insulating board ($R = 4.3$), and 0.50 inch of drywall ($R = 0.45$), all in U.S. Customary Units. (*b*) If the wall has an area of 15 m², find the heat flow per hour through it when the temperature just outside is 20 °C lower than inside.

(*a*)
$$R = R_1 + R_2 + \cdots + R_N = 1.93 + 4.3 + 0.45 = 6.7$$

in U.S. Customary Units. Using the fact that 1 U.S. Customary Unit of $R = 0.176 \text{ m}^2\cdot\text{K/W}$, we get $R = 1.18 \text{ m}^2\cdot\text{K/W}$.

(*b*)
$$\Delta Q = \frac{A\,\Delta T}{R}(\Delta t) = \frac{(15 \text{ m}^2)(20\,{}^\circ\text{C})}{1.18 \text{ m}^2\cdot\text{K/W}}(3600 \text{ s}) = 0.915 \text{ MJ} = 2.2 \times 10^2 \text{ kcal}$$

19.7 [I] A spherical body of 2.0 cm diameter is maintained at 600 °C. Assuming that it radiates as if it were a blackbody, at what rate (in watts) is energy radiated from the sphere?

$$A = \text{surface area} = 4\pi r^2 = 4\pi(0.01 \text{ m})^2 = 1.26 \times 10^{-3} \text{ m}^2$$
$$P = A\sigma T^4 = (1.26 \times 10^{-3} \text{ m}^2)(5.67 \times 10^{-8} \text{ W/m}^2\cdot\text{K}^4)(873 \text{ K})^4 = 41 \text{ W}$$

19.8 [I] An unclothed person whose body has a surface area of 1.40 m^2 with an emissivity of 0.85, has a skin temperature of 37 °C and stands in a 20 °C room. How much energy does the person lose through radiation per minute?

Energy is power (P) multiplied by time (Δt). From $P = \epsilon A\sigma(T^4 - T_e^4)$, we have the energy loss

$$\epsilon A\sigma(T^4 - T_e^4)\Delta t = (0.85)(1.40 \text{ m}^2)(\sigma)(T^4 - T_e^4)(60 \text{ s})$$

Using $\sigma = 5.67 \times 10^{-8}$ W/m$^2\cdot$K^4, $T = 273 + 37 = 310$ K, and $T_e = 273 + 20 = 293$ K gives an energy loss of

$$7.6 \text{ kJ} = 1.8 \text{ kcal}$$

Supplementary Problems

19.9 [I] What temperature gradient must exist in an aluminum rod for it to transmit 8.0 cal per second per cm^2 of cross section down the rod? k_T for aluminum is 210 W/K·m. *Ans.* 16 °C/cm

19.10 [I] A single-thickness glass window on a house actually has layers of stagnant air on its two surfaces. But if it did not, how much heat would flow out of an 80 cm × 40 cm × 3.0 mm window each hour on a day when the outside temperature was precisely 0 °C and the inside temperature was 18 °C? For glass, k_T is 0.84 W/K·m. *Ans.* 1.4×10^3 kcal/h

19.11 [I] How many grams of water at 100 °C can be evaporated per hour per cm^2 by the heat transmitted through a steel plate 0.20 cm thick, if the temperature difference between the plate faces is 100 °C? For steel, k_T is 42 W/K·m. *Ans.* 0.33 kg/h·cm^2

19.12 [II] A certain double-pane window consists of two glass sheets, each 80 cm × 80 cm × 0.30 cm, separated by a 0.30-cm stagnant air space. The indoor surface temperature is 20 °C, while the outdoor surface temperature is exactly 0 °C. How much heat passes through the window each second? $k_T = 0.84$ W/K·m for glass and about 0.080 W/K·m for air. *Ans.* 69 cal/s

19.13 [II] A small hole in a furnace acts like a blackbody. Its area is 1.00 cm^2, and its temperature is the same as that of the interior of the furnace, 1727 °C. How many calories are radiated out of the hole each second? *Ans.* 21.7 cal/s

19.14 [I] An incandescent lamp filament has an area of 50 mm^2 and operates at a temperature of 2127 °C. Assume that all the energy furnished to the bulb is radiated from it. If the filament's emissivity is 0.83, how much power must be furnished to the bulb when it is operating? *Ans.* 78 W

19.15 [I] A sphere of 3.0 cm radius acts like a blackbody. It is in equilibrium with its surroundings and absorbs 30 kW of power radiated to it from the surroundings. What is the temperature of the sphere? *Ans.* 2.6×10^3 K

19.16 [II] A 2.0 cm thick brass plate ($k_T = 105$ W/K·m) is sealed face-to-face to a glass sheet ($k_T = 0.80$ W/K·m), and both have the same area. The exposed face of the brass plate is at 80 °C, while the exposed face of the glass is at 20 °C. How thick is the glass if the glass–brass interface is at 65 °C? *Ans.* 0.46 mm

Chapter 20

First Law of Thermodynamics

HEAT (ΔQ) is the thermal energy that flows from one body or system to another, which is in contact with it, because of their temperature difference. Heat always flows from hot to cold (i.e., from the higher temperature to the lower temperature). For two objects in contact to be in thermal equilibrium with each other (i.e., for no net heat transfer from one to the other), their temperatures must be the same. If each of two objects is in thermal equilibrium with a third body, then the two are in thermal equilibrium with each other. (This fact is often referred to as the **Zeroth Law** of Thermodynamics.)

THE INTERNAL ENERGY (U) of a system is the total energy content of the system. It is the sum of all forms of energy possessed by the atoms and molecules of the system.

THE WORK DONE BY A SYSTEM (ΔW) is positive if the system thereby loses energy to its surroundings. When the surroundings do work *on* the system so as to give it energy, ΔW is a negative quantity. In a small expansion ΔV, a fluid at constant pressure P does work given by

$$\Delta W = P\Delta V$$

THE FIRST LAW OF THERMODYNAMICS is a statement of the law of conservation of energy. It maintains that if an amount of heat ΔQ flows into a system, then this energy must appear as increased internal energy ΔU for the system and/or work ΔW done *by* the system on its surroundings. As an equation, the First Law is

$$\Delta Q = \Delta U + \Delta W$$

AN ISOBARIC PROCESS is a process carried out at *constant pressure.*

AN ISOVOLUMIC PROCESS is a process carried out at *constant volume.* When a gas undergoes such a process,

$$\Delta W = P\Delta V = 0$$

and so the First Law of Thermodynamics becomes

$$\Delta Q = \Delta U$$

Any heat that flows into the system appears as increased internal energy of the system.

AN ISOTHERMAL PROCESS is a *constant-temperature* process. In the case of an ideal gas where the constituent atoms or molecules do not interact, $\Delta U = 0$ in an isothermal process. However, this is not true for many other systems. For example, $\Delta U \neq 0$ as ice melts to water at 0 °C, even though the process is isothermal.

For an ideal gas, $\Delta U = 0$ in an isothermal change and so the First Law becomes

$$\Delta Q = \Delta W \qquad \text{(ideal gas)}$$

For an ideal gas changing isothermally from (P_1, V_1) to (P_2, V_2), where $P_1V_1 = P_2V_2$,

$$\Delta Q = \Delta W = P_1V_1 \ln\left(\frac{V_2}{V_1}\right)$$

Here, ln is the logarithm to the base *e*.

AN ADIABATIC PROCESS is one in which no heat is transferred to or from the system. For such a process, $\Delta Q = 0$. Hence, in an adiabatic process, the first law becomes

$$0 = \Delta U + \Delta W$$

Any work done by the system is done at the expense of the internal energy. Any work done on the system serves to increase the internal energy.

For an ideal gas changing from conditions (P_1, V_1, T_1) to (P_2, V_2, T_2) in an adiabatic process,

$$P_1V_1^\gamma = P_2V_2^\gamma \qquad \text{and} \qquad T_1V_1^{\gamma-1} = T_2V_2^{\gamma-1}$$

where $\gamma = c_p/c_v$ is discussed below.

SPECIFIC HEATS OF GASES: When a gas is heated *at constant volume*, the heat supplied goes to increase the internal energy of the gas molecules. But when a gas is heated *at constant pressure*, the heat supplied not only increases the internal energy of the molecules but also does mechanical work in expanding the gas against the opposing constant pressure. Hence the specific heat of a gas at constant pressure c_p, is greater than its specific heat at constant volume, c_v. It can be shown that for an ideal gas of molecular mass M,

$$c_p - c_v = \frac{R}{M} \qquad \text{(ideal gas)}$$

where R is the universal gas constant. In the SI, $R = 8314$ J/kmol·K and M is in kg/kmol; then c_p and c_v must be in J/kg·K = J/kg·°C. Some people use $R = 1.98$ cal/mol·°C and M in g/mol, in which case c_p and c_v are in cal/g·°C.

SPECIFIC HEAT RATIO $(\gamma = c_p/c_v)$: As discussed above, this ratio is greater than unity for a gas. The kinetic theory of gases indicates that for monatomic gases (such as He, Ne, Ar), $\gamma = 1.67$. For most diatomic gases (the ones that are rigidly bonded such as O_2, and N_2), $\gamma = 1.40$ at ordinary temperatures.

WORK IS RELATED TO AREA in a *P–V* diagram. The work done by a fluid in an expansion is equal to the area beneath the expansion curve on a *P–V* diagram.

In a cyclic process, the work output per cycle done by a fluid is equal to the area enclosed by the *P–V* diagram representing the cycle.

THE EFFICIENCY OF A HEAT ENGINE is defined as

$$\text{eff} = \frac{\text{work output}}{\text{heat input}}$$

The **Carnot cycle** is the most efficient cycle possible for a heat engine. An engine that operates in accordance to this cycle between a hot reservoir (T_h) and a cold reservoir (T_c) has efficiency

$$\text{eff}_{\text{max}} = 1 - \frac{T_c}{T_h}$$

Kelvin temperatures must be used in this equation.

Solved Problems

20.1 [I] In a certain process, 8.00 kcal of heat is furnished to the system while the system does 6.00 kJ of work. By how much does the internal energy of the system change during the process?

We have

$$\Delta Q = (8000 \text{ cal})(4.184 \text{ J/cal}) = 33.5 \text{ kJ} \qquad \text{and} \qquad \Delta W = 6.00 \text{ kJ}$$

Therefore, from the First Law $\Delta Q = \Delta U + \Delta W$,

$$\Delta U = \Delta Q - \Delta W = 33.5 \text{ kJ} - 6.00 \text{ kJ} = 27.5 \text{ kJ}$$

20.2 [I] The specific heat of water is 4184 J/kg·K. By how many joules does the internal energy of 50 g of water change as it is heated from 21 °C to 37 °C? Assume that the expansion of the water is negligible.

The heat added to raise the temperature of the water is

$$\Delta Q = cm\,\Delta T = (4184 \text{ J/kg}\cdot\text{K})(0.050 \text{ kg})(16\,^\circ\text{C}) = 3.4 \times 10^3 \text{ J}$$

Notice that ΔT in Celsius is equal to ΔT in kelvins. If we ignore the slight expansion of the water, no work was done on the surroundings and so $\Delta W = 0$. Then, the first law, $\Delta Q = \Delta U + \Delta W$, tells us that

$$\Delta U = \Delta Q = 3.4 \text{ kJ}$$

20.3 [I] How much does the internal energy of 5.0 g of ice at precisely 0 °C increase as it is changed to water at 0 °C? Neglect the change in volume.

The heat needed to melt the ice is

$$\Delta Q = mL_f = (5.0 \text{ g})(80 \text{ cal/g}) = 400 \text{ cal}$$

No external work is done by the ice as it melts and so $\Delta W = 0$. Therefore, the First Law, $\Delta Q = \Delta U + \Delta W$, tells us that

$$\Delta U = \Delta Q = (400 \text{ cal})(4.184 \text{ J/cal}) = 1.7 \text{ kJ}$$

20.4 [II] A spring ($k = 500$ N/m) supports a 400-g mass which is immersed in 900 g of water. The specific heat of the mass is 450 J/kg·K. The spring is now stretched 15 cm and, after thermal equilibrium is reached, the mass is released so it vibrates up and down. By how much has the temperature of the water changed when the vibration has stopped?

The energy stored in the spring is dissipated by the effects of friction and goes to heat the water and mass. The energy stored in the stretched spring was

$$\mathrm{PE}_e = \tfrac{1}{2}kx^2 = \tfrac{1}{2}(500 \text{ N/m})(0.15 \text{ m})^2 = 5.625 \text{ J}$$

This energy appears as thermal energy that flows into the water and the mass. Using $\Delta Q = cm\,\Delta T$, we have

$$5.625 \text{ J} = (4184 \text{ J/kg}\cdot\text{K})(0.900 \text{ kg})\,\Delta T + (450 \text{ J/kg}\cdot\text{K})(0.40 \text{ kg})\,\Delta T$$

which gives
$$\Delta T = \frac{5.625 \text{ J}}{3950 \text{ J/K}} = 0.0014 \text{ K}$$

20.5 [II] Find ΔW and ΔU for a 6.0-cm cube of iron as it is heated from 20 °C to 300 °C at atmospheric pressure. For iron, $c = 0.11$ cal/g·°C and the volume coefficient of thermal expansion is $3.6 \times 10^{-5}\,°\text{C}^{-1}$. The mass of the cube is 1700 g.

Given that $\Delta T = 300\,°\text{C} - 20\,°\text{C} = 280\,°\text{C}$,

$$\Delta Q = cm\,\Delta T = (0.11 \text{ cal/g}\cdot°\text{C})(1700 \text{ g})(280\,°\text{C}) = 52 \text{ kcal}$$

To find that the work done by the expansion of the cube we need to determine ΔV. The volume of the cube is $V = (6.0 \text{ cm})^3 = 216 \text{ cm}^3$. Using $(\Delta V)/V = \beta\,\Delta T$ gives

$$\Delta V = V\beta\,\Delta T = (216 \times 10^{-6} \text{ m}^3)(3.6 \times 10^{-5}\,°\text{C}^{-1})(280\,°\text{C}) = 2.18 \times 10^{-6} \text{ m}^3$$

Then, assuming atmospheric pressure to be 1.0×10^5 Pa, we have

$$\Delta W = P\,\Delta V = (1.0 \times 10^5 \text{ N/m}^2)(2.18 \times 10^{-6} \text{ m}^3) = 0.22 \text{ J}$$

But the First Law tells us that

$$\Delta U = \Delta Q - \Delta W = (52\,000 \text{ cal})(4.184 \text{ J/cal}) - 0.22 \text{ J}$$
$$= 218\,000 \text{ J} - 0.22 \text{ J} \approx 2.2 \times 10^5 \text{ J}$$

Notice how very small the work of expansion against the atmosphere is in comparison to ΔU and ΔQ. Often ΔW can be neglected when dealing with liquids and solids.

20.6 [II] A motor supplies 0.4 hp to stir 5 kg of water. Assuming that all the work goes into heating the water by friction losses, how long will it take to increase the temperature of the water 6 °C?

The heat required to heat the water is

$$\Delta Q = mc\,\Delta T = (5000 \text{ g})(1 \text{ cal/g}\cdot°\text{C})(6\,°\text{C}) = 30 \text{ kcal}$$

This is actually supplied by friction work, so

$$\text{Friction work done} = \Delta Q = (30 \text{ kcal})(4.184 \text{ J/cal}) = 126 \text{ kJ}$$

and this equals the work done by the motor. But

$$\text{Work done by motor in time } t = (\text{power})(t) = (0.4 \text{ hp} \times 746 \text{ W/hp})(t)$$

Equating this to our previous value for the work done yields

$$t = \frac{1.26 \times 10^5 \text{ J}}{(0.4 \times 746) \text{ W}} = 420 \text{ s} = 7 \text{ min}$$

20.7 [I] In each of the following situations, find the change in internal energy of the system. (*a*) A system absorbs 500 cal of heat and at the same time does 400 J of work. (*b*) A system absorbs 300 cal and at the same time 420 J of work is done on it. (*c*) Twelve hundred calories is removed from a gas held at constant volume. Give your answers in kilojoules.

(*a*) $$\Delta U = \Delta Q - \Delta W = (500 \text{ cal})(4.184 \text{ J/cal}) - 400 \text{ J} = 1.69 \text{ kJ}$$

(*b*) $$\Delta U = \Delta Q - \Delta W = (300 \text{ cal})(4.184 \text{ J/cal}) - (-420 \text{ J}) = 1.68 \text{ kJ}$$

(*c*) $$\Delta U = \Delta Q - \Delta W = (-1200 \text{ cal})(4.184 \text{ J/cal}) - 0 = -5.02 \text{ kJ}$$

Notice that ΔQ is positive when heat is added to the system, and ΔW is positive when the system does work. In the reverse cases, ΔQ and ΔW must be taken negative.

20.8 [I] For each of the following adiabatic processes, find the change in internal energy. (*a*) A gas does 5 J of work while expanding adiabatically. (*b*) During an adiabatic compression, 80 J of work is done on a gas.

During an adiabatic process, no heat is transferred to or from the system.

(*a*) $$\Delta U = \Delta Q - \Delta W = 0 - 5 \text{ J} = -5 \text{ J}$$

(*b*) $$\Delta U = \Delta Q - \Delta W = 0 - (-80 \text{ J}) = +80 \text{ J}$$

20.9 [III] The temperature of 5.00 kg of N_2 gas is raised from 10.0 °C to 130.0 °C. If this is done at constant volume, find the increase in internal energy ΔU. Alternatively, if the same temperature change now occurs at constant pressure determine both ΔV and the external work ΔW done by the gas. For N_2 gas, $c_v = 0.177$ cal/g·°C and $c_p = 0.248$ cal/g·°C.

If the gas is heated at constant volume, then no work is done during the process. In that case $\Delta W = 0$, and the first law tells us that $(\Delta Q)_v = \Delta U$. Because $(\Delta Q)_v = c_v m\, \Delta T$, we would have

$$\Delta U = (\Delta Q)_v = (0.177 \text{ cal/g}\cdot{}^\circ\text{C})(5000 \text{ g})(120\,{}^\circ\text{C}) = 106 \text{ kcal} = 443 \text{ kJ}$$

The temperature change is a manifestation of the internal energy change.

When the gas is heated by 120 °C at constant pressure, the same change in internal energy occurs. In addition, however, work is done. The first law then becomes

$$(\Delta Q)_p = \Delta U + \Delta W = 443 \text{ kJ} + \Delta W$$

But $$(\Delta Q)_p = c_p m\, \Delta T = (0.248 \text{ cal/g}\cdot{}^\circ\text{C})(5000 \text{ g})(120\,{}^\circ\text{C})$$
$$= 149 \text{ kcal} = 623 \text{ kJ}$$

Hence $$\Delta W = (\Delta Q)_p - \Delta U = 623 \text{ kJ} - 443 \text{ kJ} = 180 \text{ kJ}$$

20.10 [II] One kilogram of steam at 100 °C and 101 kPa occupies 1.68 m^3. (*a*) What fraction of the observed heat of vaporization of water is accounted for by the expansion of water into stream? (*b*) Determine the increase in internal energy of 1.00 kg of water as it is vaporized at 100 °C.

(*a*) One kilogram of water expands from 1000 cm^3 to 1.68 m^3, so $\Delta V = 1.68 - 0.001 \approx 1.68$ m^3. Therefore, the work done in the expansion is

$$\Delta W = P\, \Delta V = (101 \times 10^3 \text{ N/m}^2)(1.68 \text{ m}^3) = 169 \text{ kJ}$$

The heat of vaporization of water is 540 cal/g, which is 2.26 MJ/kg. The required fraction is therefore

$$\frac{\Delta W}{mL_v} = \frac{169 \text{ kJ}}{(1.00 \text{ kg})(2260 \text{ kJ/kg})} = 0.0748$$

(*b*) From the First Law, $\Delta U = \Delta Q - \Delta W$, so

$$\Delta U = 2.26 \times 10^6 \text{ J} - 0.169 \times 10^6 \text{ J} = 2.07 \text{ MJ}$$

20.11 [I] For nitrogen gas, $c_v = 740$ J/kg·K. Assuming it to behave like an ideal gas, find its specific heat at constant pressure. (The molecular mass of nitrogen gas is 28.0 kg/kmol.)

Method 1

$$c_p = c_v + \frac{R}{M} = \frac{740 \text{ J}}{\text{kg}\cdot\text{K}} + \frac{8314 \text{ J/kmol}\cdot\text{K}}{28.0 \text{ kg/kmol}} = 1.04 \text{ kJ/kg}\cdot\text{K}$$

Method 2

Since N_2 is a diatomic gas, and since $\gamma = c_p/c_v = 1.40$ for such a gas,

$$c_p = 1.40 c_v = 1.40(740 \text{ J/kg}\cdot\text{K}) = 1.04 \text{ kJ/kg}\cdot\text{K}$$

20.12 [I] How much work is done by an ideal gas in expanding isothermally from an initial volume of 3.00 liters at 20.0 atm to a final volume of 24.0 liters?

For an isothermal expansion by an ideal gas,

$$\Delta W = P_1 V_1 \ln\left(\frac{V_2}{V_1}\right)$$

$$= (20.0 \times 1.01 \times 10^5 \text{ N/m}^2)(3.00 \times 10^{-3} \text{ m}^3) \ln\left(\frac{24.0}{3.00}\right) = 12.6 \text{ kJ}$$

20.13 [I] The P–V diagram in Fig. 20-1 applies to a gas undergoing a cyclic change in a piston–cylinder arrangement. What is the work done by the gas (*a*) during portion AB of the cycle? (*b*) During portion BC? (*c*) During portion CD? (*d*) During portion DA?

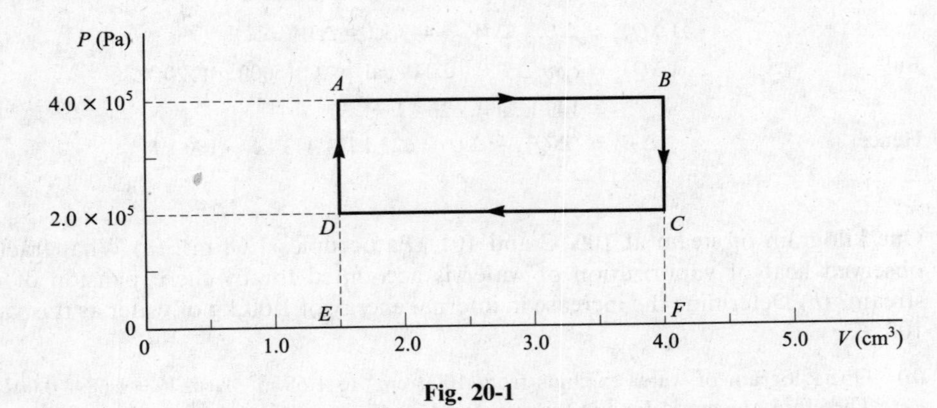

Fig. 20-1

In expansion, the work done is equal to the area under the pertinent portion of the P–V curve. In contraction, the work is numerically equal to the area but is negative.

(*a*) $$\text{Work} = \text{area } ABFEA = [(4.0 - 1.5) \times 10^{-6} \text{ m}^3](4.0 \times 10^5 \text{ N/m}^2) = 1.0 \text{ J}$$

(*b*) $$\text{Work} = \text{area under } BC = 0$$

In portion BC, the volume does not change; therefore $P\Delta V = 0$.

(*c*) This is a contraction, ΔV is negative and so the work is negative:

$$W = -(\text{area } CDEFC) = -(2.5 \times 10^{-6}\ \text{m}^3)(2.0 \times 10^5\ \text{N/m}^2) = -0.50\ \text{J}$$

(*d*)
$$W = 0$$

20.14 [I] For the thermodynamic cycle shown in Fig. 20-1, find (*a*) the net work output of the gas during the cycle and (*b*) the net heat flow into the gas per cycle.

Method 1

(*a*) From Problem 20.13, the net work done is 1.0 J − 0.50 J = 0.5 J.

Method 2

The net work done is equal to the area enclosed by the P–V diagram:

$$\text{Work} = \text{area } ABCDA = (2.0 \times 10^5\ \text{N/m}^2)(2.5 \times 10^{-6}\ \text{m}^3) = 0.50\ \text{J}$$

(*b*) Suppose the cycle starts at point A. The gas returns to this point at the end of the cycle, so there is no difference in the gas at its start and end points. For one complete cycle, ΔU is therefore zero. We have then, if the first law is applied to a complete cycle,

$$\Delta Q = \Delta U + \Delta W = 0 + 0.50\ \text{J} = 0.50\ \text{J} = 0.12\ \text{cal}$$

20.15 [I] What is the net work output per cycle for the thermodynamic cycle in Fig. 20-2?

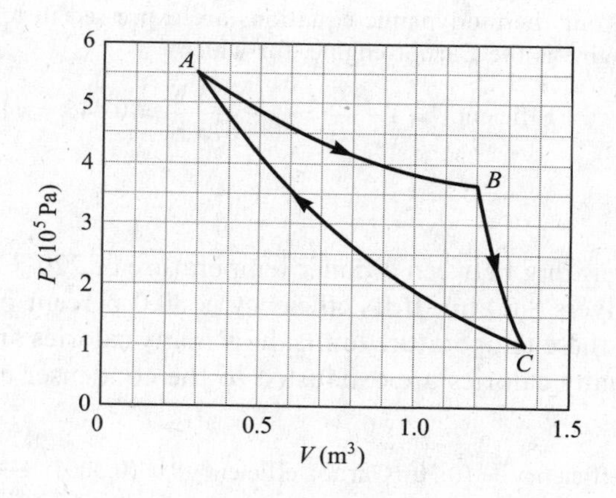

Fig. 20-2

We know that the net work output per cycle is the area enclosed by the P–V diagram. We estimate that in area $ABCA$ there are 22 squares, each of area

$$(0.5 \times 10^5\ \text{N/m}^2)(0.1\ \text{m}^3) = 5\ \text{kJ}$$

Therefore,

$$\text{Area enclosed by cycle} \approx (22)(5\ \text{kJ}) = 1 \times 10^2\ \text{kJ}$$

The net work output per cycle is 1×10^2 kJ.

20.16 [II] Twenty cubic centimeters of monatomic gas at 12 °C and 100 kPa is suddenly (and adiabatically) compressed to 0.50 cm^3. Assume that we are dealing with an ideal gas. What are its new pressure and temperature?

For an adiabatic change involving an ideal gas, $P_1 V_1^\gamma = P_2 V_2^\gamma$ where $\gamma = 1.67$ for a monatomic gas. Hence,

$$P_2 = P_1 \left(\frac{V_1}{V_2}\right)^\gamma = (1.00 \times 10^5 \text{ N/m}^2)\left(\frac{20}{0.50}\right)^{1.67} = 4.74 \times 10^7 \text{ N/m}^2 = 47 \text{ MPa}$$

To find the final temperature, we could use $P_1 V_1/T_1 = P_2 V_2/T_2$. Instead, let us use

$$T_1 V_1^{\gamma-1} = T_2 V_2^{\gamma-1}$$

or

$$T_2 = T_1 \left(\frac{V_1}{V_2}\right)^{\gamma-1} = (285 \text{ K})\left(\frac{20}{0.50}\right)^{0.67} = (285 \text{ K})(11.8) = 3.4 \times 10^3 \text{ K}$$

As a check,

$$\frac{P_1 V_1}{T_1} = \frac{P_2 V_2}{T_2}$$

$$\frac{(1 \times 10^5 \text{ N/m}^2)(20 \text{ cm}^3)}{285 \text{ K}} = \frac{(4.74 \times 10^7 \text{ N/m}^2)(0.50 \text{ cm}^3)}{3370 \text{ K}}$$

$$7000 = 7000 \checkmark$$

20.17 [I] Compute the maximum possible efficiency of a heat engine operating between the temperature limits of 100 °C and 400 °C.

Remember that our thermodynamic equations are expressed in terms of absolute temperature. The most efficient engine is the Carnot engine, for which

$$\text{Efficiency} = 1 - \frac{T_c}{T_h} = 1 - \frac{373 \text{ K}}{673 \text{ K}} = 0.446 = 44.6\%$$

20.18 [II] A steam engine operating between a boiler temperature of 220 °C and a condenser temperature of 35.0 °C delivers 8.00 hp. If its efficiency is 30.0 percent of that for a Carnot engine operating between these temperature limits, how many calories are absorbed each second by the boiler? How many calories are exhausted to the condenser each second?

$$\text{Actual efficiency} = (0.30)(\text{Carnot efficiency}) = (0.300)\left(1 - \frac{308 \text{ K}}{493 \text{ K}}\right) = 0.113$$

We can determine the input heat from the relation for the efficiency

$$\text{Efficiency} = \frac{\text{output work}}{\text{input heat}}$$

and so every second

$$\text{Input heat/s} = \frac{\text{output work/s}}{\text{efficiency}} = \frac{(8.00 \text{ hp})(746 \text{ W/hp})\left(\dfrac{1.00 \text{ cal/s}}{4.184 \text{ W}}\right)}{0.113} = 12.7 \text{ kcal/s}$$

To find the energy rejected to the condenser, we use the law of conservation of energy:

$$\text{Input energy} = (\text{output work}) + (\text{rejected energy})$$

Thus,

$$\begin{aligned}\text{Rejected energy/s} &= (\text{input energy/s}) - (\text{output work/s}) \\ &= (\text{input energy/s}) - (\text{input energy/s})(\text{efficiency}) \\ &= (\text{input energy/s})[1 - (\text{efficiency})] \\ &= (12.7\ \text{kcal/s})(1 - 0.113) = 11.3\ \text{kcal/s}\end{aligned}$$

20.19 [II] Three kilomoles (6.00 kg) of hydrogen gas at S.T.P. expands isobarically to precisely twice its volume. (*a*) What is the final temperature of the gas? (*b*) What is the expansion work done by the gas? (*c*) By how much does the internal energy of the gas change? (*d*) How much heat enters the gas during the expansion? For H_2, $c_v = 10.0\ \text{kJ/kg}\cdot\text{K}$. Assume the hydrogen will behave as an ideal gas.

(*a*) From $P_1V_1/T_1 = P_2V_2/T_2$ with $P_1 = P_2$,

$$T_2 = T_1\left(\frac{V_2}{V_1}\right) = (273\ \text{K})(2.00) = 546\ \text{K}$$

(*b*) Because 1 kmol at S.T.P. occupies 22.4 m^3, we have $V_1 = 67.2\ \text{m}^3$. Then

$$\Delta W = P\,\Delta V = P(V_2 - V_1) = (1.01 \times 10^5\ \text{N/m}^2)(67.2\ \text{m}^3) = 6.8\ \text{MJ}$$

(*c*) To raise the temperature of this ideal gas by 273 K at constant volume requires

$$\Delta Q = c_v m\,\Delta T = (10.0\ \text{kJ/kg}\cdot\text{K})(6.00\ \text{kg})(273\ \text{K}) = 16.4\ \text{MJ}$$

Because the volume is constant here, no work is done and ΔQ equals the internal energy that must be added to the 6.00 kg of H_2 to change its temperature from 273 K to 546 K. Therefore, $\Delta U = 16.4$ MJ.

(*d*) The system obeys the First Law during the process and so

$$\Delta Q = \Delta U + \Delta W = 16.4\ \text{MJ} + 6.8\ \text{MJ} = 23.2\ \text{MJ}$$

20.20 [II] A cylinder of ideal gas is closed by an 8.00 kg movable piston (area = 60.0 cm^2) as shown in Fig. 20-3. Atmospheric pressure is 100 kPa. When the gas is heated from 30.0 °C to 100.0 °C, the piston rises 20.0 cm. The piston is then fastened in place, and the gas is cooled back to 30.0 °C. Calling ΔQ_1 the heat added to the gas in the heating process, and ΔQ_2 the heat lost during cooling, find the difference between ΔQ_1 and ΔQ_2.

During the heating process, the internal energy changed by ΔU_1, and an amount of work ΔW_1 was done. The absolute pressure of the gas was

$$P = \frac{mg}{A} + P_A$$

$$P = \frac{(8.00)(9.81)\ \text{N}}{60.0 \times 10^{-4}\ \text{m}^2} + 1.00 \times 10^5\ \text{N/m}^2 = 1.13 \times 10^5\ \text{N/m}^2$$

Therefore,

$$\begin{aligned}\Delta Q_1 &= \Delta U_1 + \Delta W_1 = \Delta U_1 + P\Delta V \\ &= \Delta U_1 + (1.13 \times 10^5\ \text{N/m}^2)(0.200 \times 60.0 \times 10^{-4}\ \text{m}^3) = \Delta U_1 + 136\ \text{J}\end{aligned}$$

During the cooling process, $\Delta W = 0$ and so (since ΔQ_2 is heat *lost*)

$$-\Delta Q_2 = \Delta U_2$$

But the ideal gas returns to its original temperature, and so its internal energy is the same as at the start. Therefore $\Delta U_2 = -\Delta U_1$, or $\Delta Q_2 = \Delta U_1$. It follows that ΔQ_1 exceeds ΔQ_2 by 136 J = 32.5 cal.

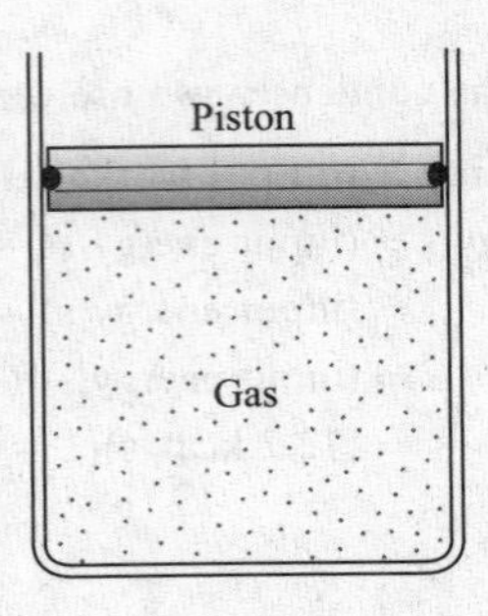

Fig. 20-3

Supplementary Problems

20.21 [I] A 2.0 kg metal block ($c = 0.137$ cal/g·°C) is heated from 15 °C to 90 °C. By how much does its internal energy change? *Ans.* 86 kJ

20.22 [I] By how much does the internal energy of 50 g of oil ($c = 0.32$ cal/g·°C) change as the oil is cooled from 100 °C to 25 °C. *Ans.* −1.2 kcal

20.23 [II] A 70-g metal block moving at 200 cm/s slides across a tabletop a distance of 83 cm before it comes to rest. Assuming 75 percent of the thermal energy developed by friction goes into the block, how much does the temperature of the block rise? For the metal, $c = 0.106$ cal/g·°C. *Ans.* 3.4×10^{-3} °C

20.24 [II] If a certain mass of water falls a distance of 854 m and all the energy is effective in heating the water, what will be the temperature rise of the water? *Ans.* 2.00 °C

20.25 [II] How many joules of heat per hour are produced in a motor that is 75.0 percent efficient and requires 0.250 hp to run it? *Ans.* 168 kJ

20.26 [II] A 100-g bullet ($c = 0.030$ cal/g·°C) is initially at 20 °C. It is fired straight upward with a speed of 420 m/s, and on returning to the starting point strikes a cake of ice at exactly 0 °C. How much ice is melted? Neglect air friction. *Ans.* 26 g

20.27 [II] To determine the specific heat of an oil, an electrical heating coil is placed in a calorimeter with 380 g of the oil at 10 °C. The coil consumes energy (and gives off heat) at the rate of 84 W. After 3.0 min, the oil temperature is 40 °C. If the water equivalent of the calorimeter and coil is 20 g, what is the specific heat of the oil? *Ans.* 0.26 cal/g·°C

20.28 [I] How much external work is done by an ideal gas in expanding from a volume of 3.0 liters to a volume of 30.0 liters against a constant pressure of 2.0 atm? *Ans.* 5.5 kJ

20.29 [I] As 3.0 liters of ideal gas at 27 °C is heated, it expands at a constant pressure of 2.0 atm. How much work is done by the gas as its temperature is changed from 27 °C to 227 °C? *Ans.* 0.40 kJ

20.30 [I] An ideal gas expands adiabatically to three times its original volume. In doing so, the gas does 720 J of work. (*a*) How much heat flows from the gas? (*b*) What is the change in internal energy of the gas? (*c*) Does its temperature rise or fall? *Ans.* (*a*) none; (*b*) −720 J; (*c*) it falls

20.31 [I] An ideal gas expands at a constant pressure of 240 cmHg from 250 cm^3 to 780 cm^3. It is then allowed to cool at constant volume to its original temperature. What is the net amount of heat that flows into the gas during the entire process? *Ans.* 40.5 cal

20.32 [I] As an ideal gas is compressed isothermally, the compressing agent does 36 J of work on the gas. How much heat flows from the gas during the compression process? *Ans.* 8.6 cal

20.33 [II] The specific heat of air at constant volume is 0.175 cal/g·°C. (*a*) By how much does the internal energy of 5.0 g of air change as it is heated from 20 °C to 400 °C? (*b*) Suppose that 5.0 g of air is adiabatically compressed so as to rise its temperature from 20 °C to 400 °C. How much work must be done on the air to compress it? *Ans.* (*a*) 0.33 kcal; (*b*) 1.4 kJ or since work done on the system is negative, −1.4 kJ

20.34 [II] Water is boiled at 100 °C and 1.0 atm. Under these conditions, 1.0 g of water occupies 1.0 cm^3, 1.0 g of steam occupies 1670 cm^3, and $L_v = 540$ cal/g. Find (*a*) the external work done when 1.0 g of steam is formed at 100 °C and (*b*) the increase in internal energy. *Ans.* (*a*) 0.17 kJ; (*b*) 0.50 kcal

20.35 [II] The temperature of 3.0 kg of krypton gas is raised from −20 °C to 80 °C. (*a*) If this is done at constant volume, compute the heat added, the work done, and the change in internal energy. (*b*) Repeat if the heating process is at constant pressure. For the monatomic gas Kr, $c_v = 0.0357$ cal/g·°C and $c_p = 0.0595$ cal/g·°C. *Ans.* (*a*) 11 kcal, 0, 45 kJ; (*b*) 18 kcal, 30 kJ, 45 kJ

20.36 [I] (*a*) Compute c_v for the monatomic gas argon, given $c_p = 0.125$ cal/g·°C and $\gamma = 1.67$. (*b*) Compute c_p for the diatomic gas nitric oxide (NO), given $c_v = 0.166$ cal/g·°C and $\gamma = 1.40$. *Ans.* (*a*) 0.0749 cal/g·°C; (*b*) 0.232 cal/g·°C

20.37 [I] Compute the work done in an isothermal compression of 30 liters of ideal gas at 1.0 atm to a volume of 3.0 liters. *Ans.* 7.0 kJ

20.38 [II] Five moles of neon gas at 2.00 atm and 27.0 °C is adiabatically compressed to one-third its initial volume. Find the final pressure, final temperature, and external work done on the gas. For neon, $\gamma = 1.67$, $c_v = 0.148$ cal/g·°C, and $M = 20.18$ kg/kmol. *Ans.* 1.27 MPa, 626 K, 20.4 kJ

20.39 [II] Determine the work done by the gas in going from *A* to *B* in the thermodynamic cycle shown in Fig. 20-2. Repeat for portion *CA*. Give answers to one significant figure. *Ans.* 0.4 MJ, −0.3 MJ

20.40 [II] Find the net work output per cycle for the thermodynamic cycle in Fig. 20-4. Give your answer to two significant figures. *Ans.* 2.1 kJ

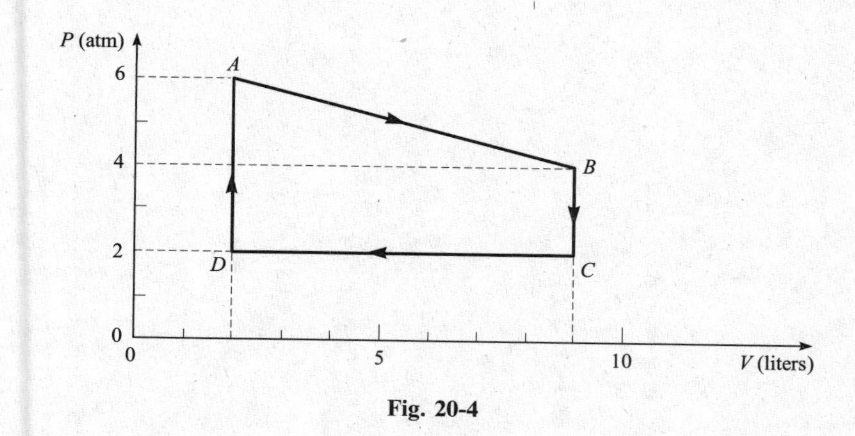

Fig. 20-4

20.41 [II] Four grams of gas, confined to a cylinder, is carried through the cycle shown in Fig. 20-4. At *A* the temperature of the gas is 400 °C. (*a*) What is its temperature at *B*? (*b*) If, in the portion from *A* to *B*, 2.20 kcal flows into the gas, what is c_v for the gas? Give your answers to two significant figures. *Ans.* (*a*) 2.0×10^3 K; (*b*) 0.25 cal/g·°C

20.42 [II] Figure 20-4 is the P–V diagram for 25.0 g of an enclosed ideal gas. At A the gas temperature is 200 °C. The value of c_v for the gas is 0.150 cal/g·°C. (*a*) What is the temperature of the gas at B? (*b*) Find ΔU for the portion of the cycle from A to B. (*c*) Find ΔW for this same portion. (*d*) Find ΔQ for this same portion. *Ans.* (*a*) 1.42×10^3 K; (*b*) 3.55 kcal = 14.9 kJ; (*c*) 3.54 kJ; (*d*) 18.4 kJ

Chapter 21

Entropy and the Second Law

THE SECOND LAW OF THERMODYNAMICS can be stated in three equivalent ways:

(1) Heat flows spontaneously from a hotter to a colder object, but not vice versa.

(2) No heat engine that cycles continuously can change all its heat-in to useful work-out.

(3) If a system undergoes spontaneous change, it will change in such a way that its entropy will increase or, at best, remain constant.

The Second Law tells us the manner in which a spontaneous change will occur, while the First Law tells us whether or not the change is possible. The First Law deals with the conservation of energy; the Second Law deals with the dispersal of energy.

ENTROPY (S) is a *state variable* for a system in equilibrium. By this is meant that S is always the same for the system when it is in a given equilibrium state. Like P, V, and U, the entropy is a characteristic of the system at equilibrium.

When heat ΔQ enters a system at an absolute temperature T, the resulting change in entropy of the system is

$$\Delta S = \frac{\Delta Q}{T}$$

provided the system changes in a reversible way. The SI unit for entropy is J/K.

A **reversible change** (or process) is one in which the values of P, V, T, and U are well-defined during the change. If the process is reversed, then P, V, T, and U will take on their original values when the system is returned to where it started. To be reversible, a process must usually be slow, and the system must be close to equilibrium during the entire change.

Another, fully equivalent, definition of entropy can be given from a detailed molecular analysis of the system. If a system can achieve a particular state (i.e., particular values of P, V, T, and U) in Ω (omega) different ways (different arrangements of the molecules, for example), then the entropy of the state is

$$S = k_B \ln \Omega$$

where ln is the logarithm to base e, and k_B is Boltzmann's constant, 1.38×10^{-23} J/K.

ENTROPY IS A MEASURE OF DISORDER: A state that can occur in only one way (one arrangement of its molecules, for example) is a state of high order. But a state that can occur in many ways is a more disordered state. One way to associate a number with disorder, is to take the disorder of a state as being proportional to Ω, the number of ways the state can occur. Because $S = k_B \ln \Omega$, entropy is a measure of disorder.

Spontaneous processes in systems that contain many molecules always occur in a direction from a

$$\begin{pmatrix}\text{state that can exist}\\ \text{in only a few ways}\end{pmatrix} \rightarrow \begin{pmatrix}\text{state that can exist}\\ \text{in many ways}\end{pmatrix}$$

Hence, when left to themselves, systems either retain their original state of order or else increase their disorder.

THE MOST PROBABLE STATE of a system is the state with the largest entropy. It is also the state with the most disorder and the state that can occur in the largest number of ways.

Solved Problems

21.1 [I] Twenty grams of ice at precisely 0 °C melts into water with no change in temperature. By how much does the entropy of the 20-g mass change in this process?

By slowly adding heat to the ice, we can melt it in a reversible way. The heat needed is

$$\Delta Q = mL_f = (20 \text{ g})(80 \text{ cal/g}) = 1600 \text{ cal}$$

so

$$\Delta S = \frac{\Delta Q}{T} = \frac{1600 \text{ cal}}{273 \text{ K}} = 5.86 \text{ cal/K} = 25 \text{ J/K}$$

Notice that melting increases the entropy (and disorder); ice is more ordered than water.

21.2 [I] As shown in Fig. 21-1, an ideal gas is confined to a cylinder by a piston. The piston is pushed down slowly so that the gas temperature remains at 20.0 °C. During the compression, 730 J of work is done on the gas. Find the entropy change of the gas.

The First Law tells us that

$$\Delta Q = \Delta U + \Delta W$$

Because the process was isothermal, the internal energy of the ideal gas did not change. Therefore, $\Delta U = 0$ and

$$\Delta Q = \Delta W = -730 \text{ J}$$

(Because the gas was compressed, the gas did negative work, hence the minus sign. In other words, work done on the gas is negative.) Now we can write

$$\Delta S = \frac{\Delta Q}{T} = \frac{-730 \text{ J}}{293 \text{ K}} = -2.49 \text{ J/K}$$

Notice that the entropy change is negative. The disorder of the gas decreased as it was pushed into a smaller volume.

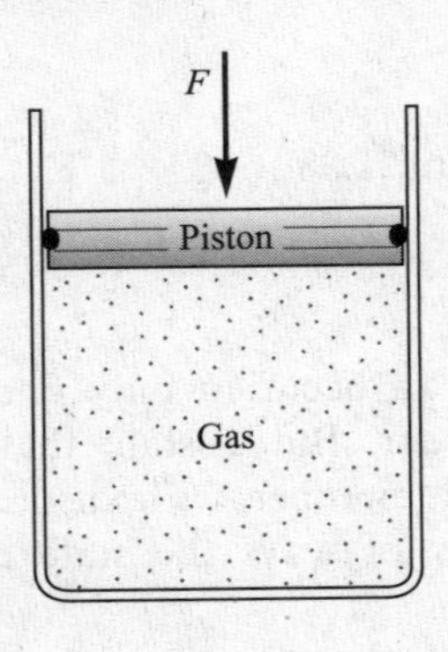

Fig. 21-1

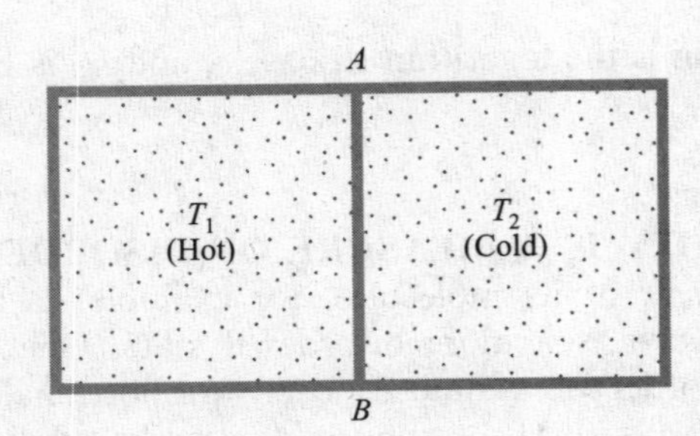

Fig. 21-2

21.3 [II] As shown in Fig. 21-2, a container is separated into two equal-volume compartments. The two compartments contain equal masses of the same gas, 0.740 g in each, and c_v for the gas is 745 J/kg·K. At the start, the hot gas is at 67.0 °C, while the cold gas is at 20.0 °C. No heat can leave or enter the compartments except slowly through the partition AB. Find the entropy change of each compartment as the hot gas cools from 67.0 °C to 65.0 °C.

The heat lost by the hot gas in the process is

$$\Delta Q = mc_v\,\Delta T = (0.000\,740\text{ kg})(745\text{ J/kg}\cdot\text{K})(-2.0\,^\circ\text{C}) = -1.10\text{ J}$$

For the hot gas (approximately the temperature as 66 °C),

$$\Delta S_h = \frac{\Delta Q}{T_h} \approx \frac{-1.10\text{ J}}{(273+66)\text{ K}} = -3.2\times10^{-3}\text{ J/K}$$

For the cold gas, since it will gain 1.10 J,

$$\Delta S_c = \frac{\Delta Q}{T_c} \approx \frac{1.10\text{ J}}{(273+21)\text{ K}} = 3.8\times10^{-3}\text{ J/K}$$

As you can see, the entropy changes were different for the two compartments; more was gained than was lost. The total entropy of the universe increased as a result of this process.

21.4 [II] The ideal gas in the cylinder in Fig. 21-1 is initially at conditions P_1, V_1, T_1. It is slowly expanded at constant temperature by allowing the piston to rise. Its final conditions are P_2, V_2, T_1, where $V_2 = 3V_1$. Find the change in entropy of the gas during this expansion. The mass of gas is 1.5 g, and $M = 28$ kg/kmol for it.

Recall from Chapter 20 that, for an isothermal expansion of an ideal gas (where $\Delta U = 0$),

$$\Delta W = \Delta Q = P_1V_1 \ln\left(\frac{V_2}{V_1}\right)$$

Consequently,
$$\Delta S = \frac{\Delta Q}{T} = \frac{P_1V_1}{T_1}\ln\left(\frac{V_2}{V_1}\right) = \frac{m}{M}R\ln\left(\frac{V_2}{V_1}\right)$$

where we have used the Ideal Gas Law. Substituting the data gives

$$\Delta S = \left(\frac{1.5\times10^{-3}\text{ kg}}{28\text{ kg/kmol}}\right)\left(8314\frac{\text{J}}{\text{kmol}\cdot\text{K}}\right)(\ln 3) = 0.49\text{ J/K}$$

21.5 [I] Two vats of water, one at 87 °C and the other at 14 °C, are separated by a metal plate. If heat flows through the plate at 35 cal/s, what is the change in entropy of the system that occurs in a time of one second?

The higher-temperature vat loses entropy, while the cooler one gains entropy:

$$\Delta S_h = \frac{\Delta Q}{T_h} = \frac{(-35\text{ cal})(4.184\text{ J/cal})}{360\text{ K}} = -0.41\text{ J/K}$$

$$\Delta S_c = \frac{\Delta Q}{T_c} = \frac{(35\text{ cal})(4.184\text{ J/cal})}{287\text{ K}} = 0.51\text{ J/K}$$

Therefore 0.51 J/K − 0.41 J/K = 0.10 J/K.

21.6 [I] A system consists of 3 coins that can come up either heads or tails. In how many different ways can the system have (*a*) all heads up? (*b*) All tails up? (*c*) One tail and two heads up? (*d*) Two tails and one head up?

(*a*) There is only one way all the coins can be heads-up: Each coin must be heads-up.

(*b*) Here, too, there is only one way.

(*c*) There are three ways, corresponding to the three choices for the coin showing the tail.

(*d*) By symmetry with (*c*), there are three ways.

21.7 [I] Find the entropy of the three-coin system described in Problem 21.6 if (*a*) all coins are heads-up, (*b*) two coins are heads-up.

We use the Boltzmann relation $S = k_B \ln \Omega$, where Ω is the number of ways the state can occur, and $k_B = 1.38 \times 10^{-23}$ J/K.

(*a*) Since this state can occur in only one way,

$$S = k_B \ln 1 = (1.38 \times 10^{-23} \text{ J/K})(0) = 0$$

(*b*) Since this state can occur in three ways,

$$S = (1.38 \times 10^{-23} \text{ J/K}) \ln 3 = 1.52 \times 10^{-23} \text{ J/K}$$

Supplementary Problems

21.8 [I] Compute the entropy change of 5.00 g of water at 100 °C as it changes to steam at 100 °C under standard pressure. *Ans.* 7.24 cal/K = 30.3 J/K

21.9 [I] By how much does the entropy of 300 g of a metal ($c = 0.093$ cal/g·°C) change as it is cooled from 90 °C to 70 °C? You may approximate $T = \frac{1}{2}(T_1 + T_2)$. *Ans.* −6.6 J/K

21.10 [II] An ideal gas was slowly expanded from 2.00 m^3 to 3.00 m^3 at a constant temperature of 30 °C. The entropy change of the gas was +47 J/K during the process. (*a*) How much heat was added to the gas during the process? (*b*) How much work did the gas do during the process? *Ans.* (*a*) 3.4 kcal; (*b*) 14 kJ

21.11 [II] Starting at standard conditions, 3.0 kg of an ideal gas ($M = 28$ kg/kmol) is isothermally compressed to one-fifth of its original volume. Find the change in entropy of the gas. *Ans.* −1.4 kJ/K

21.12 [I] Four poker chips are red on one side and white on the other. In how many different ways can (*a*) only 3 reds come up? (*b*) Only two reds come up? *Ans.* (*a*) 4; (*b*) 6

21.13 [II] When 100 coins are tossed, there is one way in which all can come up heads. There are 100 ways in which only one tail comes up. There are about 1×10^{29} ways that 50 heads can come up. One hundred coins are placed in a box with only one head up. They are shaken and then there are 50 heads up. What was the change in entropy of the coins caused by the shaking? *Ans.* 9×10^{-22} J/K

Chapter 22

Wave Motion

A PROPAGATING WAVE is a self-sustaining disturbance of a medium that travels from one point to another, carrying energy and momentum. Mechanical waves are aggregate phenomena arising from the motion of constituent particles. The wave advances, but the particles of the medium only oscillate in place. A wave has been generated on the string in Fig. 22-1 by the sinusoidal vibration of the hand at its end. Energy is carried by the wave from the source to the right, along the string. This direction, the direction of energy transport, is called the *direction of propagation* of the wave.

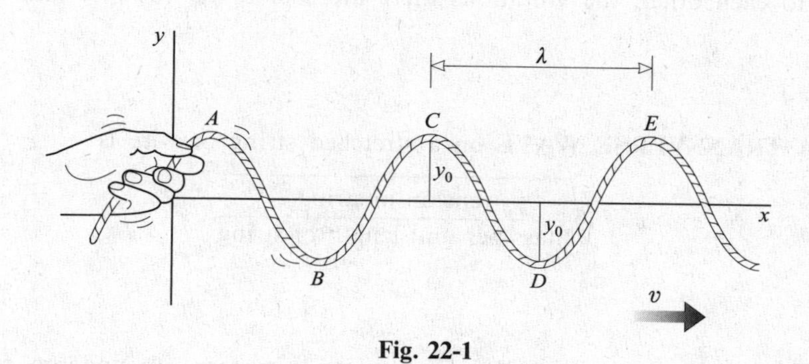

Fig. 22-1

Each particle of the string (such as the one at point-*C*) vibrates up and down, perpendicular to the direction of propagation. Any wave in which the vibration direction is perpendicular to the direction of propagation is called a **transverse wave**. Typical transverse waves, besides those on a string, are electromagnetic waves (e.g., light and radio waves). By contrast, in sound waves the vibration direction is parallel to the direction of propagation, as you will see in Chapter 23. Such a wave is called a **longitudinal** (or *compressional*) **wave.**

WAVE TERMINOLOGY: The **period** (T) of a wave is the time it takes the wave to go through one complete cycle. It is the time taken for a particle, such as the one at A, to move through one complete vibration or cycle, down from point A and then back to A. The period is the number of seconds per cycle. The **frequency** (f) of a wave is the number of cycles per second: Thus,

$$f = \frac{1}{T}$$

If T is in seconds, then f is in hertz (Hz), where 1 Hz $= 1\ \text{s}^{-1}$. The period and frequency of the wave are the same as the period and frequency of the vibration.

The top points on the wave, such as A and C, are called wave *crests*. The bottom points, such as B and D, are called *troughs*. As time goes on, the crests and troughs move to the right with speed v, the speed of the wave.

The **amplitude** of a wave is the maximum disturbance undergone during a vibration cycle, distance y_0 in Fig. 22-1.

The wavelength (λ) is the distance along the direction of propagation between corresponding points on the wave, distance AC for example. In a time T, a crest moving with speed v will move a distance λ to the right. Therefore, $s = vt$ gives

$$\lambda = vT = \frac{v}{f}$$

and

$$v = f\lambda$$

This relation holds for all waves, not just for waves on a string.

IN-PHASE VIBRATIONS exist at two points on a wave if those points undergo vibrations that are in the same direction, in step. For example, the particles of the string at points A and C in Fig. 22-1 vibrate ***in-phase***, since they move up together and down together. Vibrations are in-phase if the points are a whole number of wavelengths apart. The pieces of the string at A and B vibrate opposite to each other; the vibrations there are said to be 180°, or half a cycle, ***out-of-phase***.

THE SPEED OF A TRANSVERSE WAVE on a stretched string or wire is

$$v = \sqrt{\frac{\text{tension in string}}{\text{mass per unit length of string}}}$$

STANDING WAVES: At certain vibrational frequencies, a system can undergo resonance. That is to say, it can efficiently absorb energy from a driving source in its environment which is oscillating at that frequency (Fig. 22-2). These and similar vibration patterns are called **standing waves**, as compared to the propagating waves considered above. These might better not be called waves at all, since they do not transport energy and momentum. The stationary points (such as B and D) are called **nodes**; the points of greatest motion (such as A, C, and E) are called **antinodes**. The distance between adjacent nodes (or antinodes) is $\frac{1}{2}\lambda$. We term the portion of the string between adjacent nodes a *segment*, and the length of a segment is also $\frac{1}{2}\lambda$.

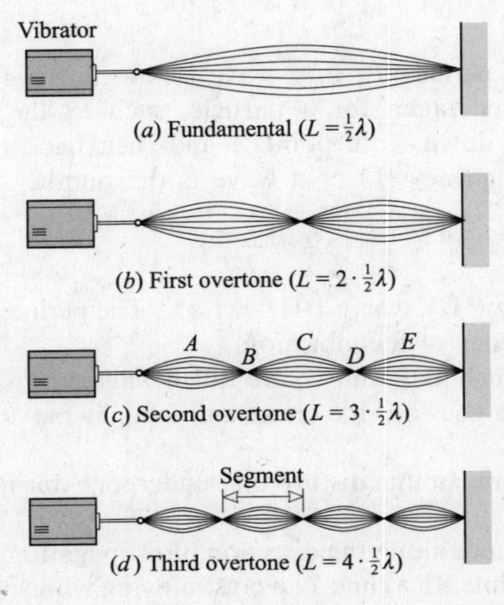

Fig. 22-2

CONDITIONS FOR RESONANCE: A string will resonate only if the vibration wavelength has certain special values: the wavelength must be such that a whole number of wave segments (each $\frac{1}{2}\lambda$ long) exactly fit on the string. A proper fit occurs when nodes and antinodes exist at positions demanded by the constraints on the string. In particular, the fixed ends of the string must be nodes. Thus, as shown in Fig. 22-2, the relation between the wavelength λ and the length L of the resonating string is $L = n(\frac{1}{2}\lambda)$, where n is any integer. Because $\lambda = vT = v/f$, the shorter the wave segments at resonance, the higher will be the resonant frequency. If we call the fundamental resonant frequency f_1, then Fig. 22-2 shows that the higher resonant frequencies are given by $f_n = nf_1$.

LONGITUDINAL (COMPRESSIONAL) WAVES occur as lengthwise vibrations of air columns, solid bars, and the like. At resonance, nodes exist at fixed points, such as the closed end of an air column in a tube, or the location of a clamp on a bar. Diagrams such as Fig. 22-2 are used to display the resonance of longitudinal waves as well as transverse waves. However, for longitudinal waves, the diagrams are mainly schematic and are used to indicate the locations of nodes and antinodes. In analyzing such diagrams, we use the fact that the distance between node and adjacent antinode is $\frac{1}{4}\lambda$.

Solved Problems

22.1 [I] Suppose that Fig. 22-1 represents a 50-Hz wave on a string. Take distance y_0 to be 3.0 mm, and distance AE to be 40 cm. Find the following for the wave: its (*a*) amplitude, (*b*) wavelength, and (*c*) speed.

(*a*) By definition, the amplitude is distance y_0 and is 3.0 mm.

(*b*) The distance between adjacent crests is the wavelength, and so $\lambda = 20$ cm.

(*c*)
$$v = \lambda f = (0.20 \text{ m})(50 \text{ s}^{-1}) = 10 \text{ m/s}$$

22.2 [I] Measurements show that the wavelength of a sound wave in a certain material is 18.0 cm. The frequency of the wave is 1900 Hz. What is the speed of the sound wave?

From $\lambda = vT = v/f$, which applies to all waves,

$$v = \lambda f = (0.180 \text{ m})(1900 \text{ s}^{-1}) = 342 \text{ m/s}$$

22.3 [I] A horizontal cord 5.00 m long has a mass of 1.45 g. What must be the tension in the cord if the wavelength of a 120 Hz wave on it is to be 60.0 cm? How large a mass must be hung from its end (say, over a pulley) to give it this tension?

We know that the speed of a wave on a rope depends on both the tension and the mass per unit length. Moreover,

$$v = \lambda f = (0.600 \text{ m})(120 \text{ s}^{-1}) = 72.0 \text{ m/s}$$

Further, since $$v = \sqrt{(\text{tension})/(\text{mass per unit length})}$$

$$\text{Tension} = (\text{mass per unit length})(v^2) = \left(\frac{1.45 \times 10^{-3}\ \text{kg}}{5.00\ \text{m}}\right)(72.0\ \text{m/s})^2 = 1.50\ \text{N}$$

The tension in the cord balances the weight of the mass hung at its end. Therefore,

$$F_T = mg \quad \text{or} \quad m = \frac{F_T}{g} = \frac{1.50\ \text{N}}{9.81\ \text{m/s}^2} = 0.153\ \text{kg}$$

22.4 [II] A uniform flexible cable is 20 m long and has a mass of 5.0 kg. It hangs vertically under its own weight and is vibrated from its upper end with a frequency of 7.0 Hz. (*a*) Find the speed of a transverse wave on the cable at its midpoint. (*b*) What are the frequency and wavelength at the midpoint?

(*a*) We shall use $v = \sqrt{(\text{tension})/(\text{mass per unit length})}$. The midpoint of the cable supports half its weight, so the tension there is

$$F_T = \tfrac{1}{2}(5.0\ \text{kg})(9.81\ \text{m/s}^2) = 24.5\ \text{N}$$

Further $$\text{Mass per unit length} = \frac{5.0\ \text{kg}}{20\ \text{m}} = 0.25\ \text{kg/m}$$

so that $$v = \sqrt{\frac{24.5\ \text{N}}{0.25\ \text{kg/m}}} = 9.9\ \text{m/s}$$

(*b*) Because wave crests do not pile up along a string or cable, the number passing one point must be the same as that for any other point. Therefore the frequency, 7.0 Hz, is the same at all points.

To find the wavelength at the midpoint, we must use the speed we found for that point, 9.9 m/s. That gives us

$$\lambda = \frac{v}{f} = \frac{9.9\ \text{m/s}}{7.0\ \text{Hz}} = 1.4\ \text{m}$$

22.5 [II] Suppose that Fig. 22-2 shows standing waves on a metal string under a tension of 88.2 N. Its length is 50.0 cm and its mass is 0.500 g. (*a*) Compute v for transverse waves on the string. (*b*) Determine the frequencies of its fundamental, first overtone, and second overtone.

(*a*) $$v = \sqrt{\frac{\text{tension}}{\text{mass per unit length}}} = \sqrt{\frac{88.2\ \text{N}}{(5.00 \times 10^{-4}\ \text{kg})/(0.500\ \text{m})}} = 297\ \text{m/s}$$

(*b*) We recall that the length of the segment is $\lambda/2$ and we use $\lambda = v/f$. For the fundamental:

$$\lambda = 1.00\ \text{m} \quad \text{and} \quad f = \frac{297\ \text{m/s}}{1.00\ \text{m}} = 297\ \text{Hz}$$

For the first overtone:

$$\lambda = 0.500\ \text{m} \quad \text{and} \quad f = \frac{297\ \text{m/s}}{0.500\ \text{m}} = 594\ \text{Hz}$$

For the second overtone:

$$\lambda = 0.333\ \text{m} \quad \text{and} \quad f = \frac{297\ \text{m/s}}{0.333\ \text{m}} = 891\ \text{Hz}$$

22.6 [II] A string 2.0 m long is driven by a 240-Hz vibrator at its end. The string resonates in four segments forming a standing wave pattern. What would be the speed of a transverse wave on such a string?

Let's first determine the wavelength of the wave from part (d) of Fig. 22-2. Since each segment is $\lambda/2$ long, we have

$$4\left(\frac{\lambda}{2}\right) = L \qquad \text{or} \qquad \lambda = \frac{L}{2} = \frac{2.0\text{ m}}{2} = 1.0\text{ m}$$

Then, using $\lambda = vT = v/f$, we have

$$v = f\lambda = (240\text{ s}^{-1})(1.0\text{ m}) = 0.24\text{ km/s}$$

22.7 [II] A banjo string 30 cm long oscillates in a standing-wave pattern. It resonates in its fundamental mode at a frequency of 256 Hz. What is the tension in the string if 80 cm of the string have a mass of 0.75 g?

First we shall find v and then the tension. We know that the string vibrates in one segment when $f = 256$ Hz. Therefore, from Fig. 22-2 (a)

$$\frac{\lambda}{2} = L \qquad \text{or} \qquad \lambda = (0.30\text{ m})(2) = 0.60\text{ m}$$

and

$$v = f\lambda = (256\text{ s}^{-1})(0.60\text{ m}) = 154\text{ m/s}$$

The mass per unit length of the string is

$$\frac{0.75 \times 10^{-3}\text{ kg}}{0.80\text{ m}} = 9.4 \times 10^{-4}\text{ kg/m}$$

Then, from $v = \sqrt{(\text{tension})/(\text{mass per unit length})}$,

$$F_T = (154\text{ m/s})^2(9.4 \times 10^{-4}\text{ kg/m}) = 22\text{ N}$$

22.8 [II] A string vibrates in five segments at a frequency of 460 Hz. (*a*) What is its fundamental frequency? (*b*) What frequency will cause it to vibrate in three segments?

Detailed Method

If the string is n segments long, then from Fig. 22-2 we have $n(\frac{1}{2}\lambda) = L$. But $\lambda = v/f_n$, so $L = n(v/2f_n)$. Solving for f_n gives

$$f_n = n\left(\frac{v}{2L}\right)$$

We are told that $f_5 = 460$ Hz, and so

$$460\text{ Hz} = 5\left(\frac{v}{2L}\right) \qquad \text{or} \qquad \frac{v}{2L} = 92.0\text{ Hz}$$

Substituting this in the above relation gives

$$f_n = (n)(92.0\text{ Hz})$$

(*a*) $f_1 = 92.0$ Hz.

(*b*) $f_3 = (3)(92\text{ Hz}) = 276$ Hz

Alternative Method

We recall that for a string held at both ends, $f_n = nf_1$. Knowing that $f_5 = 460$ Hz, we find $f_1 = 92.0$ Hz and $f_3 = 276$ Hz.

22.9 [II] A string fastened at both ends resonates at 420 Hz and 490 Hz with no resonant frequencies in between. Find its fundamental resonant frequency.

In general, $f_n = nf_1$. We are told that $f_n = 420$ Hz and $f_{n+1} = 490$ Hz. Therefore,

$$420 \text{ Hz} = nf_1 \qquad \text{and} \qquad 490 \text{ Hz} = (n+1)f_1$$

We subtract the first equation from the second to obtain $f_1 = 70.0$ Hz.

22.10 [II] A violin string resonates at its fundamental frequency of 196 Hz. Where along the string must you place your finger so its fundamental becomes 440 Hz?

In the fundamental, $L = \frac{1}{2}\lambda$. Since $\lambda = v/f$, we have $f_1 = v/2L$. Originally, the string of length L_1 resonated to a frequency of 196 Hz, and so

$$196 \text{ Hz} = \frac{v}{2L_1}$$

We want it to resonate to 440 Hz, so we have

$$440 \text{ Hz} = \frac{v}{2L_2}$$

We eliminate v from these two simultaneous equations and find

$$\frac{L_2}{L_1} = \frac{196 \text{ Hz}}{440 \text{ Hz}} = 0.445$$

To obtain the desired resonance, the finger must shorten the string to 0.445 of its original length.

22.11 [II] A 60 cm long bar, clamped at its middle, is vibrated lengthwise by an alternating force at its end. (See Fig. 22-3.) Its fundamental resonance frequency is found to be 3.0 kHz. What is the speed of longitudinal waves in the bar?

Because its ends are free, the bar must have antinodes there. The clamp point at its center must be a node. Therefore, the fundamental resonance is as shown in Fig. 22-3. Because the distance from node to antinode is always $\frac{1}{4}\lambda$, we see that $L = 2(\frac{1}{4}\lambda)$. Since $L = 0.60$ m, we find $\lambda = 1.20$ m.

Then, from the basic relation (p. 214) $\lambda = v/f$, we have

$$v = \lambda f = (1.20 \text{ m})(3.0 \text{ kHz}) = 3.6 \text{ km/s}$$

22.12 [II] Compressional waves (sound waves) are sent down an air-filled tube 90 cm long and closed at one end. The tube resonates at several frequencies, the lowest of which is 95 Hz. Find the speed of sound waves in air.

The tube and several of its resonance forms are shown in Fig. 22-4. Recall that the distance between a node and an adjacent antinode is $\lambda/4$. In our case, the top resonance form applies, since the segments are longest for it and its frequency is therefore lowest. For that form, $L = \lambda/4$, so

$$\lambda = 4L = 4(0.90 \text{ m}) = 3.6 \text{ m}$$

Using $\lambda = vT = v/f$ gives

$$v = \lambda f = (3.6 \text{ m})(95 \text{ s}^{-1}) = 0.34 \text{ km/s}$$

22.13 [II] At what other frequencies will the tube described in Problem 22.12 resonate?

The first few resonances are shown in Fig. 22-4. We see that, at resonance,

$$L = n(\tfrac{1}{4}\lambda_n)$$

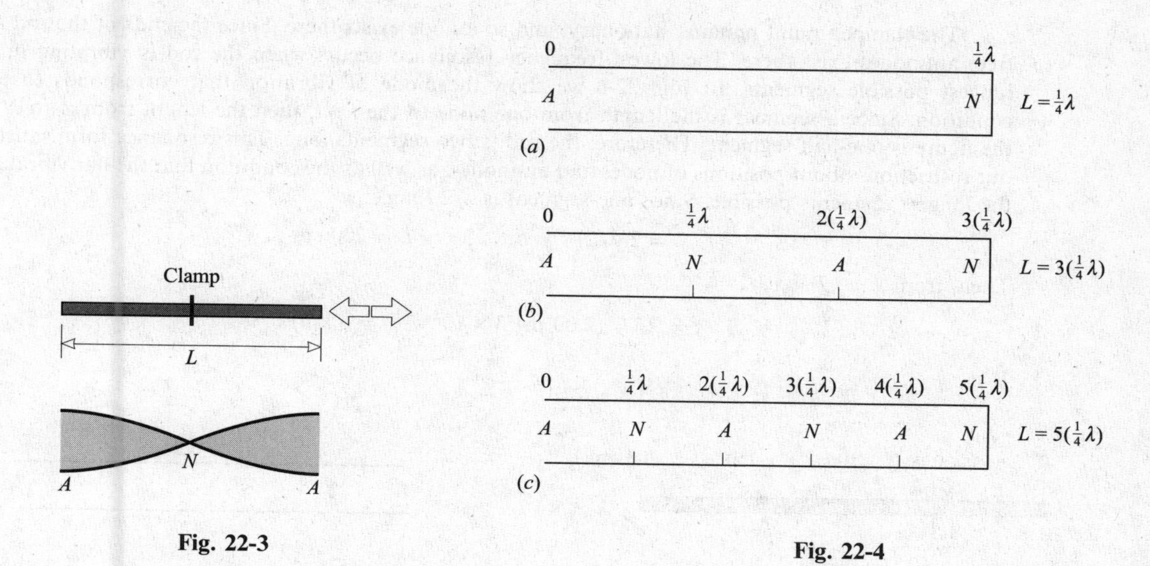

Fig. 22-3

Fig. 22-4

where $n = 1, 3, 5, 7, \ldots$, is an odd integer, and λ_n is the resonant wavelength. But $\lambda_n = v/f_n$, and so

$$L = n\frac{v}{4f_n} \qquad \text{or} \qquad f_n = n\frac{v}{4L} = nf_1$$

where, from Problem 22.12, $f_1 = 95$ Hz. The first few resonant frequencies are thus 95 Hz, 0.29 kHz, 0.48 kHz, ...

22.14 [II] A metal rod 40 cm long is dropped, end first, onto a wooden floor and rebounds into the air. Compressional waves of many frequencies are thereby set up in the bar. If the speed of compressional waves in the bar is 5500 m/s, to what lowest-frequency compressional wave will the bar resonate as it rebounds?

Both ends of the bar will be free, and so antinodes will exist there. In the lowest resonance form (i.e., the one with the longest segments), only one node will exist on the bar, at its center, as shown in Fig. 22-5. We will then have

$$L = 2\left(\frac{\lambda}{4}\right) \qquad \text{or} \qquad \lambda = 2L = 2(0.40\text{ m}) = 0.80\text{ m}$$

Then, from $\lambda = vT = v/f$,

$$f = \frac{v}{\lambda} = \frac{5500\text{ m/s}}{0.80\text{ m}} = 6875\text{ Hz} = 6.9\text{ kHz}$$

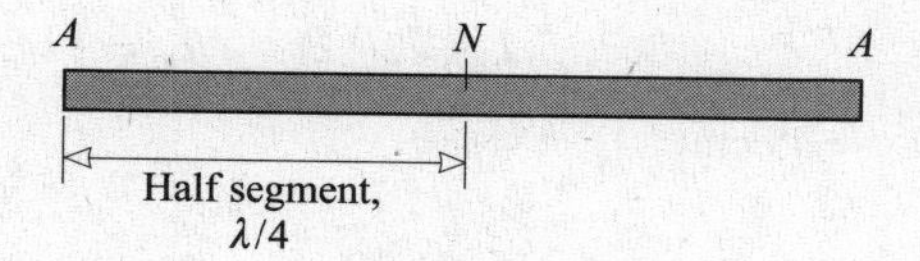

Fig. 22-5

22.15 [II] A rod 200 cm long is clamped 50 cm from one end, as shown in Fig. 22-6. It is set into longitudinal vibration by an electrical driving mechanism at one end. As the frequency of the driver is slowly increased from a very low value, the rod is first found to resonate at 3 kHz. What is the speed of sound (compressional waves) in the rod?

The clamped point remains stationary, and so a node exists there. Since the ends of the rod are free, antinodes exist there. The lowest-frequency resonance occurs when the rod is vibrating in its longest possible segments. In Fig. 22-6 we show the mode of vibration that corresponds to this condition. Since a segment is the length from one node to the next, then the length from A to N in the figure is one-half segment. Therefore, the rod is two segments long. This resonance form satisfies our restrictions about positions of nodes and antinodes, as well as the condition that the bar vibrate in the longest segments possible. Since one segment is $\lambda/2$ long,

$$L = 2(\lambda/2) \qquad \text{or} \qquad \lambda = L = 200 \text{ cm}$$

Then, from $\lambda = vT = v/f$,

$$v = \lambda f = (2.00 \text{ m})(3 \times 10^3 \text{ s}^{-1}) = 6 \text{ km/s}$$

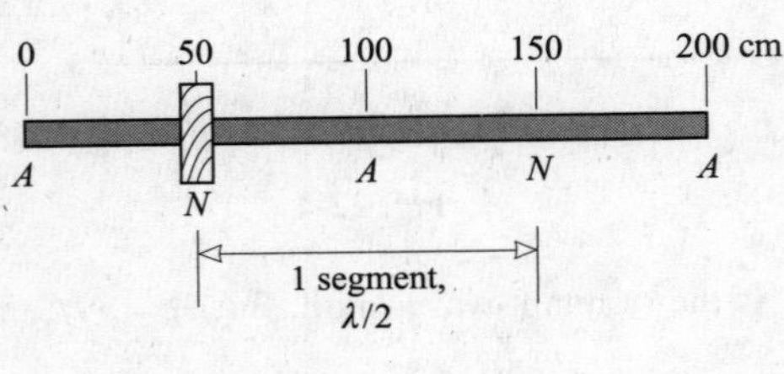

Fig. 22-6

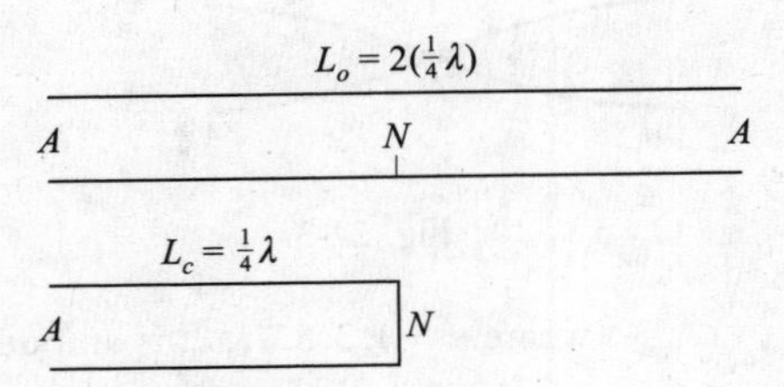

Fig. 22-7

22.16 [II] (*a*)Determine the shortest length of pipe closed at one end that will resonate in air when driven by a sound source of frequency 160 Hz. Take the speed of sound in air to be 340 m/s. (*b*) Repeat the analysis for a pipe open at both ends.

(*a*) Figure 22-4(*a*) applies in this case. The shortest pipe will be $\lambda/4$ long. Therefore,

$$L = \frac{1}{4}\lambda = \frac{1}{4}\left(\frac{v}{f}\right) = \frac{340 \text{ m/s}}{4(160 \text{ s}^{-1})} = 0.531 \text{ m}$$

(*b*) In this case the pipe will have antinodes at both ends and a node at its center. Then,

$$L = 2\left(\frac{1}{4}\lambda\right) = \frac{1}{2}\left(\frac{v}{f}\right) = \frac{340 \text{ m/s}}{2(160 \text{ s}^{-1})} = 1.06 \text{ m}$$

22.17 [II] A pipe 90 cm long is open at both ends. How long must a second pipe, closed at one end, be if it is to have the same fundamental resonance frequency as the open pipe?

The two pipes and their fundamental resonances are shown in Fig. 22-7. As we see,

$$L_o = 2(\tfrac{1}{4}\lambda) \qquad L_c = \tfrac{1}{4}\lambda$$

from which $L_c = \frac{1}{2}L_o = 45$ cm.

22.18 [II] A glass tube that is 70.0 cm long is open at both ends. Find the frequencies at which it will resonate when driven by sound waves that have a speed of 340 m/s.

A pipe that is open at both ends must have an antinode at each end. It will therefore resonate as in Fig. 22-8. We see that the resonance wavelengths λ_n are given by

$$L = n\left(\frac{\lambda_n}{2}\right) \qquad \text{or} \qquad \lambda_n = \frac{2L}{n}$$

where n is an integer. But $\lambda_n = v/f_n$, so

$$f_n = \left(\frac{n}{2L}\right)(v) = (n)\left(\frac{340 \text{ m/s}}{2 \times 0.700 \text{ m}}\right) = 243n \text{ Hz}$$

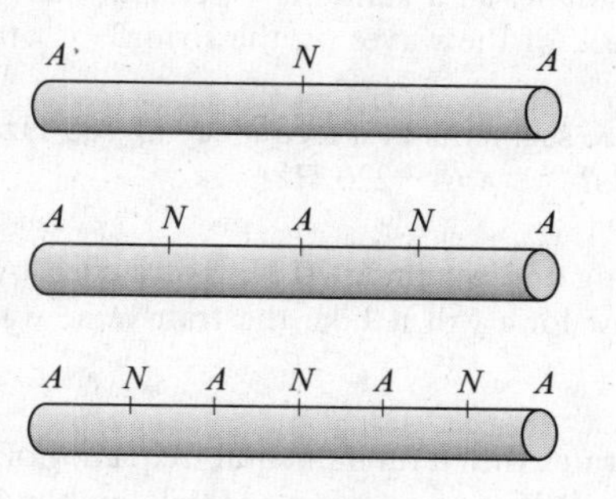

Fig. 22-8

Supplementary Problems

22.19 [I] The average person can hear sound waves ranging in frequency from about 20 Hz to 20 kHz. Determine the wavelengths at these limits, taking the speed of sound to be 340 m/s. *Ans.* 17 m, 1.7 cm

22.20 [I] Radio station WJR in Detroit broadcasts at 760 kHz. The speed of radio waves is 3.00×10^8 m/s. What is the wavelength of WJR's waves? *Ans.* 395 m

22.21 [I] Radar waves with 3.4 cm wavelength are sent out from a transmitter. Their speed is 3.00×10^8 m/s. What is their frequency? *Ans.* 8.8×10^9 Hz = 8.8 GHz

22.22 [I] When driven by a 120 Hz vibrator, a string has transverse waves of 31 cm wavelength traveling along it. (*a*) What is the speed of the waves on the string? (*b*) If the tension in the string is 1.20 N, what is the mass of 50 cm of the string? *Ans.* (*a*) 37 m/s; (*b*) 0.43 g

22.23 [I] The wave shown in Fig. 22-9 is being sent out by a 60 cycle/s vibrator. Find the following for the wave: (*a*) amplitude, (*b*) frequency, (*c*) wavelength, (*d*) speed, (*e*) period. *Ans.* (*a*) 3.0 mm; (*b*) 60 Hz; (*c*) 2.00 cm; (*d*) 1.2 m/s; (*e*) 0.017 s

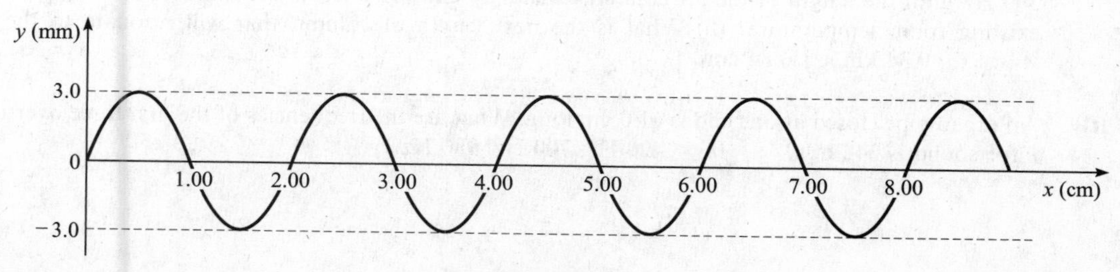

Fig. 22-9

22.24 [II] A copper wire 2.4 mm in diameter is 3.0 m long and is used to suspend a 2.0 kg mass from a beam. If a transverse disturbance is sent along the wire by striking it lightly with a pencil, how fast will the disturbance travel? The density of copper is 8920 kg/m^3. *Ans.* 22 m/s

22.25 [II] A string 180 cm long resonates in a standing wave that has three segments when driven by a 270 Hz vibrator. What is the speed of the waves on the string? *Ans.* 324 m/s

22.26 [II] A string resonates in three segments at a frequency of 165 Hz. What frequency must be used if it is to resonate in four segments? *Ans.* 220 Hz

22.27 [II] A flexible cable, 30 m long and weighing 70 N, is stretched by a force of 2.0 kN. If the cable is struck sideways at one end, how long will it take the transverse wave to travel to the other end and return? *Ans.* 0.65 s

22.28 [II] A wire under tension vibrates with a fundamental frequency of 256 Hz. What would be the fundamental frequency if the wire were half as long, twice as thick, and under one-fourth the tension? *Ans.* 128 Hz

22.29 [II] Steel and silver wires of the same diameter and same length are stretched with equal tension. Their densities are 7.80 g/cm^3 and 10.6 g/cm^3, respectively. What is the fundamental frequency of the silver wire if that of the steel is 200 Hz? *Ans.* 172 Hz

22.30 [II] A string has a mass of 3.0 gram and a length of 60 cm. What must be the tension so that when vibrating transversely its first overtone has frequency 200 Hz? *Ans.* 72 N

22.31 [II] (*a*) At what point should a stretched string be plucked to make its fundamental tone most prominent? At what point should it be plucked and then at what point touched (*b*) to make its first overtone most prominent and (*c*) to make its second overtone most prominent? *Ans.* (*a*) center; (*b*) plucked at 1/4 of its length from one end, then touched at center; (*c*) plucked at 1/6 of its length from one end, then touched at 1/3 of its length from that end

22.32 [II] What must be the length of an iron rod that has the fundamental frequency 320 Hz when clamped at its center? Assume longitudinal vibration at a speed of 5.00 km/s. *Ans.* 7.81 m

22.33 [II] A rod 120 cm long is clamped at the center and is stroked in such a way as to give its first overtone. Make a drawing showing the location of the nodes and antinodes, and determine at what other points the rod might be clamped and still emit the same tone. *Ans.* 20.0 cm from either end

22.34 [II] A metal bar 6.0 m long, clamped at its center and vibrating longitudinally in such a manner that it gives its first overtone, vibrates in unison with a tuning fork marked 1200 vibration/s. Compute the speed of sound in the metal. *Ans.* 4.8 km/s

22.35 [II] Determine the length of the shortest air column in a cylindrical jar that will strongly reinforce the sound of a tuning fork having a vibration rate of 512 Hz. Use $v = 340$ m/s for the speed of sound in air. *Ans* 16.6 cm

22.36 [II] A long, narrow pipe closed at one end does not resonate to a tuning fork having a frequency of 300 Hz until the length of the air column reaches 28 cm. (*a*) What is the speed of sound in air at the existing room temperature? (*b*) What is the next length of column that will resonate to the fork? *Ans.* (*a*) 0.34 km/s; (*b*) 84 cm

22.37 [II] An organ pipe closed at one end is 61.0 cm long. What are the frequencies of the first three overtones if v for sound is 342 m/s? *Ans.* 420 Hz, 700 Hz, 980 Hz

Chapter 23

Sound

SOUND WAVES are ***longitudinal compression waves*** in a material medium such as air, water, or steel. When the compressions and rarefactions of the waves strike the eardrum, they result in the sensation of sound, provided the frequency of the waves is between about 20 Hz and 20 000 Hz. Waves with frequencies above 20 kHz are called *ultrasonic* waves. Those with frequencies below 20 Hz are called *infrasonic* waves.

EQUATIONS FOR SOUND SPEED: In an ideal gas of molecular mass M and absolute temperature T, the speed of sound v is given by

$$v = \sqrt{\frac{\gamma RT}{M}} \qquad \text{(ideal gas)}$$

where R is the gas constant, and γ is the ratio of specific heats c_p/c_v. γ is about 1.67 for monatomic gases (He, Ne, Ar), and about 1.40 for diatomic gases (N_2, O_2, H_2).

The speed of compression waves in other materials is given by

$$v = \sqrt{\frac{\text{modulus}}{\text{density}}}$$

If the material is in the form of a solid bar, Young's modulus Y is used. For liquids, one must use the bulk modulus.

THE SPEED OF SOUND IN AIR at 0 °C is 331 m/s. The speed increases with temperature by about 0.61 m/s for each 1 °C rise. More precisely, sound speeds v_1 and v_2 at absolute temperatures T_1 and T_2 are related by

$$\frac{v_1}{v_2} = \sqrt{\frac{T_1}{T_2}}$$

The speed of sound is essentially independent of pressure, frequency, and wavelength.

THE INTENSITY (I) of any wave is the energy per unit area, per unit time; in practice, it is the average power carried by the wave through a unit area erected perpendicular to the direction of propagation of the wave. Suppose that in a time Δt an amount of energy $\Delta \mathrm{E}$ is carried through an area ΔA that is perpendicular to the propagation direction of the wave. Then

$$I = \frac{\Delta \mathrm{E}}{\Delta A\, \Delta t} = \frac{\mathrm{P}_{av}}{\Delta A}$$

It can be shown that for a sound wave with amplitude a_0 and frequency f, traveling with speed v in a material of density ρ,

$$I = 2\pi^2 f^2 \rho v a_0^2$$

If f is in Hz, ρ is in kg/m^3, v is in m/s, and a_0 (the maximum displacement of the atoms or molecules of the medium) is in m, then I is in W/m^2. Note that $I \propto a_0^2$, and that sort of relationship is true for all kinds of waves.

LOUDNESS is a measure of the human perception of sound. Although a sound wave of high intensity is perceived as louder than a wave of lower intensity, the relation is far from linear. The sensation of sound is roughly proportional to the logarithm of the sound intensity. But the exact relation between loudness and intensity is complicated and not the same for all individuals.

INTENSITY (OR SOUND) LEVEL (β) is defined by an arbitrary scale that corresponds roughly to the sensation of loudness. The zero on this scale occurs when $I_0 = 1.00 \times 10^{-12}\ \mathrm{W/m^2}$, which corresponds roughly to the weakest audible sound. The intensity level, in decibels, is then defined by

$$\beta = 10 \log_{10}\left(\frac{I}{I_0}\right)$$

Notice that when $I = I_0$ the sound level equals zero. The **decibel** (dB) is a dimensionless unit. The normal ear can distinguish between intensities that differ by an amount down to about 1 dB.

BEATS: The alternations of maximum and minimum intensity produced by the superposition of two waves of slightly different frequencies are called **beats**. The number of beats per second is equal to the difference between the frequencies of the two waves that are combined.

DOPPLER EFFECT: Suppose that a moving sound source emits a sound of frequency f_s. Let v be the speed of sound, and let the source approach the listener or observer at speed v_s, measured relative to the medium conducting the sound. Suppose further that the observer is moving toward the source at speed v_o also measured relative to the medium. Then the observer will hear a sound of frequency f_o given by $f_0 = f_s[(v + v_o)/(v - v_s)]$. In general

$$f_o = f_s \frac{v \pm v_o}{v \mp v_s}$$

Draw an arrow from the observer to the source – that's the positive direction. When the velocity of the source is in that direction we use the plus sign in front of v_s. And the same is true for v_o and the observer.

When the source and observer are approaching each other, more wave crests strike the ear each second than when both are at rest. This causes the ear to perceive a higher frequency than that emitted by the source. When the two are receding, the opposite effect occurs; the frequency appears to be lowered.

INTERFERENCE EFFECTS: Two sound waves of the same frequency and amplitude may give rise to easily observed interference effects at a point through which they both pass. If the crests of one wave fall on the crests of the other, the two waves are said to be *in-phase*. In that case, they reinforce each other and give rise to a high intensity at that point.

However, if the crests of one wave fall on the troughs of the other, the two waves will exactly cancel each other. No sound will then be heard at the point. We say that the two waves are then 180° (or a half wavelength) *out-of-phase*.

Intermediate effects are observed if the two waves are neither in-phase nor 180° out-of-phase, but have a fixed phase relationship somewhere in between.

Solved Problems

23.1 [I] An explosion occurs at a distance of 6.00 km from a person. How long after the explosion does the person hear it? Assume the temperature is 14.0 °C.

Because the speed of sound increases by 0.61 m/s for each 1.0 °C, we have

$$v = 331\ \text{m/s} + (0.61)(14)\ \text{m/s} = 340\ \text{m/s}$$

Using $s = vt$, we find that the time taken is

$$t = \frac{s}{v} = \frac{6000\ \text{m}}{340\ \text{m/s}} = 17.6\ \text{s}$$

23.2 [I] To find how far away a lightning flash is, a rough rule is the following: "Divide the time in seconds between the flash and the sound, by three. The result equals the distance in km to the flash." Justify this.

The speed of sound is $v \approx 333\ \text{m/s} \approx \frac{1}{3}\ \text{km/s}$, so the distance to the flash is

$$s = vt \approx \frac{t}{3}$$

where t, the travel time of the sound, is in seconds and s is in kilometers. The light from the flash travels so fast, 3×10^8 m/s, that it reaches the observer almost instantaneously. Hence t is essentially equal to the time between *seeing* the flash and hearing the thunder. Thus the rule.

23.3 [I] Compute the speed of sound in neon gas at 27.0 °C. For neon, $M = 20.18$ kg/kmol.

Neon, being monatomic, has $\gamma \approx 1.67$. Therefore, remembering that T is the absolute temperature,

$$v = \sqrt{\frac{\gamma RT}{M}} = \sqrt{\frac{(1.67)(8314\ \text{J/kmol}\cdot\text{K})(300\ \text{K})}{20.18\ \text{kg/kmol}}} = 454\ \text{m/s}$$

23.4 [II] Find the speed of sound in a diatomic ideal gas that has a density of 3.50 kg/m^3 and a pressure of 215 kPa.

We know that $v = \sqrt{\gamma RT/M}$ and can find the temperature from the pressure. Using the gas law $PV = (m/M)RT$,

$$\frac{RT}{M} = P\frac{V}{m}$$

Moreover, $\rho = m/V$, and so the expression for the speed becomes

$$v = \sqrt{\frac{\gamma P}{\rho}} = \sqrt{\frac{(1.40)(215 \times 10^3\ \text{Pa})}{3.50\ \text{kg/m}^3}} = 293\ \text{m/s}$$

We used the fact that $\gamma \approx 1.40$ for a diatomic ideal gas.

23.5 [II] A metal rod 60 cm long is clamped at its center. It resonates in its fundamental mode when driven by longitudinal waves of 3.00 kHz. What is Young's modulus for the material of the rod? The density of the metal is 8700 kg/m^3.

This same rod was discussed in Problem 22.11. We found there that the speed of longitudinal waves in it is 3.6 km/s. We know that $v = \sqrt{Y/\rho}$, and so

$$Y = \rho v^2 = (8700\ \text{kg/m}^3)(3600\ \text{m/s})^2 = 1.1 \times 10^{11}\ \text{N/m}^2$$

23.6 [I] What is the speed of compression waves (sound waves) in water? The bulk modulus for water is 2.2×10^9 N/m^2.

$$v = \sqrt{\frac{\text{bulk modulus}}{\text{density}}} = \sqrt{\frac{2.2 \times 10^9 \text{ N/m}^2}{1000 \text{ kg/m}^3}} = 1.5 \text{ km/s}$$

23.7 [I] A tuning fork oscillates at 284 Hz in air. Compute the wavelength of the tone emitted at 25 °C.

Remembering that the speed of sound increases by 0.61 m/s for each 1 °C increase in temperature, at 25 °C,

$$v = 331 \text{ m/s} + (0.61)(25) \text{ m/s} = 346 \text{ m/s}$$

Using $\lambda = vT = v/f$ gives

$$\lambda = \frac{v}{f} = \frac{346 \text{ m/s}}{284 \text{ s}^{-1}} = 1.22 \text{ m}$$

23.8 [II] An organ pipe whose length is held constant resonates at a frequency of 224.0 Hz when the air temperature is 15 °C. What will be its resonant frequency when the air temperature is 24 °C?

The resonant wavelength must have the same value at each temperature because it depends only on the length of the pipe. (Its nodes and antinodes must fit properly within the pipe.) But $\lambda = v/f$, and so v/f must be the same at the two temperatures. We thus have

$$\frac{v_1}{224 \text{ Hz}} = \frac{v_2}{f_2} \quad \text{or} \quad f_2 = (224 \text{ Hz})\left(\frac{v_2}{v_1}\right)$$

At temperatures near room temperature, $v = (331 + 0.61T_c)$ m/s, where T_c is the celsius temperature. Then we have

$$f_2 = (224.0 \text{ Hz})\left[\frac{331 + (0.61)(24)}{331 + (0.61)(15)}\right] = 0.228 \text{ kHz}$$

23.9 [I] An uncomfortably loud sound might have an intensity of 0.54 W/m^2. Find the maximum displacement of the molecules of air in a sound wave if its frequency is 800 Hz. Take the density of air to be 1.29 kg/m^3 and the speed of sound to be 340 m/s.

We are given I, f, ρ, and v, and have to find a_0.
From $I = 2\pi^2 f^2 \rho v a_0^2$,

$$a_0 = \frac{1}{\pi f}\sqrt{\frac{I}{2\rho v}} = \frac{1}{(800 \text{ s}^{-1}\pi)}\sqrt{\frac{0.54 \text{ W/m}^2}{(2)(1.29 \text{ kg/m}^3)(340 \text{ m/s})}} = 9.9 \times 10^{-6} \text{ m} = 9.9 \ \mu\text{m}$$

23.10 [I] A sound has an intensity of 3.00×10^{-8} W/m^2. What is the sound level in dB?

Sound level is β where $I_0 = 1.00 \times 10^{-12}$ W/m^2,

$$\beta = 10 \log_{10}\left(\frac{I}{1.00 \times 10^{-12} \text{ W/m}^2}\right)$$

$$= 10 \log_{10}\left(\frac{3.00 \times 10^{-8}}{1.00 \times 10^{-12}}\right) = 10 \log_{10}(3.00 \times 10^4) = 10(4 + \log_{10} 3.00)$$

$$= 10(4 + 0.477) = 44.8 \text{ dB}$$

23.11 [II] A noise-level meter reads the sound level in a room to be 85.0 dB. What is the sound intensity in the room?

$$\beta = 10\log_{10}\left(\frac{I}{1.00 \times 10^{-12}\ \text{W/m}^2}\right) = 85.0\ \text{dB}$$

$$\log_{10}\left(\frac{I}{1.00 \times 10^{-12}\ \text{W/m}^2}\right) = \frac{85.0}{10} = 8.50$$

$$\frac{I}{1.00 \times 10^{-12}\ \text{W/m}^2} = \text{antilog}_{10}\ 8.50 = 3.16 \times 10^8$$

$$I = (1.00 \times 10^{-12}\ \text{W/m}^2)(3.16 \times 10^8) = 3.16 \times 10^{-4}\ \text{W/m}^2$$

23.12 [II] Two sound waves have intensities of 10 and 500 μW/cm^2. What is the difference in their intensity levels?

Call the 10 μW/cm^2 sound A, and the other B. Then

$$\beta_A = 10\log_{10}\left(\frac{I_A}{I_0}\right) = 10(\log_{10} I_A - \log_{10} I_0)$$

$$\beta_B = 10\log_{10}\left(\frac{I_B}{I_0}\right) = 10(\log_{10} I_B - \log_{10} I_0)$$

Subtracting gives

$$\beta_B - \beta_A = 10(\log_{10} I_B - \log_{10} I_A) = 10\log_{10}\left(\frac{I_B}{I_A}\right)$$

$$= 10\log_{10}\left(\frac{500}{10}\right) = 10\log_{10} 50 = (10)(1.70)$$

$$= 17\ \text{dB}$$

23.13 [II] Find the ratio of the intensities of two sounds if one is 8.0 dB louder than the other.

We saw in Problem 23.12 that

$$\beta_B - \beta_A = 10\log_{10}\left(\frac{I_B}{I_A}\right)$$

In the present case this becomes

$$8.0 = 10\log_{10}\left(\frac{I_B}{I_A}\right) \quad \text{or} \quad \frac{I_B}{I_A} = \text{antilog}_{10}\ 0.80 = 6.3$$

23.14 [II] A tiny sound source emits sound uniformly in all directions. The intensity level at a distance of 2.0 m is 100 dB. How much sound power is the source emitting?

The energy emitted by a point source can be considered to flow out through a spherical surface which has the source at its center. Hence, if we find the rate of flow through such a surface, it will equal the flow from the source. Take a concentric sphere of radius 2.0 m. We know that the sound level on its surface is 100 dB. You can show that this corresponds to $I = 0.010$ W/m^2. Thus, the energy flowing each second through each m^2 of surface is 0.010 W. The total energy flow through the spherical surface is then $I(4\pi r^2)$, where $I = 0.010$ W/m^2 and $r = 2.0$ m:

$$\text{Power from source} = (0.010\ \text{W/m}^2)(4\pi)(2\ \text{m})^2 = 0.50\ \text{W}$$

Notice how little power issues as sound from even such an intense source.

23.15 [III] One typist typing furiously in a room gives rise to an average sound level of 60.0 dB. What will be the decibel level when three equally noisy typists are working?

If each typist emits the same amount of sound energy, then the final sound intensity I_f should be three times the initial intensity I_i. We have

$$\beta_f = \log_{10}\left(\frac{I_f}{I_0}\right) = \log_{10} I_f - \log_{10} I_0$$

and

$$\beta_i = \log_{10} I_i - \log_{10} I_0$$

Subtraction yields

$$\beta_f - \beta_i = \log_{10} I_f - \log_{10} I_i$$

from which

$$\beta_f = \beta_i + \log_{10}\left(\frac{I_f}{I_i}\right) = 60.0 \text{ dB} + \log 3 = 60.5 \text{ dB}$$

The sound level, being a logarithmic measure, rises very slowly with the number of typists.

23.16 [I] An automobile moving at 30.0 m/s is approaching a factory whistle that has a frequency of 500 Hz. (*a*) If the speed of sound in air is 340 m/s, what is the apparent frequency of the whistle as heard by the driver? (*b*) Repeat for the case of the car leaving the factory at the same speed.

This is a Doppler shift problem where the observer is moving in the positive direction, and $v_s = 0$. Hence we use $+v_o$ in this first part. And so

(*a*)
$$f_o = f_s \frac{v \pm v_o}{v \mp v_s} = (500 \text{ Hz}) \frac{340 \text{ m/s} + 30.0 \text{ m/s}}{340 \text{ m/s} - 0} = 544 \text{ Hz}$$

With the car leaving in the negative direction we use $-v_o$ and

(*b*)
$$f_o = f_s \frac{v \pm v_o}{v \mp v_s} = (500 \text{ Hz}) \frac{340 \text{ m/s} - 30.0 \text{ m/s}}{340 \text{ m/s} - 0} = 456 \text{ Hz}$$

23.17 [I] A car moving at 20 m/s with its horn blowing ($f = 1200$ Hz) is chasing another car going at 15 m/s in the same direction. What is the apparent frequency of the horn as heard by the driver being chased? Take the speed of sound to be 340 m/s.

This is a Doppler problem where both the source and the observer are moving in the negative direction. Hence we use $-v_o$ and $-v_s$.

$$f_o = f_s \frac{v \pm v_o}{v \mp v_s} = (1200 \text{ Hz}) \frac{340 - 15}{340 - 20} = 1.22 \text{ kHz}$$

23.18 [I] When two tuning forks are sounded simultaneously, they produce one beat every 0.30 s. (*a*) By how much do their frequencies differ? (*b*) A tiny piece of chewing gum is placed on a prong of one fork. Now there is one beat every 0.40 s. Was this tuning fork the lower- or the higher-frequency fork?

The number of beats per second equals the frequency difference.

(*a*)
$$\text{Frequency difference} = \frac{1}{0.30 \text{ s}} = 3.3 \text{ Hz}$$

(*b*)
$$\text{Frequency difference} = \frac{1}{0.40 \text{ s}} = 2.5 \text{ Hz}$$

Adding gum to the prong increases its mass and thereby decreases its vibrational frequency. This lowering of frequency caused it to come closer to the frequency of the other fork. Hence the fork in question had the higher frequency.

23.19 [II] A tuning fork of frequency 400 Hz is moved away from an observer and toward a flat wall with a speed of 2.0 m/s. What is the apparent frequency (*a*) of the unreflected sound waves coming directly to the observer, and (*b*) of the sound waves coming to the observer after reflection? (*c*) How many beats per second are heard? Assume the speed of sound in air to be 340 m/s.

(*a*) The fork, the source, is receding from the observer in the positive direction and so we use $+v_s$. It doesn't matter what the sign associated with v_o is since $v_o = 0$.

$$f_o = f_s \frac{v \pm v_o}{v \mp v_s} = (400\ \text{Hz}) \frac{340\ \text{m/s} + 0}{340\ \text{m/s} + 2.0\ \text{m/s}} = 397.7\ \text{Hz} = 398\ \text{Hz}$$

(*b*) The wave crests reaching the wall are closer together than normally because the fork is moving toward the wall. Therefore, the reflected wave appears to come from an approaching source:

$$f_o = f_s \frac{v \pm v_o}{v \mp v_s} = (400\ \text{Hz}) \frac{340\ \text{m/s} + 0}{340\ \text{m/s} - 2.0\ \text{m/s}} = 402.4\ \text{Hz} = 402\ \text{Hz}$$

(c) Beats per second = difference between frequencies = (402.4-397.7) Hz = 4.7 beats per second

23.20 [I] In Fig. 23-1, S_1 and S_2 are identical sound sources. They send out their wave crests simultaneously (the sources are in phase). For what values of $L_1 - L_2$ will constructive interference obtain and a loud sound be heard at point *P*?

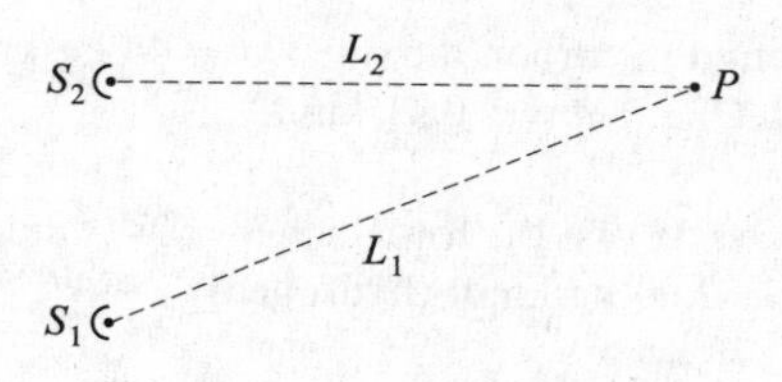

Fig. 23-1

If $L_1 = L_2$, the waves from the two sources will take equal times to reach *P*. Crests from one will arrive there at the same times as crests from the other. The waves will therefore be in phase at *P* and an interference maximum will result.

If $L_1 = L_2 + \lambda$, then the wave from S_1 will be one wavelength behind the one from S_2 when they reach *P*. But because the wave repeats each wavelength, a crest from S_1 will still reach *P* at the same time a crest from S_2 does. Once again the waves are in phase at *P* and an interference maximum will exist there.

In general, a loud sound will be heard at *P* when $L_1 - L_2 = \pm n\lambda$, where *n* is an integer.

23.21 [II] The two sound sources in Fig. 23-1 vibrate in-phase. A loud sound is heard at *P* when $L_1 = L_2$. As L_1 is slowly increased, the weakest sound is heard when $L_1 - L_2$ has the values 20.0 cm, 60.0 cm, and 100 cm. What is the frequency of the sound source if the speed of sound is 340 m/s?

The weakest sound will be heard at P when a crest from S_1 and a trough from S_2 reach there at the same time. This will happen if $L_1 - L_2$ is $\frac{1}{2}\lambda$, or $\lambda + \frac{1}{2}\lambda$, or $2\lambda + \frac{1}{2}\lambda$, and so on. Hence the increase in L_1 between weakest sounds is λ, and from the data we see that $\lambda = 0.400$ m. Then, from $\lambda = v/f$,

$$f = \frac{v}{\lambda} = \frac{340 \text{ m/s}}{0.400 \text{ m}} = 850 \text{ Hz}$$

Supplementary Problems

23.22 [I] Three seconds after a gun is fired, the person who fired the gun hears an echo. How far away was the surface that reflected the sound of the shot? Use 340 m/s for the speed of sound. *Ans.* 510 m

23.23 [I] What is the speed of sound in air when the air temperature is 31 °C? *Ans.* 0.35 km/s

23.24 [II] A shell fired at a target 800 m distant was heard to strike it 5.0 s after leaving the gun. Compute the average horizontal velocity of the shell. The air temperature is 20 °C. *Ans.* 0.30 km/s

23.25 [II] In an experiment to determine the speed of sound, two observers, A and B, were stationed 5.00 km apart. Each was equipped with gun and stopwatch. Observer A heard the report of B's gun 15.5 s after seeing its flash. Later, A fired his gun and B heard the report 14.5 s after seeing the flash. Determine the speed of sound and the component of the speed of the wind along the line joining A to B. *Ans.* 334 m/s, 11.1 m/s

23.26 [II] A disk has 40 holes around its circumference and is rotating at 1200 rpm. Determine the frequency and wavelength of the tone produced by the disk when a jet of air is blown against it. The temperature is 15 °C. *Ans.* 0.80 kHz, 0.43 m

23.27 [II] Determine the speed of sound in carbon dioxide ($M = 44$ kg/kmol, $\gamma = 1.30$) at a pressure of 0.50 atm and a temperature of 400 °C. *Ans.* 0.41 km/s

23.28 [II] Compute the molecular mass M of a gas for which $\gamma = 1.40$ and in which the speed of sound is 1260 m/s at precisely 0 °C. *Ans.* 2.00 kg/kmol (hydrogen)

23.29 [II] At S.T.P., the speed of sound in air is 331 m/s. Determine the speed of sound in hydrogen at S.T.P. if the specific gravity of hydrogen relative to air is 0.0690 and if $\gamma = 1.40$ for both gases. *Ans.* 1.26 km/s

23.30 [II] Helium is a monatomic gas that has a density of 0.179 kg/m^3 at a pressure of 76.0 cm of mercury and a temperature of precisely 0 °C. Find the speed of compression waves (sound) in helium at this temperature and pressure. *Ans.* 970 m/s

23.31 [II] A bar of dimensions 1.00 cm^2 × 200 cm and mass 2.00 kg is clamped at its center. When vibrating longitudinally it emits its fundamental tone in unison with a tuning fork making 1000 vibration/s. How much will the bar be elongated if, when clamped at one end, a stretching force of 980 N is applied at the other end? *Ans.* 0.123 m

23.32 [I] Find the speed of compression waves in a metal rod if the material of the rod has a Young's modulus of 1.20×10^{10} N/m^2 and a density of 8920 kg/m^3. *Ans.* 1.16 km/s

23.33 [II] An increase in pressure of 100 kPa causes a certain volume of water to decrease by 5×10^{-3} percent of its original volume. (*a*) What is the bulk modulus of water? (*b*) What is the speed of sound (compression waves) in water? *Ans.* (*a*) 2×10^9 N/m^2; (*b*) 1 km/s

23.34 [I] A sound has an intensity of 5.0×10^{-7} W/m^2. What is its intensity level? *Ans.* 57 dB

23.35 [I] A person riding a power mower may be subjected to a sound of intensity 2.00×10^{-2} W/m^2. What is the intensity level to which the person is subjected? *Ans.* 103 dB

23.36 [II] A rock band might easily produce a sound level of 107 dB in a room. To two significant figures, what is the sound intensity at 107 dB? *Ans.* 0.0500 W/m^2

23.37 [II] A whisper has an intensity level of about 15 dB. What is the corresponding intensity of the sound? *Ans.* 3.2×10^{-11} W/m^2

23.38 [II] What sound intensity is 3.0 dB louder than a sound of intensity of 10 μW/cm^2? *Ans.* 20 μW/cm^2

23.39 [II] Calculate the intensity of a sound wave in air at precisely 0 °C and 1.00 atm if its amplitude is 0.0020 mm and its wavelength is 66.2 cm. The density of air at S.T.P. is 1.293 kg/m^3. *Ans.* 8.4 mW/m^2

23.40 [II] What is the amplitude of vibration in a 8000 Hz sound beam if its intensity level is 62 dB? Assume that the air is at 15 °C and its density is 1.29 kg/m^3. *Ans.* 1.7×10^{-9} m

23.41 [II] One sound has an intensity level of 75.0 dB while a second has an intensity level of 72.0 dB. What is the intensity level when the two sounds are combined? *Ans.* 76.8 dB

23.42 [II] A certain organ pipe is tuned to emit a frequency of 196.00 Hz. When it and the G string of a violin are sounded together, ten beats are heard in a time of exactly 8 s. The beats become slower as the violin string is slowly tightened. What was the original frequency of the violin string? *Ans.* 194.75 Hz

23.43 [I] A locomotive moving at 30.0 m/s approaches and passes a person standing beside the track. Its whistle is emitting a note of frequency 2.00 kHz. What frequency will the person hear (*a*) as the train approaches and (*b*) as it recedes? The speed of sound is 340 m/s. *Ans.* (*a*) 2.19 kHz; (*b*) 1.84 kHz

23.44 [II] Two cars are heading straight at each other with the same speed. The horn of one ($f = 3.0$ kHz) is blowing, and is heard to have a frequency of 3.4 kHz by the people in the other car. Find the speed at which each car is moving if the speed of sound is 340 m/s. *Ans.* 21 m/s

23.45[II] To determine the speed of a harmonic oscillator, a beam of sound is sent along the line of the oscillator's motion. The sound, which is emitted at a frequency of 8000.0 Hz, is reflected straight back by the oscillator to a detector system. The detector observes that the reflected beam varies in frequency between the limits of 8003.1 Hz and 7996.9 Hz. What is the maximum speed of the oscillator? Take the speed of sound to be 340 m/s. *Ans.* 0.132 m/s

23.46 [II] In Fig. 23-1 are shown two identical sound sources sending waves to point *P*. They send out wave crests simultaneously (they are in-phase), and the wavelength of the wave is 60 cm. If $L_2 = 200$ cm, give the values of L_1 for which (*a*) maximum sound is heard at *P* and (*b*) minimum sound is heard at *P*. *Ans.* (*a*) $(200 \pm 60n)$ cm, where $n = 0, 1, 2, \ldots$; (*b*) $(230 \pm 60n)$ cm, where $n = 0, 1, 2, \ldots$.

23.47 [II] The two sources shown in Fig. 23-2 emit identical beams of sound ($\lambda = 80$ cm) toward one another. Each sends out a crest at the same time as the other (the sources are in-phase). Point *P* is a position of

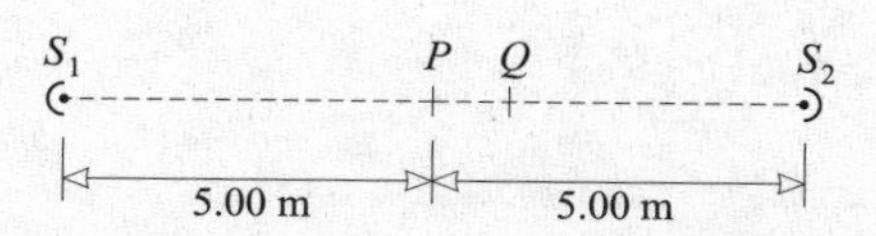

Fig. 23-2

maximum intensity, that is, loud sound. As one moves from P toward Q, the sound decreases in intensity. (*a*) How far from P will a sound minimum first be heard? (*b*) How far from P will a loud sound be heard once again? *Ans.* (*a*) 20 cm; (*b*) 40 cm

Chapter 24

Coulomb's Law and Electric Fields

COULOMB'S LAW: Suppose that two point charges, $q_\bullet$ and $q'_\bullet$, are a distance r apart in vacuum. If $q_\bullet$ and $q'_\bullet$ have the same sign, the two charges repel each other; if they have opposite signs, they attract each other. The force experienced by either charge due to the other is called a **Coulomb** or **electric force** and it is given by **Coulomb's Law**,

$$F_E = k\frac{q_\bullet q'_\bullet}{r^2} \quad \text{(in vacuum)}$$

As always in the SI, distances are measured in meters, and forces in newtons. The SI unit for charge is the *coulomb* (C). The constant k in Coulomb's Law has the value

$$k = 8.988 \times 10^9 \ \text{N}\cdot\text{m}^2/\text{C}^2$$

which we shall usually approximate as $9.0 \times 10^9 \ \text{N}\cdot\text{m}^2/\text{C}^2$. Often, k is replaced by $1/4\pi\epsilon_0$, where $\epsilon_0 = 8.85 \times 10^{-12} \ \text{C}^2/\text{N}\cdot\text{m}^2$ is called the **permittivity of free space**. Then Coulomb's Law becomes,

$$F_E = \frac{1}{4\pi\epsilon_0}\frac{q_\bullet q'_\bullet}{r^2} \quad \text{(in vacuum)}$$

When the surrounding medium is not a vacuum, forces caused by induced charges in the material reduce the force between point charges. If the material has a **dielectric constant** K, then ϵ_0 in Coulomb's Law must be replaced by $K\epsilon_o = \epsilon$, where ϵ is called the *permittivity of the material.* Then

$$F_E = \frac{1}{4\pi\epsilon}\frac{q_\bullet q'_\bullet}{r^2} = \frac{k}{K}\frac{q_\bullet q'_\bullet}{r^2}$$

For vacuum, $K = 1$; for air, $K = 1.0006$.

Coulomb's Law also applies to charged conducting spheres and spherical shells, as well as to uniform spheres of charge. This is true provided that these are all small enough, in comparison to their separations, so that the charge distribution on each doesn't become asymmetrical when two or more of them interact. In that case, r, the distance between the centers of the spheres, must be much larger than the sum of the radii of the two spheres.

CHARGE IS QUANTIZED: The magnitude of the smallest charge ever measured is denoted by e (called the **quantum of charge**), where $e = 1.60218 \times 10^{-19}$ C. All free charges, ones that can be isolated and measured, are integer multiples of e. The electron has a charge of $-e$, while the proton's charge is $+e$. Although there is good reason to believe that quarks carry charges of magnitude $e/3$ and $2e/3$, they only exist in bound systems that have a net charge equal to an integer multiple of e.

CONSERVATION OF CHARGE: The algebraic sum of the charges in the universe is constant. When a particle with charge $+e$ is created, a particle with charge $-e$ is simultaneously created in the immediate vicinity. When a particle with charge $+e$ disappears, a particle with charge $-e$ also disappears in the immediate vicinity. Hence the net charge of the universe remains constant.

THE TEST-CHARGE CONCEPT: A **test-charge** is a very small charge that can be used in making measurements on an electric system. It is assumed that such a charge, which is tiny both in magnitude and physical size, has a negligible effect on its environment.

AN ELECTRIC FIELD is said to exist at any point in space when a test charge, placed at that point, experiences an electrical force. The direction of the electric field at a point is the same as the direction of the force experienced by a *positive* test charge placed at the point.

Electric field lines can be used to sketch electric fields. The line through a point has the same direction at that point as the electric field. Where the field lines are closest together, the electric field is largest. Field lines come out of positive charges (because a positive charge repels a positive test charge) and come into negative charges (because they attract the positive test charge).

THE STRENGTH OF THE ELECTRIC FIELD ($\vec{\mathbf{E}}$) at a point is equal to the force experienced by a unit positive test charge placed at that point. Because the electric field strength is a force per unit charge, it is a vector quantity. The units of $\vec{\mathbf{E}}$ are N/C or (see Chapter 25) V/m.

If a charge q is placed at a point where the electric field due to other charges is $\vec{\mathbf{E}}$, the charge will experience a force $\vec{\mathbf{F}}_E$ given by

$$\vec{\mathbf{F}}_E = q\vec{\mathbf{E}}$$

If q is negative, $\vec{\mathbf{F}}_E$ will be opposite in direction to $\vec{\mathbf{E}}$.

ELECTRIC FIELD DUE TO A POINT CHARGE: To find E (the signed magnitude of $\vec{\mathbf{E}}$) due to a point charge $q_\bullet$, we make use of Coulomb's Law. If a point charge $q'_\bullet$ is placed at a distance r from the charge $q_\bullet$, it will experience a force

$$F_E = \frac{1}{4\pi\epsilon}\frac{q_\bullet q'_\bullet}{r^2} = q'_\bullet\left(\frac{1}{4\pi\epsilon}\frac{q_\bullet}{r^2}\right)$$

But if a point charge $q'_\bullet$ is placed at a position where the electric field is E, then the force on $q'_\bullet$ is

$$F_E = q'_\bullet E$$

Comparing these two expressions for F_E, we see that the ***electric field of a point charge*** $q_\bullet$ is

$$E_\bullet = \frac{1}{4\pi\epsilon}\frac{q_\bullet}{r^2}$$

The same relation applies at points outside of a small spherical charge q. For q positive, E is positive and $\vec{\mathbf{E}}$ is directed radially outward from q; for q negative, E is negative and $\vec{\mathbf{E}}$ is directed radially inward.

SUPERPOSITION PRINCIPLE: The force experienced by a charge due to other charges is the vector sum of the Coulomb forces acting on it due to these other charges. Similarly, the electric intensity $\vec{\mathbf{E}}$ at a point due to several charges is the vector sum of the intensities due to the individual charges.

Solved Problems

24.1 [I] Two small spheres are 1.5 m apart center-to-center. They carry identical charges. Approximately how large is the charge on each if each sphere experiences a force of 2 N?

The diameters of the spheres are small compared to the 1.5 m separation. We may therefore approximate them as point charges. Coulomb's Law, $F_E = (k/K)q_{\bullet 1}q_{\bullet 2}/r^2$, gives (with K approximated as 1.00)

$$q_{\bullet 1} q_{\bullet 2} = q^2 = \frac{F_E r^2}{k} = \frac{(2\text{ N})(1.5\text{ m})^2}{9 \times 10^9\text{ N}\cdot\text{m}^2/\text{C}^2} = 5 \times 10^{-10}\text{ C}^2$$

from which $q = 2 \times 10^{-5}$ C.

24.2 [I] Repeat Problem 24.1 if the spheres are separated by a center-to-center distance of 1.5 m in a large vat of water. The dielectric constant of water is about 80.

From Coulomb's Law,

$$F_E = \frac{k}{K}\frac{q^2}{r^2}$$

where K, the dielectric constant, is now 80. Then

$$q = \sqrt{\frac{F_E r^2 K}{k}} = \sqrt{\frac{(2\text{ N})(1.5\text{ m})^2(80)}{9 \times 10^9\text{ N}\cdot\text{m}^2/\text{C}^2}} = 2 \times 10^{-4}\text{ C}$$

24.3 [I] A helium nucleus has a charge of $+2e$, and a neon nucleus has a charge of $+10e$, where e is the quantum of charge, 1.60×10^{-19} C. Find the repulsive force exerted on one by the other when they are 3.0 nanometers (1 nm $= 10^{-9}$ m) apart. Assume the system to be in vacuum.

Nuclei have radii of order 10^{-15} m. We can assume them to be point charges in this case. Then

$$F_E = k\frac{q_\bullet q'_\bullet}{r^2} = (9.0 \times 10^9\text{ N}\cdot\text{m}^2/\text{C}^2)\frac{(2)(10)(1.6 \times 10^{-19}\text{ C})^2}{(3.0 \times 10^{-9}\text{ m})^2} = 5.1 \times 10^{-10}\text{ N} = 0.51\text{ nN}$$

24.4 [II] In the Bohr model of the hydrogen atom, an electron ($q = -e$) circles a proton ($q' = e$) in an orbit of radius 5.3×10^{-11} m. The attraction of the proton for the electron furnishes the centripetal force needed to hold the electron in orbit. Find (*a*) the force of electrical attraction between the particles and (*b*) the electron's speed. The electron mass is 9.1×10^{-31} kg.

The electron and proton are essentially point charges. Accordingly,

(*a*)
$$F_E = k\frac{q_\bullet q'_\bullet}{r^2} = (9.0 \times 10^9\text{ N}\cdot\text{m}^2/\text{C}^2)\frac{(1.6 \times 10^{-19}\text{ C})^2}{(5.3 \times 10^{-11}\text{ m})^2} = 8.2 \times 10^{-8}\text{ N} = 82\text{ nN}$$

(*b*) The force found in (*a*) is the centripetal force, mv^2/r. Therefore,

$$8.2 \times 10^{-8}\text{ N} = \frac{mv^2}{r}$$

from which

$$v = \sqrt{\frac{(8.2 \times 10^{-8}\text{ N})(r)}{m}} = \sqrt{\frac{(8.2 \times 10^{-8}\text{ N})(5.3 \times 10^{-11}\text{ m})}{9.1 \times 10^{-31}\text{ kg}}} = 2.2 \times 10^6\text{ m/s}$$

24.5 [II] Three point charges are placed on the x-axis as shown in Fig. 24-1. Find the net force on the $-5\ \mu$C charge due to the two other charges.

Because unlike charges attract, the forces on the -5μC charge are as shown. The *magnitudes* of $\vec{\mathbf{F}}_{E3}$ and $\vec{\mathbf{F}}_{E8}$ are given by Coulomb's Law:

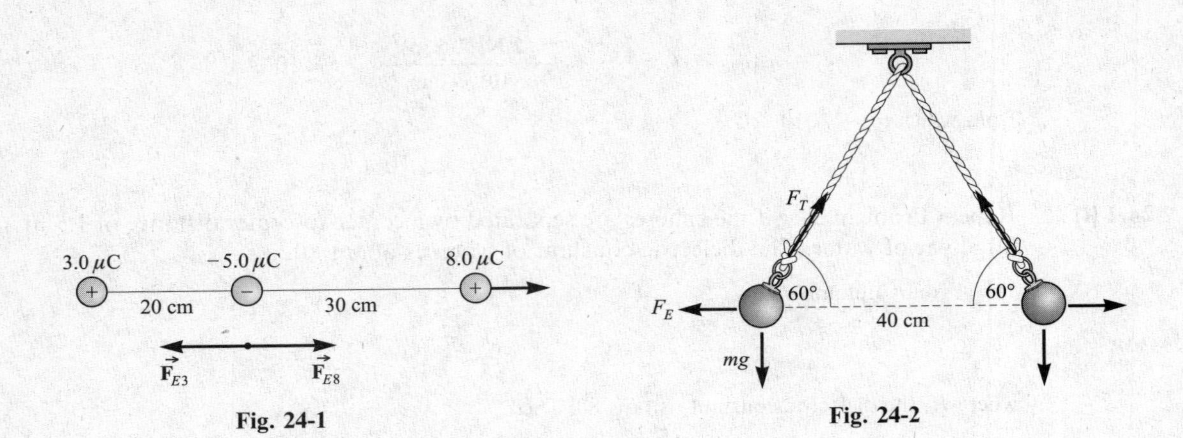

Fig. 24-1

Fig. 24-2

$$F_{E3} = (9.0 \times 10^9\ \text{N}\cdot\text{m}^2/\text{C}^2)\frac{(3.0 \times 10^{-6}\ \text{C})(5.0 \times 10^{-6}\ \text{C})}{(0.20\ \text{m})^2} = 3.4\ \text{N}$$

$$F_{E8} = (9.0 \times 10^9\ \text{N}\cdot\text{m}^2/\text{C}^2)\frac{(8.0 \times 10^{-6}\ \text{C})(5.0 \times 10^{-6}\ \text{C})}{(0.30\ \text{m})^2} = 4.0\ \text{N}$$

Notice two things about the computation: (1) Proper units (coulombs and meters) must be used. (2) Because we want only the magnitudes of the forces, *we do not carry along the signs of the charges*. (That is, we use their absolute values.) The direction of each force is given by the diagram, which we drew from inspection of the situation.

From the diagram, the resultant force on the center charge is

$$F_E = F_{E8} - F_{E3} = 4.0\ \text{N} - 3.4\ \text{N} = 0.6\ \text{N}$$

and it is in the $+x$-direction.

24.6 [II] Find the ratio of the Coulomb electric force F_E to the gravitational force F_G between two electrons in vacuum.

From Coulomb's Law and Newton's Law of gravitation,

$$F_E = k\frac{q_\bullet^2}{r^2} \quad \text{and} \quad F_G = G\frac{m^2}{r^2}$$

Therefore $$\frac{F_E}{F_G} = \frac{kq_\bullet^2/r^2}{Gm^2/r^2} = \frac{kq_\bullet^2}{Gm^2}$$

$$= \frac{(9.0 \times 10^9\ \text{N}\cdot\text{m}^2/\text{C}^2)(1.6 \times 10^{-19}\ \text{C})^2}{(6.67 \times 10^{-11}\ \text{N}\cdot\text{m}^2/\text{kg}^2)(9.1 \times 10^{-31}\ \text{kg})^2} = 4.2 \times 10^{42}$$

As you can see, the electric force is much stronger than the gravitational force.

24.7 [II] As shown in Fig. 24-2, two identical balls, each of mass 0.10 g, carry identical charges and are suspended by two threads of equal length. At equilibrium they position themselves as shown. Find the charge on either ball.

Consider the ball on the left. It is in equilibrium under three forces: (1) the tension F_T in the thread; (2) the force of gravity,

$$mg = (1.0 \times 10^{-4}\ \text{kg})(9.81\ \text{m/s}^2) = 9.8 \times 10^{-4}\ \text{N}$$

and (3) the Coulomb repulsion F_E.

Writing $\sum F_x = 0$ and $\sum F_y = 0$ for the ball on the left, we obtain

$$F_T \cos 60^\circ - F_E = 0 \qquad \text{and} \qquad F_T \sin 60^\circ - mg = 0$$

From the second equation,

$$F_T = \frac{mg}{\sin 60^\circ} = \frac{9.8 \times 10^{-4}\,\text{N}}{0.866} = 1.13 \times 10^{-3}\,\text{N}$$

Substituting into the first equation gives

$$F_E = F_T \cos 60^\circ = (1.13 \times 10^{-3}\,\text{N})(0.50) = 5.7 \times 10^{-4}\,\text{N}$$

But this is the Coulomb force, kqq'/r^2. Therefore,

$$qq' = q^2 = \frac{F_E r^2}{k} = \frac{(5.7 \times 10^{-4}\,\text{N})(0.40\,\text{m})^2}{9.0 \times 10^9\,\text{N·m}^2/\text{C}^2}$$

from which $q = 0.10\,\mu\text{C}$.

24.8 [II] The charges shown in Fig. 24-3 are stationary. Find the force on the 4.0 μC charge due to the other two.

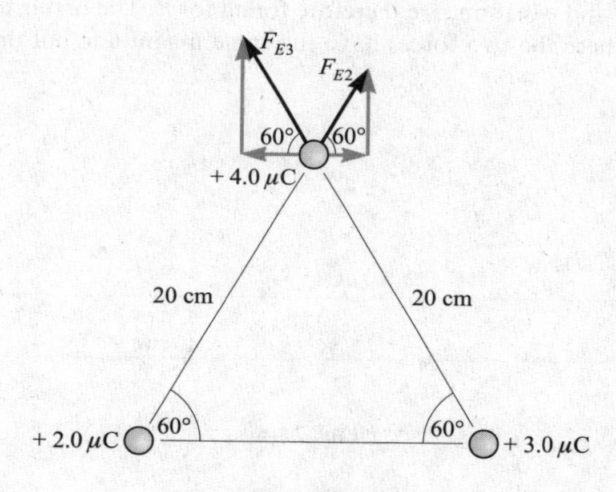

Fig. 24-3

From Coulomb's Law we have

$$F_{E2} = k\frac{qq'}{r^2} = (9.0 \times 10^9\,\text{N·m}^2/\text{C}^2)\frac{(2.0 \times 10^{-6}\,\text{C})(4.0 \times 10^{-6}\,\text{C})}{(0.20\,\text{m})^2} = 1.8\,\text{N}$$

$$F_{E3} = k\frac{qq'}{r^2} = (9.0 \times 10^9\,\text{N·m}^2/\text{C}^2)\frac{(3.0 \times 10^{-6}\,\text{C})(4.0 \times 10^{-6}\,\text{C})}{(0.20\,\text{m})^2} = 2.7\,\text{N}$$

The resultant force on the 4 μC charge has components

$$F_{Ex} = F_{E2} \cos 60^\circ - F_{E3} \cos 60^\circ = (1.8 - 2.7)(0.50)\,\text{N} = -0.45\,\text{N}$$

$$F_{Ey} = F_{E2} \sin 60^\circ + F_{E3} \sin 60^\circ = (1.8 + 2.7)(0.866)\,\text{N} = 3.9\,\text{N}$$

so

$$F_E = \sqrt{F_{Ex}^2 + F_{Ey}^2} = \sqrt{(0.45)^2 + (3.9)^2}\,\text{N} = 3.9\,\text{N}$$

The resultant makes an angle of $\tan^{-1}(0.45/3.9) = 7^\circ$ with the positive y-axis, that is, $\theta = 97^\circ$.

24.9 [II] Two small charged spheres are placed on the x-axis: $+3.0\,\mu$C at $x = 0$ and $-5.0\,\mu$C at $x = 40$ cm. Where must a third charge q be placed if the force it experiences is to be zero?

The situation is shown in Fig. 24-4. We know that q must be placed somewhere on the x-axis. (Why?) Suppose that q is positive. When it is placed in interval BC, the two forces on it are in the same direction and cannot cancel. When it is placed to the right of C, the attractive force from the $-5\,\mu$C charge is always larger than the repulsion of the $+3.0\,\mu$C charge. Therefore, the force on q cannot be zero in this region. Only in the region to the left of B can cancellation occur. (Can you show that this is also true if q is negative?)

For q placed as shown, when the net force on it is zero, we have $F_3 = F_5$ and so, for distances in meters,

$$k\frac{q(3.0\times 10^{-6}\,\text{C})}{d^2} = k\frac{q(5.0\times 10^{-6}\,\text{C})}{(0.40\,\text{m}+d)^2}$$

After canceling q, k, and 10^{-6} C from each side, we cross-multiply to obtain

$$5d^2 = 3.0(0.40+d)^2 \qquad \text{or} \qquad d^2 - 1.2d - 0.24 = 0$$

Using the quadratic formula, we find

$$d = \frac{-b\pm\sqrt{b^2-4ac}}{2a} = \frac{1.2\pm\sqrt{1.44+0.96}}{2} = 0.60 \pm 0.775\,\text{m}$$

Two values, 1.4 m and -0.18 m, are therefore found for d. The first is the correct one; the second gives the point in BC where the two forces have the same magnitude but do not cancel.

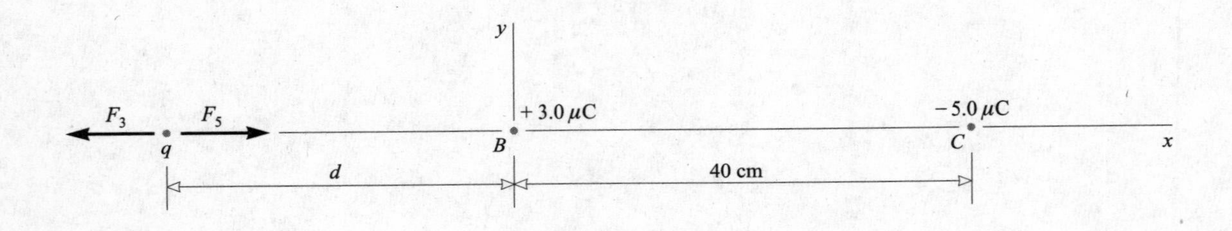

Fig. 24-4

24.10 [II] Compute (a) the electric field E in air at a distance of 30 cm from a point charge $q_{\bullet 1} = 5.0\times 10^{-9}$ C, (b) the force on a charge $q_{\bullet 2} = 4.0\times 10^{-10}$ C placed 30 cm from $q_{\bullet 1}$, and (c) the force on a charge $q_{\bullet 3} = -4.0\times 10^{-10}$ C placed 30 cm from $q_{\bullet 1}$ (in the absence of $q_{\bullet 2}$).

(a)
$$E = k\frac{q_{\bullet 1}}{r^2} = (9.0\times 10^9\,\text{N}\cdot\text{m}^2/\text{C}^2)\frac{5.0\times 10^{-9}\,\text{C}}{(0.30\,\text{m})^2} = 0.50\ \text{kN/C}$$

directed away from $q_{\bullet 1}$.

(b)
$$F_E = Eq_{\bullet 2} = (500\ \text{N/C})(4.0\times 10^{-10}\,\text{C}) = 2.0\times 10^{-7}\,\text{N} = 0.20\,\mu\text{N}$$

directed away from $q_{\bullet 1}$.

(c)
$$F_E = Eq_{\bullet 3} = (500\ \text{N/C})(-4.0\times 10^{-10}\,\text{C}) = -0.20\ \mu\text{N}$$

This force is directed toward $q_{\bullet 1}$.

24.11 [III] The situation shown in Fig. 24-5 depicts two tiny charged spheres. Find (*a*) the electric field E at point P, (*b*) the force on a -4.0×10^{-8} C charge placed at P, and (*c*) where in the region the electric field would be zero (in the absence of the -4.0×10^{-8} C charge).

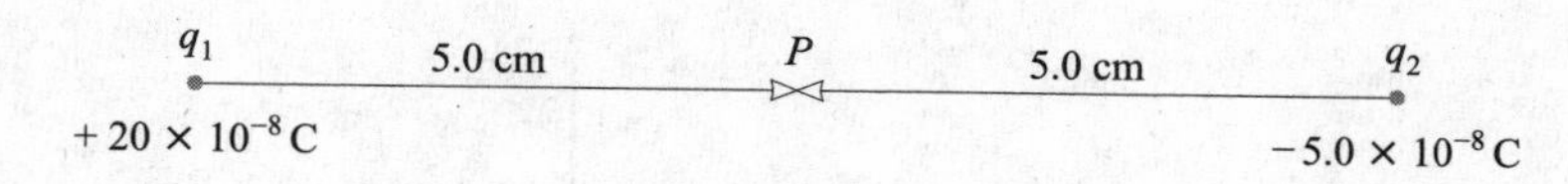

Fig. 24-5

(*a*) A positive test charge placed at P will be repelled to the right by the positive charge q_1 and attracted to the right by the negative charge q_2. Because $\vec{E}_1$ and $\vec{E}_2$ have the same direction, we can add their magnitudes to obtain the magnitude of the resultant field:

$$E = E_1 + E_2 = k\frac{|q_1|}{r_1^2} + k\frac{|q_2|}{r_2^2} = \frac{k}{r_1^2}(|q_1| + |q_2|)$$

where $r_1 = r_2 = 0.05$ m, and $|q_1|$ and $|q_2|$ are the absolute values of q_1 and q_2. Hence,

$$E = \frac{9.0 \times 10^9 \text{ N}\cdot\text{m}^2/\text{C}^2}{(0.050 \text{ m})^2}(25 \times 10^{-8} \text{ C}) = 9.0 \times 10^5 \text{ N/C}$$

directed toward the right.

(*b*) A charge q placed at P will experience a force Eq. Therefore,

$$F_E = Eq = (9.0 \times 10^5 \text{ N/C})(-4.0 \times 10^{-8} \text{ C}) = -0.036 \text{ N}$$

The negative sign tells us the force is directed toward the left. This is correct because the electric field represents the force on a positive charge. The force on a negative charge is opposite in direction to the field.

(*c*) Reasoning as in Problem 24.9, we conclude that the field will be zero somewhere to the right of the -5.0×10^{-8} C charge. Represent the distance to that point from the -5.0×10^{-8} C charge by d. At that point,

$$E_1 - E_2 = 0$$

because the field due to the positive charge is to the right, while the field due to the negative charge is to the left. Thus

$$k\left(\frac{|q_1|}{r_1^2} - \frac{|q_2|}{r_2^2}\right) = (9.0 \times 10^9 \text{ N}\cdot\text{m}^2/\text{C}^2)\left[\frac{20 \times 10^{-8} \text{ C}}{(d + 0.10 \text{ m})^2} - \frac{5.0 \times 10^{-8} \text{ C}}{d^2}\right] = 0$$

Simplifying, we obtain

$$3d^2 - 0.2d - 0.01 = 0$$

which gives $d = 0.10$ m and -0.03 m. Only the plus sign has meaning here, and therefore $d = 0.10$ m. The point in question is 10 cm to the right of the negative charge.

24.12 [II] Three charges are placed on three corners of a square, as shown in Fig. 24-6. Each side of the square is 30.0 cm. Compute $\vec{E}$ at the fourth corner. What would be the force on a 6.00 μC charge placed at the vacant corner?

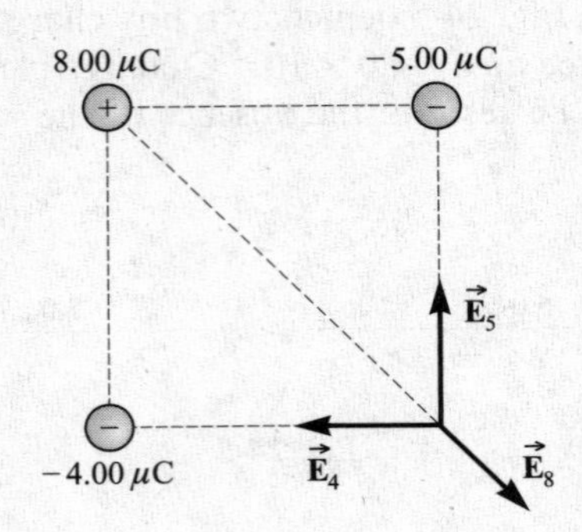

Fig. 24-6

The contributions of the three charges to the field at the vacant corner are as indicated. Notice in particular their directions. Their magnitudes are given by $E = kq/r^2$ to be

$$E_4 = 4.00 \times 10^5 \text{ N/C} \qquad E_8 = 4.00 \times 10^5 \text{ N/C} \qquad E_5 = 5.00 \times 10^5 \text{ N/C}$$

Because the E_8 vector makes an angle of 45.0° to the horizontal, we have

$$E_x = E_8 \cos 45.0^\circ - E_4 = -1.17 \times 10^5 \text{ N/C}$$

$$E_y = E_5 - E_8 \cos 45.0^\circ = 2.17 \times 10^5 \text{ N/C}$$

Using $E = \sqrt{E_x^2 + E_y^2}$ and $\tan\theta = E_y/E_x$, we find $E = 2.47 \times 10^5$ N at 118°.

The force on a charge placed at the vacant corner would be simply $F_E = Eq$. Since $q = 6.00 \times 10^{-6}$ C, we have $F_E = 1.48$ N at an angle of 118°.

24.13 [III] Two charged metal plates in vacuum are 15 cm apart as shown in Fig. 24-7. The electric field between the plates is uniform and has a strength of $E = 3000$ N/C. An electron ($q = -e$, $m_e = 9.1 \times 10^{-31}$ kg) is released from rest at point P just outside the negative plate. (*a*) How long will it take to reach the other plate? (*b*) How fast will it be going just before it hits?

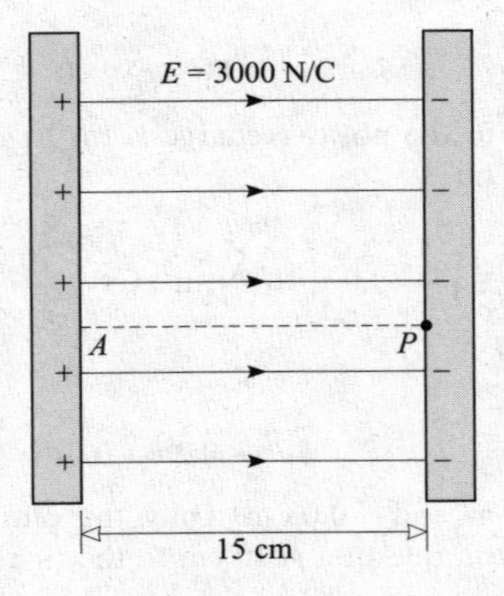

Fig. 24-7

The electric field lines show the force on a positive charge. (A positive charge would be repelled to the right by the positive plate and attracted to the right by the negative plate.) An electron, being negative, will experience a force in the opposite direction, toward the left, of magnitude

$$F_E = |q|E = (1.6 \times 10^{-19} \text{ C})(3000 \text{ N/C}) = 4.8 \times 10^{-16} \text{ N}$$

Because of this force, the electron experiences an acceleration toward the left given by

$$a = \frac{F_E}{m} = \frac{4.8 \times 10^{-16}\ \text{N}}{9.1 \times 10^{-31}\ \text{kg}} = 5.3 \times 10^{14}\ \text{m/s}^2$$

In the motion problem for the electron released at the negative plate and traveling to the positive plate,

$$v_i = 0 \qquad x = 0.15\ \text{m} \qquad a = 5.3 \times 10^{14}\ \text{m/s}^2$$

(*a*) From $x = v_i t + \frac{1}{2}at^2$ we have

$$t = \sqrt{\frac{2x}{a}} = \sqrt{\frac{(2)(0.15\ \text{m})}{5.3 \times 10^{14}\ \text{m/s}^2}} = 2.4 \times 10^{-8}\ \text{s}$$

(*b*)

$$v = v_i + at = 0 + (5.3 \times 10^{14}\ \text{m/s}^2)(2.4 \times 10^{-8}\ \text{s}) = 1.30 \times 10^7\ \text{m/s}$$

As you will see in Chapter 41, relativistic effects begin to become important at speeds above this. Therefore, this approach must be modified for very fast particles.

24.14 [I] Suppose in Fig. 24-7 an electron is shot straight upward from point P with a speed of 5.0×10^6 m/s. How far above A will it strike the positive plate?

This is a projectile problem. (Since the gravitational force is so small compared to the electrical force, we can ignore gravity.) The only force acting on the electron after its release is the horizontal electric force. We found in Problem 24.13(*a*) that under the action of this force the electron has a time-of-flight of 2.4×10^{-8} s. The vertical displacement in this time is

$$(5.0 \times 10^6\ \text{m/s})(2.4 \times 10^{-8}\ \text{s}) = 0.12\ \text{m}$$

The electron strikes the positive plate 12 cm above point A.

24.15 [II] In Fig. 24-7 a proton ($q_\bullet = +e$, $m = 1.67 \times 10^{-27}$ kg) is shot with speed 2.00×10^5 m/s toward P from A. What will be its speed just before hitting the plate at P?

Let's first calculate the acceleration knowing the electric field, and from it the force:

$$a = \frac{F_E}{m} = \frac{qE}{m} = \frac{(1.60 \times 10^{-19}\ \text{C})(3000\ \text{N/C})}{1.67 \times 10^{-27}\ \text{kg}} = 2.88 \times 10^{11}\ \text{m/s}^2$$

For the problem involving horizontal motion,

$$v_i = 2.00 \times 10^5\ \text{m/s} \qquad x = 0.15\ m \qquad a = 2.88 \times 10^{11}\ \text{m/s}^2$$

We use $v_f^2 = v_i^2 + 2ax$ to find

$$v_f = \sqrt{v_i^2 + 2ax} = \sqrt{(2.00 \times 10^5\ \text{m/s})^2 + (2)(2.88 \times 10^{11}\ \text{m/s}^2)(0.15\ \text{m})} = 356\ \text{km/s}$$

24.16 [II] Two identical tiny metal balls have charges q_1 and q_2. The repulsive force one exerts on the other when they are 20 cm apart is 1.35×10^{-4} N. After the balls are touched together and then separated once again to 20 cm, the repulsive force is found to be 1.406×10^{-4} N. Find q_1 and q_2.

Because the force is one of repulsion, q_1 and q_2 are of the same sign. After the balls are touched, they share charge equally, so each has a charge $\frac{1}{2}(q_1 + q_2)$. Writing Coulomb's Law for the two situations described, we have

$$0.000\,135\ \text{N} = k\frac{q_1 q_2}{0.040\ \text{m}^2}$$

and

$$0.000\,140\,6\ \text{N} = k\frac{[\frac{1}{2}(q_1 + q_2)]^2}{0.040\ \text{m}^2}$$

After substitution for k, these equations reduce to

$$q_1 q_2 = 6.00 \times 10^{-16}\ \text{C}^2 \qquad \text{and} \qquad q_1 + q_2 = 5.00 \times 10^{-8}\ \text{C}$$

Solving these equations simultaneously gives $q_1 = 20$ nC and $q_2 = 30$ nC (or vice versa). Alternatively, both charges could have been negative.

Supplementary Problems

24.17 [I] How many electrons are contained in 1.0 C of charge? What is the mass of the electrons in 1.0 C of charge? *Ans.* 6.2×10^{18} electrons, 5.7×10^{-12} kg

24.18 [I] If two equal point charges, each of 1 C, were separated in air by a distance of 1 km, what would be the force between them? *Ans.* 9 kN repulsion

24.19 [I] Determine the force between two free electrons spaced 1.0 angstrom (10^{-10} m) apart in vacuum. *Ans.* 23 nN repulsion

24.20 [I] What is the force of repulsion between two argon nuclei that are separated in vacuum by 1.0 nm (10^{-9} m)? The charge on an argon nucleus is $+18e$. *Ans.* 75 nN

24.21 [I] Two equally charged small balls are 3 cm apart in air and repel each other with a force of 40 μN. Compute the charge on each ball. *Ans.* 2 nC

24.22 [II] Three point charges are placed at the following points on the x-axis: $+2.0\,\mu$C at $x = 0$, $-3.0\,\mu$C at $x = 40$ cm, $-5.0\,\mu$C at $x = 120$ cm. Find the force (*a*) on the $-3.0\,\mu$C charge, (*b*) on the $-5.0\,\mu$C charge. *Ans.* (*a*) -0.55 N; (*b*) 0.15 N

24.23 [II] Four equal point charges of $+3.0\,\mu$C are placed at the four corners of a square that is 40 cm on a side. Find the force on any one of the charges. *Ans.* 0.97 N outward along the diagonal

24.24 [II] Four equal-magnitude point charges ($3.0\,\mu$C) are placed at the corners of a square that is 40 cm on a side. Two, diagonally opposite each other, are positive, and the other two are negative. Find the force on either negative charge. *Ans.* 0.46 N inward along the diagonal

24.25 [II] Charges of +2.0, +3.0, and $-8.0\,\mu$C are placed at the vertices of an equilateral triangle of side 10 cm. Calculate the magnitude of the force acting on the $-8.0\,\mu$C charge due to the other two charges. *Ans.* 31 N

24.26 [II] One charge of ($+5.0\,\mu$C) is placed at exactly $x = 0$, and a second charge ($+7.0\,\mu$C) at $x = 100$ cm. Where can a third be placed so as to experience zero net force due to the other two? *Ans.* at $x = 46$ cm

24.27 [II] Two identical tiny metal balls carry charges of +3 nC and -12 nC. They are 3 m apart. (*a*) Compute the force of attraction. (*b*) The balls are now touched together and then separated to 3 cm. Describe the forces on them now. *Ans.* (*a*) 4×10^{-4} N attraction; (*b*) 2×10^{-4} N repulsion

24.28 [II] A charge of $+6.0\,\mu C$ experiences a force of 2.0 mN in the $+x$-direction at a certain point in space. (*a*) What was the electric field at that point before the charge was placed there? (*b*) Describe the force a $-2.0\,\mu C$ charge would experience if it were used in place of the $+6.0\,\mu C$ charge. *Ans.* (*a*) 0.33 kN/C in $+x$-direction; (*b*) 0.67 mN in $-x$-direction

24.29 [I] A point charge of -3.0×10^{-5} C is placed at the origin of coordinates. Find the electric field at the point $x = 5.0$ m on the x-axis. *Ans.* 11 kN/C in $-x$-direction

24.30 [III] Four equal-magnitude ($4.0\,\mu C$) charges are placed at the four corners of a square that is 20 cm on each side. Find the electric field at the center of the square (*a*) if the charges are all positive, (*b*) if the charges alternate in sign around the perimeter of the square, (*c*) if the charges have the following sequence around the square: plus, plus, minus, minus. *Ans.* (*a*) zero; (*b*) zero; (*c*) 5.1 MN/C toward the negative side

24.31 [II] A 0.200-g ball hangs from a thread in a uniform vertical electric field of 3.00 kN/C directed upward. What is the charge on the ball if the tension in the thread is (*a*) zero and (*b*) 4.00 mN? *Ans.* (*a*) +653 nC; (*b*) −680 nC

24.32 [II] Determine the acceleration of a proton ($q = +e$, $m = 1.67 \times 10^{-27}$ kg) in an electric field of strength 0.50 kN/C. How many times is this acceleration greater than that due to gravity? *Ans.* 4.8×10^{10} m/s^2, 4.9×10^9

24.33 [II] A tiny, 0.60-g ball carries a charge of magnitude $8.0\,\mu C$. It is suspended by a vertical thread in a downward 300 N/C electric field. What is the tension in the thread if the charge on the ball is (*a*) positive, (*b*) negative? *Ans.* (*a*) 8.3 mN; (*b*) 3.5 mN

24.34 [III] The tiny ball at the end of the thread shown in Fig. 24-8 has a mass of 0.60 g and is in a horizontal electric field of strength 700 N/C. It is in equilibrium in the position shown. What are the magnitude and sign of the charge on the ball? *Ans.* $-3.1\,\mu C$

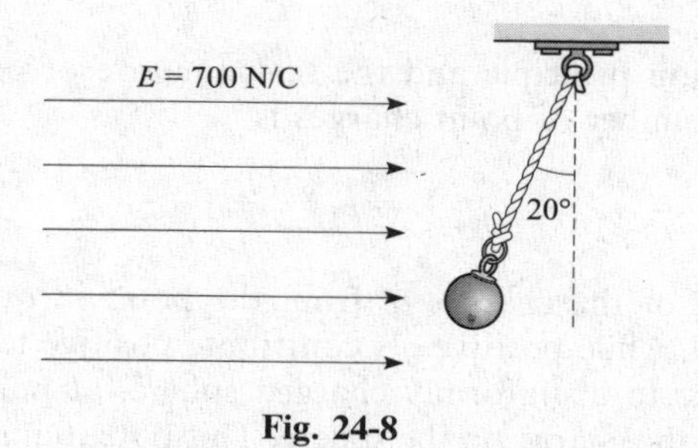

Fig. 24-8

24.35 [III] An electron ($q = -e$, $m_e = 9.1 \times 10^{-31}$ kg) is projected out along the $+x$-axis with an initial speed of 3.0×10^6 m/s. It goes 45 cm and stops due to a uniform electric field in the region. Find the magnitude and direction of the field. *Ans.* 57 N/C in $+x$-direction

24.36 [III] A particle of mass m and charge $-e$ is projected with horizontal speed v into an electric field (E) directed downward. Find (*a*) the horizontal and vertical components of its acceleration, a_x and a_y; (*b*) its horizontal and vertical displacements, x and y, after time t; (*c*) the equation of its trajectory. *Ans.* (*a*) $a_x = 0$, $a_y = Ee/m$; (*b*) $x = vt$, $y = \frac{1}{2}a_y t^2 = \frac{1}{2}(Ee/m)t^2$; (*c*) $y = \frac{1}{2}(Ee/mv^2)x^2$ (a parabola)

Chapter 25

Electric Potential; Capacitance

THE POTENTIAL DIFFERENCE between point-A and point-B is the work done against electrical forces in carrying a *unit* positive test-charge from A to B. We represent the potential difference between A and B by $V_B - V_A$ or by V. Its units are those of work per charge (joules/coulomb) and are called **volts** (V):

$$1\ \text{V} = 1\ \text{J/C}$$

Because work is a scalar quantity, so too is potential difference. Like work, potential difference may be positive or negative.

The work W done in transporting a charge q from one point-A to a second point-B is

$$W = q(V_B - V_A) = qV$$

where the appropriate sign (+ or −) must be given to the charge. If both $(V_B - V_A)$ and q are positive (or negative), the work done is positive. If $(V_B - V_A)$ and q have opposite signs, the work done is negative.

ABSOLUTE POTENTIAL: The absolute potential at a point is the work done against electric forces in carrying a unit positive test-charge from infinity to that point. Hence the absolute potential at a point-B is the difference in potential from $A = \infty$ to B.

Consider a point charge q in vacuum and a point-P at a distance r from the point charge. The absolute potential at P due to the charge q is

$$V = k\frac{q}{r}$$

where $k = 8.99 \times 10^9\ \text{N}\cdot\text{m}^2/\text{C}^2$ is the Coulomb constant. The absolute potential at infinity (at $r = \infty$) is zero.

Because of the superposition principle and the scalar nature of potential difference, the absolute potential at a point due to a number of point charges is

$$V = k\sum \frac{q_i}{r_i}$$

where the r_i are the distances of the charges q_i from the point in question. Negative q's contribute negative terms to the potential, while positive q's contribute positive terms.

The absolute potential due to a uniformly charged sphere, at points *outside* the sphere or *on* its surface is $V = kq/r$, where q is the charge on the sphere. This potential is the same as that due to a point charge q placed at the position of the sphere's center.

ELECTRICAL POTENTIAL ENERGY (PE_E)**:** To carry a charge q from infinity to a point where the absolute potential is V, work in the amount qV must be done on the charge. This work appears as electrical potential energy (PE_E) stored in the charge.

Similarly, when a charge q is carried through a **potential difference** V, work in the amount qV must be done on the charge. This work results in a change qV in the PE_E of the charge. For a potential *rise*, V will be positive and the PE_E will increase if q is positive. But for a potential *drop*, V will be negative and the PE_E of the charge will decrease if q is positive.

V RELATED TO E: Suppose that in a certain region the electric field is uniform and is in the x-direction. Call its magnitude E_x. Because E_x is the force on a unit positive test-charge, the work done in moving the test-charge through a distance x is (from $W = F_x x$)

$$V = E_x x$$

The field between two large, parallel, oppositely charged metal plates is uniform. We can therefore use this equation to relate the electric field E between the plates to the plate separation d and their potential difference V: For parallel plates,

$$V = Ed$$

ELECTRON VOLT ENERGY UNIT: The work done in carrying a charge $+e$ (coulombs) through a potential rise of exactly 1 volt is defined to be 1 **electron volt** (eV). Therefore,

$$1\ \text{eV} = (1.602 \times 10^{-19}\ \text{C})(1\ \text{V}) = 1.602 \times 10^{-19}\ \text{J}$$

Equivalently,

$$\text{Work or energy (in eV)} = \frac{\text{work (in joules)}}{e}$$

A CAPACITOR is a device that stores charge. Often, although certainly not always, it consists of two conductors separated by an insulator or dielectric. The **capacitance** (C) of a capacitor is defined as

$$\text{Capacitance} = \frac{\text{magnitude of charge on either conductor}}{\text{magnitude of potential difference between conductors}}$$

For q in coulombs and V in volts, C is in **farads** (F).

PARALLEL-PLATE CAPACITOR: The capacitance of a parallel-plate capacitor whose opposing plate faces, each of area A, are separated by a small distance d is given by

$$C = K\epsilon_0 \frac{A}{d} = \epsilon \frac{A}{d}$$

where $K = \epsilon/\epsilon_0$ is the dimensionless dielectric constant (see Chapter 24) of the nonconducting material (the *dielectric*) between the plates, and

$$\epsilon_0 = 8.85 \times 10^{-12}\ \text{C}^2/\text{N}\cdot\text{m}^2 = 8.85 \times 10^{-12}\ \text{F/m}$$

For vacuum, $K = 1$, so that a dielectric-filled parallel-plate capacitor has a capacitance K times larger than the same capacitor with vacuum between its plates. This result holds for a capacitor of arbitrary shape.

CAPACITORS IN PARALLEL AND SERIES: As shown in Fig. 25-1, capacitances add for capacitors in parallel, whereas reciprocal capacitances add for capacitors in series.

ENERGY STORED IN A CAPACITOR: The energy (PE_E) stored in a capacitor of capacitance C that has a charge q and a potential difference V is

$$\text{PE}_E = \frac{1}{2}qV = \frac{1}{2}CV^2 = \frac{1}{2}\frac{q^2}{C}$$

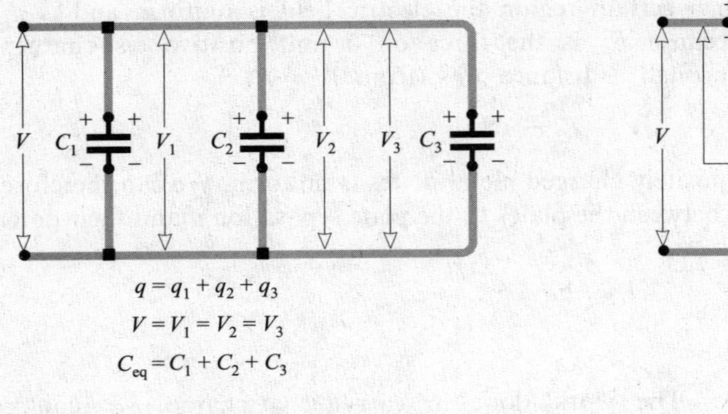

$$q = q_1 + q_2 + q_3$$
$$V = V_1 = V_2 = V_3$$
$$C_{eq} = C_1 + C_2 + C_3$$

(*a*) Capacitors in parallel

$$q = q_1 = q_2 = q_3$$
$$V = V_1 + V_2 + V_3$$
$$\frac{1}{C_{eq}} = \frac{1}{C_1} + \frac{1}{C_2} + \frac{1}{C_3}$$

(*b*) Capacitors in series

Fig. 25-1

Solved Problems

25.1 [I] In Fig. 25-2, the potential difference between the metal plates is 40 V. (*a*) Which plate is at the higher potential? (*b*) How much work must be done to carry a +3.0 C charge from *B* to *A*? From *A* to *B*? (*c*) How do we know that the electric field is in the direction indicated? (*d*) If the plate separation is 5.0 mm, what is the magnitude of $\vec{\mathbf{E}}$?

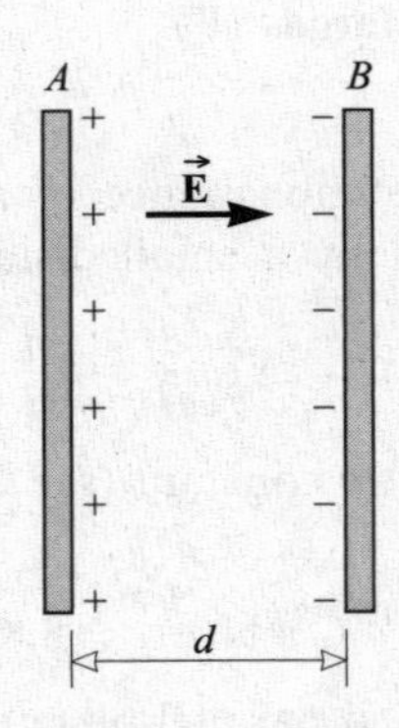

Fig. 25-2

(*a*) A positive test charge between the plates is repelled by *A* and attracted by *B*. Left to itself, the positive test charge will move from *A* to *B*, and so *A* is at the higher potential.

(*b*) The magnitude of the work done in carrying a charge *q* through a potential difference *V* is *qV*. Thus the magnitude of the work done in the present situation is

$$W = (3.0\ \text{C})(40\ \text{V}) = 0.12\ \text{kJ}$$

Because a positive charge between the plates is repelled by *A*, positive work (+120 J) must be done to drag the +3.0 C charge from *B* to *A*. To restrain the charge as it moves from *A* to *B*, negative work (−120 J) is done.

(*c*) A positive test-charge between the plates experiences a force directed from A to B and this is, by definition, the direction of the field.

(*d*) For parallel plates, $V = Ed$. Therefore,

$$E = \frac{V}{d} = \frac{40\ \text{V}}{0.0050\ \text{m}} = 8.0\ \text{kV/m}$$

Notice that the SI units for electric field, V/m and N/C, are identical.

25.2 [I] How much work is required to carry an electron from the positive terminal of a 12-V battery to the negative terminal?

Going from the positive to the negative terminal, one passes through a potential drop. In this case it is $V = -12$ V. Then

$$W = qV = (-1.6 \times 10^{-19}\ \text{C})(-12\ \text{V}) = 1.9 \times 10^{-18}\ \text{J}$$

As a check, we notice that an electron, if left to itself, will move from negative to positive because it is a negative charge. Hence positive work must be done to carry it in the reverse direction as required here.

25.3 [I] How much electrical potential energy does a proton lose as it falls through a potential drop of 5 kV?

The proton carries a positive charge. It will therefore move from regions of high potential to regions of low potential if left free to do so. Its change in potential energy as it moves through a potential difference V is Vq. In our case, $V = -5$ kV. Therefore,

$$\text{Change in PE}_E = Vq = (-5 \times 10^3\ \text{V})(1.6 \times 10^{-19}\ \text{C}) = -8 \times 10^{-16}\ \text{J}$$

25.4 [II] An electron starts from rest and falls through a potential rise of 80 V. What is its final speed?

Positive charges tend to fall through potential drops; negative charges, such as electrons, tend to fall through potential rises.

$$\text{Change in PE}_E = Vq = (80\ \text{V})(-1.6 \times 10^{-19}\ \text{C}) = -1.28 \times 10^{-17}\ \text{J}$$

This lost PE_E appears as KE of the electron:

$$\text{PE}_E \text{ lost} = \text{KE gained}$$

$$1.28 \times 10^{-17}\ \text{J} = \tfrac{1}{2}mv_f^2 - \tfrac{1}{2}mv_i^2 = \tfrac{1}{2}mv_f^2 - 0$$

$$v_f = \sqrt{\frac{(1.28 \times 10^{-17}\ \text{J})(2)}{9.1 \times 10^{-31}\ \text{kg}}} = 5.3 \times 10^6\ \text{m/s}$$

25.5 [I] (*a*) What is the absolute potential at each of the following distances from a charge of $+2.0\ \mu$C: $r = 10$ cm and $r = 50$ cm? (*b*) How much work is required to carry a $0.05\ \mu$C charge from the point at $r = 50$ cm to that at $r = 10$ cm?

(*a*)

$$V_{10} = k\frac{q}{r} = (9.0 \times 10^9\ \text{N·m}^2/\text{C}^2)\frac{2.0 \times 10^{-6}\ \text{C}}{0.10\ \text{m}} = 1.8 \times 10^5\ \text{V}$$

$$V_{50} = \frac{10}{50}V_{10} = 36\ \text{kV}$$

(*b*)

$$\text{Work} = q(V_{10} - V_{50}) = (5 \times 10^{-8}\ \text{C})(1.44 \times 10^5\ \text{V}) = 7.2\ \text{mJ}$$

25.6 [II] Suppose [in Problem 25.5(*a*) where there is a $+2.0\ \mu$C charge] that a proton is released at $r = 10$ cm. How fast will it be moving as it passes a point at $r = 50$ cm?

This is a situation where PE_E goes into KE. As the proton moves from one point to the other, there is a potential drop of

$$\text{Potential drop} = 1.80 \times 10^5\ \text{V} - 0.36 \times 10^5\ \text{V} = 1.44 \times 10^5\ \text{V}$$

The proton acquires KE as it falls through this potential drop:

$$\text{KE gained} = \text{PE}_E \text{ lost}$$
$$\tfrac{1}{2}mv_f^2 - \tfrac{1}{2}mv_i^2 = qV$$
$$\tfrac{1}{2}(1.67 \times 10^{-27}\ \text{kg})v_f^2 - 0 = (1.6 \times 10^{-19}\ \text{C})(1.44 \times 10^5\ \text{V})$$

from which $v_f = 5.3 \times 10^6$ m/s.

25.7 [II] In Fig. 25-2, which depicts two charged parallel plates, let $E = 2.0$ kV/m and $d = 5.0$ mm. A proton is shot from plate-*B* toward plate-*A* with an initial speed of 100 km/s. What will be its speed just before it strikes plate-*A*?

The proton, being positive, is repelled by plate-*A* and will therefore be slowed down. We need the potential difference between the plates, which is

$$V = Ed = (2.0\ \text{kV/m})(0.0050\ \text{m}) = 10\ \text{V}$$

Now, from the conservation of energy, for the proton,

$$\text{KE lost} = \text{PE}_E \text{ gained}$$
$$\tfrac{1}{2}mv_B^2 - \tfrac{1}{2}mv_A^2 = qV$$

Substituting $m = 1.67 \times 10^{-27}$ kg, $v_B = 1.00 \times 10^5$ m/s, $q = 1.60 \times 10^{-19}$ C, and $V = 10$ V gives $v_A = 90$ km/s. As we see, the proton is indeed slowed.

25.8 [III] The nucleus of a tin atom has a charge of $+50e$. (*a*) Find the absolute potential V at a radial distance of 1.0×10^{-12} m from the nucleus. (*b*) If a proton is released from this point, how fast will it be moving when it is 1.0 m from the nucleus?

(*a*)
$$V = k\frac{q}{r} = (9.0 \times 10^9\ \text{N}\cdot\text{m}^2/\text{C}^2)\frac{(50)(1.6 \times 10^{-19}\ \text{C})}{10^{-12}\ \text{m}} = 72\ \text{kV}$$

(*b*) The proton is repelled by the nucleus and flies out to infinity. The absolute potential at a point is the potential difference between the point in question and infinity. Hence there is a potential drop of 72 kV as the proton flies to infinity.

Usually we would simply assume that 1.0 m is far enough from the nucleus to consider it to be at infinity. But, as a check, let us compute V at $r = 1.0$ m:

$$V_{1\text{m}} = k\frac{q}{r} = (9.0 \times 10^9\ \text{N}\cdot\text{m}^2/\text{C}^2)\frac{(50)(1.6 \times 10^{-19}\text{C})}{1.0\ \text{m}} = 7.2 \times 10^{-8}\ \text{V}$$

which is essentially zero in comparison with 72 kV.

As the proton falls through 72 kV,

$$\text{KE gained} = \text{PE}_E \text{ lost}$$
$$\tfrac{1}{2}mv_f^2 - \tfrac{1}{2}mv_i^2 = qV$$
$$\tfrac{1}{2}(1.67 \times 10^{-27}\ \text{kg})v_f^2 - 0 = (1.6 \times 10^{-19}\ \text{C})(72\,000\ \text{V})$$

from which $v_f = 3.7 \times 10^6$ m/s.

25.9 [II] The following point charges are placed on the x-axis: $+2.0\,\mu\text{C}$ at $x = 20$ cm, $-3.0\,\mu\text{C}$ at $x = 30$ cm, $-4.0\,\mu\text{C}$ at $x = 40$ cm. Find the absolute potential on the axis at $x = 0$.

Potential is a scalar, and so

$$V = k\sum\frac{q_i}{r_i} = (9.0\times10^9\ \text{N}\cdot\text{m}^2/\text{C}^2)\left(\frac{2.0\times10^{-6}\ \text{C}}{0.20\ \text{m}} + \frac{-3.0\times10^{-6}\ \text{C}}{0.30\ \text{m}} + \frac{-4.0\times10^{-6}\ \text{C}}{0.40\ \text{m}}\right)$$

$$= (9.0\times10^9\ \text{N}\cdot\text{m}^2/\text{C}^2)(10\times10^{-6}\ \text{C/m} - 10\times10^{-6}\ \text{C/m} - 10\times10^{-6}\ \text{C/m}) = -90\ \text{kV}$$

25.10 [I] Two point charges, $+q$ and $-q$, are separated by a distance d. Where, besides at infinity, is the absolute potential zero?

At the point (or points) in question,

$$0 = k\frac{q}{r_1} + k\frac{-q}{r_2} \qquad \text{or} \qquad r_1 = r_2$$

This condition holds everywhere on a plane which is the perpendicular bisector of the line joining the two charges. Therefore the absolute potential is zero everywhere on that plane.

25.11 [II] Four point charges are placed at the four corners of a square that is 30 cm on each side. Find the potential at the center of the square if (*a*) the four charges are each $+2.0\,\mu\text{C}$ and (*b*) two of the four charges are $+2.0\,\mu\text{C}$ and two are $-2.0\,\mu\text{C}$.

(*a*) $$V = k\sum\frac{q_i}{r_i} = k\frac{\sum q_i}{r} = (9.0\times10^9\ \text{N}\cdot\text{m}^2/\text{C}^2)\frac{(4)(2.0\times10^{-6}\ \text{C})}{(0.30\ \text{m})(\cos 45^\circ)} = 3.4\times10^5\ \text{V}$$

(*b*) $$V = (9.0\times10^9\ \text{N}\cdot\text{m}^2/\text{C}^2)\frac{(2.0+2.0-2.0-2.0)\times10^{-6}\ \text{C}}{(0.30\ \text{m})(\cos 45^\circ)} = 0$$

25.12 [III] In Fig. 25-3, the charge at A is $+200$ pC, while the charge at B is -100 pC. (*a*) Find the absolute potentials at points-C and D. (*b*) How much work must be done to transfer a charge of $+500\,\mu\text{C}$ from point-C to point-D?

(*a*) $$V_C = k\sum\frac{q_i}{r_i} = (9.0\times10^9\ \text{N}\cdot\text{m}^2/\text{C}^2)\left(\frac{2.00\times10^{-10}\ \text{C}}{0.80\ \text{m}} - \frac{1.00\times10^{-10}\ \text{C}}{0.20\ \text{m}}\right) = -2.25\ \text{V} = -2.3\ \text{V}$$

$$V_D = (9.0\times10^9\ \text{N}\cdot\text{m}^2/\text{C}^2)\left(\frac{2.00\times10^{-10}\ \text{C}}{0.20\ \text{m}} - \frac{1.00\times10^{-10}\ \text{C}}{0.80\ \text{m}}\right) = +7.88\ \text{V} = +7.9\ \text{V}$$

(*b*) There is a potential rise from C to D of $V = V_D - V_C = 7.88\ \text{V} - (-2.25\ \text{V}) = 10.13$ V. So

$$W = Vq = (10.13\ \text{V})(5.00\times10^{-4}\ \text{C}) = 5.1\ \text{mJ}$$

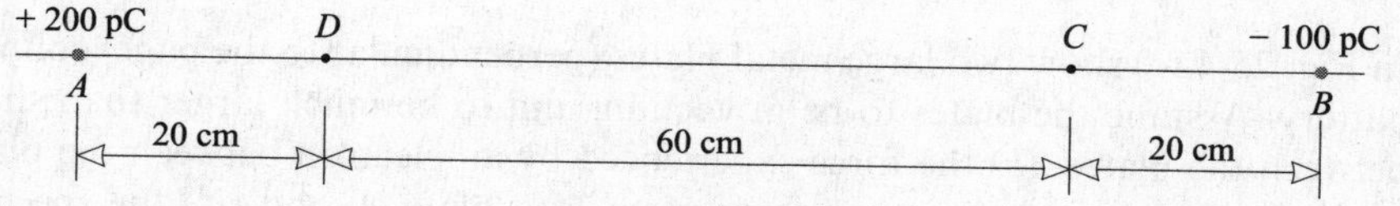

+ 200 pC D C − 100 pC
A B
20 cm 60 cm 20 cm

Fig. 25-3

25.13 [III] Find the electrical potential energy of three point charges placed as follows on the x-axis: $+2.0\,\mu$C at $x = 0$, $+3.0\,\mu$C at $x = 20$ cm, and $+6.0\,\mu$C at $x = 50$ cm. Take the PE_E to be zero when the charges are separated far apart.

Let us compute how much work must be done to bring the charges from infinity to their places on the axis. We bring in the 2.0 μC charge first; this requires no work because there are no other charges in the vicinity.

Next we bring in the 3.0 μC charge, which is repelled by the +2.0 μC charge. The potential difference between infinity and the position to which we bring it is due to the +2.0 μC charge and is

$$V_{x=0.2} = k\frac{2.0\,\mu\text{C}}{0.20\text{ m}} = (9.0\times10^9\text{ N}\cdot\text{m}^2/\text{C}^2)\left(\frac{2\times10^{-6}\text{ C}}{0.20\text{ m}}\right) = 9.0\times10^4\text{ V}$$

Therefore the work required to bring in the 3 μC charge is

$$W_{3\,\mu\text{C}} = qV_{x=0.2} = (3.0\times10^{-6}\text{ C})(9.0\times10^4\text{ V}) = 0.270\text{ J}$$

Finally we bring the 6.0 μC charge in to $x = 0.50$ m. The potential there due to the two charges already present is

$$V_{x=0.5} = k\left(\frac{2.0\times10^{-6}\text{ C}}{0.50\text{ m}} + \frac{3.0\times10^{-6}\text{ C}}{0.30\text{ m}}\right) = 12.6\times10^4\text{ V}$$

Therefore the work required to bring in the 6.0 μC charge is

$$W_{6\,\mu\text{C}} = qV_{x=0.5} = (6.0\times10^{-6}\text{ C})(12.6\times10^4\text{ V}) = 0.756\text{ J}$$

Adding the amounts of work required to assemble the charges gives the energy stored in the system:

$$PE_E = 0.270\text{ J} + 0.756\text{ J} = 1.0\text{ J}$$

Can you show that the order in which the charges are brought in from infinity does not affect this result?

25.14 [III] Two protons are held at rest, 5.0×10^{-12} m apart. When released, they fly apart. How fast will each be moving when they are far from each other?

Their original PE_E will be changed to KE. We proceed as in Problem 25.13. The potential at 5.0×10^{-12} m from the first charge due to that charge alone is

$$V = (9.0\times10^9\text{ N}\cdot\text{m}^2/\text{C}^2)\left(\frac{1.60\times10^{-19}\text{ C}}{5\times10^{-12}\text{ m}}\right) = 288\text{ V}$$

The work needed to bring in the second proton is then

$$W = qV = (1.60\times10^{-19}\text{ C})(288\text{ V}) = 4.61\times10^{-17}\text{ J}$$

and this is the PE_E of the original system. From the conservation of energy,

$$\text{Original } PE_E = \text{final KE}$$
$$4.61\times10^{-17}\text{ J} = \tfrac{1}{2}m_1v_1^2 + \tfrac{1}{2}m_2v_2^2$$

Since the particles are identical, $v_1 = v_2 = v$. Solving, we find that $v = 1.7\times10^5$ m/s when the particles are far apart.

25.15 [III] In Fig. 25-4 we show two large metal plates (perpendicular to the page) connected to a 120-V battery. Assume the plates to be in vacuum and to be much larger than shown. Find (*a*) E between the plates, (*b*) the force experienced by an electron between the plates, (*c*) the PE_E lost by an electron as it moves from plate-B to plate-A, and (*d*) the speed of the electron released from plate-B just before striking plate-A.

(*a*) E is directed from the positive plate-A to the negative plate-B. It is uniform between large parallel plates and is given by

$$E = \frac{V}{d} = \frac{120 \text{ V}}{0.020 \text{ m}} = 6000 \text{ V/m} = 6.0 \text{ kV/m}$$

directed from left to right.

(*b*)
$$F_E = qE = (-1.6 \times 10^{-19} \text{ C})(6000 \text{ V/m}) = -9.6 \times 10^{-16} \text{ N}$$

The minus sign tells us that $\vec{\mathbf{F}}_E$ is directed oppositely to $\vec{\mathbf{E}}$. Since plate-A is positive, the electron is attracted by it. The force on the electron is toward the left.

(*c*)
$$\text{Change in PE}_E = Vq = (120 \text{ V})(-1.6 \times 10^{-19} \text{ C}) = -1.92 \times 10^{-17} \text{ J} = -1.9 \times 10^{-17} \text{ J}$$

Notice that V is a potential rise from B to A.

(*d*)
$$\text{PE}_E \text{ lost} = \text{KE gained}$$
$$1.92 \times 10^{-17} \text{ J} = \tfrac{1}{2}mv_f^2 - \tfrac{1}{2}mv_i^2$$
$$1.92 \times 10^{-17} \text{ J} = \tfrac{1}{2}(9.1 \times 10^{-31} \text{ kg})v_f^2 - 0$$

from which $v_f = 6.5 \times 10^6$ m/s.

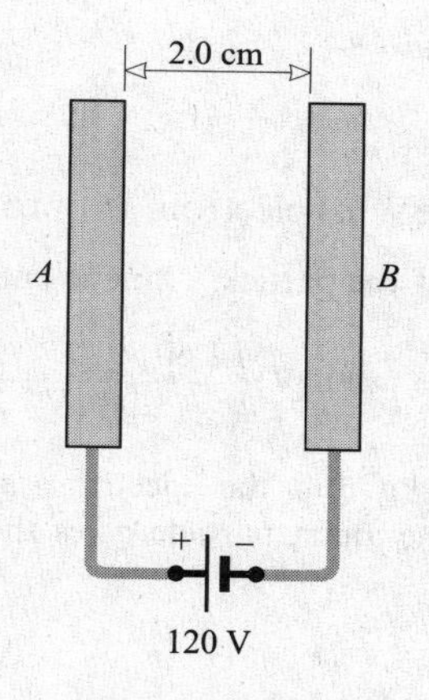

Fig. 25-4

25.16 [II] As shown in Fig. 25-5, a charged particle remains stationary between the two horizontal charged plates. The plate separation is 2.0 cm, and $m = 4.0 \times 10^{-13}$ kg and $q = 2.4 \times 10^{-18}$ C for the particle. Find the potential difference between the plates.

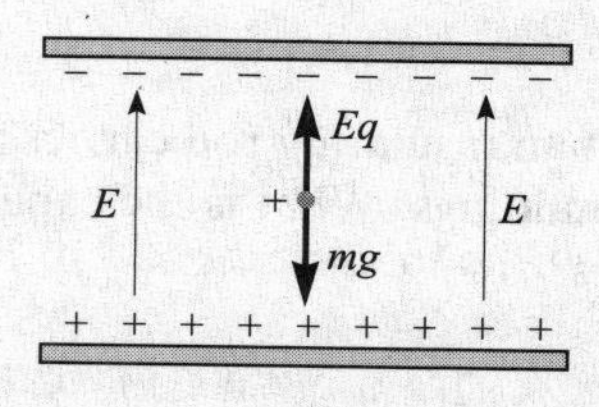

Fig. 25-5

Since the particle is in equilibrium, the weight of the particle is equal to the upward electrical force. That is,

$$mg = qE$$

or

$$E = \frac{mg}{q} = \frac{(4.0 \times 10^{-13}\ \text{kg})(9.81\ \text{m/s}^2)}{2.4 \times 10^{-18}\ \text{C}} = 1.63 \times 10^6\ \text{V/m}$$

But for a parallel-plate system,

$$V = Ed = (1.63 \times 10^6\ \text{V/m})(0.020\ \text{m}) = 33\ \text{kV}$$

25.17 [II] An alpha particle ($q = 2e$, $m = 6.7 \times 10^{-27}$ kg) falls from rest through a potential drop of 3.0×10^6 V (i.e., 3.0 MV). (*a*) What is its KE in electron volts? (*b*) What is its speed?

(*a*)

$$\text{Energy in eV} = \frac{qV}{e} = \frac{(2e)(3.0 \times 10^6)}{e} = 6.0 \times 10^6\ \text{eV} = 6.0\ \text{MeV}$$

(*b*)

$$\text{PE}_E\ \text{lost} = \text{KE gained}$$

$$qV = \tfrac{1}{2}mv_f^2 - \tfrac{1}{2}mv_i^2$$

$$(2)(1.6 \times 10^{-19}\ \text{C})(3.0 \times 10^6\ \text{V}) = \tfrac{1}{2}(6.7 \times 10^{-27}\ \text{kg})v_f^2 - 0$$

from which $v_f = 1.7 \times 10^7$ m/s.

25.18 [II] What is the speed of a 400 eV (*a*) electron, (*b*) proton, and (*c*) alpha particle?

In each case we know that the particle's kinetic energy is

$$\tfrac{1}{2}mv^2 = (400\ \text{eV})\left(\frac{1.60 \times 10^{-19}\ \text{J}}{1.00\ \text{eV}}\right) = 6.40 \times 10^{-17}\ \text{J}$$

Substituting $m_e = 9.1 \times 10^{-31}$ kg for the electron, $m_p = 1.67 \times 10^{-27}$ kg for the proton, and $m_\alpha = 4(1.67 \times 10^{-27}$ kg) for the alpha particle gives their speeds as (*a*) 1.186×10^7 m/s, (*b*) 2.77×10^5 m/s, and (*c*) 1.38×10^5 m/s.

25.19 [I] A parallel-plate capacitor has a capacitance of 8.0 μF with air between its plates. Determine its capacitance when a dielectric with dielectric constant 6.0 is placed between its plates.

$$C\ \text{with dielectric} = K(C\ \text{with air}) = (6.0)(8.0\ \mu\text{F}) = 48\ \mu\text{F}$$

25.20 [I] What is the charge on a 300 pF capacitor when it is charged to a voltage of 1.0 kV?

$$q = CV = (300 \times 10^{-12}\ \text{F})(1000\ \text{V}) = 3.0 \times 10^{-7}\ \text{C} = 0.30\ \mu\text{C}$$

25.21 [I] A metal sphere mounted on an insulating rod carries a charge of 6.0 nC when its potential is 200 V higher than its surroundings. What is the capacitance of the capacitor formed by the sphere and its surroundings?

$$C = \frac{q}{V} = \frac{6.0 \times 10^{-9}\ \text{C}}{200\ \text{V}} = 30\ \text{pF}$$

25.22 [I] A 1.2 μF capacitor is charged to 3.0 kV. Compute the energy stored in the capacitor.

$$\text{Energy} = \tfrac{1}{2}qV = \tfrac{1}{2}CV^2 = \tfrac{1}{2}(1.2 \times 10^{-6}\ \text{F})(3000\ \text{V})^2 = 5.4\ \text{J}$$

25.23 [II] The series combination of two capacitors shown in Fig. 25-6 is connected across 1000 V. Compute (*a*) the equivalent capacitance C_{eq} of the combination, (*b*) the magnitudes of the charges on the capacitors, (*c*) the potential differences across the capacitors, and (*d*) the energy stored in the capacitors.

(*a*)
$$\frac{1}{C_{eq}} = \frac{1}{C_1} + \frac{1}{C_2} = \frac{1}{3.0\ \text{pF}} + \frac{1}{6.0\ \text{pF}} = \frac{1}{2.0\ \text{pF}}$$

from which $C = 2.0$ pF.

(*b*) *In a series combination, each capacitor carries the same charge* [see Fig. 25-1(*b*)], which is the charge on the combination. Thus, using the result of (*a*), we have

$$q_1 = q_2 = q = C_{eq}V = (2.0 \times 10^{-12}\text{F})(1000\ \text{V}) = 2.0\ \text{nC}$$

(*c*)
$$V_1 = \frac{q_1}{C_1} = \frac{2.0 \times 10^{-9}\ \text{C}}{3.0 \times 10^{-12}\ \text{F}} = 667\ \text{V} = 0.67\ \text{kV}$$

$$V_2 = \frac{q_2}{C_2} = \frac{2.0 \times 10^{-9}\ \text{C}}{6.0 \times 10^{-12}\ \text{F}} = 333\ \text{V} = 0.33\ \text{kV}$$

(*d*)
$$\text{Energy in } C_1 = \tfrac{1}{2}q_1V_1 = \tfrac{1}{2}(2.0 \times 10^{-9}\ \text{C})(667\ \text{V}) = 6.7 \times 10^{-7}\ \text{J} = 0.67\ \mu\text{J}$$
$$\text{Energy in } C_2 = \tfrac{1}{2}q_2V_2 = \tfrac{1}{2}(2.0 \times 10^{-9}\ \text{C})(333\ \text{V}) = 3.3 \times 10^{-7}\ \text{J} = 0.33\ \mu\text{J}$$
$$\text{Energy in combination} = (6.7 + 3.3) \times 10^{-7}\ \text{J} = 10 \times 10^{-7}\ \text{J} = 1.0\ \mu\text{J}$$

The last result is also directly given by $\tfrac{1}{2}qV$ or $\tfrac{1}{2}C_{eq}V^2$.

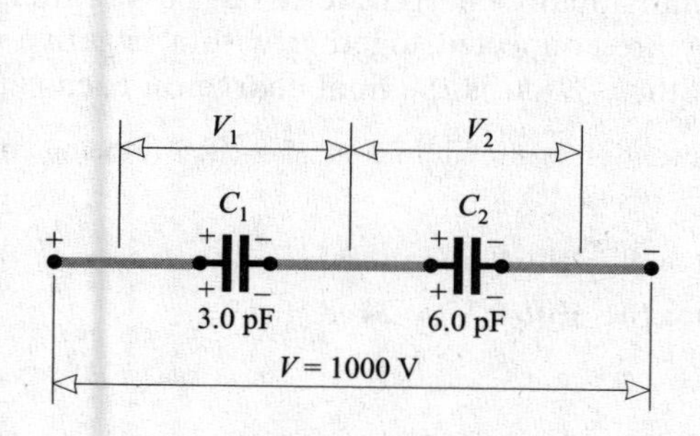

Fig. 25-6

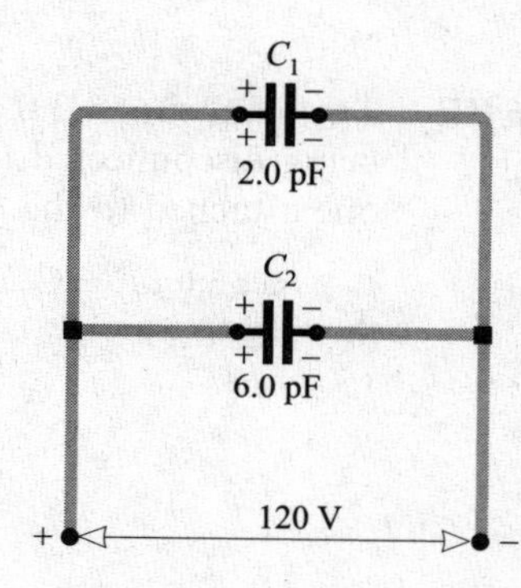

Fig. 25-7

25.24 [II] The parallel capacitor combination shown in Fig. 25-7 is connected across a 120 V source. Determine the equivalent capacitance C_{eq}, the charge on each capacitor, and the charge on the combination.

For a parallel combination,

$$C_{eq} = C_1 + C_2 = 2.0\ \text{pF} + 6.0\ \text{pF} = 8.0\ \text{pF}$$

Each capacitor has a 120 V potential difference impressed on it. Therefore,

$$q_1 = C_1V_1 = (2.0 \times 10^{-12}\ \text{F})(120\ \text{V}) = 0.24\ \text{nC}$$
$$q_2 = C_2V_2 = (6.0 \times 10^{-12}\ \text{F})(120\ \text{V}) = 0.72\ \text{nC}$$

The charge on the combination is $q_1 + q_2 = 960$ pC. Or, we could write

$$q = C_{eq}V = (8.0 \times 10^{-12}\ \text{F})(120\ \text{V}) = 0.96\ \text{nC}$$

25.25 [III] A certain capacitor consists of two parallel conducting plates, each with area 200 cm², separated by a 0.40-cm air gap. (*a*) Compute its capacitance. (*b*) If the capacitor is connected across a 500 V source, find the charge on it, the energy stored in it, and the value of E between the plates. (*c*) If a liquid with $K = 2.60$ is poured between the plates so as to fill the air gap, how much additional charge will flow onto the capacitor from the 500 V source?

(*a*) For a parallel-plate capacitor with air gap,

$$C = K\epsilon_0\frac{A}{d} = (1)(8.85 \times 10^{-12}\ \text{F/m})\frac{200 \times 10^{-4}\ \text{m}^2}{4.0 \times 10^{-3}\ \text{m}} = 4.4 \times 10^{-11}\ \text{F} = 44\ \text{pF}$$

(*b*)
$$q = CV = (4.4 \times 10^{-11}\ \text{F})(500\ \text{V}) = 2.2 \times 10^{-8}\ \text{C} = 22\ \text{nC}$$

$$\text{Energy} = \tfrac{1}{2}qV = \tfrac{1}{2}(2.2 \times 10^{-8}\ \text{C})(500\ \text{V}) = 5.5 \times 10^{-6}\ \text{J} = 5.5\ \mu\text{J}$$

$$E = \frac{V}{d} = \frac{500\ \text{V}}{4.0 \times 10^{-3}\ \text{m}} = 1.3 \times 10^5\ \text{V/m}$$

(*c*) The capacitor will now have a capacitance $K = 2.60$ times larger than before. Therefore,

$$q = CV = (2.60 \times 4.4 \times 10^{-11}\ \text{F})(500\ \text{V}) = 5.7 \times 10^{-8}\ \text{C} = 57\ \text{nC}$$

The capacitor already had a charge of 22 nC and so 57 nC − 22 nC or 35 nC must have been added to it.

25.26 [II] Two capacitors, 3.0 μF and 4.0 μF, are individually charged across a 6.0-V battery. After being disconnected from the battery, they are connected together with a negative plate of one attached to the positive plate of the other. What is the final charge on each capacitor?

Let 3.0 μF = C_1 and 4.0 μF = C_2. The situation is shown in Fig. 25-8. Before being connected, their charges are

$$q_1 = C_1V = (3.0 \times 10^{-6}\ \text{F})(6.0\ \text{V}) = 18\ \mu\text{C}$$
$$q_2 = C_2V = (4.0 \times 10^{-6}\ \text{F})(6.0\ \text{V}) = 24\ \mu\text{C}$$

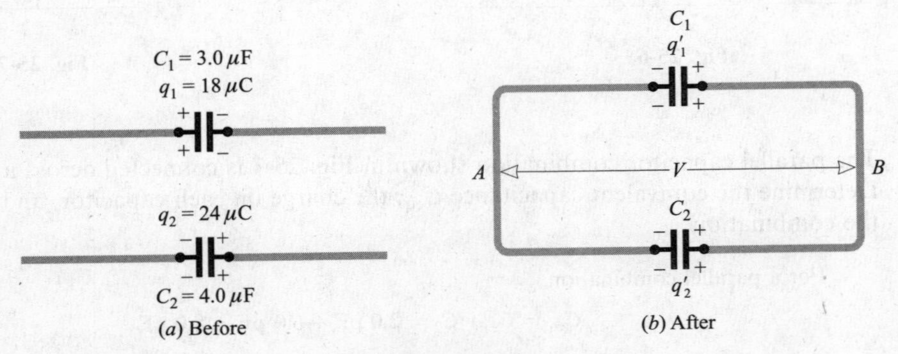

Fig. 25-8

These charges partly cancel when the capacitors are connected together. Their final charges are q_1' and q_2' where

$$q_1' + q_2' = q_2 - q_1 = 6.0\ \mu\text{C}$$

Also, the potentials across them are now the same, so that $V = q/C$ gives

$$\frac{q_1'}{3.0 \times 10^{-6}\ \text{F}} = \frac{q_2'}{4.0 \times 10^{-6}\ \text{F}} \qquad \text{or} \qquad q_1' = 0.75q_2'$$

Substitution in the previous equation gives

$$0.75q_2' + q_2' = 6.0\ \mu\text{C} \qquad \text{or} \qquad q_2' = 3.4\ \mu\text{C}$$

Then $q_1' = 0.75q_2' = 2.6\ \mu\text{C}$.

Supplementary Problems

25.27 [I] Two metal plates are attached to the two terminals of a 1.50-V battery. How much work is required to carry a $+5.0$-μC charge across the gap (*a*) from the negative to the positive plate, (*b*) from the positive to the negative plate? *Ans.* (*a*) 7.5 μJ, (*b*) $-7.5\ \mu$J

25.28 [II] The plates described in Problem 25.27 are in vacuum. An electron ($q = -e$, $m_e = 9.1 \times 10^{-31}$ kg) is released at the negative plate and falls freely to the positive plate. How fast is it going just before it strikes the plate? *Ans.* 7.3×10^5 m/s

25.29 [II] A proton ($q = e$, $m_p = 1.67 \times 10^{-27}$ kg) is accelerated from rest through a potential difference of 1.0 MV. What is its final speed? *Ans.* 1.4×10^7 m/s

25.30 [II] An electron gun shoots electrons ($q = -e$, $m_e = 9.1 \times 10^{-31}$ kg) at a metal plate that is 4.0 mm away in vacuum. The plate is 5.0 V lower in potential than the gun. How fast must the electrons be moving as they leave the gun if they are to reach the plate? *Ans.* 1.3×10^6 m/s

25.31 [I] The potential difference between two large parallel metal plates is 120 V. The plate separation is 3.0 mm. Find the electric field between the plates. *Ans.* 40 kV/m toward negative plate

25.32 [II] An electron ($q = -e$, $m_e = 9.1 \times 10^{-31}$ kg) is shot with speed 5.0×10^6 m/s parallel to a uniform electric field of strength 3.0 kV/m. How far will the electron go before it stops? *Ans.* 2.4 cm

25.33 [II] A potential difference of 24 kV maintains a downward-directed electric field between two horizontal parallel plates separated by 1.8 cm in vacuum. Find the charge on an oil droplet of mass 2.2×10^{-13} kg that remains stationary in the field between the plates. *Ans.* 1.6×10^{-18} C $= 10e$

25.34 [I] Determine the absolute potential in air at a distance of 3.0 cm from a point charge of 500 μC. *Ans.* 15 kV

25.35 [II] Compute the magnitude of the electric field and the absolute potential at a distance of 1.0 nm from a helium nucleus of charge $+2e$. What is the potential energy (relative to infinity) of a proton at this position? *Ans.* 2.9×10^9 N/C, 2.9 V, 4.6×10^{-19} J

25.36 [II] A charge of 0.20 μC is 30 cm from a point charge of 3.0 μC in vacuum. What work is required to bring the 0.20-μC charge 18 cm closer to the 3.0-μC charge? *Ans.* 0.027 J

25.37 [II] A point charge of $+2.0\,\mu C$ is placed at the origin of coordinates. A second, of $-3.0\,\mu C$, is placed on the x-axis at $x = 100$ cm. At what point (or points) on the x-axis will the absolute potential be zero? *Ans.* $x = 40$ cm and $x = -0.20$ m

25.38 [II] In Problem 25.37, what is the difference in potential between the following two points on the x-axis: point-A at $x = 0.1$ m and point-B at $x = 0.9$ m? Which point is at the higher potential? *Ans.* 4×10^5 V, point-A

25.39 [II] An electron is moving in the $+x$-direction with a speed of 5.0×10^6 m/s. There is an electric field of 3.0 kV/m in the $+x$-direction. What will be the electron's speed after it has moved 1.00 cm along the field? *Ans.* 3.8×10^6 m/s

25.40 [II] An electron has a speed of 6.0×10^5 m/s as it passes point-A on its way to point-B. Its speed at B is 12×10^5 m/s. What is the potential difference between A and B, and which is at the higher potential? *Ans.* 3.1 V, B

25.41 [I] A capacitor with air between its plates has capacitance $3.0\,\mu F$. What is its capacitance when wax of dielectric constant 2.8 is placed between the plates? *Ans.* $8.4\,\mu F$

25.42 [I] Determine the charge on each plate of a $0.050\text{-}\mu F$ parallel-plate capacitor when the potential difference between the plates is 200 V. *Ans.* $10\,\mu C$

25.43 [I] A capacitor is charged with 9.6 nC and has a 120 V potential difference between its terminals. Compute its capacitance and the energy stored in it. *Ans.* 80 pF, $0.58\,\mu J$

25.44 [I] Compute the energy stored in a 60-pF capacitor (*a*) when it is charged to a potential difference of 2.0 kV and (*b*) when the charge on each plate is 30 nC. *Ans.* (*a*) 12 mJ; (*b*) $7.5\,\mu J$

25.45 [II] Three capacitors, each of capacitance 120 pF, are each charged to 0.50 kV and then connected in series. Determine (*a*) the potential difference between the end plates, (*b*) the charge on each capacitor, and (*c*) the energy stored in the system. *Ans.* (*a*) 1.5 kV; (*b*) 60 nC; (*c*) $45\,\mu J$

25.46 [I] Three capacitors ($2.00\,\mu F$, $5.00\,\mu F$, and $7.00\,\mu F$) are connected in series. What is their equivalent capacitance? *Ans.* $1.19\,\mu F$

25.47 [I] Three capacitors ($2.00\,\mu F$, $5.00\,\mu F$, and $7.00\,\mu F$) are connected in parallel. What is their equivalent capacitance? *Ans.* $14.00\,\mu F$

25.48 [I] The capacitor combination in Problem 25.46 is connected in series with the combination in Problem 25.47. What is the capacitance of this new combination? *Ans.* $1.09\,\mu F$

25.49 [II] Two capacitors (0.30 and $0.50\,\mu F$) are connected in parallel. (*a*) What is their equivalent capacitance? A charge of $200\,\mu C$ is now placed on the parallel combination. (*b*) What is the potential difference across it? (*c*) What are the charges on the capacitors? *Ans.* (*a*) $0.80\,\mu F$; (*b*) 0.25 kV; (*c*) $75\,\mu C$, 0.13 mC

25.50 [II] A $2.0\text{-}\mu F$ capacitor is charged to 50 V and then connected in parallel (positive plate to positive plate) with a $4.0\text{-}\mu F$ capacitor charged to 100 V. (*a*) What are the final charges on the capacitors? (*b*) What is the potential difference across each? *Ans.* (*a*) 0.17 mC, 0.33 mC, (*b*) 83 V

25.51 [II] Repeat Problem 25.50 if the positive plate of one capacitor is connected to the negative plate of the other. *Ans.* (*a*) 0.10 mC, 0.20 mC; (*b*) 50 V

25.52 [II] (*a*) Calculate the capacitance of a capacitor consisting of two parallel plates separated by a layer of paraffin wax 0.50 cm thick, the area of each plate being 80 cm^2. The dielectric constant for the wax is 2.0. (*b*) If the capacitor is connected to a 100-V source, calculate the charge on the capacitor and the energy stored in the capacitor. *Ans.* (*a*) 28 pF; (*b*) 2.8 nC, $0.14\,\mu J$

Chapter 26

Current, Resistance, and Ohm's Law

A CURRENT (I) of electricity exists in a region when a net electric charge is transported from one point to another in that region. Suppose the charge is moving through a wire. If a charge q is transported through a given cross section of the wire in a time t, then the current through the wire is

$$I = \frac{q}{t}$$

Here, q is in coulombs, t is in seconds, and I is in **amperes** ($1\ \text{A} = 1\ \text{C/s}$). By custom ***the direction of the current is taken to be in the direction of flow of positive charge***. Thus, a flow of electrons to the right corresponds to a current to the left.

A BATTERY is a source of electrical energy. If no internal energy losses occur in the battery, then the potential difference (see Chapter 25) between its terminals is called the **electromotive force** (emf) of the battery. Unless otherwise stated, it will be assumed that the terminal potential difference of a battery is equal to its emf. The unit for emf is the same as the unit for potential difference, the volt.

THE RESISTANCE (R) of a wire or other object is a measure of the potential difference (V) that must be impressed across the object to cause a current of one ampere to flow through it:

$$R = \frac{V}{I}$$

The unit of resistance is the **ohm**, for which the symbol Ω (Greek omega) is used: $1\ \Omega = 1\ \text{V/A}$.

OHM'S LAW originally contained two parts. Its first part was simply the defining equation for resistance, $V = IR$. We often refer to this equation as being Ohm's Law. However, Ohm also stated that R is a constant independent of V and I. This latter part of the Law is only approximately correct.

The relation $V = IR$ can be applied to any resistor, where V is the potential difference (p.d.) between the two ends of the resistor, I is the current through the resistor, and R is the resistance of the resistor under those conditions.

MEASUREMENT OF RESISTANCE BY AMMETER AND VOLTMETER: A series circuit consisting of the resistance to be measured, an ammeter, and a battery is used. The current is measured by the (low-resistance) ammeter. The potential difference is measured by connecting the terminals of a (high-resistance) voltmeter across the resistance, i.e., in parallel with it. The resistance is computed by dividing the voltmeter reading by the ammeter reading according to Ohm's Law, $R = V/I$. (If the exact value of the resistance is required, the resistances of the voltmeter and ammeter must be considered parts of the circuit.)

THE TERMINAL POTENTIAL DIFFERENCE (or **voltage**) of a battery or generator when it delivers a current I is related to its electromotive force $\mathcal{E}$ and its **internal resistance** r as follows:

(1) When delivering current (*on discharge*):

$$\text{Terminal voltage} = (\text{emf}) - (\text{voltage drop in internal resistance})$$
$$V = \mathcal{E} - Ir$$

(2) When receiving current (*on charge*):

$$\text{Terminal voltage} = \text{emf} + (\text{voltage drop in internal resistance})$$
$$V = \mathcal{E} + Ir$$

(3) When no current exists:

$$\text{Terminal voltage} = \text{emf of battery or generator}$$

RESISTIVITY: The resistance R of a wire of length L and cross-sectional area A is

$$R = \rho \frac{L}{A}$$

where ρ is a constant called the **resistivity**. The resistivity is a characteristic of the material from which the wire is made. For L in m, A in m^2, and R in Ω, the units of ρ are $\Omega \cdot \text{m}$.

RESISTANCE VARIES WITH TEMPERATURE: If a wire has a resistance R_0 at a temperature T_0, then its resistance R at a temperature T is

$$R = R_0 + \alpha R_0 (T - T_0)$$

where α is the **temperature coefficient of resistance** of the material of the wire. Usually α varies with temperature and so this relation is applicable only over a small temperature range. The units of α are K^{-1} or $^\circ\text{C}^{-1}$.

A similar relation applies to the variation of resistivity with temperature. If ρ_0 and ρ are the resistivities at T_0 and T, respectively, then

$$\rho = \rho_0 + \alpha \rho_0 (T - T_0)$$

POTENTIAL CHANGES: The potential difference across a resistor R through which a current I flows is, by Ohm's Law, IR. The end of the resistor at which the current enters is the high-potential end of the resistor. Current always flows "downhill," from high to low potential, through a resistor.

The positive terminal of a battery is always the high-potential terminal if internal resistance of the battery is negligible or small. This is true irrespective of the direction of the current through the battery.

Solved Problems

26.1 [I] A steady current of 0.50 A flows through a wire. How much charge passes through the wire in one minute?

Because $I = q/t$, we have $q = It = (0.50\ \text{A})(60\ \text{s}) = 30\ \text{C}$. (Recall that $1\ \text{A} = 1\ \text{C/s}$.)

26.2 [I] How many electrons flow through a light bulb each second if the current through the light bulb is 0.75 A?

From $I = q/t$, the charge flowing through the bulb in 1.0 s is

$$q = It = (0.75\ \text{A})(1.0\ \text{s}) = 0.75\ \text{C}$$

But the magnitude of the charge on each electron is $e = 1.6 \times 10^{-19}$ C. Therefore,

$$\text{Number} = \frac{\text{charge}}{\text{charge/electron}} = \frac{0.75\ \text{C}}{1.6 \times 10^{-19}\ \text{C}} = 4.7 \times 10^{18}$$

26.3 [I] A light bulb has a resistance of 240 Ω when lit. How much current will flow through it when it is connected across 120 V, its normal operating voltage?

$$I = \frac{V}{R} = \frac{120\ \text{V}}{240\ \Omega} = 0.500\ \text{A}$$

26.4 [I] An electric heater uses 5.0 A when connected across 110 V. Determine its resistance.

$$R = \frac{V}{I} = \frac{110\ \text{V}}{5.0\ \text{A}} = 22\ \Omega$$

26.5 [I] What is the potential drop across an electric hot plate that draws 5.0 A when its hot resistance is 24 Ω?

$$V = IR = (5.0\ \text{A})(24\ \Omega) = 0.12\ \text{kV}$$

26.6 [II] The current in Fig. 26-1 is 0.125 A in the direction shown. For each of the following pairs of points, what is their potential difference, and which point is at the higher potential? (*a*) *A*, *B*; (*b*) *B*, *C*; (*c*) *C*, *D*; (*d*) *D*, *E*; (*e*) *C*, *E*; (*f*) *E*, *C*.

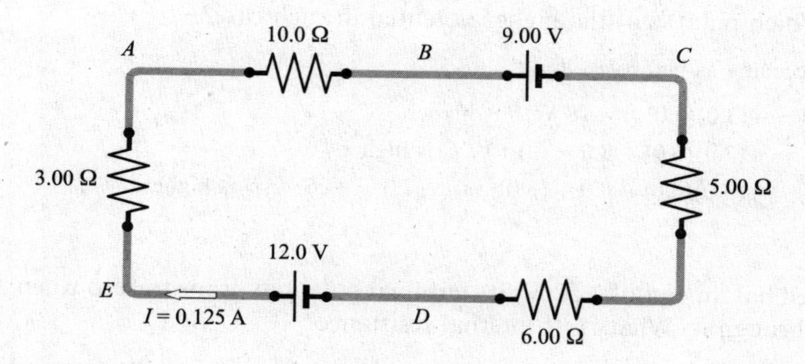

Fig. 26-1

Recall the following facts: (1) The current is the same (0.125 A) at all points in this circuit because the charge has no other place to flow. (2) Current always flows from high to low potential through a resistor. (3) The positive terminal of a pure emf (the long side of its symbol) is always the high-potential terminal. Therefore, taking potential drops as negative, we have the following:

(*a*) $V_{AB} = -IR = -(0.125\ \text{A})(10.0\ \Omega) = -1.25\ \text{V}$; *A* is higher.

(*b*) $V_{BC} = -\mathscr{E} = -9.00$ V; *B* is higher.
(*c*) $V_{CD} = -(0.125 \text{ A})(5.00\ \Omega) - (0.125 \text{ A})(6.00\ \Omega) = -1.38$ V; *C* is higher.
(*d*) $V_{DE} = +\mathscr{E} = +12.0$ V; *E* is higher.
(*e*) $V_{CE} = -(0.125 \text{ A})(5.00\ \Omega) - (0.125 \text{ A})(6.00\ \Omega) + 12.0 \text{ V} = +10.6$ V; *E* is higher.
(*f*) $V_{EC} = -(0.125 \text{ A})(3.00\ \Omega) - (0.125 \text{ A})(10.0\ \Omega) - 9.00 \text{ V} = -10.6$ V; *E* is higher.

Notice that the answers to (*e*) and (*f*) agree with each other.

26.7 [II] A current of 3.0 A flows through the wire shown in Fig. 26-2. What will a voltmeter read when connected from (*a*) *A* to *B*, (*b*) *A* to *C*, (*c*) *a* to *D*?

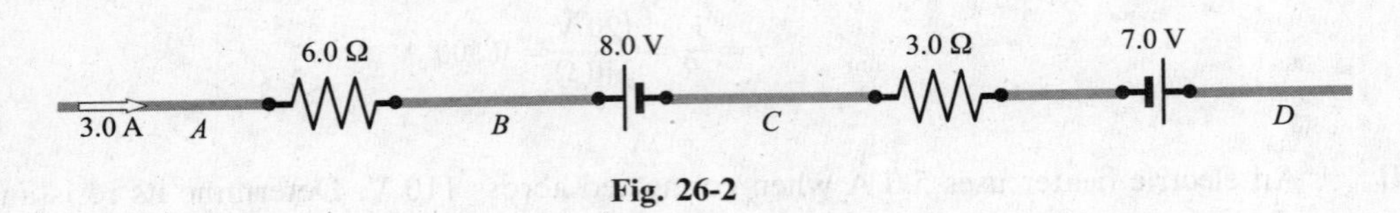

Fig. 26-2

(*a*) Point *A* is at the higher potential because current always flows "downhill" through a resistor. There is a potential drop of $IR = (3.0 \text{ A})(6.0\ \Omega) = 18$ V from *A* to *B*. The voltmeter will read −18 V.

(*b*) In going from *B* to *C* one goes from the positive to the negative side of the battery; hence there is a potential drop of 8.0 V from *B* to *C*. The drop adds to the drop of 18 V from *A* to *B*, found in (*a*), to give a 26 V drop from *A* to *C*. The voltmeter will read −26 V from *A* to *C*.

(*c*) From *C* to *D*, there is first a drop of $IR = (3.0 \text{ A})(3.0\ \Omega) = 9.0$ V through the resistor. Then, because one goes from the negative to the positive terminal of the 7.0 V battery, there is a 7.0 V rise through the battery. The voltmeter connected from *A* to *D* will read

$$-18 \text{ V} - 8.0 \text{ V} - 9.0 \text{ V} + 7.0 \text{ V} = -28 \text{ V}$$

26.8 [II] Repeat Problem 26.7 if the 3.0 A current is flowing from right to left instead of from left to right. Which point is at the higher potential in each case?

Proceeding as before, we have

(*a*) $V_{AB} = +(3.0)(6.0) = +18$ V; *B* is higher.
(*b*) $V_{AC} = +(3.0)(6.0) - 8.0 = +10$ V; *C* is higher.
(*c*) $V_{AD} = +(3.0)(6.0) - 8.0 + (3.0)(3.0) + 7.0 = +26$ V; *D* is higher.

26.9 [I] A dry cell has an emf of 1.52 V. Its terminal potential drops to zero when a current of 25 A passes through it. What is its internal resistance?

As is shown in Fig. 26-3, the battery acts like a pure emf $\mathscr{E}$ in series with a resistor *r*. We are told that, under the conditions shown, the potential difference from *A* to *B* is zero. Therefore,

$$0 = +\mathscr{E} - Ir \qquad \text{or} \qquad 0 = 1.52 \text{ V} - (25 \text{ A})r$$

from which the internal resistance is $r = 0.061\ \Omega$.

26.10 [II] A direct-current generator has an emf of 120 V; that is, its terminal voltage is 120 V when no current is flowing from it. At an output of 20 A the terminal potential is 115 V. (*a*) What is the internal resistance r of the generator? (b) What will be the terminal voltage at an output of 40 A?

The situation is much like that shown in Fig. 26-3. Now, however, $\mathscr{E} = 120$ V and I is no longer 25 A.

(a) In this case, $I = 20$ A and the p.d. from A to B is 115 V. Therefore,

$$115\text{ V} = +120\text{ V} - (20\text{ A})r$$

from which $r = 0.25\ \Omega$.

(b) Now $I = 40$ A. So

$$\text{Terminal p.d.} = \mathscr{E} - Ir = 120\text{ V} - (40\text{ A})(0.25\ \Omega) = 110\text{ V}$$

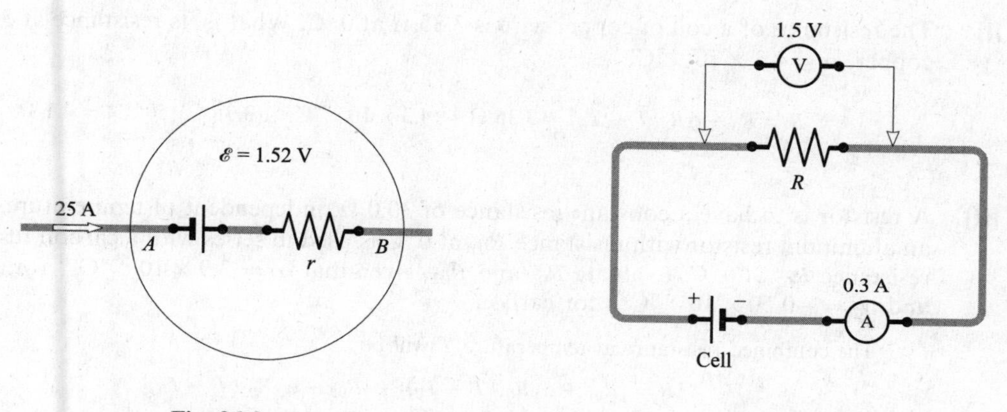

Fig. 26-3

Fig. 26-4

26.11 [I] As shown in Fig. 26-4 the ammeter–voltmeter method is used to measure an unknown resistance R. The ammeter reads 0.3 A, and the voltmeter reads 1.50 V. Compute the value of R if the ammeter and voltmeter are ideal.

$$R = \frac{V}{I} = \frac{1.50\text{ V}}{0.3\text{ A}} = 5\ \Omega$$

26.12 [I] A metal rod is 2 m long and 8 mm in diameter. Compute its resistance if the resistivity of the metal is $1.76 \times 10^{-8}\ \Omega\cdot\text{m}$.

$$R = \rho\frac{L}{A} = (1.76 \times 10^{-8}\ \Omega\cdot\text{m})\frac{2\text{ m}}{\pi(4 \times 10^{-3}\text{ m})^2} = 7 \times 10^{-4}\ \Omega$$

26.13 [I] Number 10 wire has a diameter of 2.59 mm. How many meters of number 10 aluminum wire are needed to give a resistance of 1.0 Ω? ρ for aluminum is $2.8 \times 10^{-8}\ \Omega\cdot\text{m}$.

From $R = \rho L/A$, we have

$$L = \frac{RA}{\rho} = \frac{(1.0\ \Omega)(\pi)(2.59 \times 10^{-3}\text{ m})^2/4}{2.8 \times 10^{-8}\ \Omega\cdot\text{m}} = 0.19\text{ km}$$

26.14 [II] (This problem introduces a unit sometimes used in the United States.) Number 24 copper wire has diameter 0.020 1 in. Compute (*a*) the cross-sectional area of the wire in circular mils and (*b*) the resistance of 100 ft of the wire. The resistivity of copper is 10.4 Ω· circular mils/ft.

The area of a circle in circular mils is defined as the square of the diameter of the circle expressed in mils, where 1 mil = 0.001 in.

(*a*)
$$\text{Area in circular mils} = (20.1\ \text{mil})^2 = 404\ \text{circular mils}$$

(*b*)
$$R = \rho\frac{L}{A} = \frac{(10.4\ \Omega\cdot\text{circular mil/ft})\ 100\ \text{ft}}{404\ \text{circular mils}} = 2.57\ \Omega$$

26.15 [I] The resistance of a coil of copper wire is 3.35 Ω at 0 °C. What is its resistance at 50 °C? For copper, $\alpha = 4.3 \times 10^{-3}\ {}^\circ\text{C}^{-1}$.

$$R = R_0 + \alpha R_0(T - T_0) = 3.35\ \Omega + (4.3 \times 10^{-3}\ {}^\circ\text{C}^{-1})(3.35\ \Omega)(50\ {}^\circ\text{C}) = 4.1\ \Omega$$

26.16 [II] A resistor is to have a constant resistance of 30.0 Ω, independent of temperature. For this, an aluminum resistor with resistance R_{01} at 0 °C is used in series with a carbon resistor with resistance R_{02} at 0 °C. Evaluate R_{01} and R_{02}, given that $\alpha_1 = 3.9 \times 10^{-3}\ {}^\circ\text{C}^{-1}$ for aluminum and $\alpha_2 = -0.50 \times 10^{-3}\ {}^\circ\text{C}^{-1}$ for carbon.

The combined resistance at temperature T will be

$$\begin{aligned} R &= [R_{01} + \alpha_1 R_{01}(T - T_0)] + [R_{02} + \alpha_2 R_{02}(T - T_0)] \\ &= (R_{01} + R_{02}) + (\alpha_1 R_{01} + \alpha_2 R_{02})(T - T_0) \end{aligned}$$

We thus have the two conditions

$$R_{01} + R_{02} = 30.0\ \Omega \qquad \text{and} \qquad \alpha_1 R_{01} + \alpha_2 R_{02} = 0$$

Substituting the given values of α_1 and α_2, then solving for R_{01} and R_{02}, we find

$$R_{01} = 3.4\ \Omega \qquad R_{02} = 27\ \Omega$$

26.17 [II] In the Bohr model, the electron of a hydrogen atom moves in a circular orbit of radius 5.3×10^{-11} m with a speed of 2.2×10^6 m/s. Determine its frequency f and the current I in the orbit.

$$f = \frac{v}{2\pi r} = \frac{2.2 \times 10^6\ \text{m/s}}{2\pi(5.3 \times 10^{-11}\ \text{m})} = 6.6 \times 10^{15}\ \text{rev/s}$$

Each time the electron goes around the orbit, it carries a charge e around the loop. The charge passing a point on the loop each second is

$$I = ef = (1.6 \times 10^{-19}\ \text{C})(6.6 \times 10^{15}\ \text{s}^{-1}) = 1.06 \times 10^{-3}\text{A} = 1.1\ \text{mA}$$

26.18 [II] A wire that has a resistance of 5.0 Ω is passed through an extruder so as to make it into a new wire three times as long as the original. What is the new resistance?

We shall use $R = \rho L/A$ to find the resistance of the new wire. To find ρ, we use the original data for the wire. Let L_0 and A_0 be the initial length and cross-sectional area, respectively. Then

$$5.0\ \Omega = \rho L_0/A_0 \qquad \text{or} \qquad \rho = (A_0/L_0)(5.0\ \Omega)$$

We were told that $L = 3L_0$. To find A in terms of A_0, we note that the volume of the wire cannot change. Hence,

$$V_0 = L_0 A_0 \quad \text{and} \quad V_0 = LA$$

from which

$$LA = L_0 A_0 \quad \text{or} \quad A = \left(\frac{L_0}{L_0}\right)(A_0) = \frac{A_0}{3}$$

Therefore,

$$R = \frac{\rho L}{A} = \frac{(A_0/L_0)(5.0\ \Omega)(3L_0)}{A_0/3} = 9(5.0\ \Omega) = 45\ \Omega$$

26.19 [II] It is desired to make a wire that has a resistance of 8.0 Ω from 5.0 cm^3 of metal that has a resistivity of $9.0 \times 10^{-8}\ \Omega \cdot \text{m}$. What should the length and cross-sectional area of the wire be?

We use $R = \rho L/A$ with $R = 8.0\ \Omega$ and $\rho = 9.0 \times 10^{-8}\ \Omega \cdot \text{m}$. We know further that the volume of the wire (which is LA) is $5.0 \times 10^{-6}\ \text{m}^3$. Therefore we have two equations to solve for L and A:

$$R = 8.0\ \Omega = (9.0 \times 10^{-8}\ \Omega \cdot \text{m})\left(\frac{L}{A}\right) \quad \text{and} \quad LA = 5.0 \times 10^{-6}\ \text{m}^3$$

From them, we get $L = 21$ m and $A = 2.4 \times 10^{-7}\ \text{m}^2$.

Supplementary Problems

26.20 [I] How many electrons per second pass through a section of wire carrying a current of 0.70 A? *Ans.* 4.4×10^{18} electrons/s

26.21 [I] An electron gun in a TV set shoots out a beam of electrons. The beam current is 1.0×10^{-5} A. How many electrons strike the TV screen each second? How much charge strikes the screen in a minute? *Ans.* 6.3×10^{13} electrons/s, -6.0×10^{-4} C/min

26.22 [I] What is the current through an 8.0-Ω toaster when it is operating on 120 V? *Ans.* 15 A

26.23 [I] What potential difference is required to pass 3.0 A through 28 Ω? *Ans.* 84 V

26.24 [I] Determine the potential difference between the ends of a wire of resistance 5.0 Ω if 720 C passes through it per minute. *Ans.* 60 V

26.25 [I] A copper bus bar carrying 1200 A has a potential drop of 1.2 mV along 24 cm of its length. What is the resistance per meter of the bar? *Ans.* 4.2 μΩ/m

26.26 [I] An ammeter is connected in series with an unknown resistance, and a voltmeter is connected across the terminals of the resistance. If the ammeter reads 1.2 A and the voltmeter reads 18 V, compute the value of the resistance. Assume ideal meters. *Ans.* 15 Ω

26.27 [I] An electric utility company runs two 100 m copper wires from the street mains up to a customer's premises. If the wire resistance is 0.10 Ω per 1000 m, calculate the line voltage drop for an estimated load current of 120 A. *Ans.* 2.4 V

26.28 [I] When the insulation resistance between a motor winding and the motor frame is tested, the value obtained is 1.0 megohm (10^6 Ω). How much current passes through the insulation of the motor if the test voltage is 1000 V? *Ans.* 1.0 mA

26.29 [I] Compute the internal resistance of an electric generator which has an emf of 120 V and a terminal voltage of 110 V when supplying 20 A. *Ans.* 0.50 Ω

26.30 [I] A dry cell delivering 2 A has a terminal voltage of 1.41 V. What is the internal resistance of the cell if its open-circuit voltage is 1.59 V? *Ans.* 0.09 Ω

26.31 [II] A cell has an emf of 1.54 V. When it is in series with a 1.0-Ω resistance, the reading of a voltmeter connected across the cell terminals is 1.40 V. Determine the cell's internal resistance. *Ans.* 0.10 Ω

26.32 [I] The internal resistance of a 6.4-V storage battery is 4.8 mΩ. What is the theoretical maximum current on short circuit? (In practice the leads and connections have some resistance, and this theoretical value would not be attained.) *Ans.* 1.3 kA

26.33 [I] A battery has an emf of 13.2 V and an internal resistance of 24.0 mΩ. If the load current is 20.0 A, find the terminal voltage. *Ans.* 12.7 V

26.34 [I] A storage battery has an emf of 25.0 V and an internal resistance of 0.200 Ω. Compute its terminal voltage (*a*) when it is delivering 8.00 A and (*b*) when it is being charged with 8.00 A. *Ans.* (*a*) 23.4 V; (*b*) 26.6 V

26.35 [II] A battery charger supplies a current of 10 A to charge a storage battery which has an open-circuit voltage of 5.6 V. If the voltmeter connected across the charger reads 6.8 V, what is the internal resistance of the battery at this time? *Ans.* 0.12 Ω

26.36 [II] Find the potential difference between points-*A* and *B* in Fig. 26-5 if *R* is 0.70 Ω. Which point is at the higher potential? *Ans.* −5.1 V, point-*A*

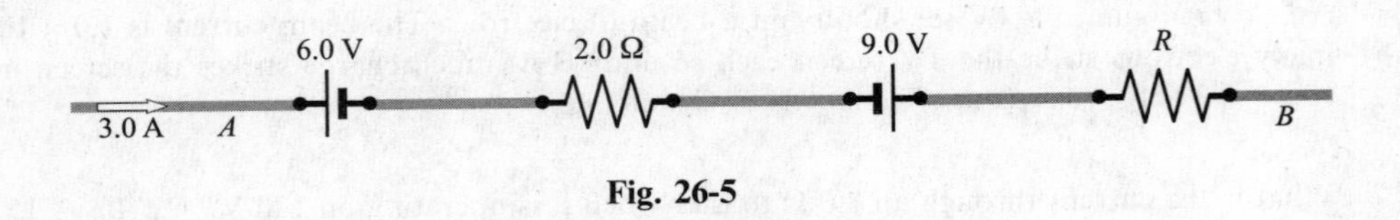

Fig. 26-5

26.37 [III] Repeat Problem 26.36 if the current flows in the opposite direction and $R = 0.70$ Ω. *Ans.* 11.1 V, point-*B*

26.38 [II] In Fig. 26-5, how large must *R* be if the potential drop from *A* to *B* is 12 V? *Ans.* 3.0 Ω

26.39 [II] For the circuit of Fig. 26-6, find the potential difference from (*a*) *A* to *B*; (*b*) *B* to *C*, and (*c*) *C* to *A*. Notice that the current is given as 2.0 A. *Ans.* (*a*) −48 V; (*b*) +28 V; (*c*) +20 V

26.40 [I] Compute the resistance of 180 m of silver wire having a cross-section of 0.30 mm^2. The resistivity of silver is 1.6×10^{-8} Ω·m. *Ans.* 9.6 Ω

26.41 [I] The resistivity of aluminum is 2.8×10^{-8} Ω·m. How long a piece of aluminum wire 1.0 mm in diameter is needed to give a resistance of 4.0 Ω? *Ans.* 0.11 km

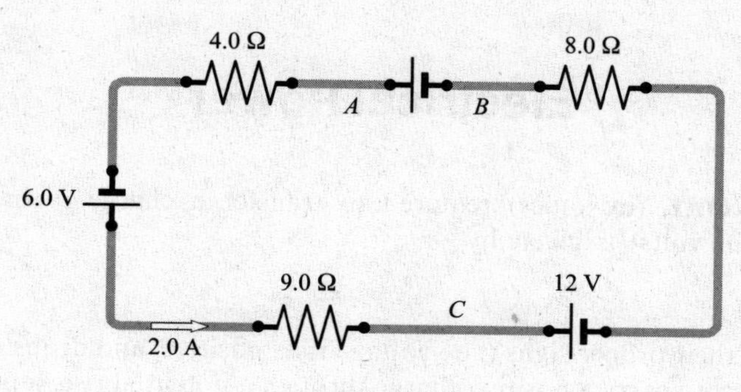

Fig. 26-6

26.42 [II] Number 6 copper wire has a diameter of 0.162 in. (*a*) Calculate its area in circular mils. (*b*) If $\rho = 10.4\ \Omega \cdot$ circular mils/ft, find the resistance of 1.0×10^3 ft of the wire. (Refer to Problem 26.14.)
Ans. (*a*) 26.0×10^3 circular mils; (*b*) 0.40 Ω

26.43 [II] A coil of wire has a resistance of 25.00 Ω at 20 °C and a resistance of 25.17 Ω at 35 °C. What is its temperature coefficient of resistance? *Ans.* $4.5 \times 10^{-4}\ {}^\circ\mathrm{C}^{-1}$

Chapter 27

Electrical Power

THE ELECTRICAL WORK (in joules) required to transfer a charge q (in coulombs) through a potential difference V (in volts) is given by

$$W = qV$$

When q and V are given their proper signs (i.e., voltage rises positive, and drops negative), the work will have its proper sign. Thus, to carry a positive charge through a potential rise, a positive amount of work must be done on the charge.

THE ELECTRICAL POWER (P), in watts, delivered by an energy source as it carries a charge q (in coulombs) through a potential rise V (in volts) in a time t (in seconds) is

$$\text{Power finished} = \frac{\text{work}}{\text{time}}$$

$$\mathrm{P} = \frac{Vq}{t}$$

Because $q/t = I$, this can be rewritten as

$$\mathrm{P} = VI$$

where I is in amperes.

THE POWER LOSS IN A RESISTOR is found by replacing V in VI by IR, or by replacing I in VI by V/R, to obtain

$$\mathrm{P} = VI = I^2R = \frac{V^2}{R}$$

THE THERMAL ENERGY GENERATED IN A RESISTOR per second is equal to the power loss in the resistor:

$$\mathrm{P} = VI = I^2R$$

CONVENIENT CONVERSIONS:

$$1\ \mathrm{W} = 1\ \mathrm{J/s} = 0.239\ \mathrm{cal/s} = 0.738\ \mathrm{ft \cdot lb/s}$$

$$1\ \mathrm{kW} = 1.341\ \mathrm{hp} = 56.9\ \mathrm{Btu/min}$$

$$1\ \mathrm{hp} = 746\ \mathrm{W} = 33\,000\ \mathrm{ft \cdot lb/min} = 42.4\ \mathrm{Btu/min}$$

$$1\ \mathrm{kW \cdot h} = 3.6 \times 10^6\ \mathrm{J} = 3.6\ \mathrm{MJ}$$

Solved Problems

27.1 [I] Compute the work and the average power required to transfer 96 kC of charge in one hour (1.0 h) through a potential rise of 50 V.

The work done equals the change in potential energy:

$$W = qV = (96\,000\ \text{C})(50\ \text{V}) = 4.8 \times 10^6\ \text{J} = 4.8\ \text{MJ}$$

Power is the rate of transferring energy:

$$\text{P} = \frac{W}{t} = \frac{4.8 \times 10^6\ \text{J}}{3600\ \text{s}} = 1.3\ \text{kW}$$

27.2 [I] How much current does a 60 W light bulb draw when connected to its proper voltage of 120 V?

From $\text{P} = VI$,

$$I = \frac{\text{P}}{V} = \frac{60\ \text{W}}{120\ \text{V}} = 0.50\ \text{A}$$

27.3 [I] An electric motor takes 5.0 A from a 110 V line. Determine the power input and the energy, in J and kW·h, supplied to the motor in 2.0 h.

$$\text{Power} = \text{P} = VI = (110\ \text{V})(5.0\ \text{A}) = 0.55\ \text{kW}$$
$$\text{Energy} = \text{P}t = (550\ \text{W})(7200\ \text{s}) = 4.0\ \text{MJ}$$
$$= (0.55\ \text{kW})(2.0\ \text{h}) = 1.1\ \text{kW}\cdot\text{h}$$

27.4 [I] An electric iron of resistance 20 Ω takes a current of 5.0 A. Calculate the thermal energy, in joules, developed in 30 s.

$$\text{Energy} = \text{P}t$$
$$\text{Energy} = I^2Rt = (5\ \text{A})^2(20\ \Omega)(30\ \text{s}) = 15\,\text{kJ}$$

27.5 [II] An electric heater of resistance 8.0 Ω draws 15 A from the service mains. At what rate is thermal energy developed, in W? What is the cost of operating the heater for a period of 4.0 h at 10 ¢/kW·h?

$$W = I^2R = (15\ \text{A})^2(8.0\ \Omega) = 1800\ \text{W} = 1.8\ \text{kW}$$
$$\text{Cost} = (1.8\ \text{kW})(4.0\ \text{h})(10\ \text{¢/kW}\cdot\text{h}) = 72\ \text{¢}$$

27.6 [II] A coil develops 800 cal/s when 20 V is supplied across its ends. Compute its resistance.

$$\text{P} = (800\ \text{cal/s})(4.184\ \text{J/cal}) = 3347\ \text{J/s}$$

Then, because $\text{P} = V^2/R$,

$$R = \frac{(20\ \text{V})^2}{3347\ \text{J/s}} = 0.12\ \Omega$$

27.7 [II] A line having a total resistance of 0.20 Ω delivers 10.00 kW at 250 V to a small factory. What is the efficiency of the transmission?

The line dissipates power due to its resistance. Consequently we'll need to find the current in the line. We use $P = VI$ to find $I = P/V$. Then

$$\text{Power lost in line} = I^2R = \left(\frac{P}{V}\right)^2 R = \left(\frac{10\,000\ \text{W}}{250\ \text{V}}\right)^2 (0.20\ \Omega) = 0.32\ \text{kW}$$

$$\text{Efficiency} = \frac{\text{power delivered by line}}{\text{power supplied to line}} = \frac{10.00\ \text{kW}}{(10.00 + 0.32)\ \text{kW}} = 0.970 = 97.0\%$$

27.8 [II] A hoist motor supplied by a 240-V source requires 12.0 A to lift an 800-kg load at a rate of 9.00 m/min. Determine the power input to the motor and the power output, both in horsepower, and the overall efficiency of the system.

$$\text{Power input} = IV = (12.0\ \text{A})(240\ \text{V}) = 2880\ \text{W} = (2.88\ \text{kW})(1.34\ \text{hp/kW}) = 3.86\ \text{hp}$$

$$\text{Power output} = Fv = (800 \times 9.81\ \text{N})\left(\frac{9.00\ \text{m}}{\text{min}}\right)\left(\frac{1.00\ \text{min}}{60.0\ \text{s}}\right)\left(\frac{1.00\ \text{hp}}{746\ \text{J/s}}\right) = 1.58\ \text{hp}$$

$$\text{Efficiency} = \frac{1.58\ \text{hp output}}{3.86\ \text{hp input}} = 0.408 = 40.8\%$$

27.9 [II] The lights on a car are inadvertently left on. They dissipate 95.0 W. About how long will it take for the fully charged 12.0 V car battery to run down if the battery is rated at 150 ampere-hours (A · h)?

As an approximation, assume the battery maintains 12.0 V until it goes dead. Its 150 A · h rating means it can supply the energy equivalent of a 150-A current that flows for 1.00 h (3600 s). Therefore, the total energy the battery can supply is

$$\text{Total output energy} = (\text{power})(\text{time}) = (VI)t = (12.0\ \text{V} \times 150\ \text{A})(3600\ \text{s}) = 6.48 \times 10^6\ \text{J}$$

The energy consumed by the lights in a time t is

$$\text{Energy dissipated} = (95\ \text{W})(t)$$

Equating these two energies and solving for t, we find $t = 6.82 \times 10^4\ \text{s} = 18.9\ \text{h}$.

27.10 [II] What is the cost of electrically heating 50 liters of water from 40 °C to 100 °C at 8.0 ¢/kW · h?

$$\text{Heat gained by water} = (\text{mass}) \times (\text{specific heat}) \times (\text{temperature rise})$$
$$= (50\ \text{kg}) \times (1000\ \text{cal/kg}\cdot{}^\circ\text{C}) \times (60\ {}^\circ\text{C}) = 3.0 \times 10^6\ \text{cal}$$

$$\text{Cost} = (3.0 \times 10^6\ \text{cal})\left(\frac{4.184\ \text{J}}{1\ \text{cal}}\right)\left(\frac{1\ \text{kW}\cdot\text{h}}{3.6 \times 10^6\ \text{J}}\right)\left(\frac{8.0\ ¢}{1\ \text{kW}\cdot\text{h}}\right) = 28\ ¢$$

Supplementary Problems

27.11 [I] A resistive heater is labeled 1600 W/120 V. How much current does the heater draw from a 120-V source? *Ans.* 13.3 A

27.12 [I] A bulb is stamped 40 W/120 V. What is its resistance when lighted by a 120-V source? *Ans.* 0.36 kΩ

27.13 [II] A spark of artificial 10.0-MV lightning had an energy output of 0.125 MW·s. How many coulombs of charge flowed? *Ans.* 0.0125 C

27.14 [II] A current of 1.5 A exists in a conductor whose terminals are connected across a potential difference of 100 V. Compute the total charge transferred in one minute, the work done in transferring this charge, and the power expended in heating the conductor if all the electrical energy is converted into heat. *Ans.* 90 C, 9.0 kJ, 0.15 kW

27.15 [II] An electric motor takes 15.0 A at 110 V. Determine (*a*) the power input and (*b*) the cost of operating the motor for 8.00 h at 10.0 ¢/kW·h. *Ans.* (*a*) 1.65 kW; (*b*) $1.32

27.16 [I] A current of 10 A exists in a line of 0.15 Ω resistance. Compute the rate of production of thermal energy in watts. *Ans.* 15 W

27.17 [II] An electric broiler develops 400 cal/s when the current through it is 8.0 A. Determine the resistance of the broiler. *Ans.* 26 Ω

27.18 [II] A 25.0-W, 120-V bulb has a cold resistance of 45.0 Ω. When the voltage is switched on, what is the instantaneous current? What is the current under normal operation? *Ans.* 2.67 A, 0.208 A

27.19 [II] While carrying a current of 400 A, a defective switch becomes overheated due to faulty surface contact. A millivoltmeter connected across the switch shows a 100-mV drop. What is the power loss due to the contact resistance? *Ans.* 40.0 W

27.20 [II] How much power does a 60 W/120 V incandescent light bulb dissipate when operated at a voltage of 115 V? Neglect the bulb's decrease in resistance with lowered voltage. *Ans.* 55 W

27.21 [II] A house wire is to carry a current of 30 A while dissipating no more than 1.40 W of heat per meter of its length. What is the minimum diameter of the wire if its resistivity is 1.68×10^{-8} Ω·m? *Ans.* 3.7 mm

27.22 [II] A 10.0-Ω electric heater operates on a 110-V line. Compute the rate at which it develops thermal energy in W and in cal/s. *Ans.* 1.21 kW = 290 cal/s

27.23 [III] An electric motor, which has 95 percent efficiency, uses 20 A at 110 V. What is the horsepower output of the motor? How many watts are lost in thermal energy? How many calories of thermal energy are developed per second? If the motor operates for 3.0 h, what energy, in MJ and in kW·h, is dissipated? *Ans.* 2.8 hp, 0.11 kW, 26 cal/s, 24 MJ = 6.6 kW·h

27.24 [II] An electric crane uses 8.0 A at 150 V to raise a 450-kg load at the rate of 7.0 m/min. Determine the efficiency of the system. *Ans.* 43%

27.25 [III] What should be the resistance of a heating coil which will be used to raise the temperature of 500 g of water from 28 °C to the boiling point in 2.0 minutes, assuming that 25 percent of the heat is lost? The heater operates on a 110-V line. *Ans.* 7.2 Ω

27.26 [II] Compute the cost per hour at 8.0 ¢/kW·h of electrically heating a room, if it requires 1.0 kg/h of anthracite coal having a heat of combustion of 8000 kcal/kg. *Ans.* 74 ¢/h

27.27 [II] Power is transmitted at 80 kV between two stations. If the voltage can be increased to 160 kV without a change in cable size, how much additional power can be transmitted for the same current? What effect does the power increase have on the line heating loss? *Ans.* additional power = original power, no effect

27.28 [II] A storage battery, of emf 6.4 V and internal resistance 0.080 Ω, is being charged by a current of 15 A. Calculate (*a*) the power loss in internal heating of the battery, (*b*) the rate at which energy is stored in the battery, and (*c*) its terminal voltage. *Ans.* (*a*) 18 W; (*b*) 96 W; (*c*) 7.6 V

27.29 [II] A tank containing 200 kg of water was used as a constant-temperature bath. How long would it take to heat the bath from 20 °C to 25 °C with a 250-W immersion heater? Neglect the heat capacity of the tank frame and any heat losses to the air. *Ans.* 4.6 h

Chapter 28

Equivalent Resistance; Simple Circuits

RESISTORS IN SERIES: When current can follow only one path as it flows through two or more resistors connected in line, the resistors are in **series**. In other words, when one and only one terminal of a resistor is connected directly to one and only one terminal of another resistor, the two are in series and the same current passes through both. A **node** is a point where three or more current-carrying wires or branches meet. There are no nodes between circuit elements (such as capacitors, resistors, and batteries) that are connected in series. A typical case is shown in Fig. 28-1(*a*). For several resistors in series, their equivalent resistance R_{eq} is given by

$$R_{eq} = R_1 + R_2 + R_3 + \cdots \qquad \text{(series combination)}$$

where $R_1, R_2, R_3, \ldots,$ are the resistances of the several resistors. Observe that resistances in series combine like capacitances in parallel (see Chapter 25). It i s assumed that all connection wire is effectively resistanceless.

In a series combination, the current through each resistance is the same as that through all the others. The potential drop (p.d.) across the combination is equal to the sum of the individual potential drops. *The equivalent resistance in series is always greater than the largest of the individual resistances.*

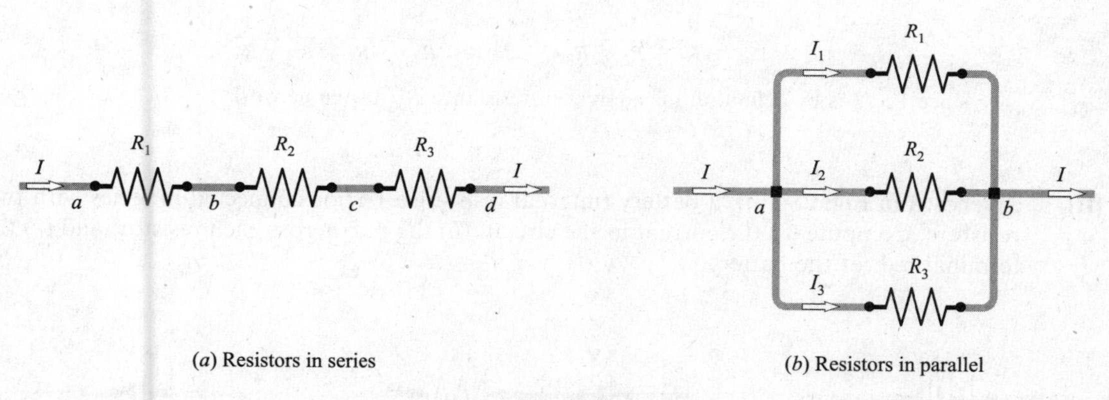

(*a*) Resistors in series

(*b*) Resistors in parallel

Fig. 28-1

RESISTORS IN PARALLEL: Several resistors are connected in **parallel** between two nodes if one end of each resistor is connected to one node and the other end of each is connected to the other node. A typical case is shown in Fig. 28-1(*b*), where points *a* and *b* are nodes. Their equivalent resistance R_{eq} is given by

$$\frac{1}{R_{eq}} = \frac{1}{R_1} + \frac{1}{R_2} + \frac{1}{R_3} + \cdots \qquad \text{(parallel combination)}$$

The equivalent resistance in parallel is always less than the smallest of the individual resistances. Connecting additional resistances in parallel decreases R_{eq} for the combination. Observe that resistances in parallel combine like capacitances in series (see Chapter 25).

The potential drop V across any one resistor in a parallel combination is the same as the potential drop across each of the others. The current through the nth resistor is $I_n = V/R_n$ and the total current entering the combination is equal to the sum of the individual branch currents [see Fig. 28-1(*b*)].

Solved Problems

28.1 [II] Derive the formula for the equivalent resistance R_{eq} of resistors R_1, R_2, and R_3 (*a*) in series and (*b*) in parallel, as shown in Fig. 28-1(*a*) and (*b*).

(*a*) For the series network,

$$V_{ad} = V_{ab} + V_{bc} + V_{cd} = IR_1 + IR_2 + IR_3$$

since the current I is the same in all three resistors. Dividing by I gives

$$\frac{V_{ad}}{I} = R_1 + R_2 + R_3 \qquad \text{or} \qquad R_{eq} = R_1 + R_2 + R_3$$

since V_{ad}/I is by definition the equivalent resistance R_{eq} of the network.

(*b*) The p.d. is the same for all three resistors, whence

$$I_1 = \frac{V_{ab}}{R_1} \qquad I_2 = \frac{V_{ab}}{R_2} \qquad I_3 = \frac{V_{ab}}{R_3}$$

Since the line current I is the sum of the branch currents,

$$I = I_1 + I_2 + I_3 = \frac{V_{ab}}{R_1} + \frac{V_{ab}}{R_2} + \frac{V_{ab}}{R_3}$$

Dividing by V_{ab} gives

$$\frac{I}{V_{ab}} = \frac{1}{R_1} + \frac{1}{R_2} + \frac{1}{R_3} \qquad \text{or} \qquad \frac{1}{R_{eq}} = \frac{1}{R_1} + \frac{1}{R_2} + \frac{1}{R_3}$$

since V_{ab}/I is by definition the equivalent resistance R_{eq} of the network.

28.2 [II] As shown in Fig. 28-2(*a*), a battery (internal resistance 1 Ω) is connected in series with two resistors. Compute (*a*) the current in the circuit, (*b*) the p.d. across each resistor, and (*c*) the terminal p.d. of the battery.

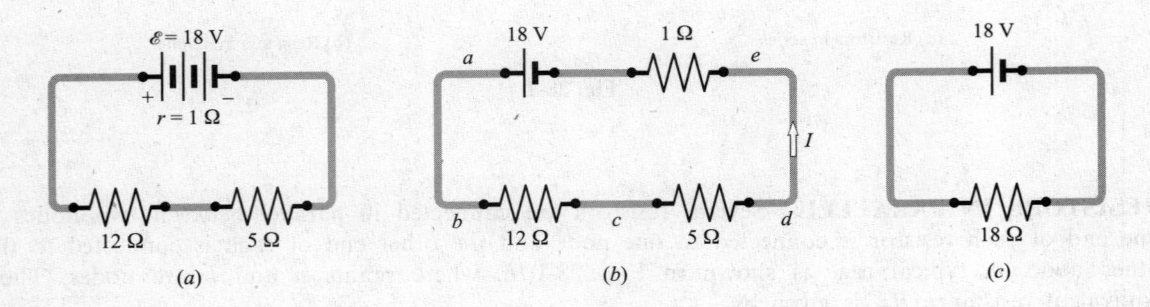

Fig. 28-2

The circuit is redrawn in Fig. 28-2(*b*) so as to show the battery resistance. We have

$$R_{eq} = 5\ \Omega + 12\ \Omega + 1\ \Omega = 18\ \Omega$$

Hence the circuit is equivalent to the one shown in Fig. 28-2(*c*). Applying $V = IR$ to it, we have:

(*a*)
$$I = \frac{V}{R} = \frac{18\ \text{V}}{18\ \Omega} = 1.0\ \text{A}$$

(*b*) Since $I = 1.0$ A, we can find the p.d. from point *b* to point *c* as

$$V_{bc} = IR_{bc} = (1.0\ \text{A})(12\ \Omega) = 12\ \text{V}$$

and that from *c* to *d* as

$$V_{cd} = IR_{cd} = (1.0\ \text{A})(5\ \Omega) = 5\ \text{V}$$

Notice that I is the same at all points in a series circuit.

(*c*) The terminal p.d. of the battery is the p.d. from *a* to *e*. Therefore,

$$\text{Terminal p.d.} = V_{bc} + V_{cd} = 12 + 5 = 17\ \text{V}$$

Or, we could start at *e* and keep track of the voltage changes as we go through the battery from *e* to *a*. Taking voltage drops as negative, we have

$$\text{Terminal p.d.} = -Ir + \mathscr{E} = -(1.0\ \text{A})(1\ \Omega) + 18\ \text{V} = 17\ \text{V}$$

28.3 [II] A 120-V house circuit has the following light bulbs turned on: 40.0 W, 60.0 W, and 75.0 W. Find the equivalent resistance of these lights.

House circuits are so constructed that each device is connected in parallel with the others. From $\text{P} = VI = V^2/R$, we have for the first bulb

$$R_1 = \frac{V^2}{\text{P}_1} = \frac{(120\ \text{V})^2}{40.0\ \text{W}} = 360\ \Omega$$

Similarly, $R_2 = 240\ \Omega$ and $R_3 = 192\ \Omega$. Because devices in a house circuit are in parallel,

$$\frac{1}{R_{\text{eq}}} = \frac{1}{360\ \Omega} + \frac{1}{240\ \Omega} + \frac{1}{192\ \Omega} \quad \text{or} \quad R_{\text{eq}} = 82.3\ \Omega$$

As a check, we note that the total power drawn from the line is 40.0 W + 60.0 W + 75.0 W = 175.0 W. Then, using $\text{P} = V^2/R$, we have

$$R_{\text{eq}} = \frac{V^2}{\text{total power}} = \frac{(120\ \text{V})^2}{175.0\ \text{W}} = 82.3\ \Omega$$

28.4 [I] What resistance must be placed in parallel with 12 Ω to obtain a combined resistance of 4 Ω?

From

$$\frac{1}{R_{\text{eq}}} = \frac{1}{R_1} + \frac{1}{R_2}$$

we have

$$\frac{1}{4\ \Omega} = \frac{1}{12\ \Omega} + \frac{1}{R_2}$$

so

$$R_2 = 6\ \Omega$$

28.5 [II] Several 40-Ω resistors are to be connected so that 15 A flows from a 120-V source. How can this be done?

The equivalent resistance must be such that 15 A flows from 120 V. Thus,

$$R_{\text{eq}} = \frac{V}{I} = \frac{120\ \text{V}}{15\ \text{A}} = 8\ \Omega$$

The resistors must be in parallel, since the combined resistance is to be smaller than any of them. If the required number of 40-Ω resistors is n, then we have

$$\frac{1}{80\ \Omega} = n\left(\frac{1}{40\ \Omega}\right) \quad \text{or} \quad n = 5$$

28.6 [II] For each circuit shown in Fig. 28-3, determine the current I through the battery.

(*a*) The 3.0-Ω and 7.0-Ω resistors are in parallel; their joint resistance R_1 is found from

$$\frac{1}{R_1} = \frac{1}{3.0\ \Omega} + \frac{1}{7.0\ \Omega} = \frac{10}{21\ \Omega} \qquad \text{or} \qquad R_1 = 2.1\ \Omega$$

Then the equivalent resistance of the entire circuit is

$$R_{eq} = 2.1\ \Omega + 5.0\ \Omega + 0.4\ \Omega = 7.5\ \Omega$$

and the battery current is

$$I = \frac{\mathscr{E}}{R_{eq}} = \frac{30\ \text{V}}{7.5\ \Omega} = 4.0\ \text{A}$$

(*b*) The 7.0-Ω, 1.0-Ω, and 10.0-Ω resistors are in series; their joint resistance is 18.0 Ω. Then 18.0 Ω is in parallel with 6.0 Ω; their combined resistance R_1 is given by

$$\frac{1}{R_1} = \frac{1}{18.0\ \Omega} + \frac{1}{6.0\ \Omega} \qquad \text{or} \qquad R_1 = 4.5\ \Omega$$

Hence, the equivalent resistance of the entire circuit is

$$R_{eq} = 4.5\ \Omega + 2.0\ \Omega + 8.0\ \Omega + 0.3\ \Omega = 14.8\ \Omega$$

and the battery current is

$$I = \frac{\mathscr{E}}{R_{eq}} = \frac{20\ \text{V}}{14.8\ \Omega} = 1.4\ \text{A}$$

(*c*) The 5.0-Ω and 19.0-Ω resistors are in series; their joint resistance is 24.0 Ω. Then 24.0 Ω is in parallel with 8.0 Ω; their joint resistance R_1 is given by

$$\frac{1}{R_1} = \frac{1}{24.0\ \Omega} + \frac{1}{8.0\ \Omega} \qquad \text{or} \qquad R_1 = 6.0\ \Omega$$

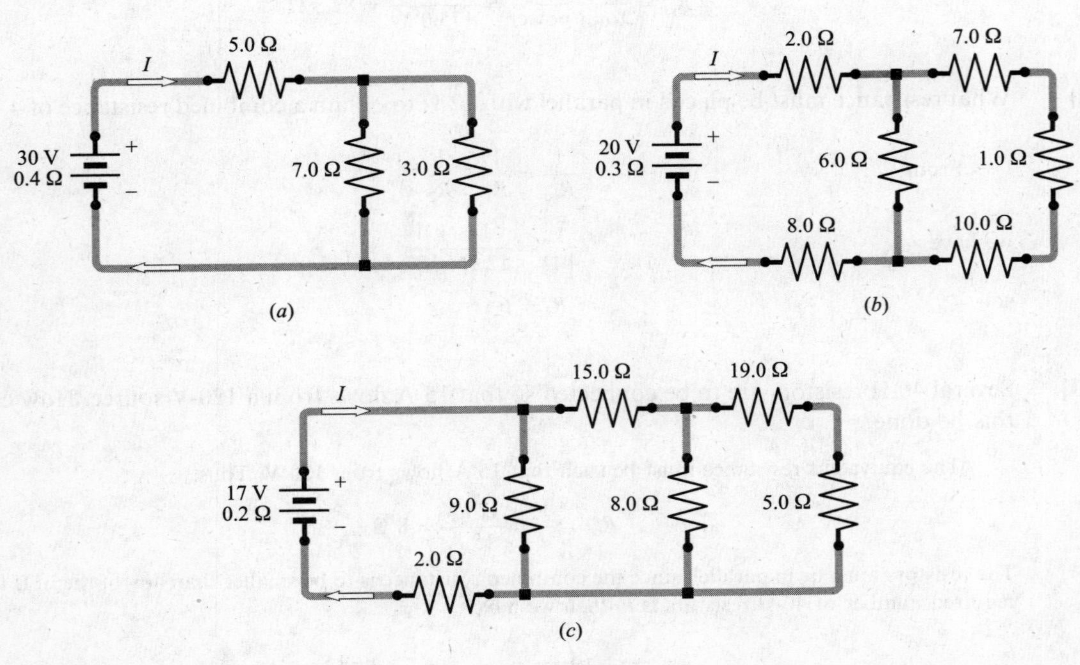

Fig. 28-3

Now $R_1 = 6.0\ \Omega$ is in series with 15.0 Ω; their joint resistance is $6.0\ \Omega + 15.0\ \Omega = 21.0\ \Omega$. Thus 21.0 Ω is in parallel with 9.0 Ω; their combined resistance is found from

$$\frac{1}{R_2} = \frac{1}{21.0\ \Omega} + \frac{1}{9.0\ \Omega} \qquad \text{or} \qquad R_2 = 6.3\ \Omega$$

Hence the equivalent resistance of the entire circuit is

$$R_{eq} = 6.3\ \Omega + 2.0\ \Omega + 0.2\ \Omega = 8.5\ \Omega$$

and the battery current is

$$I = \frac{\mathscr{E}}{R_{eq}} = \frac{17\ \text{V}}{8.5\ \Omega} = 2.0\ \text{A}$$

28.7 [II] For the circuit shown in Fig. 28-4, find the current in each resistor and the current drawn from the 40-V source.

Notice that the p.d. from a to b is 40 V. Therefore, the p.d. across each resistor is 40 V. Then,

$$I_2 = \frac{40\ \text{V}}{2.0\ \Omega} = 20\ \text{A} \qquad I_5 = \frac{40\ \text{V}}{5.0\ \Omega} = 8.0\ \text{A} \qquad I_8 = \frac{40\ \text{V}}{8.0\ \Omega} = 5.0\ \text{A}$$

Because I splits into three currents.

$$I = I_2 + I_5 + I_8 = 20\ \text{A} + 8.0\ \text{A} + 5.0\ \text{A} = 33\ \text{A}$$

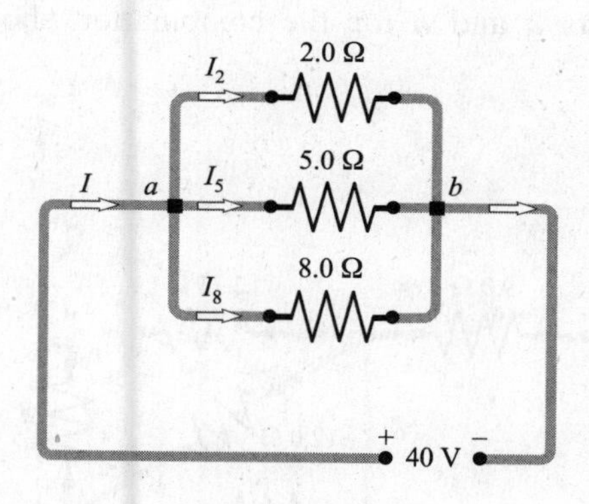

Fig. 28-4

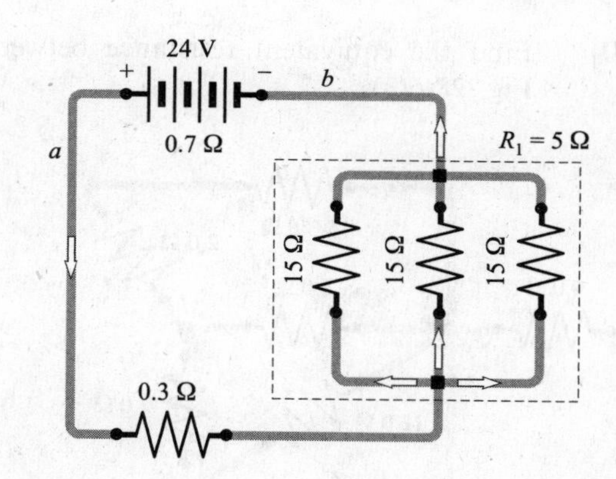

Fig. 28-5

28.8 [II] In Fig. 28-5, the battery has an internal resistance of 0.7 Ω. Find (*a*) the current drawn from the battery, (*b*) the current in each 15-Ω resistor, and (*c*) the terminal voltage of the battery.

(*a*) First we'll have to find the equivalent resistance of the entire circuit, and with that and Ohm's Law, determine the current. For parallel group resistance R_1 we have

$$\frac{1}{R_1} = \frac{1}{15\ \Omega} + \frac{1}{15\ \Omega} + \frac{1}{15\ \Omega} = \frac{3}{15\ \Omega} \qquad \text{or} \qquad R_1 = 5.0\ \Omega$$

Then

$$R_{eq} = 5.0\ \Omega + 0.3\ \Omega + 0.7\ \Omega = 6.0\ \Omega$$

and

$$I = \frac{\mathscr{E}}{R_{eq}} = \frac{24\ \text{V}}{6.0\ \Omega} = 4.0\ \text{A}$$

(*b*) **Method 1**

The three-resistor combination is equivalent to $R_1 = 5.0\ \Omega$. A current of 4.0 A flows through it. Hence, the p.d. across the combination is

$$IR_1 = (4.0\ \text{A})(5.0\ \Omega) = 20\ \text{V}$$

This is also the p.d. across each 15-Ω resistor. Therefore, the current through each 15-Ω resistor is

$$I_{15} = \frac{V}{R} = \frac{20\ \text{V}}{15\ \Omega} = 1.3\ \text{A}$$

Method 2

In this special case, we know that one-third of the current will go through each 15-Ω resistor. Hence

$$I_{15} = \frac{4.0\ \text{A}}{3} = 1.3\ \text{A}$$

(*c*) We start at *a* and go to *b* outside the battery:

$$V \text{ from } a \text{ to } b = -(4.0\ \text{A})(0.3\ \Omega) - (4.0\ \text{A})(5.0\ \Omega) = -21.2\ \text{V}$$

The terminal p.d. of the battery is 21.2 V. Or, we could write for this case of a discharging battery,

$$\text{Terminal p.d.} = \mathscr{E} - Ir = 24\ \text{V} - (4.0\ \text{A})(0.7\ \Omega) = 21.2\ \text{V}$$

28.9 [II] Find the equivalent resistance between points *a* and *b* for the combination shown in Fig. 28-6(*a*).

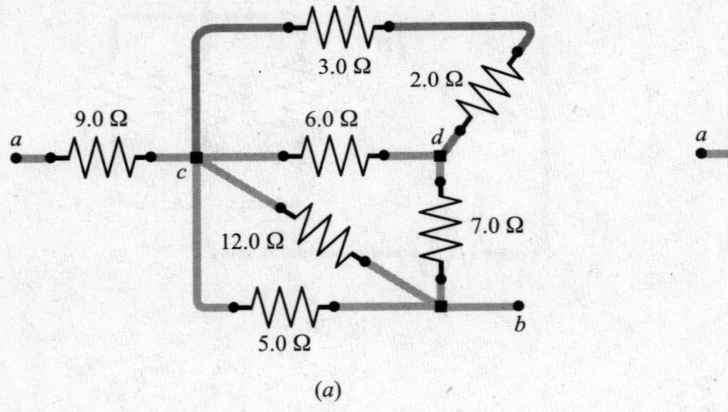

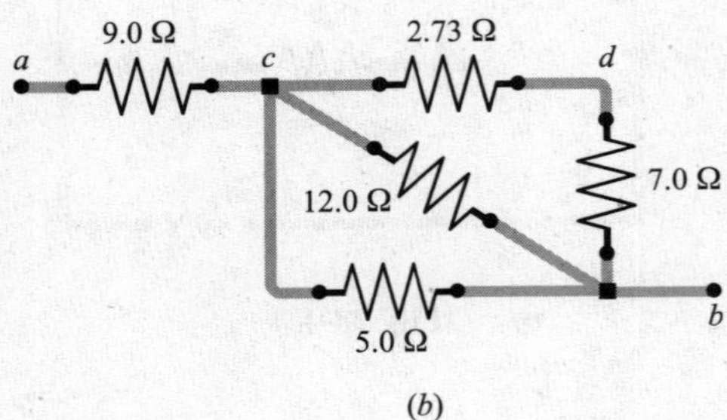

Fig. 28-6

The 3.0-Ω and 2.0-Ω resistors are in series and are equivalent to a 5.0-Ω resistor. The equivalent 5.0 Ω is in parallel with the 6.0 Ω, and their equivalent, R_1, is

$$\frac{1}{R_1} = \frac{1}{5.0\ \Omega} + \frac{1}{6.0\ \Omega} = 0.20 + 0.167 = 0.367\ \Omega^{-1} \qquad \text{or} \qquad R_1 = 2.73\ \Omega$$

The circuit thus far reduced is shown in Fig. 28-6(*b*).

The 7.0 Ω and 2.73 Ω are equivalent to 9.73 Ω. Now the 5.0 Ω, 12.0 Ω, and 9.73 Ω are in parallel, and their equivalent, R_2, is

$$\frac{1}{R_2} = \frac{1}{5.0\ \Omega} + \frac{1}{12.0\ \Omega} + \frac{1}{9.73\ \Omega} = 0.386\ \Omega^{-1} \qquad \text{or} \qquad R_2 = 2.6\ \Omega$$

This 2.6 Ω is in series with the 9.0-Ω resistor. Therefore, the equivalent resistance of the combination is 9.0 Ω + 2.6 Ω = 11.6 Ω.

28.10 [II] A current of 5.0 A flows into the circuit in Fig. 28-6(a) at point-a and out at point-b. (a) What is the potential difference from a to b? (b) How much current flows through the 12.0-Ω resistor?

In Problem 28.9, we found that the equivalent resistance for this combination is 11.6 Ω, and we are told the current through it is 5.0 A.

(a)
$$\text{Voltage drop from } a \text{ to } b = IR_{\text{eq}} = (5.0\text{ A})(11.6\ \Omega) = 58\text{ V}$$

(b) The voltage drop from a to c is (5.0 A)(9.0 Ω) = 45 V. Hence, from part (a), the voltage drop from c to b is

$$58\text{ V} - 45\text{ V} = 13\text{ V}$$

and the current in the 12.0-Ω resistor is

$$I_{12} = \frac{V}{R} = \frac{13\text{ V}}{12\ \Omega} = 1.1\text{ A}$$

28.11 [II] As shown in Fig. 28-7, the current I divides into I_1 and I_2. Find I_1 and I_2 in terms of I, R_1, and R_2.

The potential drops across R_1 and R_2 are the same because the resistors are in parallel, so

$$I_1R_1 = I_2R_2$$

But $I = I_1 + I_2$ and so $I_2 = I - I_1$. Substituting in the first equation gives

$$I_1R_1 = (I - I_1)R_2 = IR_2 - I_1R_2 \qquad \text{or} \qquad I_1 = \frac{R_2}{R_1 + R_2}I$$

Using this result together with the first equation gives

$$I_2 = \frac{R_1}{R_2}I_1 = \frac{R_1}{R_1 + R_2}I$$

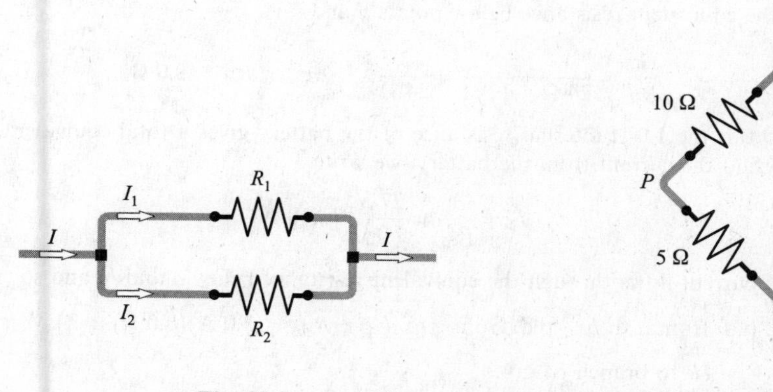

Fig. 28-7

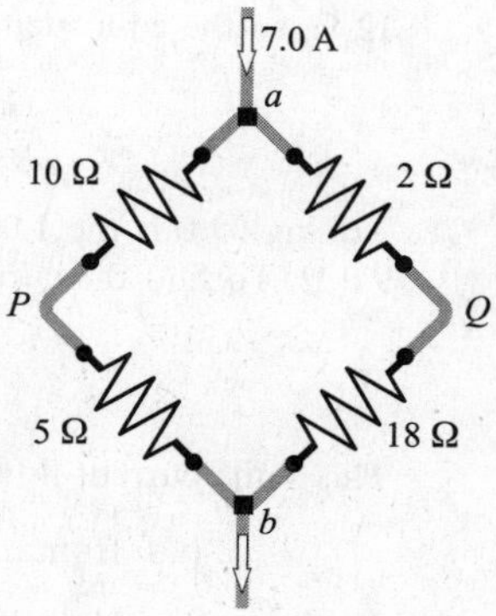

Fig. 28-8

28.12 [II] Find the potential difference between points P and Q in Fig. 28-8. Which point is at the higher potential?

From the result of Problem 28.11, the currents through P and Q are

$$I_P = \frac{2\ \Omega + 18\ \Omega}{10\ \Omega + 5\ \Omega + 2\ \Omega + 18\ \Omega}(7.0\ \text{A}) = 4.0\ \text{A}$$

$$I_Q = \frac{10\ \Omega + 5\ \Omega}{10\ \Omega + 5\ \Omega + 2\ \Omega + 18\ \Omega}(7.0\ \text{A}) = 3.0\ \text{A}$$

Now we start at point P and go through point a to point Q, to find

$$\text{Voltage change from } P \text{ to } Q = +(4.0\ \text{A})(10\ \Omega) - (3.0\ \text{A})(2\ \Omega) = +34\ \text{V}$$

(Notice that we go through a potential rise from P to a because we are going against the current. From a to Q there is a drop.) Therefore, the voltage difference between P and Q is 34 V, with Q being at the higher potential.

28.13 [II] For the circuit of Fig. 28-9(*a*), find (*a*) I_1 I_2, and I_3; (*b*) the current in the 12-Ω resistor.

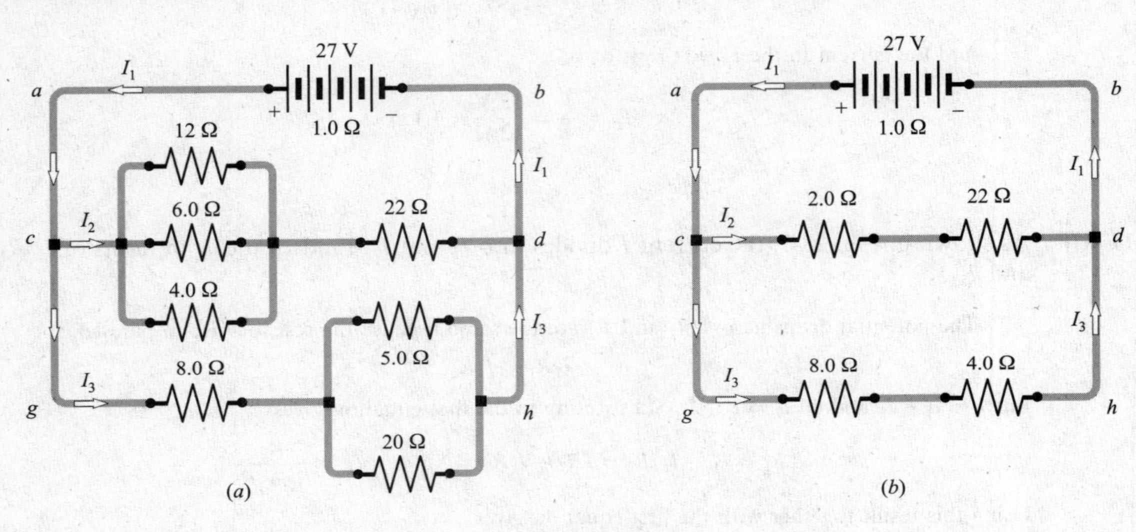

Fig. 28-9

(*a*) The circuit reduces at once to that shown in Fig. 28-9(*b*). There we have 24 Ω in parallel with 12 Ω, so the equivalent resistance below points a and b is

$$\frac{1}{R_{ab}} = \frac{1}{24\ \Omega} + \frac{1}{12\ \Omega} = \frac{3}{24\ \Omega} \qquad \text{or} \qquad R_{ab} = 8.0\ \Omega$$

Adding to this the 1.0-Ω internal resistance of the battery gives a total equivalent resistance of 9.0 Ω. To find the current from the battery, we write

$$I_1 = \frac{\mathscr{E}}{R_{\text{eq}}} = \frac{27\ \text{V}}{9.0\ \Omega} = 3.0\ \text{A}$$

This same current flows through the equivalent resistance below a and b, and so

$$\text{p.d. from } a \text{ to } b = \text{p.d. from } c \text{ to } d = I_1 R_{ab} = (3.0\ \text{A})(8.0\ \Omega) = 24\ \text{V}$$

Applying $V = IR$ to branch cd gives

$$I_2 = \frac{V_{cd}}{R_{cd}} = \frac{24\ \text{V}}{24\ \Omega} = 1.0\ \text{A}$$

Similarly,
$$I_3 = \frac{V_{gh}}{R_{gh}} = \frac{24\ \text{V}}{12\ \Omega} = 2.0\ \text{A}$$

As a check, we note that $I_2 + I_3 = 3.0\ \text{A} = I_1$, as should be.

(*b*) Because $I_2 = 1.0$ A, the p.d. across the 2.0-Ω resistor in Fig. 28-9(*b*) is $(1.0\ \text{A})(2.0\ \Omega) = 2.0$ V. But this is also the p.d. across the 12-Ω resistor in Fig. 28-9(*a*). Applying $V = IR$ to the 12 Ω gives

$$I_{12} = \frac{V_{12}}{R} = \frac{2.0\ \text{V}}{12\ \Omega} = 0.17\ \text{A}$$

28.14 [II] A galvanometer has a resistance of 400 Ω and deflects full scale for a current of 0.20 mA through it. How large a shunt resistor is required to change it to a 3.0 A ammeter?

In Fig. 28-10 we label the galvanometer G and the shunt resistance R_s. At full scale deflection, the currents are as shown.

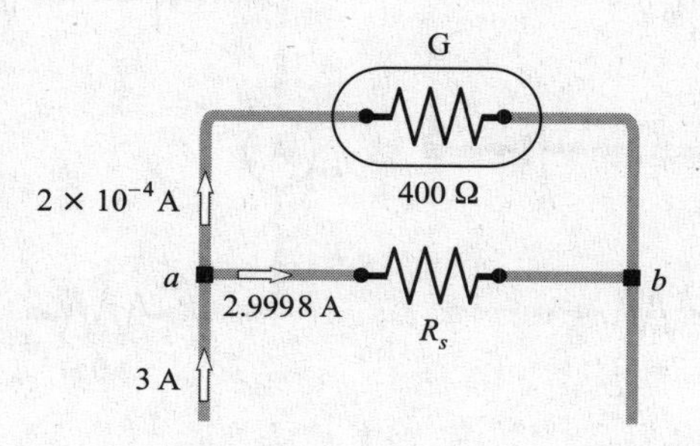

Fig. 28-10

The voltage drop from *a* to *b* across G is the same as that across R_s. Therefore,

$$(2.9998\ \text{A})R_s = (2.0 \times 10^{-4}\ \text{A})(400\ \Omega)$$

from which $R_s = 0.027\ \Omega$.

28.15 [II] A voltmeter is to deflect full scale for a potential difference of 5.000 V across it and is to be made by connecting a resistor R_x in series with a galvanometer. The 80.00-Ω galvanometer deflects full scale for a potential of 20.00 mV across it. Find R_x.

When the galvanometer is deflecting full scale, the current through it is

$$I = \frac{V}{R} = \frac{20.00 \times 10^{-3}\ \text{V}}{80.00\ \Omega} = 2.500 \times 10^{-4}\ \text{A}$$

When R_x is connected in series with the galvanometer, we wish I to be 2.500×10^{-4} A for a potential difference of 5.000 V across the combination. Hence, $V = IR$ becomes

$$5.000\ \text{V} = (2.500 \times 10^{-4}\ \text{A})(80.00\ \Omega + R_x)$$

from which $R_x = 19.92\ \text{k}\Omega$.

28.16 [III] The currents in the circuit in Fig. 28-11 are steady. Find I_1, I_2, I_3, and the charge on the capacitor.

When a capacitor has a constant charge, as it does here, the current flowing to it is zero. Therefore $I_2 = 0$, and the circuit behaves just as though the center wire were missing.

With the center wire missing, the remaining circuit is simply 12 Ω connected across a 15-V battery. Therefore,

$$I_1 = \frac{\mathscr{E}}{R} = \frac{15\ \text{V}}{12\ \Omega} = 1.25\ \text{A}$$

In addition, because $I_2 = 0$, we have $I_3 = I_1 = 1.3$ A.

To find the charge on the capacitor, we first find the voltage difference between points a and b. We start at a and go around the upper path.

$$\text{Voltage change from } a \text{ to } b = -(5.0\ \Omega)I_3 + 6.0\ \text{V} + (3.0\ \Omega)I_2$$
$$= -(5.0\ \Omega)(1.25\ \text{A}) + 6.0\ \text{V} + (3.0\ \Omega)(0) = -0.25\ \text{V}$$

Therefore b is at the lower potential and the capacitor plate at b is negative. To find the charge on the capacitor, we write

$$Q = CV_{ab} = (2 \times 10^{-6}\ \text{F})(0.25\ \text{V}) = 0.5\ \mu\text{C}$$

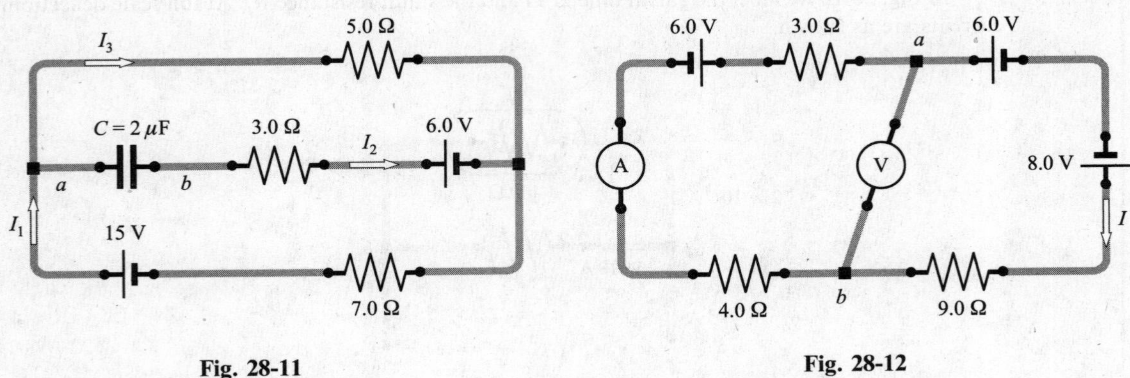

Fig. 28-11

Fig. 28-12

28.17 [II] Find the ammeter reading and the voltmeter reading in the circuit in Fig. 28-12. Assume both meters to be ideal.

The ideal voltmeter has infinite resistance and so its wire can be removed without altering the circuit. The ideal ammeter has zero resistance. It can be shown (see Chapter 29) that batteries in series simply add or subtract. The two 6.0-V batteries cancel each other because they tend to push current in opposite directions. As a result, the circuit behaves as though it had a single 8.0-V battery that causes a clockwise current.

The equivalent resistance is $3.0\ \Omega + 4.0\ \Omega + 9.0\ \Omega = 16.0\ \Omega$, and the equivalent battery is 8.0 V. Therefore,

$$I = \frac{\mathscr{E}}{R} = \frac{8.0\ \text{V}}{16\ \Omega} = 0.50\ \text{A}$$

and this is what the ammeter will read.

Adding up the voltage changes from a to b around the right-hand side of the circuit gives

$$\text{Voltage change from } a \text{ to } b = -6.0\ \text{V} + 8.0\ \text{V} - (0.50\ \text{A})(9.0\ \Omega) = -2.5\ \text{V}$$

Therefore, a voltmeter connected from a to b will read 2.5 V, with b being at the lower potential.

Supplementary Problems

28.18 [I] Compute the equivalent resistance of 4.0 Ω and 8.0 Ω (*a*) in series and (*b*) in parallel. *Ans.* (*a*) 12 Ω; (*b*) 2.7 Ω

28.19 [I] Compute the equivalent resistance of (*a*) 3.0 Ω, 6.0 Ω, and 9.0 Ω in parallel; (*b*) 3.0 Ω, 4.0 Ω, 7.0 Ω, 10.0 Ω, and 12.0 Ω in parallel; (*c*) three 33-Ω heating elements in parallel; (*d*) twenty 100-Ω lamps in parallel. *Ans.* (*a*) 1.6 Ω; (*b*) 1.1 Ω; (*c*) 11 Ω; (*d*) 5.0 Ω

28.20 [I] What resistance must be placed in parallel with 20 Ω to make the combined resistance 15 Ω? *Ans.* 60 Ω

28.21 [II] How many 160-Ω resistors (in parallel) are required to carry a total of 5.0 A on a 100-V line? *Ans.* 8

28.22 [II] Three resistors, of 8.0 Ω, 12 Ω, and 24 Ω, are in parallel, and a current of 20 A is drawn by the combination. Determine (*a*) the potential difference across the combination and (*b*) the current through each resistance. *Ans.* (*a*) 80 V; (*b*) 10 A, 6.7 A, 3.3 A

28.23 [II] By use of one or more of the three resistors 3.0 Ω, 5.0 Ω, and 6.0 Ω, a total of 18 resistances can be obtained. What are they? *Ans.* 0.70 Ω, 1.4 Ω, 1.9 Ω, 2.0 Ω, 2.4 Ω, 2.7 Ω, 3.0 Ω, 3.2 Ω, 3.4 Ω, 5.0 Ω, 5.7 Ω, 6.0 Ω, 7.0 Ω, 7.9 Ω, 8.0 Ω, 9.0 Ω, 11 Ω, 14 Ω

28.24 [II] Two resistors, of 4.00 Ω and 12.0 Ω, are connected in parallel across a 22-V battery having internal resistance 1.00 Ω. Compute (*a*) the battery current, (*b*) the current in the 4.00-Ω resistor, (*c*) the terminal voltage of the battery, (*d*) the current in the 12.0-Ω resistor. *Ans.* (*a*) 5.5 A; (*b*) 4.1 A; (*c*) 17 V; (*d*) 1.4 A

28.25 [II] Three resistors, of 40 Ω, 60 Ω, and 120 Ω, are connected in parallel, and this parallel group is connected in series with 15 Ω in series with 25 Ω. The whole system is then connected to a 120-V source. Determine (*a*) the current in the 25 Ω, (*b*) the potential drop across the parallel group, (*c*) the potential drop across the 25 Ω, (*d*) the current in the 60 Ω, (*e*) the current in the 40 Ω. *Ans.* (*a*) 2.0 A; (*b*) 40 V; (*c*) 50 V; (*d*) 0.67 A; (*e*) 1.0 A

28.26 [II] What shunt resistance should be connected in parallel with an ammeter having a resistance of 0.040 Ω so that 25 percent of the total current will pass through the ammeter? *Ans.* 0.013 Ω

28.27 [II] A 36-Ω galvanometer is shunted by a resistor of 4.0 Ω. What part of the total current will pass through the instrument? *Ans.* 1/10

28.28 [II] A relay having a resistance of 6.0 Ω operates with a minimum current of 0.030 A. It is required that the relay operate when the current in the line reaches 0.240 A. What resistance should be used to shunt the relay? *Ans.* 0.86 Ω

28.29 [II] Show that if two resistors are connected in parallel, the rates at which they produce thermal energy vary inversely as their resistances.

28.30 [II] For the circuit shown in Fig. 28-13, find the current through each resistor and the potential drop across each resistor. *Ans.* for 20 Ω, 3.0 A and 60 V; for 75 Ω, 2.4 A and 180 V; for 300 Ω, 0.6 A and 180 V

28.31 [II] For the circuit shown in Fig. 28-14, find (*a*) its equivalent resistance; (*b*) the current drawn from the power source; (*c*) the potential differences across *ab*, *cd*, and *de*; (*d*) the current in each resistor. *Ans.* (*a*) 15 Ω; (*b*) 20 A; (*c*) $V_{ab} = 80$ V, $V_{cd} = 120$ V, $V_{de} = 100$ V; (*d*) $I_4 = 20$ A, $I_{10} = 12$ A, $I_{15} = 8$ A, $I_9 = 11.1$ A, $I_{18} = 5.6$ A, $I_{30} = 3.3$ A

28.32 [II] It is known that the potential difference across the 6.0-Ω resistance in Fig. 28-15 is 48 V. Determine (*a*) the entering current *I*, (*b*) the potential difference across the 8.0-Ω resistance, (*c*) the potential difference across the 10-Ω resistance, (*d*) the potential difference from *a* to *b*. (*Hint*: The wire connecting *c* and *d* can be shrunk to zero length without altering the currents or potentials.) *Ans.* (*a*) 12 A; (*b*) 96 V; (*c*) 60 V; (*d*) 204 V

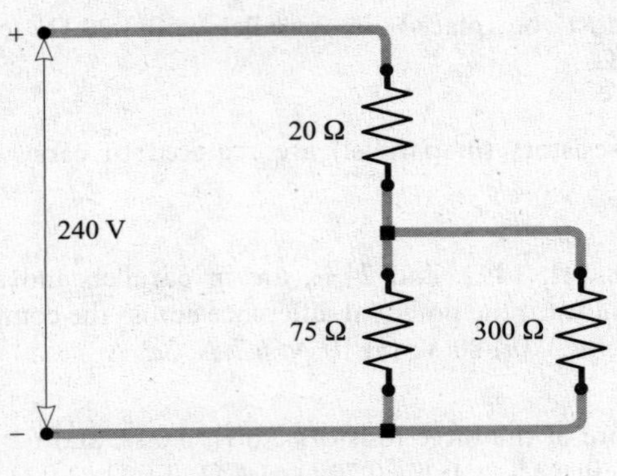

Fig. 28-13

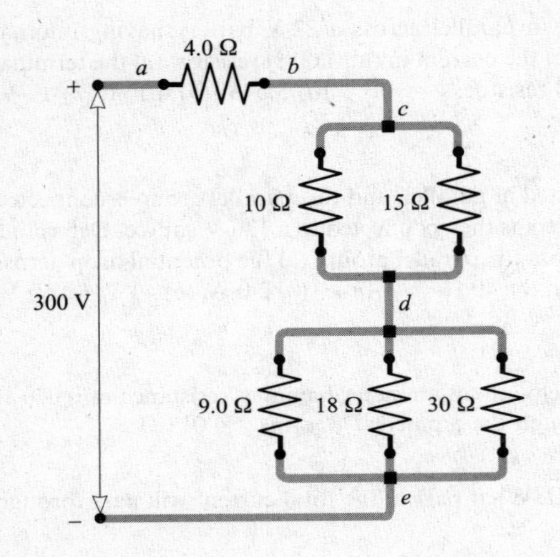

Fig. 28-14

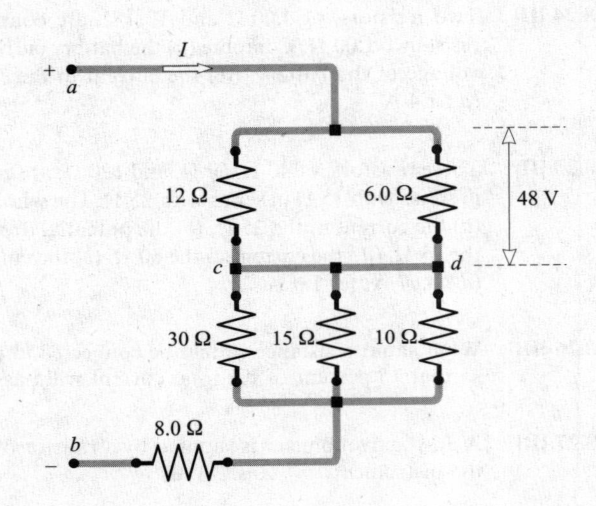

Fig. 28-15

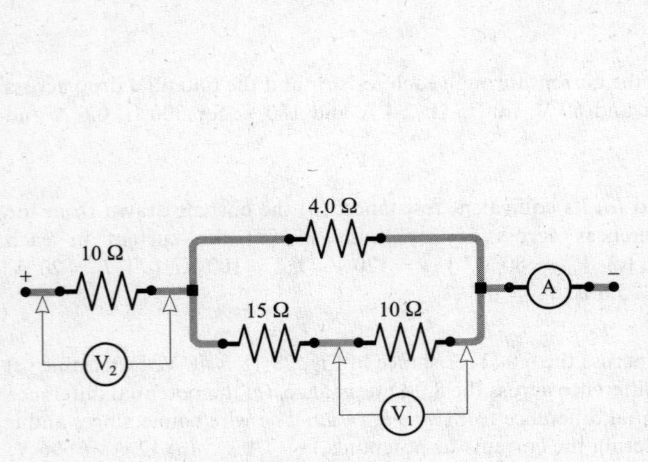

Fig. 28-16

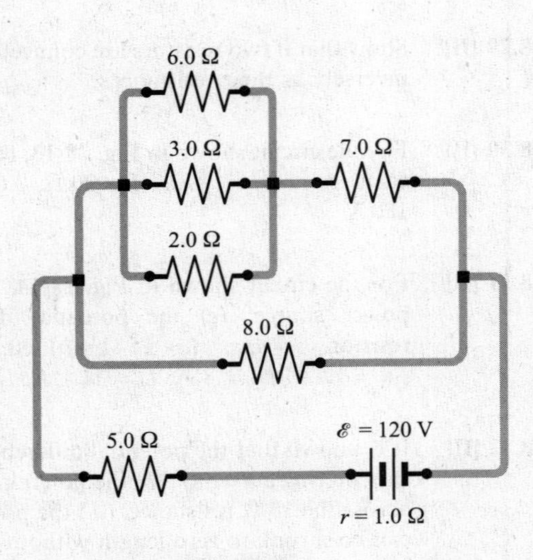

Fig. 28-17

28.33 [II] In the circuit shown in Fig. 28-16, 23.9 calories of thermal energy are produced each second in the 4.0-Ω resistor. Assuming the ammeter and two voltmeters to be ideal, what will be their readings?
Ans. 5.8 A, 8.0 V, 58 V

28.34 [II] For the circuit shown in Fig. 28-17, find (*a*) the equivalent resistance; (*b*) the currents through the 5.0-Ω, 7.0-Ω, and 3.0-Ω resistors; (*c*) the total power delivered by the battery to the circuit. *Ans.* (*a*) 10 Ω; (*b*) 12 A, 6.0 A, 2.0 A; (*c*) 1.3 kW

28.35 [II] In the circuit shown in Fig. 28-18, the ideal ammeter registers 2.0 A. (*a*) Assuming XY to be a resistance, find its value. (*b*) Assuming XY to be a battery (with 2.0-Ω internal resistance) that is being charged, find its emf. (*c*) Under the conditions of part (*b*), what is the potential change from point Y to point X? *Ans.* (*a*) 5.0 Ω; (*b*) 6.0 V; (*c*) −10 V

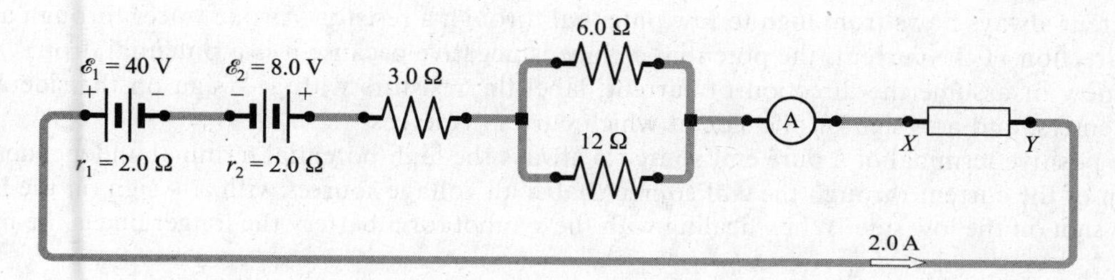

Fig. 28-18

28.36 II] The *Wheatstone bridge* shown in Fig. 28-19 is being used to measure resistance X. At balance, the current through the galvanometer G is zero and resistances L, M, and N are 3.0 Ω, 2.0 Ω, and 10 Ω, respectively. Find the value of X. *Ans.* 15 Ω

28.37 [II] The slidewire Wheatstone bridge shown in Fig. 28-20 is balanced (refer back to Problem 28.36) when the uniform resistive slide wire AB is divided as shown. Find the value of the resistance X.
Ans. 2 Ω

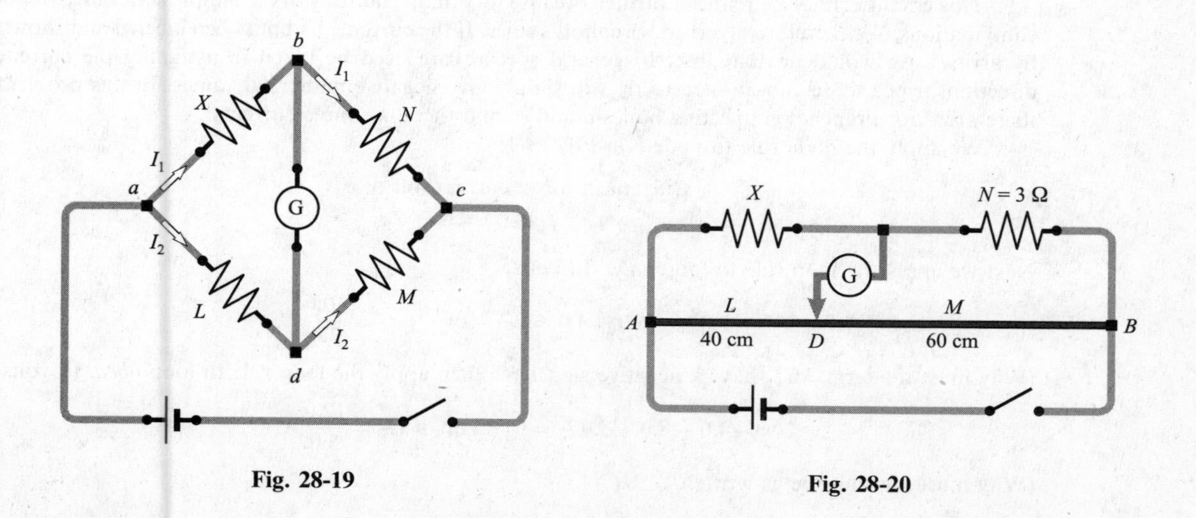

Fig. 28-19 **Fig. 28-20**

Chapter 29

Kirchhoff's Laws

KIRCHHOFF'S NODE (OR JUNCTION) RULE: The sum of all the currents coming into a ***node*** (i.e., a junction where three or more current-carrying leads or ***branches*** attach) must equal the sum of all the currents leaving that node. If we designate the currents-in as positive and the currents-out as negative, then *the sum of the currents equals zero* is a common alternative statement of the rule.

KIRCHHOFF'S LOOP (OR CIRCUIT) RULE: As one traces around any closed path (or ***loop***) in a circuit, the algebraic sum of the potential changes encountered is zero. In this sum, a potential (i.e., voltage) rise is positive and a potential drop is negative.

Current always flows from high to low potential through a resistor. As one traces through a resistor in the direction of the current, the potential change is negative because it is a potential drop. Once you either know or assume the direction of current, label the resistors with a + sign on the side at which current enters, and a − sign on the side at which current emerges.

The positive terminal of a pure emf source is always the high-potential terminal, independent of the direction of the current through the emf source. Label all voltage sources with a + sign on the high side and a − sign on the low side. When dealing with the symbol for a battery the longer line is the high side.

THE SET OF EQUATIONS OBTAINED by use of Kirchhoff's loop rule will be independent provided that each new loop equation contains at least one voltage change not included in a previous equation.

Solved Problems

29.1 [II] Find the currents in the circuit shown in Fig. 29-1.

Notice that the signs of the voltage drops have been provided in the circuit diagram. You will not need them in this solution but it's a good habit to put them in as a first step.

This circuit cannot be reduced further because it contains no resistors in simple series or parallel combinations. We therefore revert to Kirchhoff's rules. If the currents had not been labeled and shown by arrows, we would do that first. In general special care need be taken in assigning the current directions, since those chosen incorrectly will simply give negative numerical values. In this problem there are three branches connecting nodes-*a* and -*b* and therefore three currents.

We apply the node rule to node-*b* in Fig. 29-1:

$$\text{Current into } b = \text{current out of } b$$

$$I_1 + I_2 + I_3 = 0 \qquad (1)$$

Next we apply the loop rule to loop *adba*. In volts,

$$-7.0\,I_1 + 6.0 + 4.0 = 0 \qquad \text{or} \qquad I_1 = \frac{10.0}{7.0}\,\text{A}$$

(Why must the term $7.0\,I_1$ have a negative sign?) We then apply the loop rule to loop *abca*. In volts,

$$-4.0 - 8.0 + 5.0\,I_2 = 0 \qquad \text{or} \qquad I_2 = \frac{12.0}{5.0}\,\text{A}$$

(Why must the signs be as written?)

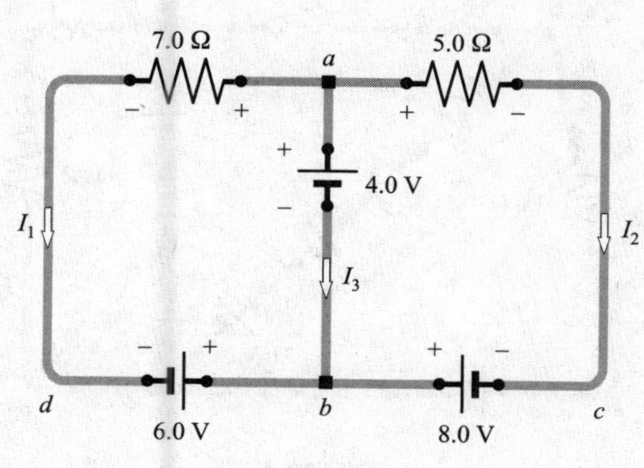

Fig. 29-1

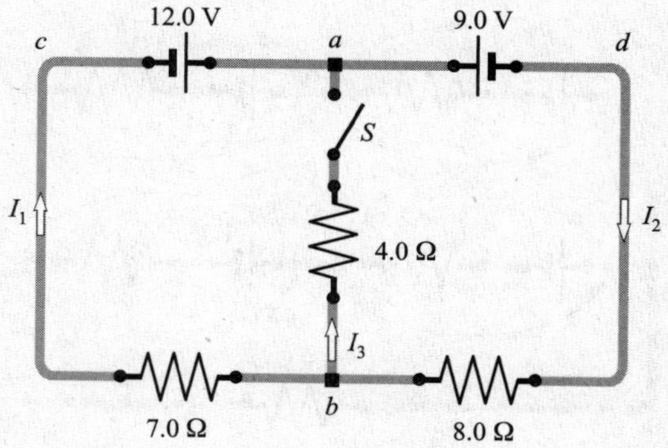

Fig. 29-2

Now we return to Eq. (*1*) to find

$$I_3 = -I_1 - I_2 = -\frac{10.0}{7.0} - \frac{12.0}{5.0} = \frac{-50 - 84}{35} = -3.8 \text{ A}$$

The minus sign tells us that I_3 is opposite in direction to that shown in the figure.

29.2 [II] For the circuit shown in Fig. 29-2, find I_1, I_2, and I_3 if switch S is (*a*) open and (*b*) closed.

(*a*) When S is open, $I_3 = 0$, because no current can flow through the middle branch. Applying the node rule to point-*a* we get

$$I_1 + I_3 = I_2 \qquad \text{or} \qquad I_2 = I_1 + 0 = I_1$$

Applying the loop rule to the outer loop *acbda* yields

$$-12.0 + 7.0\,I_1 + 8.0\,I_2 + 9.0 = 0 \tag{1}$$

To understand the use of signs, remember that current always flows from high to low potential through a resistor.

Because $I_2 = I_1$, Eq. (*1*) becomes

$$15.0\,I_1 = 3.0 \qquad \text{or} \qquad I_1 = 0.20 \text{ A}$$

Also, $I_2 = I_1 = 0.20$ A. Notice that this is the same result that one would obtain by replacing the two batteries by a single 3.0-V battery.

(*b*) With S closed, I_3 is no longer necessarily zero. Applying the node rule to point-*a* gives

$$I_1 + I_3 = I_2 \tag{2}$$

Applying the loop rule to loop *acba* we get

$$-12.0 + 7.0\,I_1 - 4.0\,I_3 = 0 \tag{3}$$

and to loop *adba* gives

$$-9.0 - 8.0\,I_2 - 4.0\,I_3 = 0 \tag{4}$$

Applying the loop rule to the remaining loop, *acbda*, would yield a redundant equation, because it would contain no new voltage change.

We must now solve Eqs. (*2*), (*3*), and (*4*) for I_1, I_2, and I_3. From Eq. (*4*),

$$I_3 = -2.0\,I_2 - 2.25$$

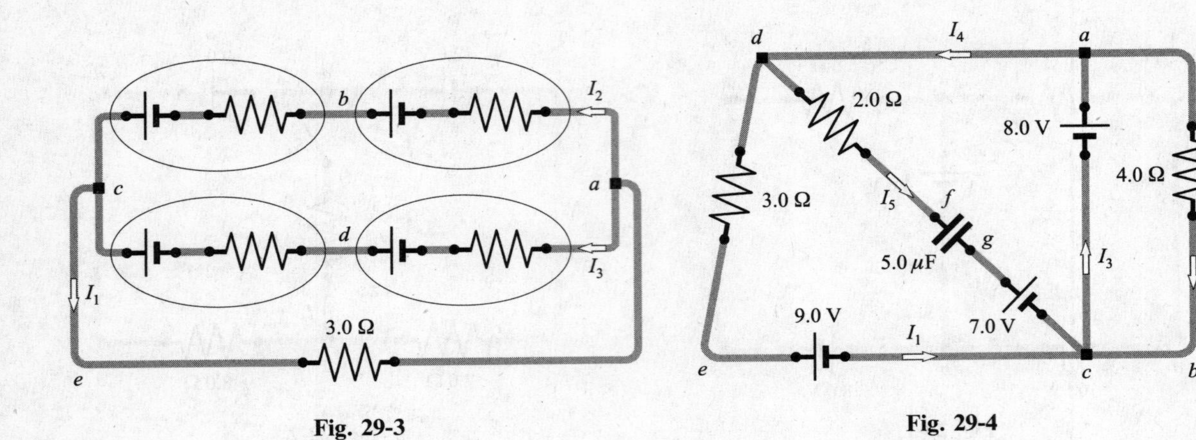

Fig. 29-3 **Fig. 29-4**

Substituting this in Eq. (*3*) yields

$$-12.0 + 7.0\,I_1 + 9.0 + 8.0\,I_2 = 0 \qquad \text{or} \qquad 7.0\,I_1 + 8.0\,I_2 = 3.0$$

Substituting for I_3 in Eq. (*2*) also gives

$$I_1 - 2.0\,I_2 - 2.25 = I_2 \qquad \text{or} \qquad I_1 = 3.0\,I_2 + 2.25$$

Substituting this value in the previous equation finally gives

$$21.0\,I_2 + 15.75 + 8.0\,I_2 = 3.0 \qquad \text{or} \qquad I_2 = -0.44 \text{ A}$$

Using this in the equation for I_1 gives

$$I_1 = 3.0(-0.44) + 2.25 = -1.32 + 2.25 = 0.93 \text{ A}$$

Notice that the minus sign is a part of the value we have found for I_2. It must be carried along with its numerical value. Now we can use (*2*) to find

$$I_3 = I_2 - I_1 = (-0.44) - 0.93 = -1.37 \text{ A}$$

29.3 [II] Each of the cells shown in Fig. 29-3 has an emf of 1.50 V and a 0.0750-Ω internal resistance. Find I_1, I_2, and I_3.

Applying the node rule to point-*a* gives

$$I_1 = I_2 + I_3 \tag{1}$$

Applying the loop rule to loop *abcea* yields, in volts,

$$-(0.0750)I_2 + 1.50 - (0.0750)I_2 + 1.50 - 3.00\,I_1 = 0$$

or

$$3.00\,I_1 + 0.150\,I_2 = 3.00 \tag{2}$$

Also, for loop *adcea*,

$$-(0.0750)I_3 + 1.50 - (0.0750)I_3 + 1.50 - 3.00\,I_1 = 0$$

or

$$3.00\,I_1 + 0.150\,I_3 = 3.00 \tag{3}$$

We solve Eq. (*2*) for $3.00\,I_1$ and substitute in Eq. (*3*) to get

$$3.00 - 0.150\,I_3 + 0.150\,I_2 = 3.00 \qquad \text{or} \qquad I_2 = I_3$$

as we might have guessed from the symmetry of the problem. Then Eq. (*1*) yields

$$I_1 = 2I_2$$

and substituting this in Eq. (*2*) we get

$$6.00\,I_2 + 0.150\,I_2 = 3.00 \qquad \text{or} \qquad I_2 = 0.488 \text{ A}$$

Then, $I_3 = I_2 = 0.488$ A and $I_1 = 2I_2 = 0.976$ A.

29.4 [III] The currents are steady in the circuit of Fig. 29-4. Find I_1, I_2, I_3, I_4, I_5, and the charge on the capacitor.

The capacitor passes no current when charged, and so $I_5 = 0$. Consider loop *acba*. The loop rule leads to

$$-8.0 + 4.0\,I_2 = 0 \qquad \text{or} \qquad I_2 = 2.0 \text{ A}$$

Using loop *adeca* gives

$$-3.0\,I_1 - 9.0 + 8.0 = 0 \qquad \text{or} \qquad I_1 = -0.33 \text{ A}$$

Applying the node rule at point-*c* results in

$$I_1 + I_5 + I_2 = I_3 \qquad \text{or} \qquad I_3 = 1.67 \text{ A} = 1.7 \text{ A}$$

and at point-*a*, it yields

$$I_3 = I_4 + I_2 \qquad \text{or} \qquad I_4 = -0.33 \text{ A}$$

(We should have realized this at once, because $I_5 = 0$ and so $I_4 = I_1$.)

To find the charge on the capacitor, we need the voltage V_{fg} across it. Put in all the signs on the resistors, batteries and capacitor. Applying the loop rule to loop *dfgced* gives

$$-2.0\,I_5 + V_{fg} - 7.0 + 9.0 + 3.0\,I_1 = 0 \qquad \text{or} \qquad 0 + V_{fg} - 7.0 + 9.0 - 1.0 = 0$$

from which $V_{fg} = -1.0$ V. The minus sign tells us that plate g is negative. The capacitor's charge is

$$Q = CV = (5.0\,\mu\text{F})(1.0 \text{ V}) = 5.0\,\mu\text{C}$$

29.5 [III] For the circuit shown in Fig. 29-5, the resistance R is 5.0 Ω and $\mathscr{E} = 20$ V. Find the readings of the ammeter and the voltmeter. Assume the meters to be ideal.

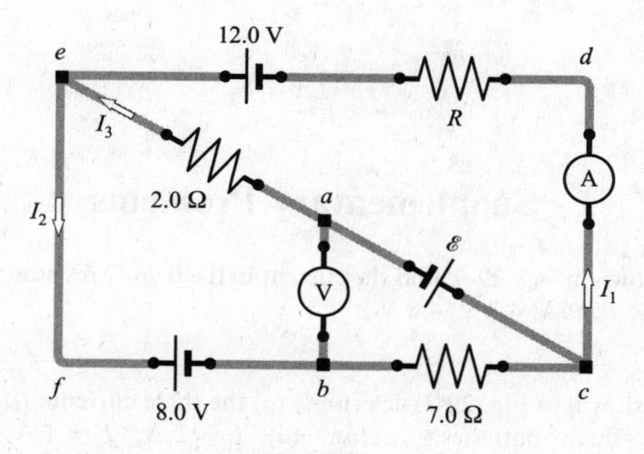

Fig. 29-5

The ideal voltmeter has infinite resistance (no current passes through it) and so it can be removed from the circuit with no effect. Write the loop equation for loop *cdefc*:

$$-RI_1 + 12.0 - 8.0 - 7.0\,I_2 = 0$$

which becomes

$$5.0\,I_1 + 7.0\,I_2 = 4.0 \tag{1}$$

Next write the loop equation for loop *cdeac*. It is

$$-5.0\,I_1 + 12.0 + 2.0\,I_3 + 20.0 = 0$$

$$5.0\,I_1 - 2.0\,I_3 = 32.0 \tag{2}$$

But the node rule applied at e gives

$$I_1 + I_3 = I_2 \tag{3}$$

Substituting Eq. (*3*) in Eq. (*1*) yields

$$5.0\,I_1 + 7.0\,I_1 + 7.0\,I_3 = 4.0$$

We solve this for I_3 and substitute in (*2*) to get

$$5.0\,I_1 - 2.0\left(\frac{4.0 - 12.0\,I_1}{7.0}\right) = 32.0$$

which yields $I_1 = 3.9$ A, which is the ammeter reading. Then Eq. (*1*) gives $I_2 = -2.2$ A.

To find the voltmeter reading V_{ab}, we write the loop equation for loop *abca*:

$$V_{ab} - 7.0\,I_2 - \mathscr{E} = 0$$

Substituting the known values of I_2 and $\mathscr{E}$, then solving, we obtain $V_{ab} = 4.3$ V. Since this is the potential difference between a to b, point b must be at the higher potential.

29.6 [III] In the circuit in Fig. 29-5, $I_1 = 0.20$ A and $R = 5.0\ \Omega$. Find $\mathscr{E}$.

We write the loop equation for loop *cdefc*:

$$-RI_1 + 12.0 - 8.0 - 7.0\,I_2 = 0 \qquad \text{or} \qquad -(5.0)(0.20) + 12.0 - 8.0 - 7.0\,I_2 = 0$$

from which $I_2 = 0.43$ A. We can now find I_3 by applying the node rule at e:

$$I_1 + I_3 = I_2 \qquad \text{or} \qquad I_3 = I_2 - I_1 = 0.23 \text{ A}$$

Now we apply the loop rule to loop *cdeac*:

$$-(5.0)(0.20) + 12.0 + (2.0)(0.23) + \mathscr{E} = 0$$

from which $\mathscr{E} = -11.5$ V. The minus sign tells us that the polarity of the battery is actually the reverse of that shown.

Supplementary Problems

29.7 [II] For the circuit shown in Fig. 29-6, find the current in the 0.96-Ω resistor and the terminal voltages of the batteries. *Ans.* 5.0 A, 4.8 V, 4.8 V

29.8 [III] For the network shown in Fig. 29-7, determine (*a*) the three currents I_1, I_2, and I_3, and (*b*) the terminal voltages of the three batteries. *Ans.* (*a*) $I_1 = 2$ A, $I_2 = 1$ A, $I_3 = -3$ A; (*b*) $V_{16} = 14$ V, $V_4 = 3.8$ V, $V_{10} = 8.5$ V

29.9 [II] Refer back to Fig. 29-5. If the voltmeter reads 16.0 V (with point-b at the higher potential) and $I_2 = 0.20$ A, find $\mathscr{E}$, R and the ammeter reading. *Ans.* 14.6 V, 0.21 Ω, 12 A

29.10 [III] Find I_1, I_2, I_3, and the potential difference between point-b to point-e in Fig. 29-8. *Ans.* 2.0 A, −8.0 A, 6.0 A, −13.0 V

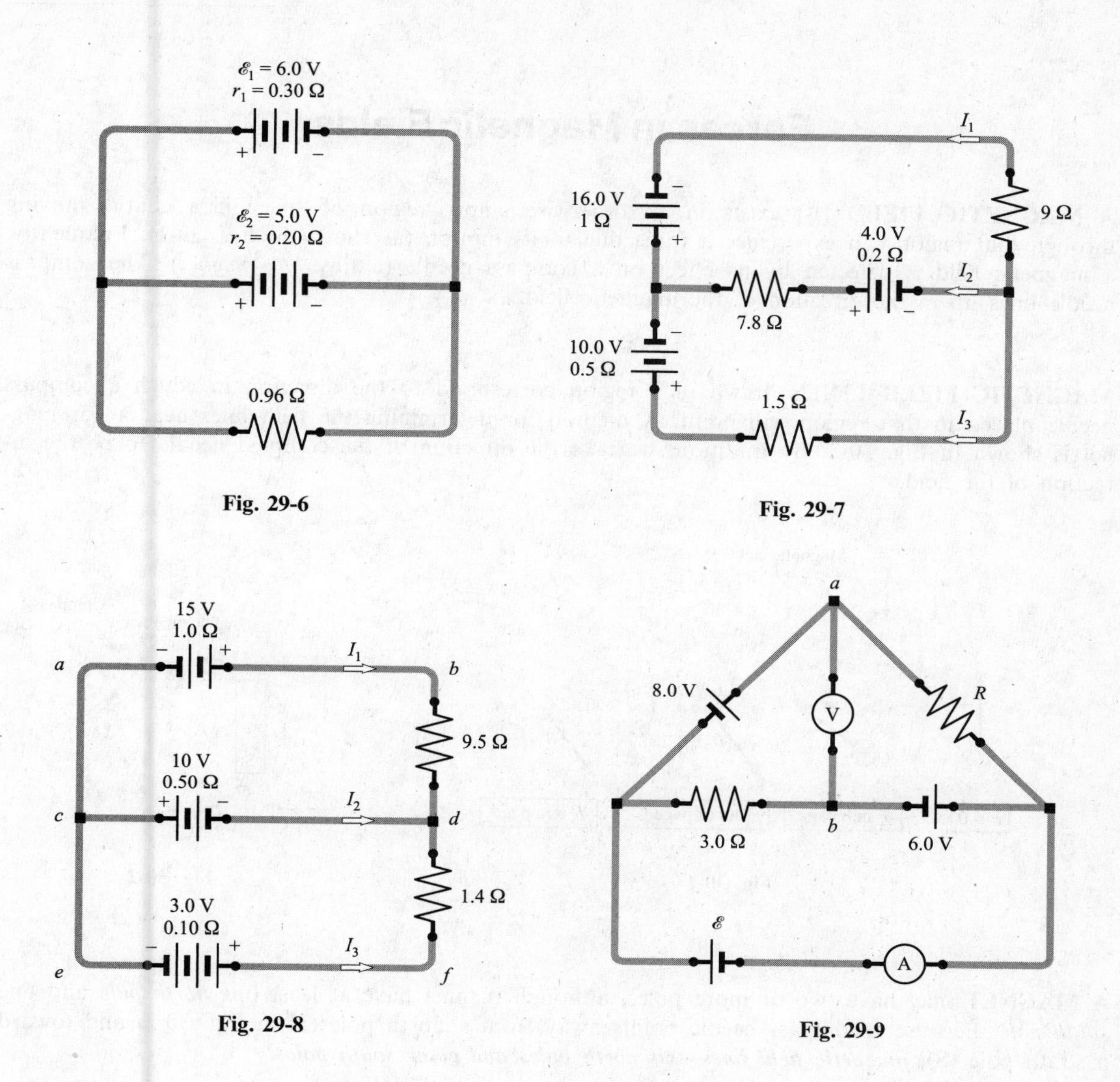

Fig. 29-6

Fig. 29-7

Fig. 29-8

Fig. 29-9

29.11 [II] In Fig. 29-9, $R = 10.0\ \Omega$ and $\mathscr{E} = 13$ V. Find the readings of the ideal ammeter and voltmeter. *Ans.* 8.4 A, 27 V with point-*a* positive

29.12 [II] In Fig. 29-9, the voltmeter reads 14 V (with point-*a* at the higher potential) and the ammeter reads 4.5 A. Find $\mathscr{E}$ and R. *Ans.* $\mathscr{E} = 0$, $R = 3.2\ \Omega$

Chapter 30

Forces in Magnetic Fields

A MAGNETIC FIELD ($\vec{\mathbf{B}}$) exists in an otherwise empty region of space if a charge moving through that region can experience a force due to its motion (as shown in Fig. 30-1). Frequently, a magnetic field is detected by its effect on a compass needle (a tiny *bar magnet*). The compass needle lines up in the direction of the magnetic field.

MAGNETIC FIELD LINES drawn in a region correspond to the direction in which a compass needle placed in that region will point. A method for determining the field lines near a bar magnet is shown in Fig. 30-2. By tradition, we take the direction of the compass needle to be the direction of the field.

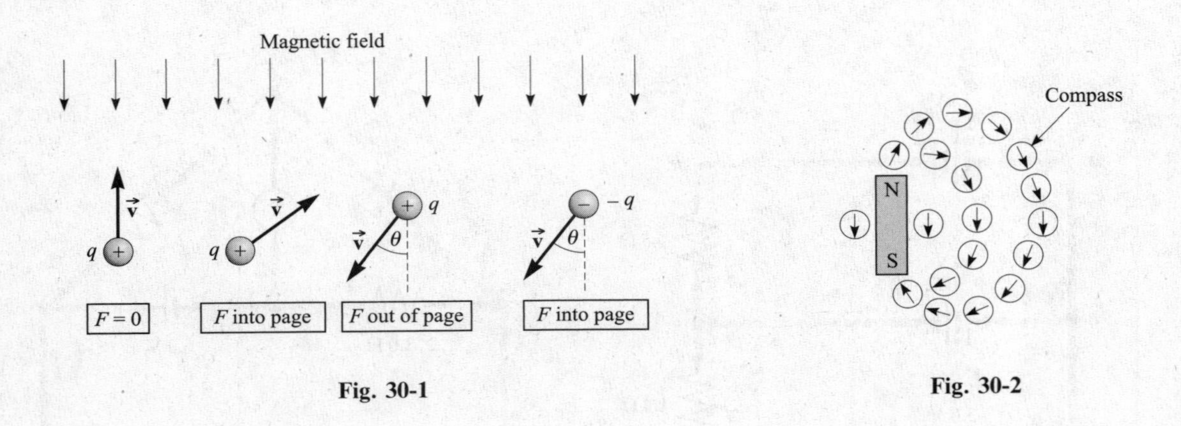

Fig. 30-1 **Fig. 30-2**

A MAGNET may have two or more poles, although it must have at least one *north pole* and one *south pole*. Because a compass needle points away from a north pole (N in Fig. 30-2) and toward a south pole (S), ***magnetic field lines exit north poles and enter south poles.***

MAGNETIC POLES of the same type (north or south) repel each other, while unlike poles attract each other.

A CHARGE MOVING THROUGH A MAGNETIC FIELD experiences a force due to the field, provided its velocity vector is not along a magnetic field line. In Fig. 30-1, charge (q) is moving with velocity $\vec{\mathbf{v}}$ in a magnetic field directed as shown. The direction of the force $\vec{\mathbf{F}}$ on each charge is indicated. Notice that ***the direction of the force on a negative charge is opposite to that on a positive charge*** with the same velocity.

THE DIRECTION OF THE FORCE acting on a charge $+q$ moving in a magnetic field can be found from a **right-hand rule** (Fig. 30-3):

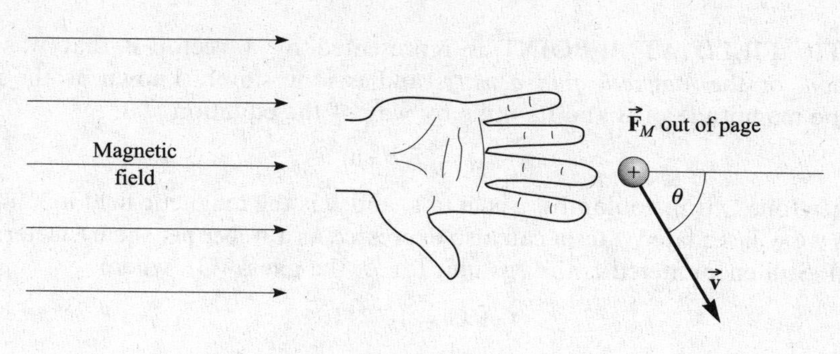

Fig. 30-3

> Hold the right hand flat. Point its fingers in the direction of the field. Orient the thumb along the direction of the velocity of the positive charge. Then the palm of the hand pushes in the direction of the force on the charge. The force direction on a negative charge is opposite to that on a positive charge.

It is often helpful to note that the field line through the particle and the velocity vector of the particle determine a plane (the plane of the page in Fig. 30-3). The force vector is always perpendicular to this plane. An alternative rule is based on the vector cross product: put the fingers of the right hand in the direction of $\vec{v}$, rotate your hand until the fingers can naturally close toward $\vec{B}$ through the smallest angle and your thumb then points in the direction of $\vec{F}_M$ (see Fig. 30-4). We say that $\vec{F}_M$ is in the direction of $\vec{v}$ cross $\vec{B}$.

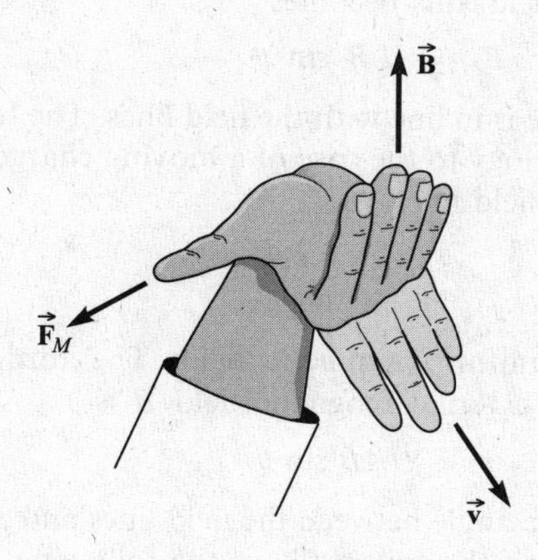

Fig. 30-4

THE MAGNITUDE OF THE FORCE (F_M) on a charge moving in a magnetic field depends upon the product of four factors:

(1) q, the charge (in C)

(2) v, the magnitude of the velocity of the charge (in m/s)

(3) B, the strength of the magnetic field

(4) $\sin\theta$, where θ is the angle between the field lines and the velocity $\vec{v}$.

THE MAGNETIC FIELD AT A POINT is represented by a vector $\vec{\mathbf{B}}$ that was once called the *magnetic induction*, or the *magnetic flux density*, and is now simply known as the **magnetic field**.

We define the magnitude of $\vec{\mathbf{B}}$ and its units by way of the equation

$$F_M = qvB \sin\theta$$

where F_M is in newtons, q is in coulombs, v is in m/s, and B is the magnetic field in a unit called the *tesla* (T). For reasons we will see later, a tesla can also expressed as a **weber per square meter**: $1\ \text{T} = 1\ \text{Wb/m}^2$ (see Chapter 32). Still encountered is the cgs unit for B, the **gauss** (G), where

$$1\ \text{G} = 10^{-4}\ \text{T}$$

The Earth's magnetic field is a few tenths of a gauss. Also note that

$$1\ \text{T} = 1\ \text{Wb/m}^2 = 1\frac{\text{N}}{\text{C}\cdot(\text{m/s})} = 1\frac{\text{N}}{\text{A}\cdot\text{m}}$$

FORCE ON A CURRENT IN A MAGNETIC FIELD: Since a current is simply a stream of positive charges, a current experiences a force due to a magnetic field. The direction of the force is found by the right-hand rule shown in Fig. 30-3 or 30-4, with the direction of the current used in place of the velocity vector.

The magnitude ΔF_M of the force on a small length ΔL of wire carrying current I is given by

$$\Delta F_M = I(\Delta L)B \sin\theta$$

where θ is the angle between the direction of the current I and the direction of the field. For a straight wire of length L in a uniform magnetic field, this becomes

$$F_M = ILB \sin\theta$$

Notice that the force is zero if the wire is in line with the field lines. The force is maximum if the field lines are perpendicular to the wire. In analogy to the case of a moving charge, the force is perpendicular to the plane defined by the wire and the field lines.

TORQUE ON A FLAT COIL in a uniform magnetic field: The torque τ on a flat coil of N loops, each carrying a current I, in an external magnetic field B is

$$\tau = NIAB \sin\theta$$

where A is the area of the coil, and θ is the angle between the field lines and a perpendicular to the plane of the coil. For the direction of rotation of the coil, we have the following right-hand rule:

> Orient the right thumb perpendicular to the plane of the coil, such that the fingers run in the direction of the current flow. Then the torque acts to rotate the thumb into alignment with the external field (at which orientation the torque will be zero).

Solved Problems

30.1 [I] A uniform magnetic field, $B = 3.0$ G, exists in the $+x$-direction. A proton ($q = +e$) shoots through the field in the $+y$-direction with a speed of 5.0×10^6 m/s. (*a*) Find the magnitude and direction of the force on the proton. (*b*) Repeat with the proton replaced by an electron.

(*a*) The situation is shown in Fig. 30-5. We have, after changing 3.0 G to 3.0×10^{-4} T,

$$F_M = qvB \sin \theta = (1.6 \times 10^{-19}\ \text{C})(5.0 \times 10^6\ \text{m/s})(3.0 \times 10^{-4}\ \text{T}) \sin 90° = 2.4 \times 10^{-16}\ \text{N}$$

The force is perpendicular to the *xy*-plane, the plane defined by the field lines and $\vec{v}$. The right-hand rule tells us that the force is in the $-z$-direction.

(*b*) The magnitude of the force is the same as in (*a*), 2.4×10^{-16} N. But, because the electron is negative, the force direction is reversed. The force is in the $+z$-direction.

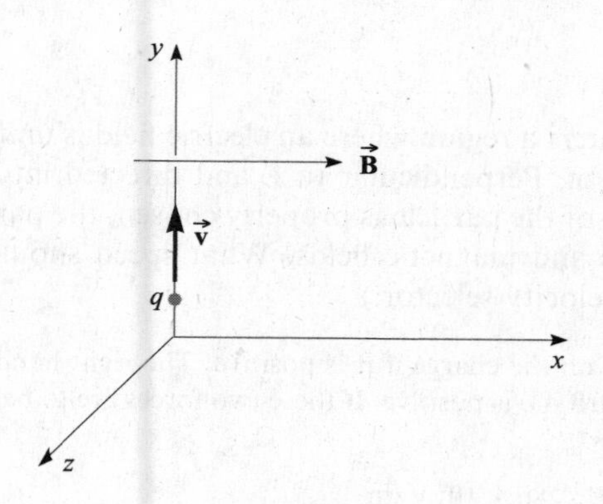

Fig. 30-5

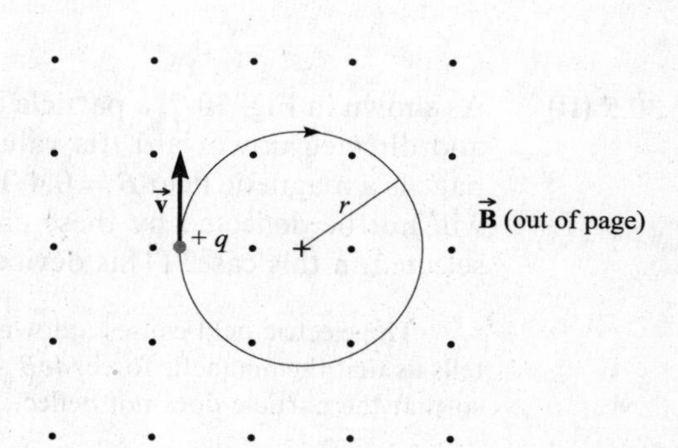

Fig. 30-6

30.2 [II] The charge shown in Fig. 30-6 is a proton ($q = +e$, $m_p = 1.67 \times 10^{-27}$ kg) with speed 5.0×10^6 m/s. It is passing through a uniform magnetic field directed up out of the page; *B* is 30 G. Describe the path followed by the proton.

Because the proton's velocity is perpendicular to $\vec{B}$, the force on the proton is

$$qvB \sin 90° = qvB$$

This force is perpendicular to $\vec{v}$ and so it does no work on the proton. It simply deflects the proton and causes it to follow the circular path shown, as you can verify using the right-hand rule. The force qvB is radially inward and supplies the centripetal force for the circular motion: $F_M = qvB = ma = mv^2/r$ and

$$r = \frac{mv}{qB} \qquad (1)$$

For the given data,

$$r = \frac{(1.67 \times 10^{-27}\ \text{kg})(5.0 \times 10^6\ \text{m/s})}{(1.6 \times 10^{-19}\ \text{C})(30 \times 10^{-4}\ \text{T})} = 17\ \text{m}$$

Observe from Eq. (*1*) that the momentum of the charged particle is directly proportional to the radius of its circular orbit.

30.3 [I] A proton enters a magnetic field of flux density 1.5 Wb/m² with a velocity of 2.0×10^7 m/s at an angle of 30° with the field. Compute the magnitude of the force on the proton.

$$F_M = qvB \sin \theta = (1.6 \times 10^{-19}\ \text{C})(2.0 \times 10^7\ \text{m/s})(1.5\ \text{Wb/m}^2) \sin 30° = 2.4 \times 10^{-12}\ \text{N}$$

30.4 [I] A cathode ray beam (i.e., an electron beam; $m_e = 9.1 \times 10^{-31}$ kg, $q = -e$) is bent in a circle of radius 2.0 cm by a uniform field with $B = 4.5 \times 10^{-3}$ T. What is the speed of the electrons?

To describe a circle like this, the particles must be moving perpendicular to $\vec{\mathbf{B}}$. From Eq. (*1*) of Problem 30.2,

$$v = \frac{rqB}{m} = \frac{(0.020\ \text{m})(1.6 \times 10^{-19}\ \text{C})(4.5 \times 10^{-3}\ \text{T})}{9.1 \times 10^{-31}\ \text{kg}} = 1.58 \times 10^7\ \text{m/s} = 1.6 \times 10^4\ \text{km/s}$$

30.5 [II] As shown in Fig. 30-7, a particle of charge q enters a region where an electric field is uniform and directed downward. Its value E is 80 kV/m. Perpendicular to $\vec{\mathbf{E}}$ and directed into the page is a magnetic field $B = 0.4$ T. If the speed of the particle is properly chosen, the particle will not be deflected by these crossed electric and magnetic fields. What speed should be selected in this case? (This device is called a velocity selector.)

The electric field causes a downward force Eq on the charge if it is positive. The right-hand rule tells us that the magnetic force, $qvB \sin 90°$, is upward if q is positive. If these two forces are to balance so that the particle does not deflect, then

$$Eq = qvB \sin 90° \qquad \text{or} \qquad v = \frac{E}{B} = \frac{80 \times 10^3\ \text{V/m}}{0.4\ \text{T}} = 2 \times 10^5\ \text{m/s}$$

When q is negative, both forces are reversed, so the result $v = E/B$ still holds.

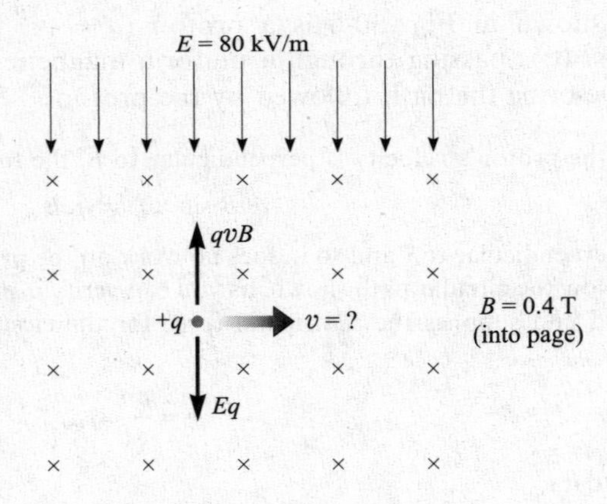

Fig. 30-7

30.6 [III] In Fig. 30-8(*a*), a proton ($q = +e$, $m_p = 1.67 \times 10^{-27}$ kg) is shot with a speed of 8.0×10^6 m/s at an angle of 30.0° to an x-directed field $B = 0.15$ T. Describe the path followed by the proton.

We resolve the particle velocity into components parallel to and perpendicular to the magnetic field. The magnetic force in the direction of $v_{\parallel}$ is zero ($\sin \theta = 0$); the magnetic force in the direction of $v_{\perp}$ has no x-component. Therefore, the motion in the x-direction is uniform, at speed

$$v_{\parallel} = (0.866)(8.0 \times 10^6\ \text{m/s}) = 6.93 \times 10^6\ \text{m/s}$$

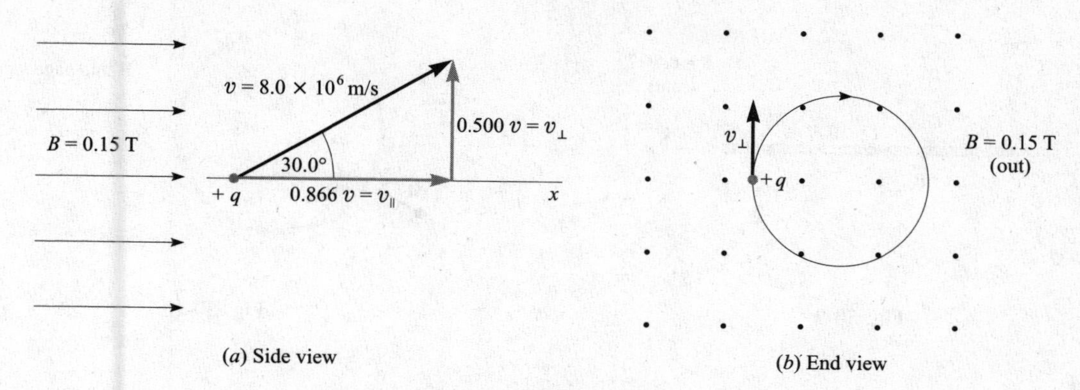

Fig. 30-8

while the transverse motion is circular (see Problem 30.2), with radius

$$r = \frac{mv_\perp}{qB} = \frac{(1.67 \times 10^{-27}\ \text{kg})(0.500 \times 8.0 \times 10^6\ \text{m/s})}{(1.6 \times 10^{-19}\ \text{C})(0.15\ \text{T})} = 0.28\ \text{m}$$

The proton will spiral along the x-axis; the radius of the spiral (or helix) will be 28 cm.

To find the **pitch** of the helix (the x-distance traveled during one revolution), we note that the time taken to complete one circle is

$$\text{Period} = \frac{2\pi r}{v_\perp} = \frac{2\pi(0.28\ \text{m})}{(0.500)(8.0 \times 10^6\ \text{m/s})} = 4.4 \times 10^{-7}\ \text{s}$$

During that time, the proton will travel an x-distance of

$$\text{Pitch} = (v_\parallel)(\text{period}) = (6.93 \times 10^6\ \text{m/s})(4.4 \times 10^{-7}\ \text{s}) = 3.0\ \text{m}$$

30.7 [II] Alpha particles ($m_\alpha = 6.68 \times 10^{-27}$ kg, $q = +2e$) are accelerated from rest through a p.d. of 1.0 kV. They then enter a magnetic field $B = 0.20$ T perpendicular to their direction of motion. Calculate the radius of their path.

Their final KE is equal to the electric potential energy they lose during acceleration, Vq:

$$\tfrac{1}{2}mv^2 = Vq \qquad \text{or} \qquad v = \sqrt{\frac{2Vq}{m}}$$

From Problem 30.2, they follow a circular path in which

$$r = \frac{mv}{qB} = \frac{m}{qB}\sqrt{\frac{2Vq}{m}} = \frac{1}{B}\sqrt{\frac{2Vm}{q}}$$

$$= \frac{1}{0.20\ \text{T}}\sqrt{\frac{2(1000\ \text{V})(6.68 \times 10^{-27}\ \text{kg})}{3.2 \times 10^{-19}\ \text{C}}} = 0.032\ \text{m}$$

30.8 [I] In Fig. 30-9, the magnetic field is up out of the page and $B = 0.80$ T. The wire shown carries a current of 30 A. Find the magnitude and direction of the force on a 5.0 cm length of the wire.

We know that

$$\Delta F_M = I(\Delta L)B \sin\theta = (30\ \text{A})(0.050\ \text{m})(0.80\ \text{T})(1) = 1.2\ \text{N}$$

By the right-hand rule, the force is perpendicular to both the wire and the field and is directed toward the bottom of the page.

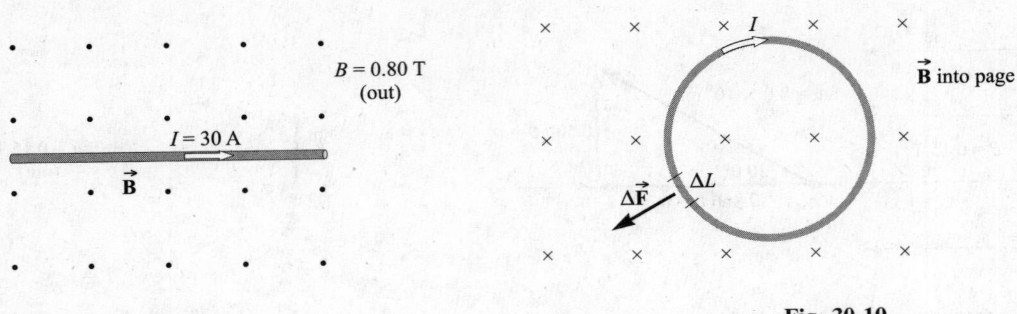

Fig. 30-9

Fig. 30-10

30.9 [I] As shown in Fig. 30-10, a loop of wire carries a current I and its plane is perpendicular to a uniform magnetic field $\vec{\mathbf{B}}$. What are the resultant force and torque on the loop?

Consider the length ΔL shown. The force $\Delta\vec{\mathbf{F}}$ on it has the direction indicated. A point directly opposite this on the loop has an equal, but opposite, force acting on it. Hence the forces on the loop cancel and the resultant force on it is zero.

We see from the figure that the $\Delta\vec{\mathbf{F}}$'s acting on the loop are trying to expand it, not rotate it. Therefore the torque (τ) on the loop is zero. Or, we can make use of the torque equation,

$$\tau = NIAB \sin \theta$$

where θ is the angle between the field lines and the perpendicular to the plane of the loop. We see that $\theta = 0$. Therefore $\sin \theta = 0$ and the torque is zero.

30.10 [I] The 40-loop coil shown in Fig. 30-11 carries a current of 2.0 A in a magnetic field $B = 0.25$ T. Find the torque on it. How will it rotate?

Method 1

$$\tau = NIAB \sin \theta = (40)(2.0 \text{ A})(0.10 \text{ m} \times 0.12 \text{ m})(0.25 \text{ T})(\sin 90^\circ) = 0.24 \text{ N}\cdot\text{m}$$

(Remember that θ is the angle between the field lines and the perpendicular to the loop.) By the right-hand rule, the coil will turn about a vertical axis in such a way that side *ad* moves out of the page.

Method 2

Because sides *dc* and *ab* are in line with the field, the force on each of them is zero, while the force on each vertical wire is

$$F_M = ILB = (2.0 \text{ A})(0.12 \text{ m})(0.25 \text{ T}) = 0.060 \text{ N}$$

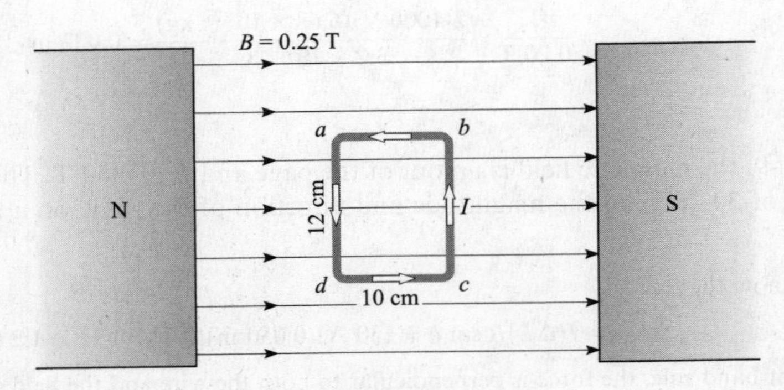

Fig. 30-11

out of the page on side *ab* and into the page on side *bc*. If we take torques about side *bc* as axis, only the force on side *ad* gives a nonzero torque. It is

$$\tau = (40 \times 0.060 \text{ N})(0.10 \text{ m}) = 0.24 \text{ N} \cdot \text{m}$$

and it tends to rotate side *ad* out of the page.

30.11 [I] In Fig. 30-12 is shown one-quarter of a single circular loop of wire that carries a current of 14 A. Its radius is $a = 5.0$ cm. A uniform magnetic field, $B = 300$ G, is directed in the $+x$-direction. Find the torque on the loop and the direction in which it will rotate.

The normal to the loop, $\overline{OP}$, makes an angle $\theta = 60°$ with the $+x$-direction, the field direction. Hence,

$$\tau = NIAB \sin\theta = (1)(14 \text{ A})(\pi \times 25 \times 10^{-4} \text{ m}^2)(0.0300 \text{ T}) \sin 60° = 2.9 \times 10^{-3} \text{ N} \cdot \text{m}$$

The right-hand rule shows that the loop will rotate about the y-axis so as to decrease the angle labeled 60°.

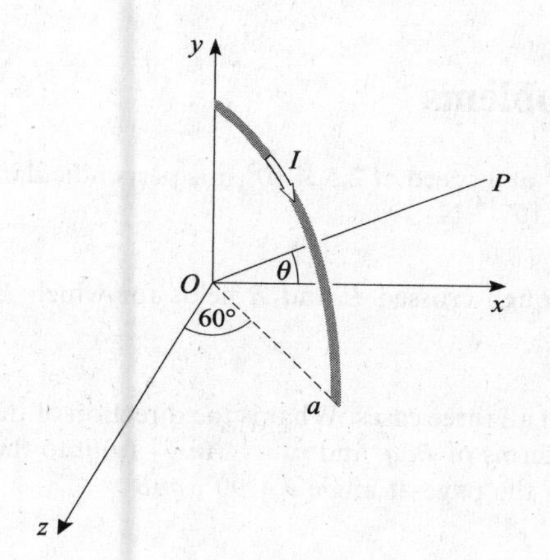

Fig. 30-12

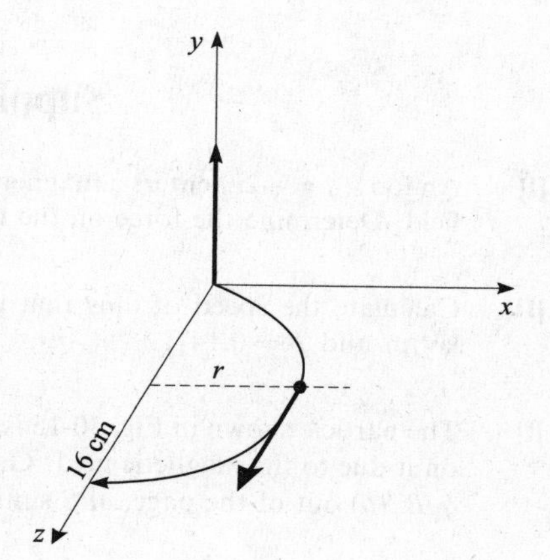

Fig. 30-13

30.12 [II] Two electrons, both with speed 5.0×10^6 m/s, are shot into a uniform magnetic field $\vec{\mathbf{B}}$. The first is shot from the origin out along the $+x$-axis, and it moves in a circle that intersects the $+z$-axis at $z = 16$ cm. The second is shot out along the $+y$-axis, and it moves in a straight line. Find the magnitude and direction of $\vec{\mathbf{B}}$.

The situation is shown in Fig. 30-13. Because a charge experiences no force when moving along a field line, the field must be in either the $+y$- or $-y$-direction. Use of the right-hand rule for the motion shown in the diagram for the *negative* electron charge leads us to conclude that the field is in the $-y$-direction.

To find the magnitude of $\vec{\mathbf{B}}$, we notice that $r = 8$ cm. The magnetic force Bqv provides the needed centripetal force mv^2/r, and so

$$B = \frac{mv}{qr} = \frac{(9.1 \times 10^{-31} \text{ kg})(5.0 \times 10^6 \text{ m/s})}{(1.6 \times 10^{-19} \text{ C})(0.080 \text{ m})} = 3.6 \times 10^{-4} \text{ T}$$

30.13 [I] At a certain place on the planet, the Earth's magnetic field is 5.0×10^{-5} T, directed 40° below the horizontal. Find the force per meter of length on a horizontal wire that carries a current of 30 A northward.

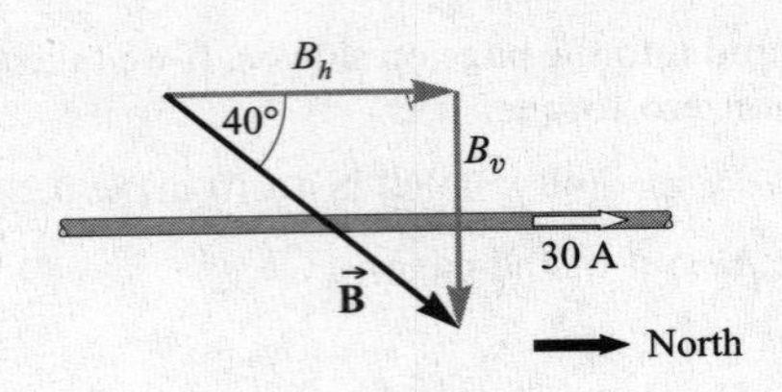

Fig. 30-14

Nearly everywhere, the Earth's field is directed northward. (That is the direction in which a compass needle points.) Therefore, the situation is that shown in Fig. 30-14. The force on the wire is

$$F_M = (30\ \text{A})(L)(5.0 \times 10^{-5}\ \text{T}) \sin 40° \qquad \text{so that} \qquad \frac{F_M}{L} = 9.6 \times 10^{-4}\ \text{N/m}$$

The right-hand rule indicates that the force is into the page, which is west.

Supplementary Problems

30.14 [I] An ion ($q = +2e$) enters a magnetic field of 1.2 Wb/m² at a speed of 2.5×10^5 m/s perpendicular to the field. Determine the force on the ion. *Ans.* 9.6×10^{-14} N

30.15 [II] Calculate the speed of ions that pass undeflected through crossed E and B fields for which $E = 7.7$ kV/m and $B = 0.14$ T. *Ans.* 55 km/s

30.16 [I] The particle shown in Fig. 30-15 is positively charged in all three cases. What is the direction of the force on it due to the magnetic field? Give its magnitude in terms of B, q, and v. *Ans.* (*a*) into the page, qvB; (*b*) out of the page, $qvB \sin\theta$; (*c*) in the plane of the page at angle $\theta + 90°$, qvB

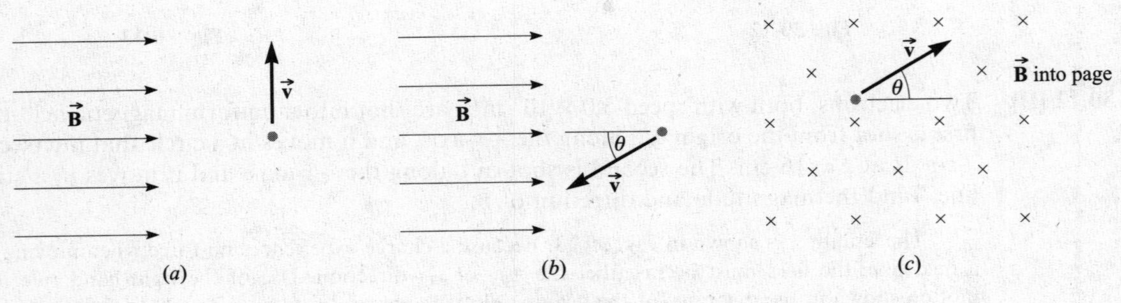

Fig. 30-15

30.17 [II] What might be the mass of a positive ion that is moving at 1.0×10^7 m/s and is bent into a circular path of radius 1.55 m by a magnetic field of 0.134 Wb/m²? (There are several possible answers.) *Ans.* $n(3.3 \times 10^{-27}$ kg), where ne is the ion's charge

30.18 [II] An electron is accelerated from rest through a potential difference of 3750 V. It enters a region where $B = 4.0 \times 10^{-3}$ T perpendicular to its velocity. Calculate the radius of the path it will follow. *Ans.* 5.2 cm

30.19 [II] An electron is shot with speed 5.0×10^6 m/s out from the origin of coordinates. Its initial velocity makes an angle of 20° to the $+x$-axis. Describe its motion if a magnetic field $B = 2.0$ mT exists in the $+x$-direction. *Ans.* helix, $r = 0.49$ cm, pitch = 8.5 cm

30.20 [II] A beam of electrons passes undeflected through two mutually perpendicular electric and magnetic fields. If the electric field is cut off and the same magnetic field maintained, the electrons move in the magnetic field in a circular path of radius 1.14 cm. Determine the ratio of the electronic charge to the electron mass if $E = 8.00$ kV/m and the magnetic field has flux density 2.00 mT. *Ans.* $e/m_e = 175$ GC/kg

30.21 [I] A straight wire 15 cm long, carrying a current of 6.0 A, is in a uniform field of 0.40 T. What is the force on the wire when it is (*a*) at right angles to the field and (*b*) at 30° to the field? *Ans.* (*a*) 0.36 N; (*b*) 0.18 N

30.22 [I] What is the direction of the force, due to the Earth's magnetic field, on a wire carrying current vertically downward? *Ans.* horizontally toward east

30.23 [I] Find the force on each segment of the wire shown in Fig. 30-16 if $B = 0.15$ T. Assume the current in the wire to be 5.0 A. *Ans.* In sections AB and DE, the force is zero; in section BC, 0.12 N into page; in section CD, 0.12 N out of page

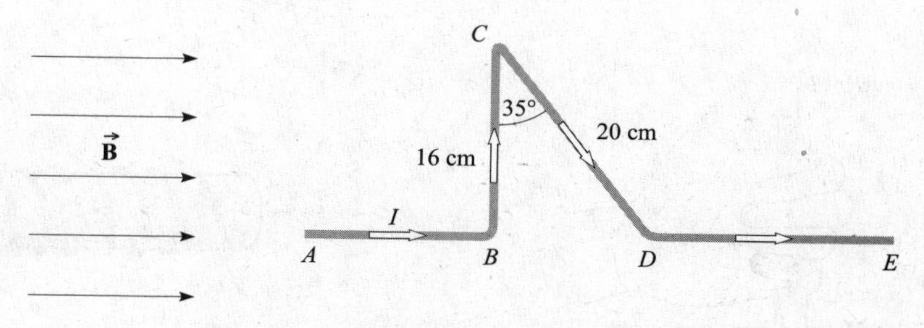

Fig. 30-16

30.24 [II] A flat rectangular coil of 25 loops is suspended in a uniform magnetic field of 0.20 Wb/m^2. The plane of the coil is parallel to the direction of the field. The dimensions of the coil are 15 cm perpendicular to the field lines and 12 cm parallel to them. What is the current in the coil if there is a torque of 5.4 N·m acting on it? *Ans.* 60 A

30.25 [II] An electron is accelerated from rest through a potential difference of 800 V. It then moves perpendicularly to a magnetic field of 30 G. Find the radius of its orbit and its orbital frequency. *Ans.* 3.2 cm, 84 MHz

30.26 [II] A proton and a deuteron ($m_d \approx 2m_p$, $q_d = e$) are both accelerated through the same potential difference and enter a magnetic field along the same line. If the proton follows a path of radius R_p, what will be the radius of the deuteron's path? *Ans.* $R_d = R_p\sqrt{2}$

Chapter 31

Sources of Magnetic Fields

MAGNETIC FIELDS ARE PRODUCED by moving charges, and of course that includes electric currents. Figure 31-1 shows the nature of the magnetic fields produced by several current configurations. Below each is given the value of B at the indicated point P. The constant $\mu_0 = 4\pi \times 10^{-7}\ \mathrm{T \cdot m/A}$ is called the *permeability of free space*. It is assumed that the surrounding material is either vacuum or air.

THE DIRECTION OF THE MAGNETIC FIELD of a current-carrying wire can be found by using a right-hand rule, as illustrated in Fig. 31-1(*a*):

> Grasp the wire in the right hand, with the thumb pointing in the direction of the current. The fingers then circle the wire in the same direction as the magnetic field does.

This same rule can be used to find the direction of the field for a current loop such as that shown in Fig. 31-1(*b*).

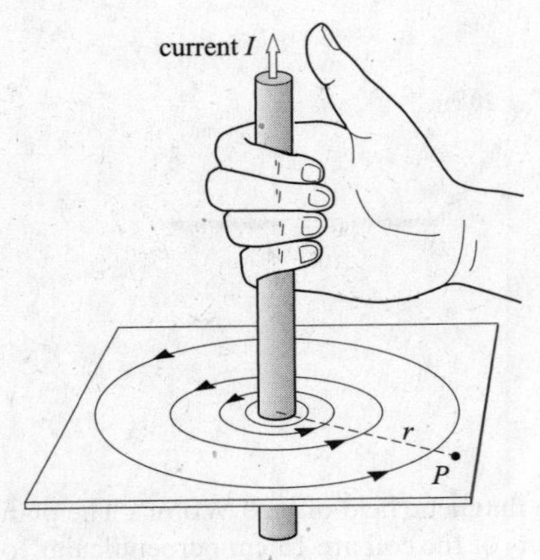

(*a*) Long straight wire:

$$B = \frac{\mu_0 I}{2\pi r}$$

where r is distance to P from the axis of the wire

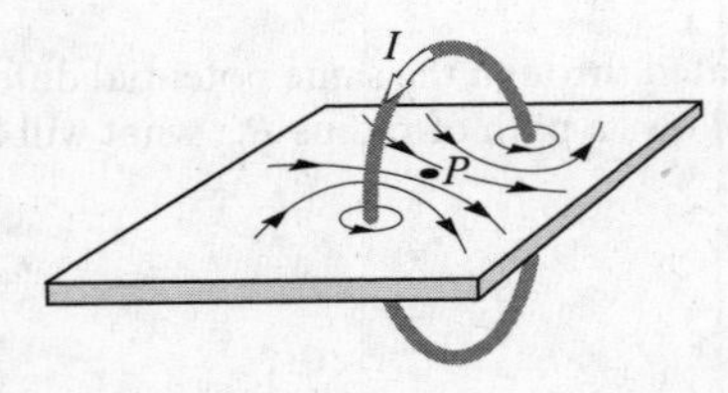

(*b*) Center of a circular coil with radius r and N loops:

$$B = \frac{\mu_0 N I}{2r}$$

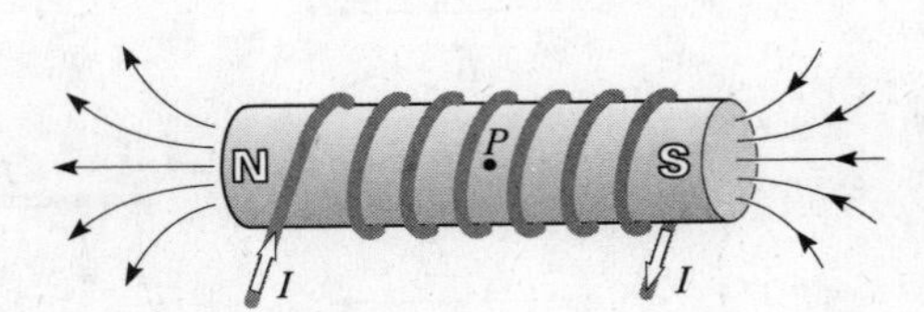

(*c*) Interior point of long solenoid with n loops per meter:

$$B = \mu_0 n I$$

It is constant in the interior

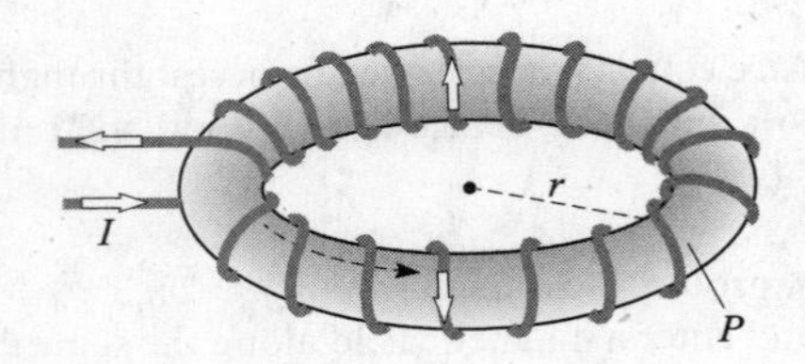

(*d*) Interior point of toroid having N loops:

$$B = \frac{\mu_0 N I}{2\pi r}$$

where r is the radius of the circle on which P lies

Fig. 31-1

FERROMAGNETIC MATERIALS, primarily iron and the other transition elements, greatly enhance magnetic fields. Other materials influence B-fields only slightly. The ferromagnetic materials contain *domains*, or regions of aligned atoms, that act as tiny bar magnets. When the domains within an object are aligned with each other, the object becomes a magnet. The alignment of domains in permanent magnets is not easily disrupted.

THE MAGNETIC MOMENT of a flat current-carrying loop (current $= I$, area $= A$) is IA. The magnetic moment is a vector quantity that points along the field line perpendicular to the plane of the loop. In terms of the magnetic moment, the torque on a flat coil with N loops in a magnetic field B is $\tau = N(IA)B \sin\theta$, where θ is the angle between the field and the magnetic moment vector.

MAGNETIC FIELD OF A CURRENT ELEMENT: The current element of length ΔL shown in Fig. 31-2 contributes $\Delta\vec{\mathbf{B}}$ to the field at P. The magnitude of $\Delta\vec{\mathbf{B}}$ is given by the *Biot–Savart Law*:

$$\Delta B = \frac{\mu_0 I\,\Delta L}{4\pi r^2}\sin\theta$$

where r and θ are defined in the figure. The direction of $\Delta\vec{\mathbf{B}}$ is perpendicular to the plane determined by ΔL and r (the plane of the page). In the case shown, the right-hand rule tells us that $\Delta\vec{\mathbf{B}}$ is out of the page.

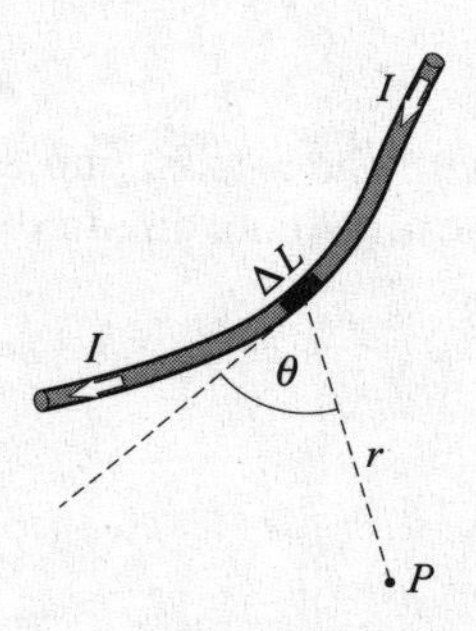

Fig. 31-2

When r is in line with ΔL, then $\theta = 0$ and thus $\Delta B = 0$. This means that the field due to a straight wire at a point on the line of the wire is zero.

Solved Problems

31.1 [I] Compute the value of B in air at a point 5 cm from a long straight wire carrying a current of 15 A.

From Fig. 31-1(*a*),

$$B = \frac{\mu_0 I}{2\pi r} = \frac{(4\pi \times 10^{-7}\ \mathrm{T\cdot m/A})(15\ \mathrm{A})}{2\pi(0.05\ \mathrm{m})} = 6 \times 10^{-5}\ \mathrm{T}$$

31.2 [I] A flat circular coil with 40 loops of wire has a diameter of 32 cm. What current must flow in its wires to produce a field of 3.0×10^{-4} Wb/m^2 at its center?

From Fig. 31-1(*b*),

$$B = \frac{\mu_0 NI}{2r} \qquad \text{or} \qquad 3.0 \times 10^{-4}\ \text{T} = \frac{(4\pi \times 10^{-7}\ \text{T}\cdot\text{m/A})(40)(I)}{2(0.16\ \text{m})}$$

which gives $I = 1.9$ A.

31.3 [I] An air-core solenoid with 2000 loops is 60 cm long and has a diameter of 2.0 cm. If a current of 5.0 A is sent through it, what will be the flux density within it?

From Fig. 31-3(*c*),

$$B = \mu_0 nI = (4\pi \times 10^{-7}\ \text{T}\cdot\text{m/A})\left(\frac{2000}{0.60\ \text{m}}\right)(5.0\ \text{A}) = 0.021\ \text{T}$$

31.4 [II] In Bohr's model of the hydrogen atom, the electron travels with speed 2.2×10^6 m/s in a circle ($r = 5.3 \times 10^{-11}$ m) about the nucleus. Find the value of B at the nucleus due to the electron's motion.

In Problem 26.17 we found that the orbiting electron corresponds to a current loop with $I = 1.06$ mA. The field at the center of the current loop is

$$B = \frac{\mu_0 I}{2r} = \frac{(4\pi \times 10^{-7}\ \text{T}\cdot\text{m/A})(1.06 \times 10^{-3}\ \text{A})}{2(5.3 \times 10^{-11}\ \text{m})} = 13\ \text{T}$$

31.5 [III] A long straight wire coincides with the x-axis, and another coincides with the y-axis. Each carries a current of 5 A in the positive coordinate direction. (See Fig. 31-3.) Where is their combined field equal to zero?

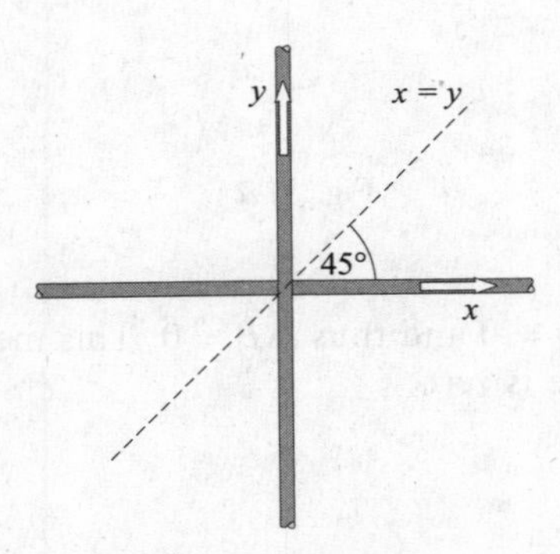

Fig. 31-3

Use of the right-hand rule should convince you that their fields tend to cancel in the first and third quadrants. A line at $\theta = 45°$ passing through the origin is equidistant from the two wires in these quadrants. Hence the fields exactly cancel along the line $x = y$, the 45° line.

31.6 [II] A long wire carries a current of 20 A along the axis of a long solenoid. The field due to the solenoid is 4.0 mT. Find the resultant field at a point 3.0 mm from the solenoid axis.

The situation is shown in Fig. 31-4. The field of the solenoid, $\vec{\mathbf{B}}_s$, is directed parallel to the wire. The field of the long straight wire, $\vec{\mathbf{B}}_w$, circles the wire and is perpendicular to $\vec{\mathbf{B}}_s$. We have $B_s = 4.0$ mT and

$$B_w = \frac{\mu_0 I}{2\pi r} = \frac{(4\pi \times 10^{-7}\ \text{T}\cdot\text{m/A})(20\ \text{A})}{2\pi(3.0 \times 10^{-3}\ \text{m})} = 1.33\ \text{mT}$$

Since $\vec{\mathbf{B}}_s$ and $\vec{\mathbf{B}}_w$ are perpendicular, their resultant $\vec{\mathbf{B}}$ has magnitude

$$B = \sqrt{(4.0\ \text{mT})^2 + (1.33\ \text{mT})^2} = 4.2\ \text{mT}$$

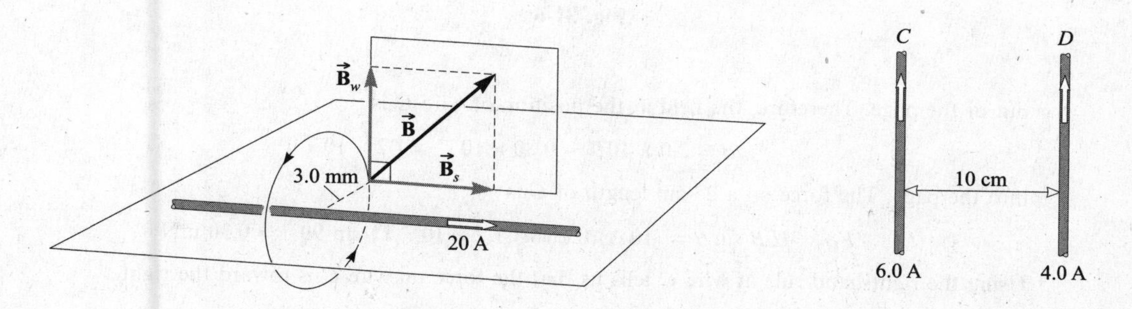

Fig. 31-4

Fig. 31-5

31.7 [II] As shown in Fig. 31-5, two long parallel wires are 10 cm apart and carry currents of 6.0 A and 4.0 A. Find the force on a 1.0 m length of wire D if the currents are (*a*) parallel and (*b*) antiparallel.

(*a*) This is the situation shown in Fig. 31-5. The field at wire D due to wire C is directed into the page and has the value

$$B = \frac{\mu_0 I}{2\pi r} = \frac{(4\pi \times 10^{-7}\ \text{T}\cdot\text{m/A})(6.0\ \text{A})}{2\pi(0.10\ \text{m})} = 1.2 \times 10^{-5}\ \text{T}$$

The force on 1 m of wire D due to this field is

$$F_M = ILB \sin\theta = (4.0\ \text{A})(1.0\ \text{m})(1.2 \times 10^{-5}\ \text{T})(\sin 90^\circ) = 48\ \mu\text{N}$$

The right-hand rule applied to wire D tells us the force on D is toward the left. The wires attract each other.

(*b*) If the current in D flows in the reverse direction, the force direction will be reversed. The wires will repel each other. The force per meter of length is still 48 μN.

31.8 [III] Consider the three long, straight, parallel wires shown in Fig. 31-6. Find the force experienced by a 25-cm length of wire C.

The fields due to wires D and G at wire C are

$$B_D = \frac{\mu_0 I}{2\pi r} = \frac{(4\pi \times 10^{-7}\ \text{T}\cdot\text{m/A})(30\ \text{A})}{2\pi(0.030\ \text{m})} = 2.0 \times 10^{-4}\ \text{T}$$

into the page, and

$$B_G = \frac{(4\pi \times 10^{-7}\ \text{T}\cdot\text{m/A})(20\ \text{A})}{2\pi(0.050\ \text{m})} = 0.80 \times 10^{-4}\ \text{T}$$

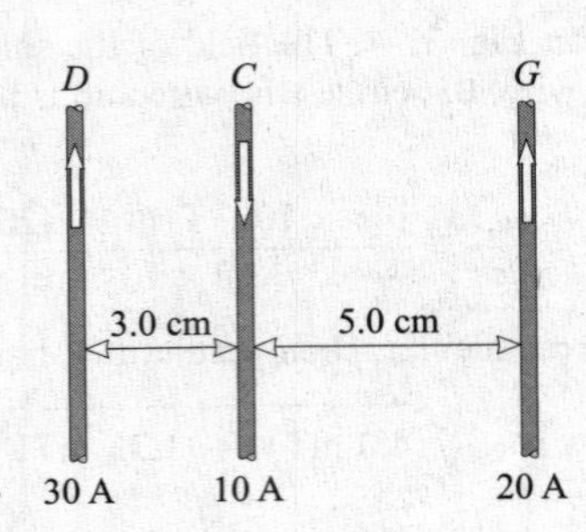

Fig. 31-6

out of the page. Therefore, the field at the position of wire C is

$$B = 2.0 \times 10^{-4} - 0.80 \times 10^{-4} = 1.2 \times 10^{-4}\ \text{T}$$

into the page. The force on a 25-cm length of C is

$$F_M = ILB \sin\theta = (10\ \text{A})(0.25\ \text{m})(1.2 \times 10^{-4}\ \text{T})(\sin 90^\circ) = 0.30\ \text{mN}$$

Using the right-hand rule at wire C tells us that the force on wire C is toward the right.

31.9 [III] A flat circular coil having 10 loops of wire has a diameter of 2.0 cm and carries a current of 0.50 A. It is mounted inside a long solenoid that has 200 loops on its 25-cm length. The current in the solenoid is 2.4 A. Compute the torque required to hold the coil with its central axis perpendicular to that of the solenoid.

Let the subscripts s and c refer to the solenoid and coil respectively. Then

$$\tau = N_c I_c A_c B_s \sin 90^\circ$$

But $B_s = \mu_0 n I_s = \mu_0 (N_s/L_s) I_s$, which gives

$$\tau = \frac{\mu_0 N_c N_s I_c I_s (\pi r_c^2)}{L_s}$$

$$= \frac{(4\pi \times 10^{-7}\ \text{T}\cdot\text{m/A})(10)(200)(0.50\ \text{A})(2.4\ \text{A})\pi(0.010\ \text{m})^2}{0.25\ \text{m}}$$

$$= 3.8 \times 10^{-6}\ \text{N}\cdot\text{m}$$

31.10 [III] The wire shown in Fig. 31-7 carries a current of 40 A. Find the field at point P.

Since P lies on the lines of the straight wires, those wires contribute no field at P. A circular loop of radius r gives a field of $B = \mu_0 I/2r$ at its center point. Here we have only three-fourths of a loop, and so we can assume that

$$B \text{ at point-}P = \left(\frac{3}{4}\right)\left(\frac{\mu_0 I}{2r}\right) = \frac{(3)(4\pi \times 10^{-7}\ \text{T}\cdot\text{m/A})(40\ \text{A})}{(4)(2)(0.020\ \text{m})}$$

$$= 9.4 \times 10^{-4}\ \text{T} = 0.94\ \text{mT}$$

The field is out of the page.

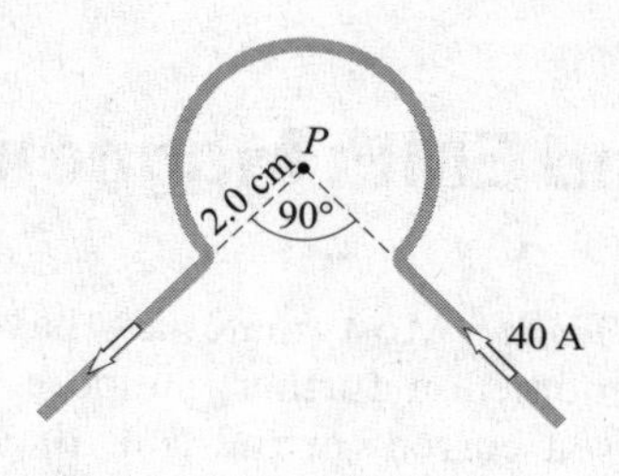

Fig. 31-7

Supplementary Problems

31.11 [I] Compute the magnitude of the magnetic field in air at a point 6.0 cm from a long straight wire carrying a current of 9.0 A. *Ans.* 30 μT

31.12 [I] A closely wound, flat, circular coil of 25 turns of wire has a diameter of 10 cm and carries a current of 4.0 A. Determine the value of B at its center. *Ans.* 1.3×10^{-3} Wb/m^2

31.13 [I] An air-core solenoid 50 cm long has 4000 turns of wire wound on it. Compute B in its interior when a current of 0.25 A exists in the winding. *Ans.* 2.5 mT

31.14 [I] A uniformly wound air-core toroid has 750 loops on it. The radius of the circle through the center of its windings is 5 cm. What current in the winding will produce a field of 1.8 mT on this central circle? *Ans.* 0.6 A

31.15 [II] Two long parallel wires are 4 cm apart and carry currents of 2 A and 6 A in the same direction. Compute the force between the wires per meter of wire length. *Ans.* 6×10^{-5} N/m, attraction

31.16 [II] Two long fixed parallel wires, A and B, are 10 cm apart in air and carry 40 A and 20 A respectively, in opposite directions. Determine the resultant field (*a*) on a line midway between the wires and parallel to them and (*b*) on a line 8.0 cm from wire A and 18 cm from wire B. (*c*) What is the force per meter on a third long wire, midway between A and B and in their plane, when it carries a current of 5.0 A in the same direction as the current in A? *Ans.* (*a*) 2.4×10^{-4} T; (*b*) 7.8×10^{-5} T; (*c*) 1.2×10^{-3} N/m, toward A

31.17 [II] The long straight wires in Fig. 31-3 both carry a current of 12 A, in the directions shown. Find B at the points (*a*) $x = -5.0$ cm, $y = 5.0$ cm and (*b*) $x = -7.0$ cm, $y = -6.0$ cm. *Ans.* (*a*) 96 μT, out; (*b*) 5.7 μT, in

31.18 [II] A certain electromagnet consists of a solenoid (5.0 cm long with 200 turns of wire) wound on a soft-iron core that intensifies the field 130 times. (We say that the *relative permeability* of the iron is 130.) Find B within the iron when the current in the solenoid is 0.30 A. *Ans.* 0.20 T

31.19 [III] A particular solenoid (50 cm long with 2000 turns of wire) carries a current of 0.70 A and is in vacuum. An electron is shot at an angle of 10° to the solenoid axis from a point on the axis. (*a*) What must be the speed of the electron if it is to just miss hitting the inside of the 1.6 cm diameter solenoid? (*b*) What is then the pitch of the electron's helical path? *Ans.* (*a*) 1.4×10^{7} m/s; (*b*) 14 cm

Chapter 32

Induced EMF; Magnetic Flux

MAGNETIC EFFECTS OF MATTER: Most materials have only a slight effect on a steady magnetic field. To explore that phenomenon further, suppose that a very long solenoid or a toroid is located in vacuum. With a fixed current in the coil, the magnetic field at a certain point inside the solenoid or toroid is B_0, where the subscript $_0$ stands for vacuum. If now the solenoid or toroid core is filled with a material, the field at that point will be changed to a new value B. We define:

$$\textit{Relative permeability} \text{ of the material} = k_M = \frac{B}{B_0}$$

$$\textit{Permeability} \text{ of the material} = \mu = k_M \mu_0$$

Recall that μ_0 is the permeability of free space, $4\pi \times 10^{-7}\ \mathrm{T \cdot m/A}$.

Diamagnetic materials have values for k_M slightly below unity (0.999 984 for solid lead, for example). They slightly decrease the value of B in the solenoid or toroid.

Paramagnetic materials have values for k_M slightly larger than unity (1.000 021 for solid aluminum, for example). They slightly increase the value of B in the solenoid or toroid.

Ferromagnetic materials, such as iron and its alloys, have k_M values of about 50 or larger. They greatly increase the value of B in the toroid or solenoid.

MAGNETIC FIELD LINES: A magnetic field may be represented pictorially using lines, to which $\vec{\mathbf{B}}$ is everywhere tangential. These magnetic field lines are constructed in such a way that the number of lines piercing a unit area perpendicular to them is proportional to the local value of B.

THE MAGNETIC FLUX (Φ_M) through an area A is defined to be the product of $B_\perp$ and A where $B_\perp$ is the component of $\vec{\mathbf{B}}$ perpendicular to the surface of area A:

$$\Phi_M = B_\perp A = BA\cos\theta$$

Here θ is the angle between the direction of the magnetic field and the perpendicular to the area. The flux is expressed in **webers** (Wb).

AN INDUCED EMF exists in a loop of wire whenever there is a change in the magnetic flux passing through the area surrounded by the loop. The induced emf exists only during the time that the flux through the area is changing, either increasing or decreasing.

FARADAY'S LAW FOR INDUCED EMF: Suppose that a coil with N loops is subject to a changing magnetic flux passing through the coil. If a change in flux $\Delta\Phi_M$ occurs in a time Δt, then the average emf induced between the two terminals of the coil is given by

$$\mathscr{E} = -N\frac{\Delta\Phi_M}{\Delta t}$$

The emf e is measured in volts if $\Delta\Phi_M/\Delta t$ is in Wb/s. The minus sign indicates that the induced emf opposes the change which produces it, as stated generally in **Lenz's Law.**

LENZ'S LAW: An induced emf always has such a direction as to oppose the change in magnetic flux that produced it. For example, if the flux is increasing through a coil, the current produced by the induced emf will generate a flux that tends to cancel the increasing flux (though it generally does not succeed at doing it completely). Or, if the flux is decreasing through the coil, that current will produce a flux that tends to restore the decreasing flux (though it generally does not succeed at doing it completely). Lenz's Law is a consequence of Conservation of Energy. If this were not the case, the induced currents would enhance the flux change that caused them to begin with and the process would build endlessly.

MOTIONAL EMF: When a conductor moves through a magnetic field so as to cut field lines, an induced emf will exist in it, in accordance with Faraday's Law. In this case,

$$|\mathscr{E}| = \frac{\Delta\Phi_M}{\Delta t}$$

The symbol $|\mathscr{E}|$ means that we are concerned here only with the magnitude of the average induced emf; its direction will be considered below.

The induced emf in a straight conductor of length L moving with velocity $\vec{\mathbf{v}}$ perpendicular to a field $\vec{\mathbf{B}}$ is given by

$$|\mathscr{E}| = BLv$$

where $\vec{\mathbf{B}}$, $\vec{\mathbf{v}}$, and the wire must be mutually perpendicular.

In this case, Lenz's Law still tells us that the induced emf opposes the process. But now the opposition is produced by way of the force exerted by the magnetic field on the induced current in the conductor. The current direction must be such that the force opposes the motion of the conductor (though it generally does not completely cancel it). Knowing the current direction, we also know the direction of $\mathscr{E}$.

Solved Problems

32.1 [II] A solenoid is 40 cm long, has cross-sectional area 8.0 cm^2, and is wound with 300 turns of wire that carry a current of 1.2 A. The relative permeability of its iron core is 600. Compute (*a*) B for an interior point and (*b*) the flux through the solenoid.

(*a*) From Fig. 31-1(*c*),

$$B_0 = \frac{\mu_0 NI}{L} = \frac{(4\pi \times 10^{-7}\ \mathrm{T\cdot m/A})(300)(1.2\ \mathrm{A})}{0.40\ \mathrm{m}} = 1.13\ \mathrm{mT}$$

and so $$B = k_M B_0 = (600)(1.13 \times 10^{-3}\ \mathrm{T}) = 0.68\ \mathrm{T}$$

(*b*) Because the field lines are perpendicular to the cross-section of the solenoid,

$$\Phi_M = B_\perp A = BA = (0.68\ \mathrm{T})(8.0 \times 10^{-4}\ \mathrm{m^2}) = 54\ \mu\mathrm{Wb}$$

32.2 [I] The flux through a current-carrying toroidal coil changes from 0.65 mWb to 0.91 mWb when the air core is replaced by another material. What are the relative permeability and the permeability of the material?

The air core is essentially the same as a vacuum core. Since $k_M = B/B_0$ and $\Phi_M = B_\perp A$,

$$k_M = \frac{0.91 \text{ mWb}}{0.65 \text{ mWb}} = 1.40$$

This is the relative permeability. The magnetic permeability is

$$\mu = k_M \mu_0 = (1.40)(4\pi \times 10^{-7} \text{ T·m/A}) = 5.6\pi \times 10^{-7} \text{ T·m/A}$$

32.3 [I] The quarter-circle loop shown in Fig. 32-1 has an area of 15 cm^2. A constant magnetic field, $B = 0.16$ T, pointing in the $+x$-direction, fills the space independent of the loop. Find the flux through the loop in each orientation shown.

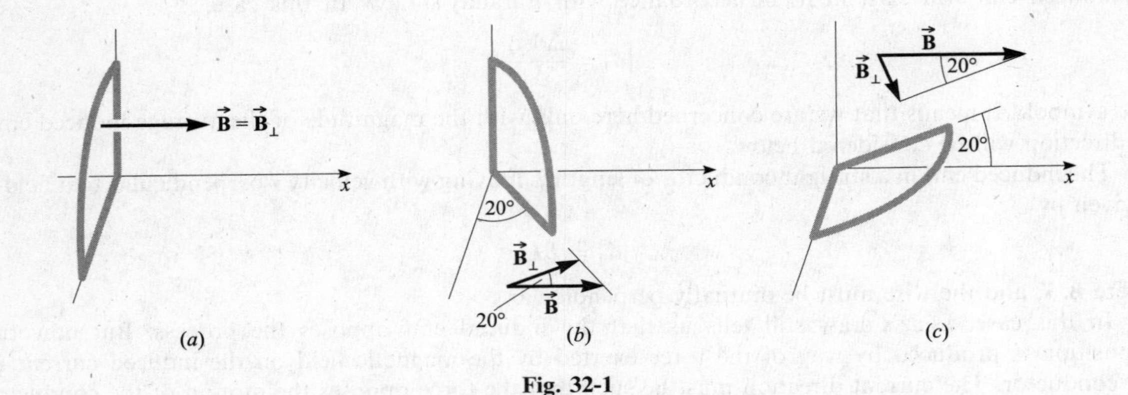

Fig. 32-1

The magnetic flux is determined by the amount of $\vec{\mathbf{B}}$-field passing perpendicularly through the particular area, times that area. That is, $\Phi_M = B_\perp A$.

(a) $$\Phi_M = B_\perp A = BA = (0.16 \text{ T})(15 \times 10^{-4} \text{ m}^2) = 2.4 \times 10^{-4} \text{ Wb}$$

(b) $$\Phi_M = (B \cos 20°)A = (2.4 \times 10^{-4} \text{ Wb})(\cos 20°) = 2.3 \times 10^{-4} \text{ Wb}$$

(c) $$\Phi_M = (B \sin 20°)A = (2.4 \times 10^{-4} \text{ Wb})(\sin 20°) = 8.2 \times 10^{-5} \text{ Wb}$$

32.4 [II] A hemispherical surface of radius R is placed in a uniform magnetic field $\vec{\mathbf{B}}$ as shown in Fig. 32-2. What is the magnetic flux through the hemispherical surface?

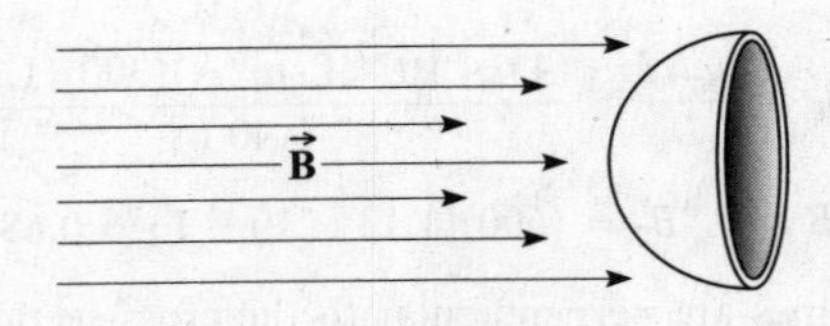

Fig. 32-2

The same number of field lines pass through the curved surface as through the shaded flat circular cross-section. Therefore,

$$\text{Flux through curved surface} = \text{flux through flat surface} = B_\perp A$$

where in this case $B_\perp = B$ and $A = \pi R^2$. Then $\Phi_M = \pi B R^2$.

32.5 [I] A 50-loop circular coil has a radius of 3.0 cm. It is oriented so that the field lines of a magnetic field are normal to the area of the coil. Suppose that the magnetic field is varied so that B increases from 0.10 T to 0.35 T in a time of 2.0 milliseconds. Find the average induced emf in the coil.

$$\Delta\Phi_M = B_{\text{final}}A - B_{\text{initial}}A = (0.25\ \text{T})(\pi r^2) = (0.25\ \text{T})\pi(0.030\ \text{m})^2 = 7.1 \times 10^{-4}\ \text{Wb}$$

$$|\mathscr{E}| = N\left|\frac{\Delta\Phi_M}{\Delta t}\right| = (50)\left(\frac{7.1 \times 10^{-4}\ \text{Wb}}{2 \times 10^{-3}\ \text{s}}\right) = 18\ \text{V}$$

32.6 [II] The cylindrical permanent magnet in the center of Fig. 32-3 induces an emf in the coils as the magnet moves toward the right or the left. Find the directions of the induced currents through both resistors when the magnet is moving (*a*) toward the right and (*b*) toward the left. In each case discuss the voltage across the resistor:

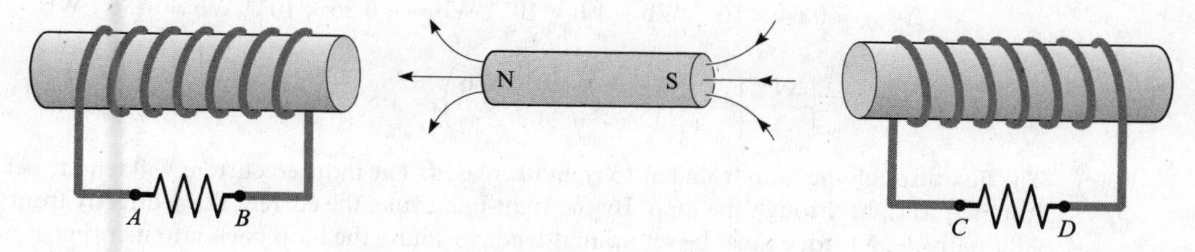

Fig. 32-3

(*a*) Consider first the coil on the left. As the magnet moves to the right, the flux through that coil, which is directed more or less to the left, decreases. To compensate for this, the induced current in the coil on the left will flow so as to produce a flux toward the left through itself. Apply the right-hand rule to the loop on the left end. For it to produce flux inside the coil toward the left, the current must flow directly through the resistor from B to A. The voltage at B is higher than at A.

Now consider the coil on the right. As the magnet moves toward the right, the flux inside that coil on the right, which is more or less directed to the left, increases. The induced current in the coil will produce a flux toward the right to cancel this increased flux. Applying the right-hand rule to the loop on the right end, we find that the loop generates flux to the right inside itself if the current flows from D to C directly through the resistor. The voltage at D is higher than at C.

(*b*) In this case the flux change caused by the magnet's motion toward the left is opposite to what it was in (*a*). Using the same type of reasoning, we find that the induced currents flow through the resistors directly from A to B and from C to D. The voltage at A is higher than at B, and it's higher at C than at D.

32.7 [III] In Fig. 32-4(*a*) there is a uniform magnetic field in the $+x$-direction, with a value of $B = 0.20$ T. The circular loop of wire is in the yz-plane. The loop has an area of 5.0 cm^2 and rotates about line CD as axis. Point A rotates toward positive x-values from the position shown. If the loop rotates through 50° from its indicated position, as shown in Fig. 32-4(*b*), in a time of 0.20 s, (*a*) what is the change in flux through the coil, (*b*) what is the average induced emf in it, and (*c*) does the induced current flow directly from A to C or C to A in the upper part of the coil?

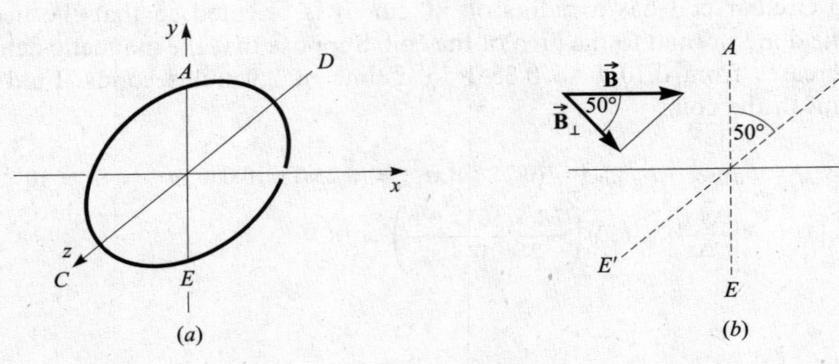

Fig. 32-4

(a) $$\text{Initial flux} = B_{\perp}A = BA = (0.20\text{ T})(5.0 \times 10^{-4}\text{ m}^2) = 1.0 \times 10^{-4}\text{ Wb}$$

$$\text{Final flux} = (B\cos 50°)A = (1.0 \times 10^{-4}\text{ Wb})(\cos 50°) = 0.64 \times 10^{-4}\text{ Wb}$$

$$\Delta\Phi_M = 0.64 \times 10^{-4}\text{ Wb} - 1.0 \times 10^{-4}\text{ Wb} = -0.36 \times 10^{-4}\text{ Wb} = -36\ \mu\text{Wb}$$

(b) $$|\mathscr{E}| = N\left|\frac{\Delta\Phi_M}{\Delta t}\right| = (1)\left(\frac{0.36 \times 10^{-4}\text{ Wb}}{0.20\text{ s}}\right) = 1.8 \times 10^{-4}\text{ V} = 0.18\text{ mV}$$

(c) The flux through the loop from left to right decreased. The induced current will tend to set up flux from left to right through the loop. By the right-hand rule, the current flows directly from A to C. Alternatively, a torque must be set up that tends to rotate the loop back into its original position. The appropriate right-hand rule from Chapter 30 again gives a current flow directly from A to C.

32.8 [I] A coil having 50 turns of wire is removed in 0.020 s from between the poles of a magnet, where its area intercepted a flux of 3.1×10^{-4} Wb, to a place where the intercepted flux is 0.10×10^{-4} Wb. Determine the average emf induced in the coil.

$$|\mathscr{E}| = N\left|\frac{\Delta\Phi_M}{\Delta t}\right| = 50\,\frac{(3.1 - 0.10) \times 10^{-4}\text{ Wb}}{0.020\text{ s}} = 0.75\text{ V}$$

32.9 [I] A copper bar 30 cm long is perpendicular to a uniform magnetic field of 0.80 Wb/m^2 and moves at right angles to the field with a speed of 0.50 m/s. Determine the emf induced in the bar.

$$|\mathscr{E}| = BLv = (0.80\text{ Wb/m}^2)(0.30\text{ m})(0.50\text{ m/s}) = 0.12\text{ V}$$

32.10 [II] As shown in Fig. 32-5, a metal rod makes contact with two parallel wires and completes the circuit. The circuit is perpendicular to a magnetic field with $B = 0.15$ T. If the resistance is 3.0 Ω, how large a force is needed to move the rod to the right with a constant speed of 2.0 m/s? At what rate is energy dissipated in the resistor?

As the wire moves, the downward flux through the loop increases. Accordingly, the induced emf in the rod causes a current to flow counterclockwise in the circuit so as to produce an upward induced $\vec{\mathbf{B}}$-field in the loop that opposes the downward flux increase. Because of this current in the rod, it experiences a force to the left due to the magnetic field. To pull the rod to the right with a constant speed, this force must be balanced.

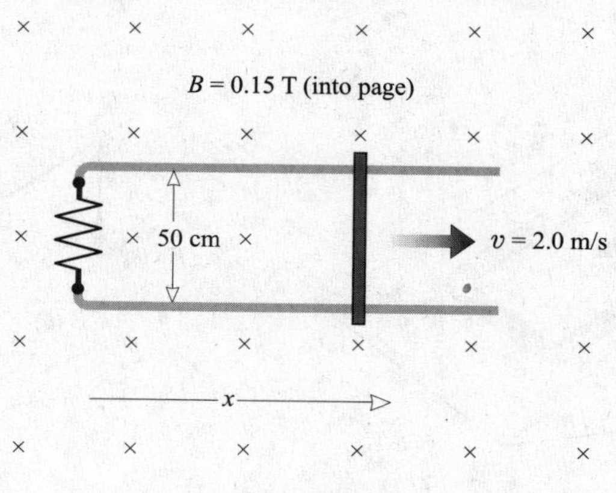

Fig. 32-5

Method 1

The emf induced in the rod is

$$|\mathscr{E}| = BLv = (0.15\ \text{T})(0.50\ \text{m})(2.0\ \text{m/s}) = 0.15\ \text{V}$$

and

$$I = \frac{|\mathscr{E}|}{R} = \frac{0.15\ \text{V}}{3.0\ \Omega} = 0.050\ \text{A}$$

from which

$$F_M = ILB \sin 90^\circ = (0.050\ \text{A})(0.50\ \text{m})(0.15\ \text{T})(1) = 3.8\ \text{mN}$$

Method 2

The emf induced in the loop is

$$|\mathscr{E}| = N\left|\frac{\Delta\Phi_M}{\Delta t}\right| = (1)\frac{B\Delta A}{\Delta t} = \frac{B(L\Delta x)}{\Delta t} = BLv$$

as before. Now proceed as in Method 1.

To find the power loss in the resistor, we can use

$$\text{P} = I^2R = (0.050\ \text{A})^2(3.0\ \Omega) = 7.5\ \text{mW}$$

Alternatively,

$$\text{P} = Fv = (3.75 \times 10^{-3}\ \text{N})(2.0\ \text{m/s}) = 7.5\ \text{mW}$$

32.11 [III] The metal bar of length L, mass m, and resistance R shown in Fig. 32-6(a) slides without friction on a rectangular circuit composed of resistanceless wire resting on an inclined plane. There is a vertical uniform magnetic field $\vec{\mathbf{B}}$. Find the terminal speed of the bar (that is, the constant speed it attains).

Gravity pulls the bar down the incline as shown in Fig. 32-6(b). Induced current flowing in the bar interacts with the field so as to retard this motion.

Because of the motion of the bar in the magnetic field, an emf is induced in the bar:

$$\mathscr{E} = (Blv)_\perp = BL(v \cos\theta)$$

This causes a current

$$I = \frac{\text{emf}}{R} = \left(\frac{BLv}{R}\right)\cos\theta$$

in the loop. A wire carrying a current in a magnetic field experiences a force that is perpendicular to the plane defined by the wire and the magnetic field lines. The bar thus experiences a horizontal force $\vec{\mathbf{F}}_h$ (perpendicular to the plane of $\vec{\mathbf{B}}$ and the bar) given by

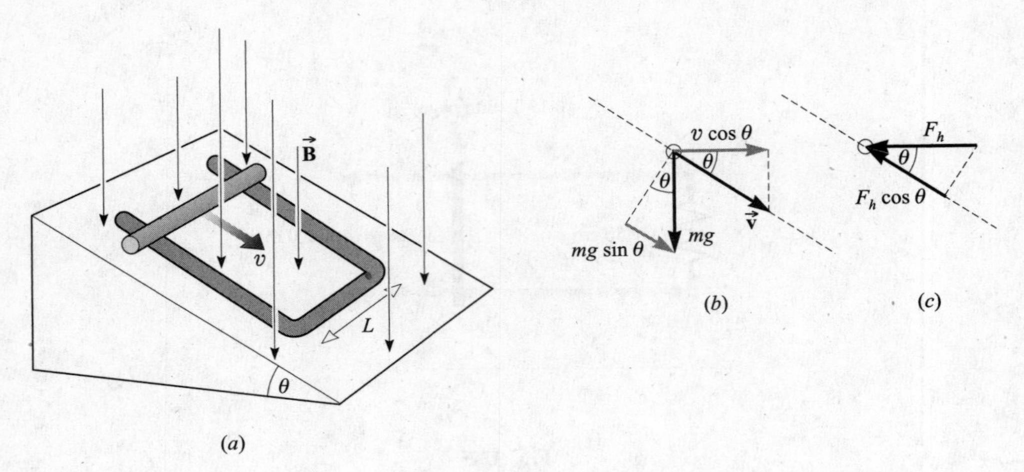

Fig. 32-6

$$F_h = BIL = \left(\frac{B^2L^2v}{R}\right)\cos\theta$$

and shown in Fig. 32-6(*c*). However, we want the force component along the plane, which is

$$F_{\text{up plane}} = F_h \cos\theta = \left(\frac{B^2L^2v}{R}\right)\cos^2\theta$$

When the bar reaches its terminal velocity, this force equals the gravitational force down the plane. Therefore,

$$\left(\frac{B^2L^2v}{R}\right)\cos^2\theta = mg\sin\theta$$

from which the terminal speed is

$$v = \left(\frac{Rmg}{B^2L^2}\right)\left(\frac{\sin\theta}{\cos^2\theta}\right)$$

Can you show that this answer is reasonable in the limiting cases $\theta = 0$, $B = 0$, and $\theta = 90°$, and for R very large or very small?

32.12 [III] The rod shown in Fig. 32-7 rotates about point-C as pivot with a constant frequency of 5.0 rev/s. Find the potential difference between its two ends, which are 80 cm apart, due to the magnetic field $B = 0.30$ T directed into the page.

Consider an imaginary loop $CADC$. As time goes on, its area and the flux through it will both increase. The induced emf in this loop will equal the potential difference we seek.

$$|\mathscr{E}| = N\left|\frac{\Delta\Phi_M}{\Delta t}\right| = (1)\left(\frac{B\,\Delta A}{\Delta t}\right)$$

It takes one-fifth second for the area to change from zero to that of a full circle, πr^2. Therefore,

$$|\mathscr{E}| = B\frac{\Delta A}{\Delta t} = B\frac{\pi r^2}{0.20\text{ s}} = (0.30\text{ T})\frac{\pi(0.80\text{ m})^2}{0.20\text{ s}} = 3.0\text{ V}$$

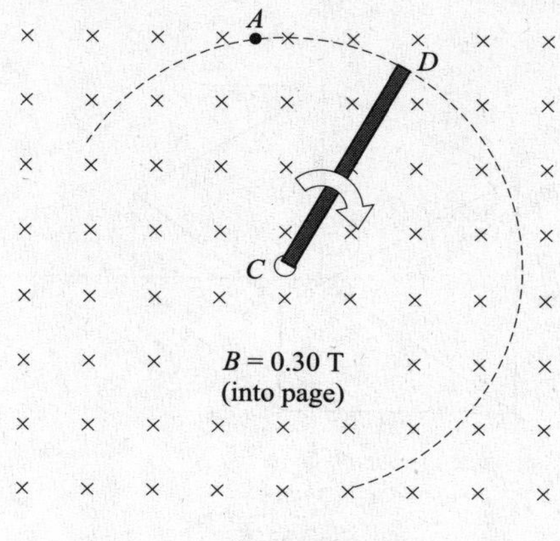

Fig. 32-7

32.13 [III] A 5.0-Ω coil, of 100 turns and diameter 6.0 cm, is placed between the poles of a magnet so that the magnetic flux is maximum through the coil's cross-sectional area. When the coil is suddenly removed from the field of the magnet, a charge of 1.0×10^{-4} C flows through a 595-Ω galvanometer connected to the coil. Compute B between the poles of the magnet.

As the coil is removed, the flux changes from BA, where A is the coil's cross-sectional area, to zero. Therefore,

$$|\mathscr{E}| = N\left|\frac{\Delta\Phi_M}{\Delta t}\right| = N\frac{BA}{\Delta t}$$

We are told that $\Delta q = 1.0 \times 10^{-4}$ C. But, by Ohm's Law,

$$|\mathscr{E}| = IR = \frac{\Delta q}{\Delta t}R$$

where $R = 600\ \Omega$ is the total resistance. If we now equate these two expressions for $|\mathscr{E}|$ and solve for B, we find

$$B = \frac{R\Delta q}{NA} = \frac{(600\ \Omega)(1.0 \times 10^{-4}\ \text{C})}{(100)(\pi \times 9.0 \times 10^{-4}\ \text{m}^2)} = 0.21\ \text{T}$$

Supplementary Problems

32.14 [II] A flux of 9.0×10^{-4} Wb is produced in the iron core of a solenoid. When the core is removed, a flux (in air) of 5.0×10^{-7} Wb is produced in the same solenoid by the same current. What is the relative permeability of the iron? *Ans.* 1.8×10^3

32.15 [I] In Fig. 32-8 there is a $+x$-directed uniform magnetic field of 0.2 T filling the space. Find the magnetic flux through each face of the box shown. *Ans.* Zero through bottom and rear and front sides; through top, 1 mWb; through left side, 2 mWb; through right side, 0.8 mWb.

32.16 [II] A solenoid 60 cm long has 5000 turns of wire and is wound on an iron rod having a 0.75 cm radius. Find the flux inside the solenoid when the current through the wire is 3.0 A. The relative permeability of the iron is 300. *Ans.* 1.7 m/Wb

32.17 [II] A room has its walls aligned accurately with respect to north, south, east, and west. The north wall has an area of 15 m², the east wall has an area of 12 m², and the floor's area is 35 m². At the site the Earth's

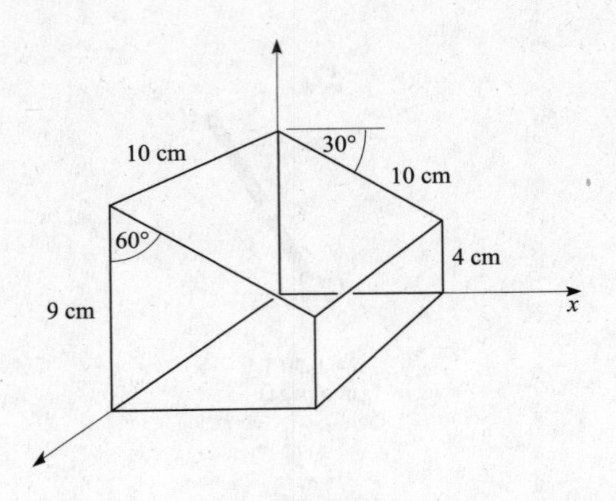

Fig. 32-8

magnetic field has a value of 0.60 G and is directed 50° below the horizontal and 7.0° east of north. Find the magnetic flux through the north wall, the east wall, and the floor. *Ans.* 0.57 mWb, 56 μWb, 1.6 mWb

32.18 [I] The flux through the solenoid of Problem 32.16 is reduced to a value of 1.0 mWb in a time of 0.050 s. Find the induced emf in the solenoid. *Ans.* 67 V

32.19 [II] A flat coil with a radius of 8.0 mm has 50 turns of wire. It is placed in a magnetic field $B = 0.30$ T in such a way that the maximum flux goes through it. Later, it is rotated in 0.020 s to a position such that no flux goes through it. Find the average emf induced between the terminals of the coil. *Ans.* 0.15 V

32.20 [II] The square coil shown in Fig. 32-9 is 20 cm on a side and has 15 turns of wire. It is moving to the right at 3.0 m/s. Find the induced emf (magnitude and direction) in it (*a*) at the instant shown and (*b*) when the entire coil is in the field region. The uniform magnetic field is 0.40 T into the page. *Ans.* (*a*) 3.6 V counterclockwise; (*b*) zero

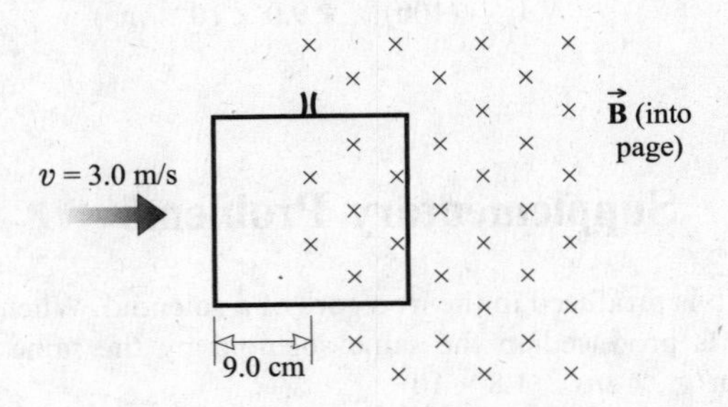

Fig. 32-9

32.21 [I] The cylindrical magnet at the center of Fig. 32-10 rotates as shown on a pivot through its center. At the instant shown, in what direction is the induced current flowing (*a*) in resistor *AB*? (*b*) in resistor *CD*? *Ans.* (*a*) directly from *B* to *A*; (*b*) directly from *C* to *D*

32.22 [II] A train is moving directly south at a constant speed of 10 m/s. If the downward vertical component of the Earth's magnetic field is 0.54 G, compute the magnitude and direction of the emf induced in a rail car axle 1.2 m long. *Ans.* 0.65 mV from west to east

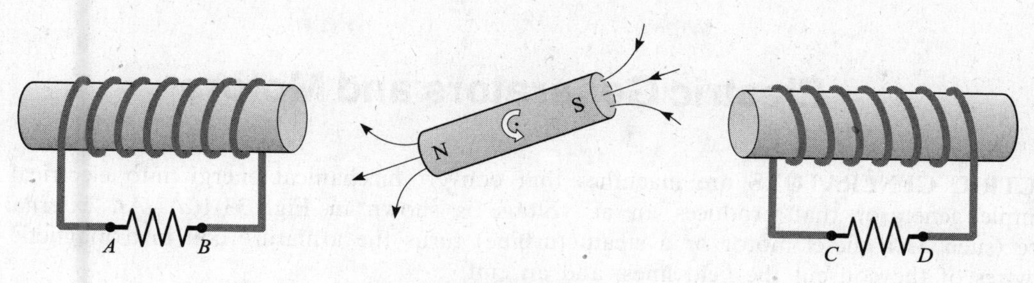

Fig. 32-10

32.23 [III] A copper disk of 10 cm radius is rotating at 20 rev/s about its central symmetry axis. The plane of the disk is perpendicular to a uniform magnetic field $B = 0.60$ T. What is the potential difference between the center and rim of the disk? (*Hint*: There is some similarity with Problem 32.12.) *Ans.* 0.38 V

32.24 [II] How much charge will flow through a 200-Ω galvanometer connected to a 400-Ω circular coil of 1000 turns wound on a wooden stick 2.0 cm in diameter, if a uniform magnetic field $B = 0.011\,3$ T parallel to the axis of the stick is decreased suddenly to zero? *Ans.* 5.9 μC

32.25 [III] In Fig. 32-6, described in Problem 32.11, what is the acceleration of the rod when its speed down the incline is v? *Ans.* $g \sin\theta - (B^2L^2v/Rm)\cos^2\theta$

Chapter 33

Electric Generators and Motors

ELECTRIC GENERATORS are machines that convert mechanical energy into electrical energy. A simple generator that produces an ac voltage is shown in Fig. 33-1(a). An external energy source (such as a diesel motor or a steam turbine) turns the armature coil in a magnetic field $\vec{\mathbf{B}}$. The wires of the coil cut the field lines, and an emf

$$\mathcal{E} = 2\pi NABf \cos 2\pi ft$$

is induced between the terminals of the coil. In this relation, N is the number of loops (each of area A) on the coil, and f is the frequency of its rotation. Figure 33-1(b) shows the emf in graphical form.

As current is drawn from the generator, the wires of its coil experience a retarding force because of the interaction between current and field. Thus the work required to rotate the coil is the source of the electrical energy supplied by the generator. For any generator,

$$\text{(input mechanical energy)} = \text{(output electrical energy)} + \text{(friction and heat losses)}$$

Usually the losses are only a very small fraction of the input energy.

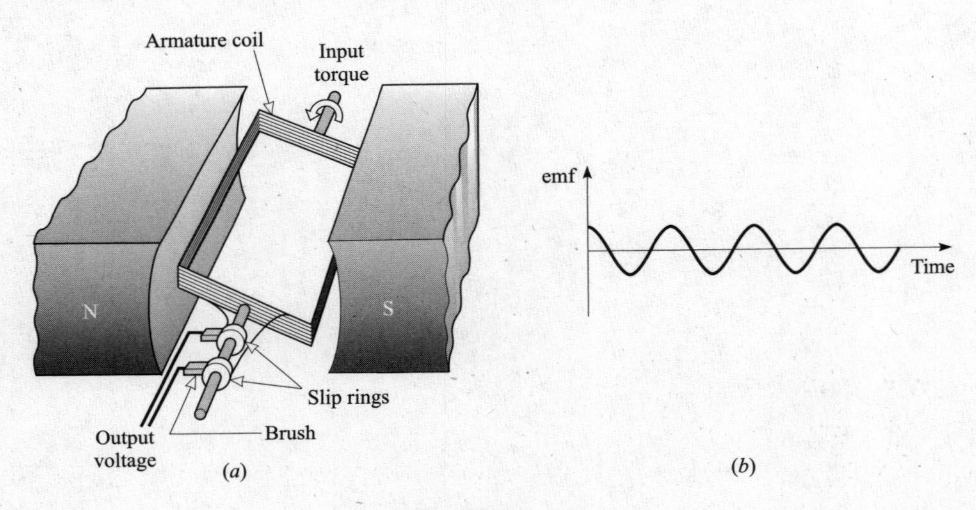

Fig. 33-1

ELECTRIC MOTORS convert electrical energy into mechanical energy. A simple dc motor (i.e., one that runs on a constant voltage) is shown in Fig. 33-2. The current through the armature coil interacts with the magnetic field to cause a torque

$$\tau = NIAB \sin \theta$$

on the coil (see Chapter 30), which rotates the coil and shaft. Here, θ is the angle between the field lines and the perpendicular to the plane of the coil. The split-ring commutator reverses I each time $\sin \theta$ changes sign, thereby ensuring that the torque always rotates the coil in the same sense. For such a motor,

$$\text{Average torque} = \text{(constant)}\ |NIAB|$$

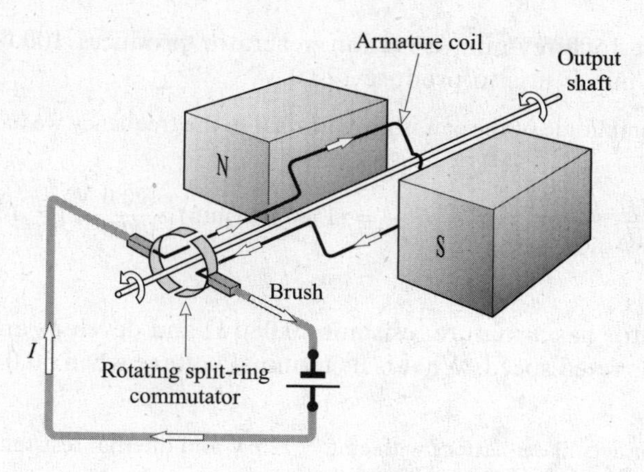

Fig. 33-2

Because the rotating armature coil of the motor acts as a generator, a **back** (or **counter**) **emf** is induced in the coil. The back emf opposes the voltage source that drives the motor. Hence, the net potential difference that causes current through the armature is

$$\text{Net p.d. across armature} = (\text{line voltage}) - (\text{back emf})$$

and

$$\text{Armature current} = \frac{(\text{line voltage}) - (\text{back emf})}{\text{armature resistance}}$$

The mechanical power P developed within the armature of a motor is

$$\text{P} = (\text{armature current})(\text{back emf})$$

The useful mechanical power delivered by the motor is slightly less, due to friction, windage, and iron losses.

Solved Problems

ELECTRIC GENERATORS

33.1 [I] An ac generator produces an output voltage of $\mathscr{E} = 170 \sin 377t$ volts, where t is in seconds. What is the frequency of the ac voltage?

A sine curve plotted as a function of time is no different from a cosine curve, except for the location of $t = 0$. Since $\mathscr{E} = 2\pi NABf \cos 2\pi ft$, we have $377t = 2\pi ft$, from which we find that the frequency $f = 60$ Hz.

33.2 [II] How fast must a 1000-turn coil (each with a 20 cm^2 area) turn in the Earth's magnetic field of 0.70 G to generate a voltage that has a maximum value (i.e., an amplitude) of 0.50 V?

We assume the coil's axis to be oriented in the field so as to give maximum flux change when rotated. Then $B = 7.0 \times 10^{-5}$ T in the expression

$$\mathscr{E} = 2\pi NABf \cos 2\pi ft$$

Because $\cos 2\pi ft$ has a maximum value of unity, the amplitude of the voltage is $2\pi NABf$. Therefore,

$$f = \frac{0.50\ \text{V}}{2\pi NAB} = \frac{0.50\ \text{V}}{(2\pi)(1000)(20 \times 10^{-4}\ \text{m}^2)(7.0 \times 10^{-5}\ \text{T})} = 0.57\ \text{kHz}$$

33.3 [I] When turning at 1500 rev/min, a certain generator produces 100.0 V. What must be its frequency in rev/min if it is to produce 120.0 V?

Because the amplitude of the emf is proportional to the frequency we have, for two frequencies f_1 and f_2,

$$\frac{\mathscr{E}_1}{\mathscr{E}_2} = \frac{f_1}{f_2} \qquad \text{or} \qquad f_2 = f_1 \frac{\mathscr{E}_2}{\mathscr{E}_1} = (1500 \text{ rev/min})\left(\frac{120.0 \text{ V}}{100.0 \text{ V}}\right) = 1800 \text{ rev/min}$$

33.4 [II] A certain generator has armature resistance 0.080 Ω and develops an induced emf of 120 V when driven at its rated speed. What is its terminal voltage when 50.0 A is being drawn from it?

The generator acts like a battery with emf = 120 V and internal resistance $r = 0.080\ \Omega$. As with a battery,

$$\text{Terminal p.d.} = (\text{emf}) - Ir = 120 \text{ V} - (50.0 \text{ A})(0.080\ \Omega) = 116 \text{ V}$$

33.5 [III] Some generators, called *shunt generators*, use electromagnets in place of permanent magnets, with the field coils for the electromagnets activated by the induced voltage. The magnet coil is in parallel with the armature coil (it shunts the armature). As shown in Fig. 33-3, a certain shunt generator has an armature resistance of 0.060 Ω and a shunt resistance of 100 Ω. What power is developed in the armature when it delivers 40 kW at 250 V to an external circuit?

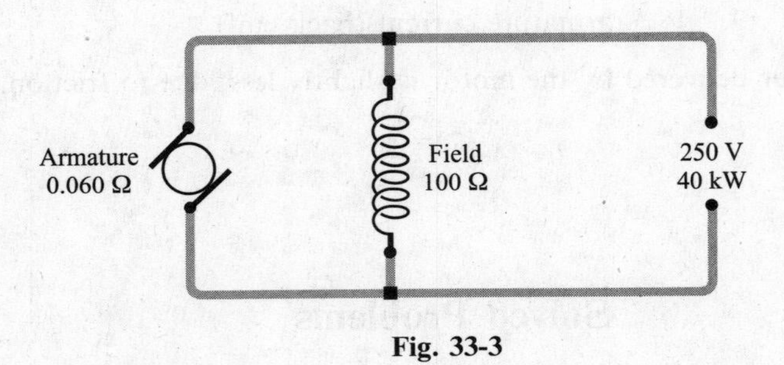

Fig. 33-3

From $\mathrm{P} = VI$,

$$\text{Current to the external circuit} = I_x = \frac{\mathrm{P}}{V} = \frac{40\,000 \text{ W}}{250 \text{ V}} = 160 \text{ A}$$

$$\text{Field current} = I_f = \frac{V_f}{r_f} = \frac{250 \text{ V}}{100\ \Omega} = 2.5 \text{ A}$$

$$\text{Armature current} = I_a = I_x + I_f = 162.5 \text{ A}$$

$$\text{Total induced emf} = |\mathscr{E}| = (250 \text{ V} + I_a r_a \text{ drop in armature})$$

$$= 250 \text{ V} + (162.5 \text{ A})(0.06\ \Omega) = 260 \text{ V}$$

$$\text{Armature power} = I_a|\mathscr{E}| = (162.5 \text{ A})(260 \text{ V}) = 42 \text{ kW}$$

Alternative Method

$$\text{Power loss in the armature} = I_a^2 r_a = (162.5 \text{ A})^2(0.06\ \Omega) = 1.6 \text{ kW}$$

$$\text{Power loss in the field} = I_f^2 r_f = (2.5 \text{ A})^2(100\ \Omega) = 0.6 \text{ kW}$$

$$\text{Power developed} = (\text{power delivered}) + (\text{power loss in armature}) + (\text{power loss in field})$$

$$= 40 \text{ kW} + 1.6 \text{ kW} + 0.6 \text{ kW} = 42 \text{ kW}$$

ELECTRIC MOTORS

33.6 [II] The resistance of the armature in the motor shown in Fig. 33-2 is 2.30 Ω. It draws a current of 1.60 A when operating on 120 V. What is its back emf under these circumstances?

The motor acts like a back emf in series with an IR drop through its internal resistance. Therefore,

$$\text{Line voltage} = \text{back emf} + Ir$$

or

$$\text{Back emf} = 120\ \text{V} - (1.60\ \text{A})(2.30\ \Omega) = 116\ \text{V}$$

33.7 [II] A 0.250-hp motor (like that in Fig. 33-2) has a resistance of 0.500 Ω. (*a*) How much current does it draw on 110 V when its output is 0.250 hp? (*b*) What is its back emf?

(*a*) Assume the motor to be 100 percent efficient so that the input power VI equals its output power (0.250 hp). Then

$$(110\ \text{V})(I) = (0.250\ \text{hp})(746\ \text{W/hp}) \qquad \text{or} \qquad I = 1.695\ \text{A}$$

(*b*)

$$\text{Back emf} = (\text{line voltage}) - Ir = 110\ \text{V} - (1.695\ \text{A})(0.500\ \Omega) = 109\ \text{V}$$

33.8 [III] In a *shunt motor*, the permanent magnet is replaced by an electromagnet activated by a field coil that shunts the armature. The shunt motor shown in Fig. 33-4 has an armature resistance of 0.050 Ω and is connected to a 120 V line. (*a*) What is the armature current at the starting instant, i.e., before the armature develops any back emf? (*b*) What starting rheostat resistance R, in series with the armature, will limit the starting current to 60 A? (*c*) With no starting resistance, what back emf is generated when the armature current is 20 A? (*d*) If this machine were running as a generator, what would be the total induced emf developed by the armature when the armature is delivering 20 A at 120 V to the shunt field and external circuit?

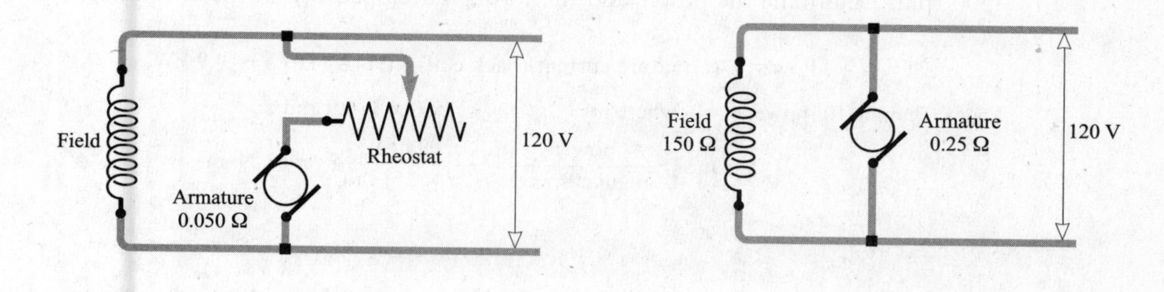

Fig. 33-4

Fig. 33-5

(*a*)

$$\text{Armature current} = \frac{\text{impressed voltage}}{\text{armature resistance}} = \frac{120\ \text{V}}{0.050\ \Omega} = 2.4\ \text{kA}$$

(*b*)

$$\text{Armature current} = \frac{\text{impressed voltage}}{0.050\ \Omega + R} \qquad \text{or} \qquad 60\ \text{A} = \frac{120\ \text{V}}{0.050\ \Omega + R}$$

from which $R = 2.0\ \Omega$.

(*c*)

$$\begin{aligned}\text{Back emf} &= (\text{impressed voltage}) - (\text{voltage drop in armature resistance})\\ &= 120\ \text{V} - (20\ \text{A})(0.050\ \Omega) = 119\ \text{V} = 0.12\ \text{kV}\end{aligned}$$

(*d*)

$$\begin{aligned}\text{Induced emf} &= (\text{terminal voltage}) + (\text{voltage drop in armature resistance})\\ &= 120\ \text{V} + (20\ \text{A})(0.050\ \Omega) = 121\ \text{V} = 0.12\ \text{kV}\end{aligned}$$

33.9 [III] The shunt motor shown in Fig. 33-5 has an armature resistance of 0.25 Ω and a field resistance of 150 Ω. It is connected across 120-V mains and is generating a back emf of 115 V. Compute: (*a*) the armature current I_a, the field current I_f, and the total current I_t taken by the motor; (*b*) the total power taken by the motor; (*c*) the power lost in heat in the armature and field circuits; (*d*) the electrical efficiency of this machine (when only heat losses in the armature and field are considered).

(*a*)
$$I_a = \frac{(\text{impressed voltage}) - (\text{back emf})}{\text{armature resistance}} = \frac{(120 - 115)}{0.25\ \Omega} = 20\ \text{A}$$
$$I_f = \frac{\text{impressed voltage}}{\text{field resistance}} = \frac{120\ \text{V}}{150\ \Omega} = 0.80\ \text{A}$$
$$I_t = I_a + I_f = 20.80\ \text{A} = 21\ \text{A}$$

(*b*)
$$\text{Power input} = (120\ \text{V})(20.80\ \text{A}) = 2.5\ \text{kW}$$

(*c*)
$$I_a^2 r_a \text{ loss in armature} = (20\ \text{A})^2(0.25\ \Omega) = 0.10\ \text{kW}$$
$$I_f^2 r_f \text{ loss in field} = (0.80\ \text{A})^2(150\ \Omega) = 96\ \text{W}$$

(*d*)
$$\text{Power output} = (\text{power input}) - (\text{power losses}) = 2496 - (100 + 96) = 2.3\ \text{kW}$$

Alternatively,

$$\text{Power output} = (\text{armature current})(\text{back emf}) = (20\ \text{A})(115\ \text{V}) = 2.3\ \text{kW}$$

Then
$$\text{Efficiency} = \frac{\text{power output}}{\text{power input}} = \frac{2300\ \text{W}}{2496\ \text{W}} = 0.921 = 92\%$$

33.10 [II] A motor has a back emf of 110 V and an armature current of 90 A when running at 1500 rpm. Determine the power and the torque developed within the armature.

$$\text{Power} = (\text{armature current})(\text{back emf}) = (90\ \text{A})(110\ \text{V}) = 9.9\ \text{kW}$$

From Chapter 10, power $= \tau\omega$ where $\omega = 2\pi f = 2\pi(1500 \times 1/60)$ rad/s

$$\text{Torque} = \frac{\text{power}}{\text{angular speed}} = \frac{9900\ \text{W}}{(2\pi \times 25)\ \text{rad/s}} = 63\ \text{N} \cdot \text{m}$$

33.11 [III] A motor armature develops a torque of 100 N·m when it draws 40 A from the line. Determine the torque developed if the armature current is increased to 70 A and the magnetic field strength is reduced to 80 percent of its initial value.

The torque developed by the armature of a given motor is proportional to the armature current and to the field strength (see Chapter 30). In other words, the ratio of the torques equals the ratio of the two sets of values of $|NIAB|$. Using subscripts i and f for *initial* and *final* values, $\tau_f/\tau_i = I_f B_f / I_i B_i$ hence,

$$\tau_f = (100\ \text{N} \cdot \text{m})\left(\frac{70}{40}\right)(0.80) = 0.14\ \text{kN} \cdot \text{m}$$

Supplementary Problems

ELECTRIC GENERATORS

33.12 [I] Determine the separate effects on the induced emf of a generator if (*a*) the flux per pole is doubled, and (*b*) the speed of the armature is doubled. *Ans.* (*a*) doubled; (*b*) doubled

33.13 [II] The emf induced in the armature of a shunt generator is 596 V. The armature resistance is 0.100 Ω. (*a*) Compute the terminal voltage when the armature current is 460 A. (*b*) The field resistance is 110 Ω. Determine the field current, and the current and power delivered to the external circuit. *Ans.* (*a*) 550 V; (*b*) 5 A, 455 A, 250 kW

33.14 [II] A dynamo (generator) delivers 30.0 A at 120 V to an external circuit when operating at 1200 rpm. What torque is required to drive the generator at this speed if the total power losses are 400 W? *Ans.* 31.8 N·m

33.15 [II] A 75.0-kW, 230-V shunt generator has a generated emf of 243.5 V. If the field current is 12.5 A at rated output, what is the armature resistance? *Ans.* 0.0399 Ω

33.16 [III] A 120-V generator is run by a windmill that has blades 2.0 m long. The wind, moving at 12 m/s, is slowed to 7.0 m/s after passing the windmill. The density of air is 1.29 kg/m^3. If the system has no losses, what is the largest current the generator can produce? (*Hint*: How much energy does the wind lose per second?) *Ans.* 77 A

ELECTRIC MOTORS

33.17 [II] A generator has an armature with 500 turns, which cut a flux of 8.00 mWb during each rotation. Compute the back emf it develops when run as a motor at 1500 rpm. *Ans.* 100 V

33.18 [I] The active length of each armature conductor of a motor is 30 cm, and the conductors are in a field of 0.40 Wb/m^2. A current of 15 A flows in each conductor. Determine the force acting on each conductor. *Ans.* 1.8 N

33.19 [II] A shunt motor with armature resistance 0.080 Ω is connected to 120 V mains. With 50 A in the armature, what are the back emf and the mechanical power developed within the armature? *Ans.* 0.12 kV, 5.8 kW

33.20 [II] A shunt motor is connected to a 110-V line. When the armature generates a back emf of 104 V, the armature current is 15 A. Compute the armature resistance. *Ans.* 0.40 Ω

33.21 [II] A shunt dynamo has an armature resistance of 0.120 Ω. (*a*) If it is connected across 220-V mains and is running as a motor, what is the induced (back) emf when the armature current is 50.0 A? (*b*) If this machine is running as a generator, what is the induced emf when the armature is delivering 50.0 A at 220 V to the shunt field and external circuit? *Ans.* (*a*) 214 V; (*b*) 226 V

33.22 [II] A shunt motor has a frequency of 900 rpm when it is connected to 120-V mains and delivering 12 hp. The total losses are 1048 W. Compute the power input, the line current, and the motor torque. *Ans.* 10 kW, 83 A, 93 N·m

33.23 [II] A shunt motor has armature resistance 0.20 Ω and field resistance 150 Ω, and draws 30 A when connected to a 120-V supply line. Determine the field current, the armature current, the back emf, the mechanical power developed within the armature, and the electrical efficiency of the machine. *Ans.* 0.80 A, 29 A, 0.11 kV, 3.3 kW, 93%

33.24 [II] A shunt motor develops 80 N·m of torque when the flux density in the air gap is 1.0 $\mathrm{Wb/m^2}$ and the armature current is 15 A. What is the torque when the flux density is 1.3 $\mathrm{Wb/m^2}$ and the armature current is 18 A? *Ans.* 0.13 kN·m

33.25 [II] A shunt motor has a field resistance of 200 Ω and an armature resistance of 0.50 Ω and is connected to 120-V mains. The motor draws a current of 4.6 A when running at full speed. What current will be drawn by the motor if the speed is reduced to 90 percent of full speed by application of a load? *Ans.* 28 A

Chapter 34

Inductance; *R-C* and *R-L* Time Constants

SELF-INDUCTANCE (L): A coil can induce an emf in itself. If the current in a coil changes, the flux through the coil due to the current also changes. As a result, the changing current in a coil induces an emf in that same coil.

Because an induced emf e is proportional to $\Delta\Phi_M/\Delta t$ and because $\Delta\Phi_M$ is proportional to Δi, where i is the current that causes the flux,

$$\mathscr{E} = -(\text{constant})\frac{\Delta i}{\Delta t}$$

Here i is the current through the same coil in which e is induced. (We shall denote a time-varying current by i instead of I.) The minus sign indicates that the self-induced emf $\mathscr{E}$ is a back emf and opposes the change in current.

The proportionality constant depends upon the geometry of the coil. We represent it by L and call it the **self-inductance** of the coil. Then

$$\mathscr{E} = -L\frac{\Delta i}{\Delta t}$$

For $\mathscr{E}$ in units of V, i in units of A, and t in units of s, L is in **henries** (H).

MUTUAL INDUCTANCE (M): When the flux from one coil threads through another coil, an emf can be induced in either one by the other. The coil that contains the power source is called the *primary coil.* The other coil, in which an emf is induced by the changing current in the primary, is called the *secondary coil.* The induced secondary emf $\mathscr{E}_s$ is proportional to the time rate of change of the primary current, $\Delta i_p/\Delta t$:

$$\mathscr{E}_s = M\frac{\Delta i_p}{\Delta t}$$

where M is a constant called the **mutual inductance** of the two-coil system.

ENERGY STORED IN AN INDUCTOR: Because of its self-induced back emf, work must be done to increase the current through an inductor from zero to I. The energy furnished to the coil in the process is stored in the coil and can be recovered as the coil's current is decreased once again to zero. If a current I is flowing in an inductor of self-inductance L, then the energy stored in the inductor is

$$\text{Stored energy} = \tfrac{1}{2}LI^2$$

For L in units of H and I in units of A, the energy is in J.

***R-C* TIME CONSTANT:** Consider the *R-C* circuit shown in Fig. 34-1(a). The capacitor is initially uncharged. If the switch is now closed, the current i in the circuit and the charge q on the capacitor vary as shown in Fig. 34-1(b). If we call the p.d. across the capacitor v_c, writing the loop rule for this circuit gives

$$-iR - v_c + \mathscr{E} = 0 \qquad \text{or} \qquad i = \frac{\mathscr{E} - v_c}{R}$$

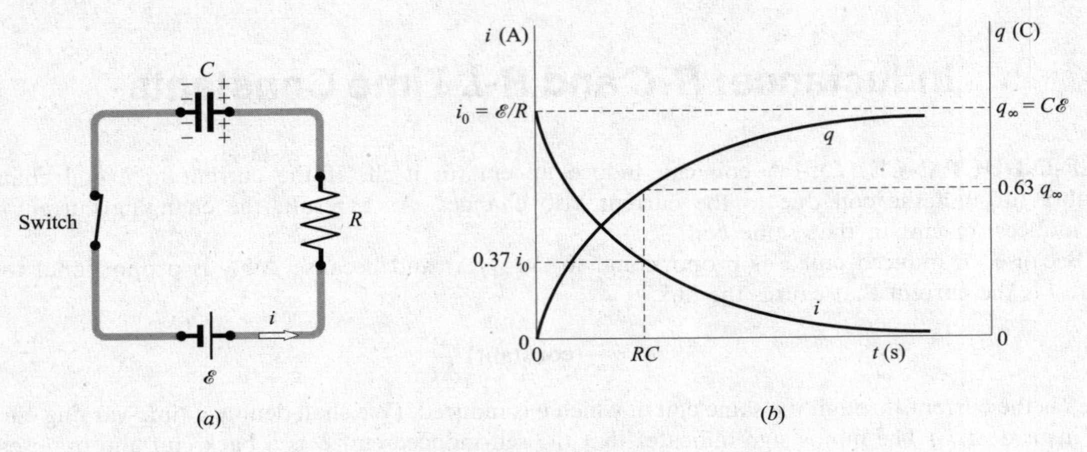

Fig. 34-1

At the first instant after the switch is closed, $v_c = 0$ and $i = \mathscr{E}/R$. As time goes on, v_c increases and i decreases. The time, in seconds, taken for the current to drop to 1/2.718 or 0.368 of its initial value is RC, which is called the **time constant** of the *R-C* circuit.

Also shown in Fig. 34-1(*b*) is the variation of q, the charge on the capacitor, with time. At $t = RC$, q has attained 0.632 of its final value.

When a charged capacitor C with initial charge q_0 is discharged through a resistor R, its discharge current follows the same curve as for charging. The charge q on the capacitor follows a curve similar to that for the discharge current. At time RC, $i = 0.368 i_0$ and $q = 0.368 q_0$ during discharge.

***R-L* TIME CONSTANT:** Consider the circuit in Fig. 34-2(*a*). The symbol (coil symbol) represents a coil having a self-inductance of L henries. When the switch in the circuit is first closed, the current in the circuit rises as shown in Fig. 34-2(*b*). The current does not jump to its final value because the changing flux through the coil induces a back emf in the coil, which opposes the rising current. After L/R seconds, the current has risen to 0.632 of its final value i_∞. This time, $t = L/R$, is called the **time constant** of the *R-L* circuit. After a long time, the current is changing so slowly that the back emf in the inductor, $L(\Delta i/\Delta t)$, is negligible. Then $i = i_\infty = \mathscr{E}/R$.

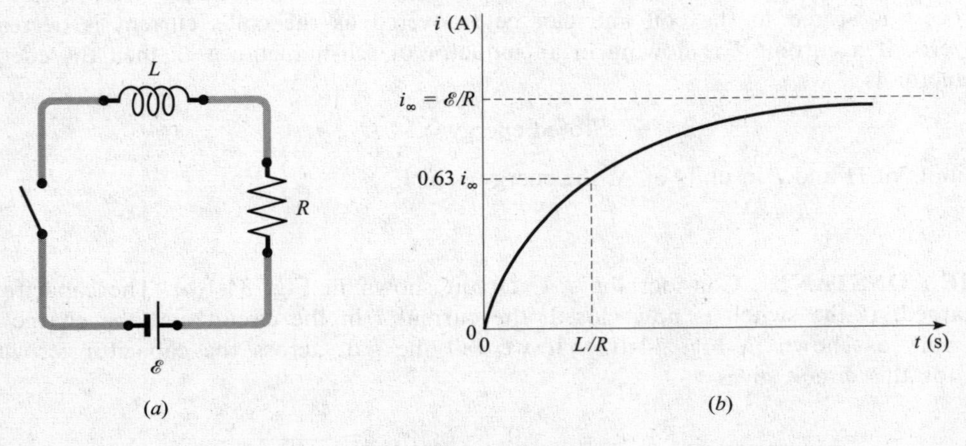

Fig. 34-2

EXPONENTIAL FUNCTIONS are used as follows to describe the curves of Figs 34-1 and 34-2:

$i = i_0\, e^{-t/RC}$	capacitor charging and discharging
$q = q_\infty(1 - e^{-t/RC})$	capacitor charging
$q = q_\infty\, e^{-t/RC}$	capacitor discharging
$i = i_\infty(1 - e^{-t/(L/R)})$	inductor current buildup

where $e = 2.718$ is the base of the natural logarithms.

When t is equal to the time constant, the relations for a capacitor give $i = 0.368 i_0$ and $q = 0.632 q_\infty$ for charging, and $q = 0.368 q_\infty$ for discharging. The equation for current in an inductor gives $i = 0.632 i_\infty$ when t equals the time constant.

The equation for i in the capacitor circuit (as well as for q in the capacitor discharge case) has the following property: After n time constants have passed,

$$i = i_0(0.368)^n \qquad \text{and} \qquad q = q_\infty(0.368)^n$$

For example, after four time constants have passed,

$$i = i_0(0.368)^4 = 0.018\,3 i_0$$

Solved Problems

34.1 [II] A steady current of 2 A in a coil of 400 turns causes a flux of 10^{-4} Wb to link (pass through) the loops of the coil. Compute (*a*) the average back emf induced in the coil if the current is stopped in 0.08 s, (*b*) the inductance of the coil, and (*c*) the energy stored in the coil.

(*a*)
$$|\mathscr{E}| = N\left|\frac{\Delta\Phi_M}{\Delta t}\right| = 400\,\frac{(10^{-4} - 0)\ \text{Wb}}{0.08\ \text{s}} = 0.5\ \text{V}$$

(*b*)
$$|\mathscr{E}| = L\left|\frac{\Delta i}{\Delta t}\right| \qquad \text{or} \qquad L = \left|\frac{\mathscr{E}\Delta t}{\Delta i}\right| = \frac{(0.5\ \text{V})(0.08\ \text{s})}{(2-0)\ \text{A}} = 0.02\ \text{H}$$

(*c*)
$$\text{Energy} = \tfrac{1}{2}LI^2 = \tfrac{1}{2}(0.02\ \text{H})(2\ \text{A})^2 = 0.04\ \text{J}$$

34.2 [III] A long air-core solenoid has cross-sectional area A and N loops of wire on its length d. (*a*) Find its self-inductance. (*b*) What is its inductance if the core material has a permeability of μ?

(*a*) We can write

$$|\mathscr{E}| = N\left|\frac{\Delta\Phi_M}{\Delta t}\right| \qquad \text{and} \qquad |\mathscr{E}| = L\left|\frac{\Delta i}{\Delta t}\right|$$

Equating these two expressions for $|\mathscr{E}|$ yields

$$L = N\left|\frac{\Delta\Phi_M}{\Delta i}\right|$$

If the current changes from zero to I, then the flux changes from zero to Φ_M. Therefore, $\Delta i = I$ and $\Delta\Phi_M = \Phi_M$ in this case. The self-inductance, assumed constant for all cases, is then

$$L = N\frac{\Phi_M}{I} = N\frac{BA}{I}$$

But, for an air-core solenoid, $B = \mu_0 nI = \mu_0(N/d)I$. Substitution gives $L = \mu_0 N^2 A/d$.

(*b*) If the material of the core has permeability μ instead of μ_0, then B, and therefore L, will be increased by the factor μ/μ_0. In that case, $L = \mu N^2 A/d$. An iron-core solenoid has a much higher self-inductance than an air-core solenoid has.

34.3 [II] A solenoid 30 cm long is made by winding 2000 turns of wire on an iron rod whose cross-sectional area is 1.5 cm^2. If the relative permeability of the iron is 600, what is the self-inductance of the solenoid? What average emf is induced in the solenoid as the current in it is decreased from 0.60 A to 0.10 A in a time of 0.030 s? Refer back to Problem 34.2.

From Problem 34.2(*b*) with $k_M = \mu/\mu_0$,

$$L = \frac{k_m \mu_0 N^2 A}{d} = \frac{(600)(4\pi \times 10^{-7}\ \mathrm{T \cdot m/A})(2000)^2(1.5 \times 10^{-4}\ \mathrm{m^2})}{0.30\ \mathrm{m}} = 1.51\ \mathrm{H}$$

and

$$|\mathscr{E}| = L\left|\frac{\Delta i}{\Delta t}\right| = (1.51\ \mathrm{H})\frac{0.50\ \mathrm{A}}{0.030\ \mathrm{s}} = 25\ \mathrm{V}$$

34.4 [II] At a certain instant, a coil with a resistance of 0.40 Ω and a self-inductance of 200 mH carries a current of 0.30 A that is increasing at the rate of 0.50 A/s. (*a*) What is the potential difference across the coil at that instant? (*b*) Repeat if the current is decreasing at 0.50 A/s.

We can represent the coil by a resistance in series with an emf (the induced emf), as shown in Fig. 34.3.

(*a*) Because the current is increasing, $\mathscr{E}$ will oppose the current and therefore have the polarity shown. We write the loop equation for the circuit:

$$V_{ba} - iR - \mathscr{E} = 0$$

Since V_{ba} is the voltage across the coil, and since $\mathscr{E} = L|\Delta i/\Delta t|$, we have

$$V_{\mathrm{coil}} = iR + \mathscr{E} = (0.30\ \mathrm{A})(0.40\ \Omega) + (0.200\ \mathrm{H})(0.50\ \mathrm{A/s}) = 0.22\ \mathrm{V}$$

(*b*) With i decreasing, the induced emf must be reversed in Fig. 34-3. This gives $V_{\mathrm{coil}} = iR - \mathscr{E} = 0.020$ V.

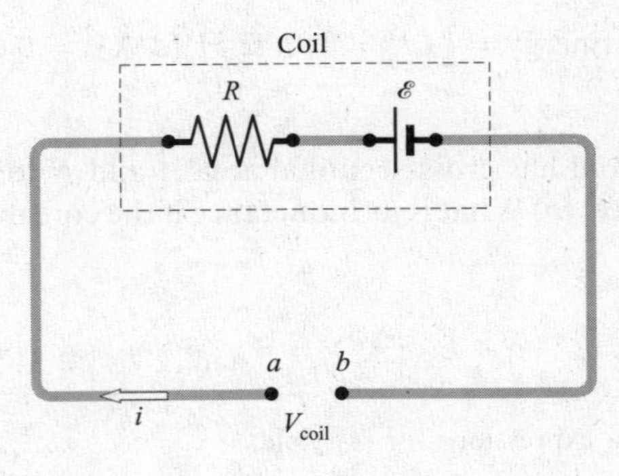

Fig. 34-3

34.5 [II] A coil of resistance 15 Ω and inductance 0.60 H is connected to a steady 120-V power source. At what rate will the current in the coil rise (*a*) at the instant the coil is connected to the power source, and (*b*) at the instant the current reaches 80 percent of its maximum value?

The effective driving voltage in the circuit is the 120 V power supply minus the induced back emf, $L(\Delta i/\Delta t)$. This equals the p.d. in the resistance of the coil:

$$120\text{ V} - L\frac{\Delta i}{\Delta t} = iR$$

[This same equation can be obtained by writing the loop equation for the circuit of Fig. 34-2(*a*). In doing so, remember that the inductance acts as a back emf of value $L\ \Delta i/\Delta t$.]

(*a*) At the first instant, i is essentially zero. Then

$$\frac{\Delta i}{\Delta t} = \frac{120\text{ V}}{L} = \frac{120\text{ V}}{0.60\text{ H}} = 0.20\text{ mA/s}$$

(*b*) The current reaches a maximum value of (120 V)/R when the current finally stops changing (i.e., when $\Delta i/\Delta t = 0$). We are interested in the case when

$$i = (0.80)\left(\frac{120\text{ V}}{R}\right)$$

Substitution of this value for i in the loop equation gives

$$120\text{ V} - L\frac{\Delta i}{\Delta t} = (0.80)\left(\frac{120\text{ V}}{R}\right)R$$

from which

$$\frac{\Delta i}{\Delta t} = \frac{(0.20)(120\text{ V})}{L} = \frac{(0.20)(120\text{ V})}{0.60\text{ H}} = 40\text{ A/s}$$

34.6 [II] When the current in a certain coil is changing at a rate of 3.0 A/s, it is found that an emf of 7.0 mV is induced in a nearby coil. What is the mutual inductance of the combination?

$$\mathscr{E}_s = M\frac{\Delta i_p}{\Delta t} \qquad \text{or} \qquad M = \mathscr{E}_s\frac{\Delta t}{\Delta i_p} = (7.0\times10^{-3}\text{ V})\frac{1.0\text{ s}}{3.0\text{ A}} = 2.3\text{ mH}$$

34.7 [II] Two coils are wound on the same iron rod so that the flux generated by one passes through the other also. The primary coil has N_p loops and, when a current of 2.0 A flows through it, the flux in it is 2.5×10^{-4} Wb. Determine the mutual inductance of the two coils if the secondary coil has N_s loops.

$$|\mathscr{E}_s| = N_s\left|\frac{\Delta\Phi_{Ms}}{\Delta t}\right| \qquad \text{and} \qquad |\mathscr{E}_s| = M\left|\frac{\Delta i_p}{\Delta t}\right|$$

give

$$M = N_s\left|\frac{\Delta\Phi_{Ms}}{\Delta i_p}\right| = N_s\frac{(2.5\times10^{-4} - 0)\text{ Wb}}{(2.0-0)\text{ A}} = (1.3\times10^{-4}\ N_s)\text{ H}$$

34.8 [II] A 2000-loop solenoid is wound uniformly on a long rod with length d and cross-section A. The relative permeability of the iron is k_m. On top of this is wound a 50-loop coil which is used as a secondary. Find the mutual inductance of the system.

The flux through the solenoid is

$$\Phi_M = BA = (k_M\mu_0 nI_p)A = (k_M\mu_0 I_p A)\left(\frac{2000}{d}\right)$$

This same flux goes through the secondary. We have, then,

$$|\mathscr{E}_s| = N_s\left|\frac{\Delta\Phi_M}{\Delta t}\right| \qquad \text{and} \qquad |\mathscr{E}_s| = M\left|\frac{\Delta i_p}{\Delta t}\right|$$

from which

$$M = N_s \left| \frac{\Delta \Phi_M}{\Delta i_p} \right| = N_s \frac{\Phi_M - 0}{I_p - 0} = 50 \frac{k_M \mu_0 I_p A(2000/d)}{I_p} = \frac{10 \times 10^4 \, k_M \mu_0 A}{d}$$

34.9 [II] A certain series circuit consists of a 12-V battery, a switch, a 1.0-MΩ resistor, and a 2.0-μF capacitor, initially uncharged. If the switch is now closed, find (*a*) the initial current in the circuit, (*b*) the time for the current to drop to 0.37 of its initial value, (*c*) the charge on the capacitor then, and (*d*) the final charge on the capacitor.

(*a*) The loop rule applied to the circuit of Fig. 34-1(*a*) at any instant gives

$$12 \text{ V} - iR - v_c = 0$$

where v_c is the p.d. across the capacitor. At the first instant, q is essentially zero and so $v_c = 0$. Then

$$12 \text{ V} - iR - 0 = 0 \qquad \text{or} \qquad i = \frac{12 \text{ V}}{1.0 \times 10^6 \; \Omega} = 12 \; \mu\text{A}$$

(*b*) The current drops to 0.37 of its initial value when

$$t = RC = (1.0 \times 10^6 \; \Omega)(2.0 \times 10^{-6} \text{ F}) = 2.0 \text{ s}$$

(*c*) At $t = 2.0$ s the charge on the capacitor has increased to 0.63 of its final value. [See part (*d*) below.]

(*d*) The charge ceases to increase when $i = 0$ and $v_c = 12$ V. Therefore,

$$q_{\text{final}} = Cv_c = (2.0 \times 10^{-6} \text{ F})(12 \text{ V}) = 24 \; \mu\text{C}$$

34.10 [II] A 5.0-μF capacitor is charged to a potential difference of 20 kV across its plates. After being disconnected from the power source, it is connected across a 7.0-MΩ resistor to discharge. What is the initial discharge current, and how long will it take for the capacitor voltage to decrease to 37 percent of the 20 kV?

The loop equation for the discharging capacitor is

$$v_c - iR = 0$$

where v_c is the p.d. across the capacitor. At the first instant, $v_c = 20$ kV, so

$$i = \frac{v_c}{R} = \frac{20 \times 10^3 \text{ V}}{7.0 \times 10^6 \; \Omega} = 2.9 \text{ mA}$$

The potential across the capacitor, as well as the charge on it, will decrease to 0.37 of its original value in one time constant. The required time is

$$RC = (7.0 \times 10^6 \; \Omega)(5.0 \times 10^{-6} \text{ F}) = 35 \text{ s}$$

34.11 [II] A coil has an inductance of 1.5 H and a resistance of 0.60 Ω. If the coil is suddenly connected across a 12-V battery, find the time required for the current to rise to 0.63 of its final value. What will be the final current through the coil?

The time required is the time constant of the circuit:

$$\text{Time constant} = \frac{L}{R} = \frac{1.5 \text{ H}}{0.60 \; \Omega} = 2.5 \text{ s}$$

After a long time, the current will be steady and so no back emf will exist in the coil. Under those conditions,

$$I = \frac{\mathscr{E}}{R} = \frac{12\ \text{V}}{0.60\ \Omega} = 20\ \text{A}$$

34.12 [I] A capacitor that has been charged to 2.0×10^5 V is allowed to discharge through a resistor. What will be the voltage across the capacitor after five time constants have elapsed?

We know (p. 333) that after n time constants, $q = q_\infty(0.368)^n$. Because v is proportional to q (that is, $v = q/C$), we may write

$$v_{n=5} = (2.0 \times 10^5\ \text{V})(0.368)^5 = 1.4\ \text{kV}$$

34.13 [II] A 2.0-μF capacitor is charged through a 30-MΩ resistor by a 45-V battery. Find (*a*) the charge on the capacitor and (*b*) the current through the resistor, both determined 83 s after the charging process starts.

The time constant of the circuit is $RC = 60$ s. Also,

$$q_\infty = V_\infty C = (45\ \text{V})(2.0 \times 10^{-6}\ \text{F}) = 9.0 \times 10^{-5}\ \text{C}$$

(*a*)

$$q = q_\infty(1 - e^{-t/RC}) = (9.0 \times 10^{-5}\ \text{C})(1 - e^{-83/60})$$

But

$$e^{-83/60} = e^{-1.383} = 0.25$$

Then substitution gives

$$q = (9.0 \times 10^{-5}\ \text{C})(1 - 0.25) = 67\ \mu\text{C}$$

(*b*)

$$i = i_0 e^{-t/RC} = \left(\frac{45\ \text{V}}{30 \times 10^6\ \Omega}\right)(e^{-1.383}) = 0.38\ \mu\text{A}$$

34.14 [II] If, in Fig. 34-2, $R = 20\ \Omega$, $L = 0.30$ H, and $\mathscr{E} = 90$ V, what will be the current in the circuit 0.050 s after the switch is closed?

We are going to use the exponential equation for i given on p.333.
The time constant for this circuit is $L/R = 0.015$ s, and $i_\infty = \mathscr{E}/R = 4.5$ A. Then

$$i = i_\infty(1 - e^{-t/(L/R)}) = (4.5\ \text{A})(1 - e^{-3.33}) = (4.5\ \text{A})(1 - 0.0357) = 4.3\ \text{A}$$

Supplementary Problems

34.15 [I] An emf of 8.0 V is induced in a coil when the current in it changes at the rate of 32 A/s. Compute the inductance of the coil. *Ans.* 0.25 H

34.16 [I] A steady current of 2.5 A creates a flux of 1.4×10^{-4} Wb in a coil of 500 turns. What is the inductance of the coil? *Ans.* 28 mH

34.17 [I] The mutual inductance between the primary and secondary of a transformer is 0.30 H. Compute the induced emf in the secondary when the primary current changes at the rate of 4.0 A/s. *Ans.* 1.2 V

34.18 [II] A coil of inductance 0.20 H and 1.0-Ω resistance is connected to a constant 90-V source. At what rate will the current in the coil grow (*a*) at the instant the coil is connected to the source, and (*b*) at the instant the current reaches two-thirds of its maximum value? *Ans.* (*a*) 0.45 kA/s; (*b*) 0.15 kA/s

34.19 [II] Two neighboring coils, *A* and *B*, have 300 and 600 turns, respectively. A current of 1.5 A in *A* causes 1.2×10^{-4} Wb to pass through *A* and 0.90×10^{-4} Wb to pass through *B*. Determine (*a*) the self-inductance of *A*, (*b*) the mutual inductance of *A* and *B*, and (*c*) the average induced emf in *B* when the current in *A* is interrupted in 0.20 s. *Ans.* (*a*) 24 mH; (*b*) 36 mH; (*c*) 0.27 V

34.20 [I] A coil of 0.48 H carries a current of 5 A. Compute the energy stored in it. *Ans.* 6 J

34.21 [I] The iron core of a solenoid has a length of 40 cm and a cross-section of 5.0 cm^2, and is wound with 10 turns of wire per cm of length. Compute the inductance of the solenoid, assuming the relative permeability of the iron to be constant at 500. *Ans.* 0.13 H

34.22 [I] Show that (*a*) $1\ N/A^2 = 1\ T \cdot m/A = 1\ Wb/A \cdot m = 1\ H/m$, and (*b*) $1\ C^2/N \cdot m^2 = 1\ F/m$.

34.23 [II] A series circuit consisting of an uncharged 2.0-μF capacitor and a 10-MΩ resistor is connected across a 100-V power source. What are the current in the circuit and the charge on the capacitor (*a*) after one time constant, and (*b*) when the capacitor has acquired 90 percent of its final charge? *Ans.* (*a*) 3.7 μA, 0.13 mC; (*b*) 1.0 μA, 0.18 mC

34.24 [II] A charged capacitor is connected across a 10-kΩ resistor and allowed to discharge. The potential difference across the capacitor drops to 0.37 of its original value after a time of 7.0 s. What is the capacitance of the capacitor? *Ans.* 0.70 mF

34.25 [II] When a long iron-core solenoid is connected across a 6-V battery, the current rises to 0.63 of its maximum value after a time of 0.75 s. The experiment is then repeated with the iron core removed. Now the time required to reach 0.63 of the maximum is 0.0025 s. Calculate (*a*) the relative permeability of the iron and (*b*) *L* for the air-core solenoid if the maximum current is 0.5 A. *Ans.* (*a*) 0.3×10^3; (*b*) 0.03 H

34.26 [I] What fraction of the initial current still flows in the circuit of Fig. 34-1 seven time constants after the switch has been closed? *Ans.* 0.000 91

34.27 [II] By what fraction does the current in Fig. 34-2 differ from i_∞ three time constants after the switch is first closed? *Ans.* $(i_\infty - i)/i_\infty = 0.050$

34.28 [II] In Fig. 34-2, $R = 5.0\ \Omega$, $L = 0.40$ H, and $\mathscr{E} = 20$ V. Find the current in the circuit 0.20 s after the switch is first closed. *Ans.* 3.7 A

34.29 [II] The capacitor in Fig. 34-1 is initially uncharged when the switch is closed. Find the current in the circuit and the charge on the capacitor five seconds later. Use $R = 7.00\ M\Omega$, $C = 0.300\ \mu F$, and $\mathscr{E} = 12.0$ V. *Ans.* 159 nA, 3.27 μC

Chapter 35

Alternating Current

THE EMF GENERATED BY A ROTATING COIL in a magnetic field has a graph similar to the one shown in Fig. 35-1. It is called an *ac voltage* because there is a reversal of polarity (i.e., the voltage changes sign); ac voltages need not be sinusoidal. If the coil rotates with a frequency of f revolutions per second, then the emf has a frequency of f in hertz (cycles per second). The instantaneous voltage v that is generated has the form

$$v = v_0 \sin \omega t = v_0 \sin 2\pi f t$$

where v_0 is the amplitude (maximum value) of the voltage in volts, and $\omega = 2\pi f$ is the angular velocity in rad/s. The frequency f of the voltage is related to its period T by

$$T = \frac{1}{f}$$

where T is in seconds.

Rotating coils are not the only source of ac voltages; electronic devices for generating ac voltages are very common. Alternating voltages produce alternating currents.

An alternating current produced by a typical generator has a graph much like that for the voltage shown in Fig. 35-1. Its instantaneous value is i, and its amplitude is i_0. Often the current and voltage do not reach a maximum at the same time, even though they both have the same frequency.

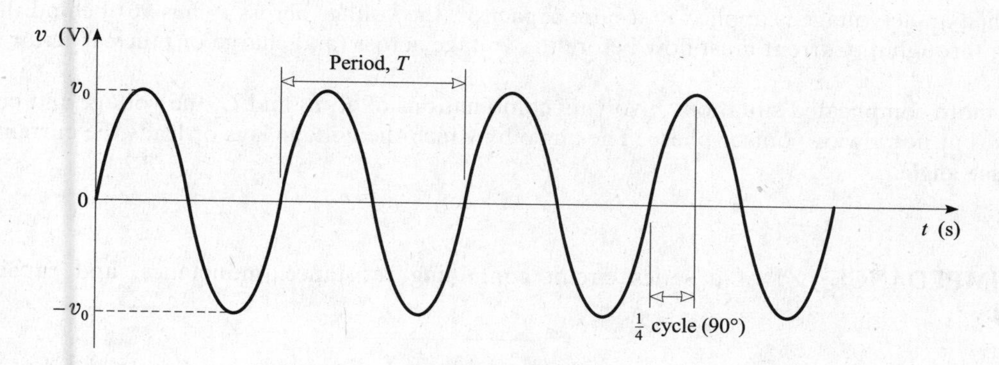

Fig. 35-1

METERS (i.e., measuring devices) for use in ac circuits read the **effective**, or **root mean square** (rms), values of the current and voltage. These values are always positive and are related to the amplitudes of the instantaneous sinusoidal values through

$$V = V_{\text{rms}} = \frac{v_0}{\sqrt{2}} = 0.707 v_0$$

$$I = I_{\text{rms}} = \frac{i_0}{\sqrt{2}} = 0.707 i_0$$

It is customary to represent meter readings by capital letters (V, I), while instantaneous values are represented by small letters (v, i).

THE THERMAL ENERGY GENERATED OR POWER LOST by an rms current I in a resistor R is given by I^2R.

FORMS OF OHM'S LAW: Suppose that a sinusoidal current of frequency f with rms value I flows through a pure resistor R, or a pure inductor L, or a pure capacitor C. Then an ac voltmeter placed across the element in question will read an rms voltage V as follows:

$$\text{Pure resistor:} \quad V = IR$$

$$\text{Pure inductor:} \quad V = IX_L$$

where $X_L = 2\pi fL$ is called the **inductive reactance**. Its unit is ohms when L is in henries and f is in hertz.

$$\text{Pure capacitor:} \quad V = IX_C$$

where $X_C = 1/2\pi fC$ is called the **capacitive reactance**. Its unit is ohms when C is in farads.

PHASE: When an ac voltage is applied to a pure resistance, the voltage across the resistance and the current through it attain their maximum values at the same instant and their zero values at the same instant; the voltage and current are said to be *in-phase*.

When an ac voltage is applied to a pure inductance, the voltage across the inductance reaches its maximum value one-quarter cycle ahead of the current, i.e., when the current is zero. The back emf of the inductance causes the current through the inductance to lag behind the voltage by one-quarter cycle (or 90°), and the two are 90° *out-of-phase*.

When an ac voltage is applied to a pure capacitor, the voltage across it lags 90° behind the current flowing through it. Current must flow before the voltage across (and charge on) the capacitor can build up.

In more complicated situations involving combinations of R, L, and C, the voltage and current are usually (but not always) out-of-phase. The angle by which the voltage lags or leads the current is called the **phase angle.**

THE IMPEDANCE (Z) of a series circuit containing resistance, inductance, and capacitance is given by

$$Z = \sqrt{R^2 + (X_L - X_C)^2}$$

with Z in ohms. If a voltage V is applied to such a series circuit, then a form of Ohm's Law relates V to the current I through it:

$$V = IZ$$

The phase angle ϕ between V and I is given by

$$\tan\phi = \frac{X_L - X_C}{R} \qquad \text{or} \qquad \cos\phi = \frac{R}{Z}$$

PHASORS: A **phasor** is a quantity that behaves, in many regards, like a vector. Phasors are used to describe series R-L-C circuits because the above expression for the impedance can be associated with the Pythagorean theorem for a right triangle. As shown in Fig. 35-2(a), Z is the hypotenuse of the right triangle, while R and $(X_L - X_C)$ are its two legs. The angle labeled ϕ is the phase angle between the current and the voltage.

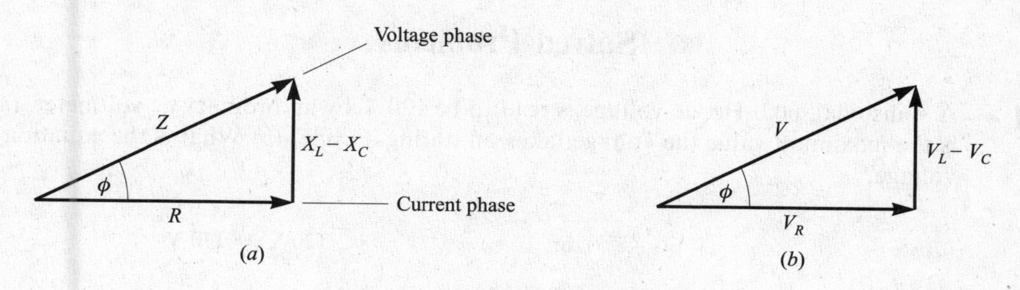

Fig. 35-2

A similar relation applies to the voltages across the elements in the series circuit. As shown in Fig. 35-2(*b*), it is

$$V^2 = V_R^2 + (V_L - V_C)^2$$

Because of the phase differences a measurement of the voltage across a series circuit is not equal to the algebraic sum of the individual voltage readings across its elements. Instead, the above relation must be used.

RESONANCE occurs in a series *R-L-C* circuit when $X_L = X_C$. Under this condition $Z = R$ is minimum, so that I is maximum for a given value of V. Equating X_L to X_C, we find for the **resonant** (or **natural**) **frequency** of the circuit

$$f_0 = \frac{1}{2\pi\sqrt{LC}}$$

POWER LOSS: Suppose that an ac voltage V is impressed across an impedance of any type. It gives rise to a current I through the impedance, and the phase angle between V and I is ϕ. The power loss in the impedance is given by

$$\text{Power loss} = VI \cos \phi$$

The quantity $\cos \phi$ is called the **power factor**. It is unity for a pure resistor; but it is zero for a pure inductor or capacitor (no power loss occurs in a pure inductor or capacitor).

A TRANSFORMER is a device used to raise or lower the voltage in an ac circuit. It consists of a primary and a secondary coil wound on the same iron core. An alternating current in one coil creates a continuously changing magnetic flux through the core. This change of flux induces an alternating emf in the other coil.

The efficiency of a transformer is usually very high. Thus, we may often *neglect losses* and write

$$\text{Power in primary} = \text{power in secondary}$$
$$V_1 I_1 = V_2 I_2$$

The voltage ratio equals the ratio of the numbers of turns on the two coils; the current ratio equals the inverse ratio of the numbers of turns:

$$\frac{V_1}{V_2} = \frac{N_1}{N_2} \quad \text{and} \quad \frac{I_1}{I_2} = \frac{N_2}{N_1}$$

Solved Problems

35.1 [I] A sinusoidal, 60.0-Hz, ac voltage is read to be 120 V by an ordinary ac voltmeter. (*a*) What is the maximum value the voltage takes on during a cycle? (*b*) What is the equation for the voltage?

(*a*) $$V = \frac{v_0}{\sqrt{2}} \quad \text{or} \quad v_0 = \sqrt{2}V = \sqrt{2}(120\text{ V}) = 170\text{ V}$$

(*b*) $$v = v_0 \sin 2\pi ft = (170\text{ V})\sin 120\pi t$$

where t is in s, and v_0 is the maximum voltage.

35.2 [I] A voltage $v = (60.0\text{ V})\sin 120\pi t$ is applied across a 20.0-Ω resistor. What will an ac ammeter in series with the resistor read?

The rms voltage across the resistor is

$$V = 0.707v_0 = (0.707)(60.0\text{ V}) = 42.4\text{ V}$$

Then $$I = \frac{V}{R} = \frac{42.4\text{ V}}{20.0\ \Omega} = 2.12\text{ A}$$

35.3 [II] A 120-V ac voltage source is connected across a 2.0-μF capacitor. Find the current to the capacitor if the frequency of the source is (*a*) 60 Hz and (*b*) 60 kHz. (*c*) What is the power loss in the capacitor?

(*a*) $$X_C = \frac{1}{2\pi fC} = \frac{1}{2\pi(60\text{ s}^{-1})(2.0\times 10^{-6}\text{ F})} = 1.33\text{ k}\Omega$$

Then $$I = \frac{V}{X_C} = \frac{120\text{ V}}{1330\ \Omega} = 0.090\text{ A}$$

(*b*) Now $X_C = 1.33\ \Omega$, so $I = 90$ A. Notice that the impedance of a capacitor varies inversely with the frequency.

(*c*) Inasmuch as $\cos\phi = R/Z$ and $R = 0$;

$$\text{Power loss} = VI\cos\phi = VI\cos 90^\circ = 0$$

35.4 [II] A 120-V ac voltage source is connected across a pure 0.700-H inductor. Find the current through the inductor if the frequency of the source is (*a*) 60.0 Hz and (*b*) 60.0 kHz. (*c*) What is the power loss in the inductor?

(*a*) $$X_L = 2\pi fL = 2\pi(60.0\text{ s}^{-1})(0.700\text{ H}) = 264\ \Omega$$

Then $$I = \frac{V}{X_L} = \frac{120\text{ V}}{264\ \Omega} = 0.455\text{ A}$$

(*b*) Now $X_L = 264\times 10^3\ \Omega$, so $I = 0.455\times 10^{-3}$ A. Notice that the impedance of an inductor varies directly with the frequency.

(*c*) Inasmuch as $\cos\phi = R/Z$ and $R = 0$;

$$\text{Power loss} = VI\cos\phi = VI\cos 90^\circ = 0$$

35.5 [II] A coil having inductance 0.14 H and resistance of 12 Ω is connected across a 110-V, 25-Hz line. Compute (*a*) the current in the coil, (*b*) the phase angle between the current and the supply voltage, (*c*) the power factor, and (*d*) the power loss in the coil.

(*a*)
$$X_L = 2\pi fL = 2\pi(25)(0.14) = 22.0\ \Omega$$

and
$$Z = \sqrt{R^2 + (X_L - X_C)^2} = \sqrt{(12)^2 + (22 - 0)^2} = 25.1\ \Omega$$

so
$$I = \frac{V}{Z} = \frac{110\ \text{V}}{25.1\ \Omega} = 4.4\ \text{A}$$

(*b*)
$$\tan\phi = \frac{X_L - X_C}{R} = \frac{22 - 0}{12} = 1.83 \quad \text{or} \quad \phi = 61.3^\circ$$

The voltage leads the current by 61°.

(*c*)
$$\text{Power factor} = \cos\phi = \cos 61.3^\circ = 0.48$$

(*d*)
$$\text{Power loss} = VI\cos\phi = (110\ \text{V})(4.4\ \text{A})(0.48) = 0.23\ \text{kW}$$

Or, since power loss occurs only because of the resistance of the coil,

$$\text{Power loss} = I^2R = (4.4\ \text{A})^2(12\ \Omega) = 0.23\ \text{kW}$$

35.6 [II] A capacitor is in series with a resistance of 30 Ω and is connected to a 220-V ac line. The reactance of the capacitor is 40 Ω. Determine (*a*) the current in the circuit, (*b*) the phase angle between the current and the supply voltage, and (*c*) the power loss in the circuit.

(*a*)
$$Z = \sqrt{R^2 + (X_L - X_C)^2} = \sqrt{(30)^2 + (0 - 40)^2} = 50\ \Omega$$

so
$$I = \frac{V}{Z} = \frac{220\ \text{V}}{50\ \Omega} = 4.4\ \text{A}$$

(*b*)
$$\tan\phi = \frac{X_L - X_C}{R} = \frac{0 - 40}{30} = -1.33 \quad \text{or} \quad \phi = -53^\circ$$

The minus sign tells us that the voltage *lags* the current by 53°. The angle ϕ in Fig. 35-2 would lie below the horizontal axis.

(*c*) **Method 1**

$$\text{Power loss} = VI\cos\phi = (220)(4.4)\cos(-53^\circ) = (220)(4.4)\cos 53^\circ = 0.58\ \text{kW}$$

Method 2

Because the power loss occurs only in the resistor, and not in the pure capacitor,

$$\text{Power loss} = I^2R = (4.4\ \text{A})^2(30\ \Omega) = 0.58\ \text{kW}$$

35.7 [III] A series circuit consisting of a 100-Ω noninductive resistor, a coil with a 0.10-H inductance and negligible resistance, and a 20-μF capacitor is connected across a 110-V, 60-Hz power source. Find (*a*) the current, (*b*) the power loss, (*c*) the phase angle between the current and the source voltage, and (*d*) the voltmeter readings across the three elements.

(*a*) For the entire circuit, $Z = \sqrt{R^2 + (X_L - X_C)^2}$, with

$$R = 100\ \Omega$$
$$X_L = 2\pi fL = 2\pi(60\ \text{s}^{-1})(0.10\ \text{H}) = 37.7\ \Omega$$
$$X_C = \frac{1}{2\pi fC} = \frac{1}{2\pi(60\ \text{s}^{-1})(20 \times 10^{-6}\ \text{F})} = 132.7\ \Omega$$

from which

$$Z = \sqrt{(100)^2 + (38 - 133)^2} = 138\ \Omega \qquad \text{and} \qquad I = \frac{V}{Z} = \frac{110\ \text{V}}{138\ \Omega} = 0.79\ \text{A}$$

(*b*) The power loss all occurs in the resistor, so

$$\text{Power loss} = I^2 R = (0.79\ \text{A})^2(100\ \Omega) = 63\ \text{W}$$

(*c*)
$$\tan\phi = \frac{X_L - X_C}{R} = \frac{-95\ \Omega}{100\ \Omega} = -0.95 \qquad \text{or} \qquad \phi = -44°$$

The voltage lags the current.

(*d*)
$$V_R = IR = (0.79\ \text{A})(100\ \Omega) = 79\ \text{V}$$
$$V_C = IX_C = (0.79\ \text{A})(132.7\ \Omega) = 0.11\ \text{kV}$$
$$V_L = IX_L = (0.79\ \text{A})(37.7\ \Omega) = 30\ \text{V}$$

Notice that $V_C + V_L + V_R$ does not equal the source voltage. From Fig. 35-2(*b*), the correct relationship is

$$V = \sqrt{V_R^2 + (V_L - V_C)^2} = \sqrt{(79)^2 + (-75)^2} = 109\ \text{V}$$

which checks within the limits of rounding-off errors.

35.8 [III] A 5.00-Ω resistance is in a series circuit with a 0.200-H pure inductance and a 40.0-nF pure capacitance. The combination is placed across a 30.0-V, 1780-Hz power supply. Find (*a*) the current in the circuit, (*b*) the phase angle between source voltage and current, (*c*) the power loss in the circuit, and (*d*) the voltmeter reading across each element of the circuit.

(*a*)
$$X_L = 2\pi fL = 2\pi(1780\ \text{s}^{-1})(0.200\ \text{H}) = 2.24\ \text{k}\Omega$$
$$X_C = \frac{1}{2\pi fC} = \frac{1}{2\pi(1780\ \text{s}^{-1})(4.00 \times 10^{-8}\ \text{F})} = 2.24\ \text{k}\Omega$$

and
$$Z = \sqrt{R^2 + (X_L - X_C)^2} = R = 5.00\ \Omega$$

Then
$$I = \frac{V}{Z} = \frac{30.0\ \text{V}}{5.00\ \Omega} = 6.00\ \text{A}$$

(*b*)
$$\tan\phi = \frac{X_L - X_C}{R} = 0 \qquad \text{or} \qquad \phi = 0°$$

(*c*)
$$\text{Power loss} = VI\cos\phi = (30.0\ \text{V})(6.00\ \text{A})(1) = 180\ \text{W}$$

or
$$\text{Power loss} = I^2 R = (6.00\ \text{A})^2(5.00\ \Omega) = 180\ \text{W}$$

(*d*)
$$V_R = IR = (6.00\ \text{A})(5.00\ \Omega) = 30.00\ \text{V}$$
$$V_C = IX_C = (6.00\ \text{A})(2240\ \Omega) = 13.4\ \text{kV}$$
$$V_L = IX_L = (6.00\ \text{A})(2240\ \Omega) = 13.4\ \text{kV}$$

This circuit is in resonance because $X_C = X_L$. Notice how very large the voltages across the inductor and capacitor become, even though the source voltage is low.

35.9 [III] As shown in Fig. 35-3, a series circuit connected across a 200-V, 60-Hz line consists of a capacitor of capacitive reactance 30 Ω, a noninductive resistor of 44 Ω, and a coil of inductive reactance 90 Ω and resistance 36 Ω. Determine (*a*) the current in the circuit, (*b*) the potential difference across each element, (*c*) the power factor of the circuit, and (*d*) the power absorbed by the circuit.

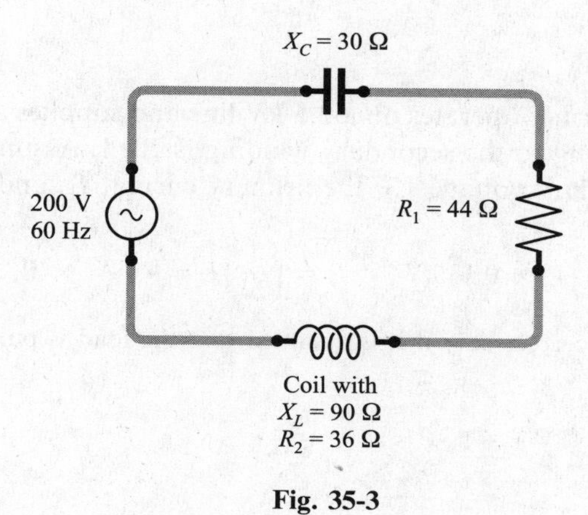

Fig. 35-3

(*a*)
$$Z = \sqrt{(R_1 + R_2)^2 + (X_L - X_C)^2} = \sqrt{(44 + 36)^2 + (90 - 30)^2} = 0.10\ \text{k}\Omega$$

so
$$I = \frac{V}{Z} = \frac{200\ \text{V}}{100\ \Omega} = 2.0\ \text{A}$$

(*b*)
$$\text{p.d. across capacitor} = IX_C = (2.0\ \text{A})(30\ \Omega) = 60\ \text{V}$$
$$\text{p.d. across resistor} = IR_1 = (2.0\ \text{A})(44\ \Omega) = 88\ \text{V}$$
$$\text{Impedance of coil} = \sqrt{R_2^2 + X_L^2} = \sqrt{(36)^2 + (90)^2} = 97\ \Omega$$
$$\text{p.d. across coil} = (2.0\ \text{A})(97\ \Omega) = 0.19\ \text{kV}$$

(*c*)
$$\text{Power factor} = \cos\phi = \frac{R}{Z} = \frac{80}{100} = 0.80$$

(*d*)
$$\text{Power used} = VI\cos\phi = (200\ \text{V})(2\ \text{A})(0.80) = 0.32\ \text{kW}$$

or
$$\text{Power used} = I^2R = (2\ \text{A})^2(80\ \Omega) = 0.32\ \text{kW}$$

35.10 [I] Calculate the resonant frequency of a circuit of negligible resistance containing an inductance of 40.0 mH and a capacitance of 600 pF.

$$f_0 = \frac{1}{2\pi\sqrt{LC}} = \frac{1}{2\pi\sqrt{(40.0 \times 10^{-3}\ \text{H})(600 \times 10^{-12}\ \text{F})}} = 32.5\ \text{kHz}$$

35.11 [I] A step-up transformer is used on a 120-V line to furnish 1800 V. The primary has 100 turns. How many turns are on the secondary?

$$\frac{V_1}{V_2} = \frac{N_1}{N_2} \qquad \text{or} \qquad \frac{120\ \text{V}}{1800\ \text{V}} = \frac{100\ \text{turns}}{N_2}$$

from which $N_2 = 1.50 \times 10^3$ turns.

35.12 [I] A transformer used on a 120-V line delivers 2.0 A at 900 V. What current is drawn from the line? Assume 100 percent efficiency.

$$\text{Power in primary} = \text{power in secondary}$$
$$I_1(120\text{ V}) = (2.0\text{ A})(900\text{ V})$$
$$I_1 = 15\text{ A}$$

35.13 [I] A step-down transformer operates on a 2.5-kV line and supplies a load with 80 A. The ratio of the primary winding to the secondary winding is 20 : 1. Assuming 100 percent efficiency, determine the secondary voltage V_2, the primary current I_1, and the power output P_2.

$$V_2 = \left(\frac{1}{20}\right)V_1 = 0.13\text{ kV} \qquad I_1 = \left(\frac{1}{20}\right)I_2 = 4.0\text{ A} \qquad P_2 = V_2 I_2 = 10\text{ kW}$$

The last expression is correct only if it is assumed that the load is pure resistive, so that the power factor is unity.

Supplementary Problems

35.14 [I] A voltmeter reads 80.0 V when it is connected across the terminals of a sinusoidal power source with $f = 1000$ Hz. Write the equation for the instantaneous voltage provided by the source.
Ans. $v = (113\text{ V})\sin 2000\pi t$ for t in seconds

35.15 [I] An ac current in a 10 Ω resistance produces thermal energy at the rate of 360 W. Determine the effective values of the current and voltage. *Ans.* 6.0 A, 60 V

35.16 [I] A 40.0-Ω resistor is connected across a 15.0-V variable-frequency electronic oscillator. Find the current through the resistor when the frequency is (*a*) 100 Hz and (*b*) 100 kHz. *Ans.* (*a*) 0.375 A; (*b*) 0.375 A

35.17 [I] Solve Problem 35.16 if the 40.0-Ω resistor is replaced by a 2.00-mH inductor. *Ans.* (*a*) 11.9 A; (*b*) 11.9 mA

35.18 [I] Solve Problem 35.16 if the 40.0-Ω resistor is replaced by 0.300-μF capacitor. *Ans.* (*a*) 2.83 mA; (*b*) 2.83 A

35.19 [II] A coil has resistance 20 Ω and inductance 0.35 H. Compute its reactance and its impedance to an alternating current of 25 cycles/s. *Ans.* 55 Ω, 59 Ω

35.20 [II] A current of 30 mA is supplied to a 4.0-μF capacitor connected across an alternating current line having a frequency of 500 Hz. Compute the reactance of the capacitor and the voltage across the capacitor.
Ans. 80 Ω, 2.4 V

35.21 [II] A coil has an inductance of 0.100 H and a resistance of 12.0 Ω. It is connected to a 110-V, 60.0-Hz line. Determine (*a*) the reactance of the coil, (*b*) the impedance of the coil, (*c*) the current through the coil, (*d*) the phase angle between current and supply voltage, (*e*) the power factor of the circuit, and (*f*) the reading of a wattmeter connected in the circuit. *Ans.* (*a*) 37.7 Ω; (*b*) 39.6 Ω; (*c*) 2.78 A; (*d*) voltage leads by 72.3°; (*e*) 0.303; (*f*) 92.6 W

35.22 [III] A 10.0-μF capacitor is in series with a 40.0-Ω resistance, and the combination is connected to a 110-V, 60.0-Hz line. Calculate (*a*) the capacitive reactance, (*b*) the impedance of the circuit, (*c*) the current in the circuit, (*d*) the phase angle between current and supply voltage, and (*e*) the power factor for the circuit. *Ans.* (*a*) 266 Ω; (*b*) 269 Ω; (*c*) 0.409 A; (*d*) voltage lags by 81.4°; (*e*) 0.149

35.23 [III] A circuit having a resistance, an inductance, and a capacitance in series is connected to a 110-V ac line. For the circuit, $R = 9.0\ \Omega$, $X_L = 28\ \Omega$, and $X_C = 16\ \Omega$. Compute (*a*) the impedance of the circuit, (*b*) the current, (*c*) the phase angle between the current and the supply voltage, and (*d*) the power factor of the circuit. *Ans.* (*a*) 15 Ω; (*b*) 7.3 A; (*c*) voltage leads by 53°; (*d*) 0.60

35.24 [II] An experimenter has a coil of inductance 3.0 mH and wishes to construct a circuit whose resonant frequency is 1.0 MHz. What should be the value of the capacitor used? *Ans.* 8.4 pF

35.25 [II] A circuit has a resistance of 11 Ω, a coil of inductive reactance 120 Ω, and a capacitor with a 120-Ω reactance, all connected in series with a 110-V, 60-Hz power source. What is the potential difference across each circuit element? *Ans.* $V_R = 0.11$ kV, $V_L = V_C = 1.2$ kV

35.26 [II] A 120-V, 60-Hz power source is connected across an 800-Ω noninductive resistance and an unknown capacitance in series. The voltage drop across the resistor is 102 V. (*a*) What is the voltage drop across the capacitor? (*b*) What is the reactance of the capacitor? *Ans.* (*a*) 63 V; (*b*) 0.50 kΩ

35.27 [II] A coil of negligible resistance is connected in series with a 90-Ω resistor across a 120-V, 60-Hz line. A voltmeter reads 36 V across the resistance. Find the voltage across the coil and the inductance of the coil. *Ans.* 0.11 kV, 0.76 H

35.28 [I] A step-down transformer is used on a 2.2-kV line to deliver 110 V. How many turns are on the primary winding if the secondary has 25 turns? *Ans.* 5.0×10^2

35.29 [I] A step-down transformer is used on a 1650-V line to deliver 45 A at 110 V. What current is drawn from the line? Assume 100 percent efficiency. *Ans.* 3.0 A

35.30 [II] A step-up transformer operates on a 110-V line and supplies a load with 2.0 A. The ratio of the primary and secondary windings is 1 : 25. Determine the secondary voltage, the primary current, and the power output. Assume a resistive load and 100 percent efficiency. *Ans.* 2.8 kV, 50 A, 5.5 kW

Chapter 36

Reflection of Light

THE NATURE OF LIGHT: Light (along with all other forms of electromagnetic radiation) is a fundamental entity and physics is still struggling to understand it. On an observable level, light manifests two seemingly contradictory behaviors, crudely pictured via wave and particle models. Usually the amount of energy present is so large that light behaves as if it were an ideal continuous wave, a wave of interdependent electric and magnetic fields. The interaction of light with lenses, mirrors, prisms, slits, and so forth, can satisfactorily be understood via the wave model (provided we don't probe too deeply into what's happening on a microscopic level). On the other hand, when light is emitted or absorbed by the atoms of a system, these processes occur as if the radiant energy is in the form of minute, localized, well-directed blasts; that is, as if light is a stream of "particles". Fortunately, without worrying about the very nature of light, we can predict its behavior in a wide range of practical situations.

LAW OF REFLECTION: A ray is a mathematical line drawn perpendicular to the wavefronts of a lightwave. It shows the direction of propagation of electromagnetic energy. In *specular* (or *mirror*) reflection, the angle of incidence (θ_i) equals the angle of reflection (θ_r), as shown in Fig. 36-1. Furthermore, the incident ray, reflected ray, and normal to the surface all lie in the same plane, called the **plane-of-incidence.**

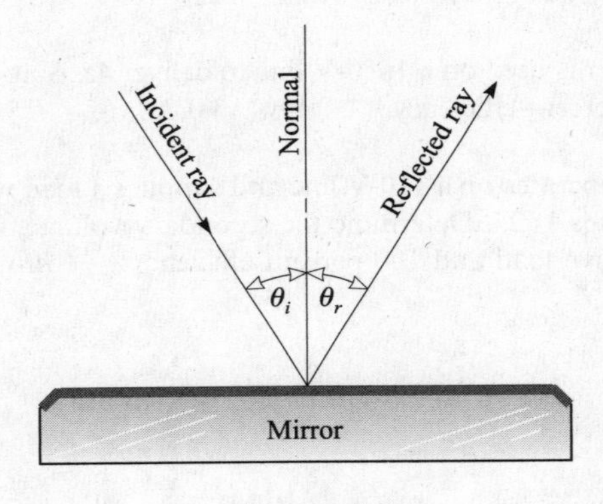

Fig. 36-1

PLANE MIRRORS form images that are erect, of the same size as the object, and as far behind the reflecting surface as the object is in front of it. Such an image is **virtual**; i.e., the image will not appear on a screen located at the position on the image because the light does not converge there.

SPHERICAL MIRRORS: The **principal focus** of a spherical mirror, such as the ones shown in Fig. 36-2, is the point F where rays parallel to and very close to the *central* or **optical axis** of the mirror are focused. This focus is real for a concave mirror and virtual for a convex mirror. It is located on the optical axis and midway between the center of curvature C and the mirror.

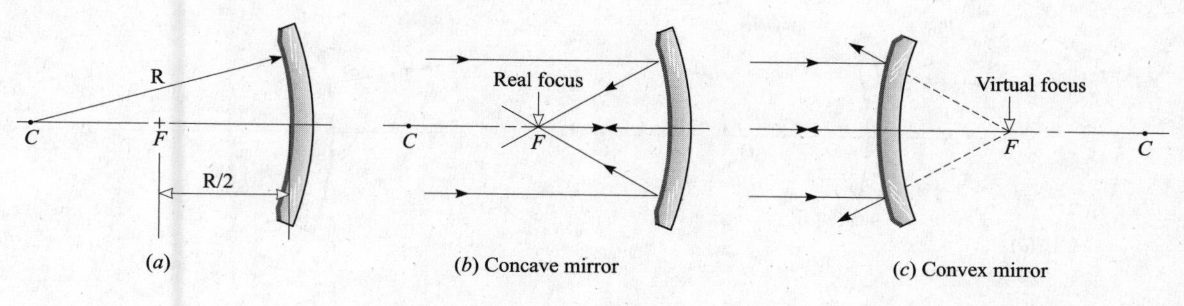

Fig. 36-2

Concave mirrors form inverted real images of objects placed beyond the principal focus. If the object is between the principal focus and the mirror, the image is virtual, erect, and enlarged.

Convex mirrors produce only erect virtual images of objects placed in front of them. The images are diminished (smaller than the object) in size.

RAY TRACING: We can locate the image of any point on an object by tracing at least two rays from that point through the optical system that forms the image—in this case the system is a mirror. There are four especially convenient rays to use because we know, without making any calculations, exactly how they will reflect from the mirror. These rays are shown for a concave spherical mirror in Fig. 36-3, and for a convex spherical mirror in Fig. 36-4. Notice that a line drawn from C to the point of reflection is a radius and therefore normal to the mirror's surface. That line always bisects the angle formed by the incident and reflected rays (i.e., $\theta_i = \theta_r$).

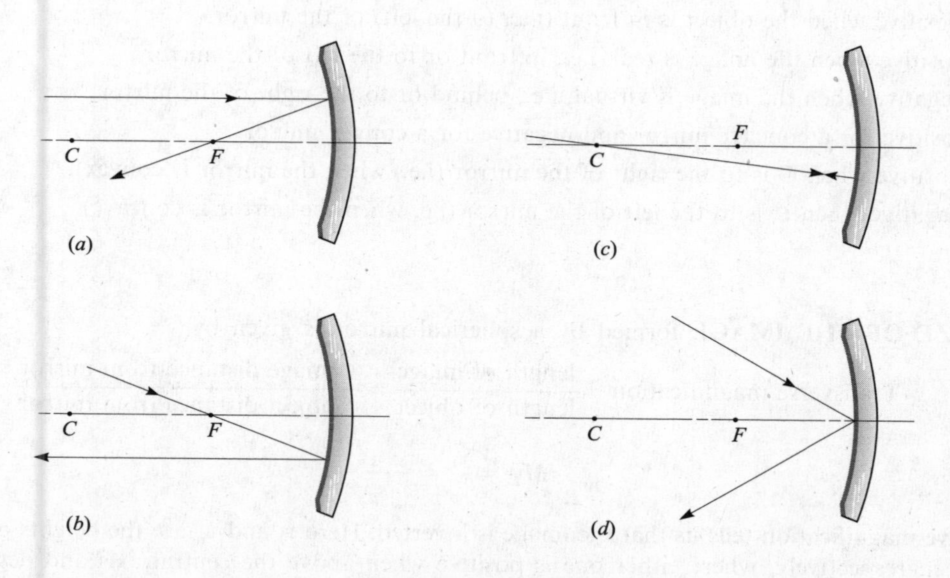

Fig. 36-3

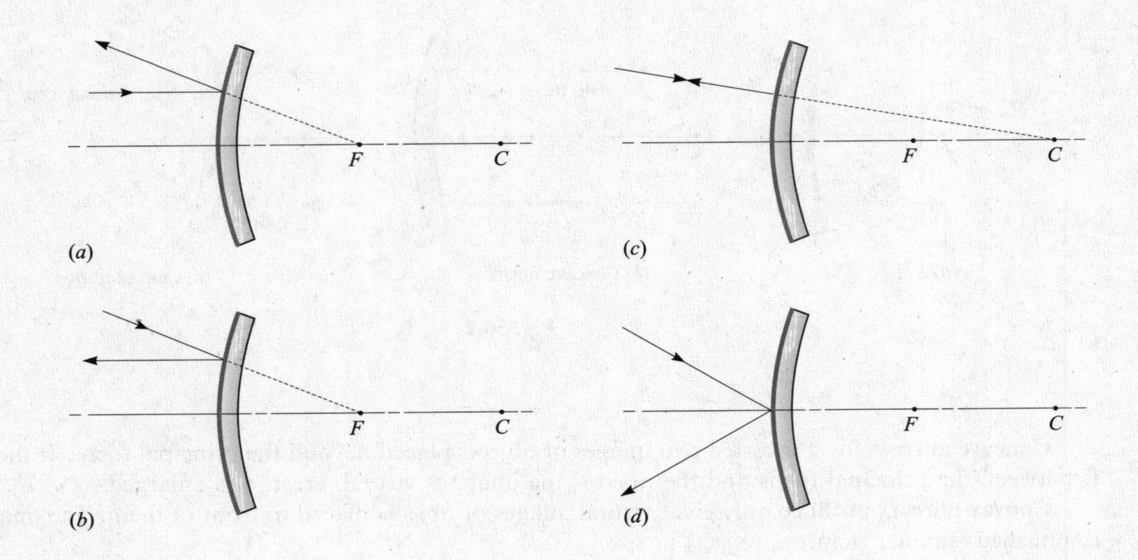

Fig. 36-4

MIRROR EQUATION for both concave and convex spherical mirrors:

$$\frac{1}{s_o} + \frac{1}{s_i} = -\frac{2}{R} = \frac{1}{f}$$

where s_o = object distance from the mirror
s_i = image distance from the mirror
R = radius of curvature of the mirror
f = focal length of the mirror $= -R/2$.

There are several sign conventions; the following is the most widely used one. With light entering from the left:

- s_o is positive when the object is in front (i.e., to the left) of the mirror.
- s_i is positive when the image is real, i.e., in front or to the left of the mirror.
- s_i is negative when the image is virtual, i.e., behind or to the right of the mirror.
- f is positive for a concave mirror and negative for a convex mirror.
- R is positive when C is to the right of the mirror (i.e., when the mirror is convex).
- R is negative when C is to the left of the mirror (i.e, when the mirror is concave).

THE SIZE OF THE IMAGE formed by a spherical mirror is given by

$$\text{Transverse magnification} = \frac{\text{length of image}}{\text{length of object}} = -\frac{\text{image distance from mirror}}{\text{object distance from mirror}}$$

$$M_T = \frac{y_i}{y_o} = -\frac{s_i}{s_o}$$

A negative magnification tells us that the image is inverted. Here y_i and y_o are the heights of the image and object, respectively, where either one is positive when above the central axis and negative when below it.

Solved Problems

36.1 [II] Two plane mirrors make an angle of 30° with each other. Locate graphically four images of a luminous point A placed between the two mirrors. (See Fig. 36-5.)

From A draw normals AA' and AB' to mirrors OY and OX, respectively, making $\overline{AL} = \overline{LA'}$ and $\overline{AM} = \overline{MB'}$. Then A' and B' are images of A.

Next, from A' and B' draw normals to OX and OY, making $\overline{A'N} = \overline{NA''}$ and $\overline{B'P} = \overline{PB''}$. Then A'' is the image of A' in OX and B'' is the image of B' in OY.

The four images of A are A', B', A'', B''. Additional images also exist, for example, images of A'' and B''.

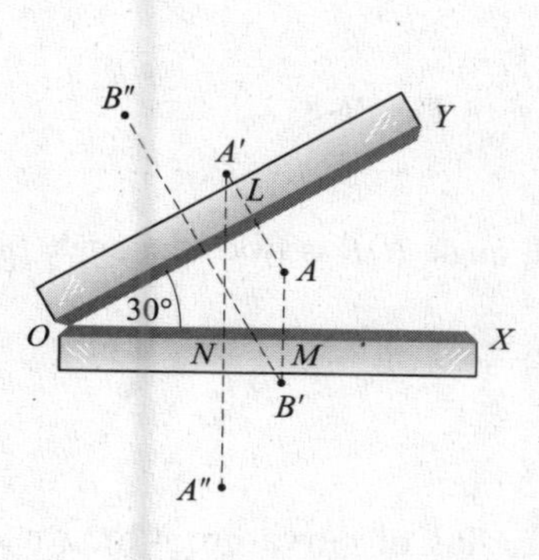

Fig. 36-5

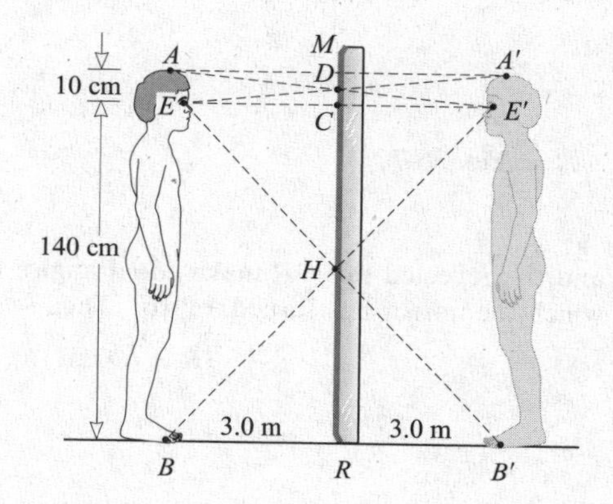

Fig. 36-6

36.2 [II] A boy is 1.50 m tall and can just see his image in a vertical plane mirror 3.0 m away. His eyes are 1.40 m from the floor level. Determine the vertical dimension and elevation of the shortest mirror in which he could see his full image.

In Fig. 36-6, let AB represent the boy. His eyes are at E. Then $A'B'$ is the image of AB in mirror MR, and DH represents the shortest mirror necessary for the eye to view the image $A'B'$.

Triangles DEC and $DA'M$ are congruent and so

$$\overline{CD} = \overline{DM} = 5.0 \text{ cm}$$

Triangles HRB' and HCE are congruent and so

$$\overline{RH} = \overline{HC} = 70 \text{ cm}$$

The dimension of the mirror is $\overline{HC} + \overline{CD} = 75$ cm and its elevation is $\overline{RH} = 70$ cm.

36.3 [II] As shown in Fig. 36-7, a light ray IO is incident on a small plane mirror. The mirror reflects this ray back onto a straight ruler SC which is 1 m away from and parallel to the undeflected mirror MM. When the mirror turns through an angle of 8.0° and assumes the position $M'M'$, across what distance on the scale will the spot of light move? (This device, called an *optical lever*, is useful in measuring small deflections.)

When the mirror turns through 8.0° the normal to it also turns through 8.0°, and the incident ray makes an angle of 8.0° with the normal NO to the deflected mirror $M'M'$. Because the incident ray IO

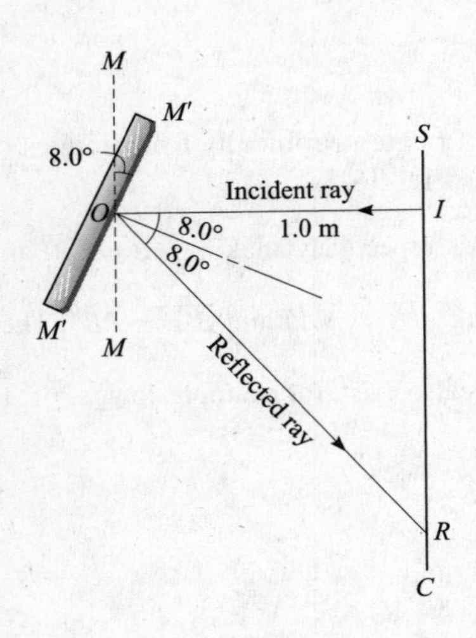

Fig. 36-7

Fig. 36-8

and the reflected ray OR make equal angles with the normal, angle IOR is twice the angle through which the mirror has turned, or 16°. Then

$$\overline{IR} = \overline{IO} \tan 16° = (1.0 \text{ m})(0.287) = 29 \text{ cm}$$

36.4 [II] The concave spherical mirror shown in Fig. 36-8 has radius of curvature 4 m. An object OO', 5 cm high, is placed 3 m in front of the mirror. By (*a*) construction and (*b*) computation, determine the position and height of the image II'.

In Fig. 36-8, C is the center of curvature, 4 m from the mirror, and F is the principal focus, 2 m from the mirror.

(*a*) Two of the following three convenient rays from O will locate the image.

(1) The ray OA, parallel to the principal axis. This ray, like all parallel rays, is reflected through the principal focus F in the direction AFA'.

(2) The ray OB, drawn as if it passed through the center of curvature C. This ray is normal to the mirror and is reflected back on itself in the direction BCB'.

(3) The ray OFD which passes through the principal focus F and, like all rays passing through F, is reflected parallel to the principal axis in the direction DD'.

The intersection I of any two of these reflected rays is the image of O. Thus II' represents the position and size of the image of OO'. The image is real, inverted, magnified, and at a greater distance from the mirror than the object. (*Note*: If the object were at II', the image would be at OO' and would be real, inverted, and smaller.)

(*b*) Using the mirror equation in which $R = -4$ m,

$$\frac{1}{s_o} + \frac{1}{s_i} = -\frac{2}{R} \qquad \text{or} \qquad \frac{1}{3} + \frac{1}{s_i} = -\frac{2}{-4} \qquad \text{or} \qquad s_i = 6 \text{ m}$$

The image is real (since s_i is positive) and located 6 m from the mirror. Also, since the image is inverted, both the magnification and y_i are negative:

$$M_T = -\frac{s_i}{s_o} = -\frac{6 \text{ m}}{3 \text{ m}} = -2 \qquad \text{and so} \qquad y_i = (-2)(5 \text{ cm}) = -0.10 \text{ m}$$

36.5 [II] An object OO' is 25 cm from a concave spherical mirror of radius 80 cm (Fig. 36-9). Determine the position and relative size of its image II' (*a*) by construction and (*b*) by use of the mirror equation.

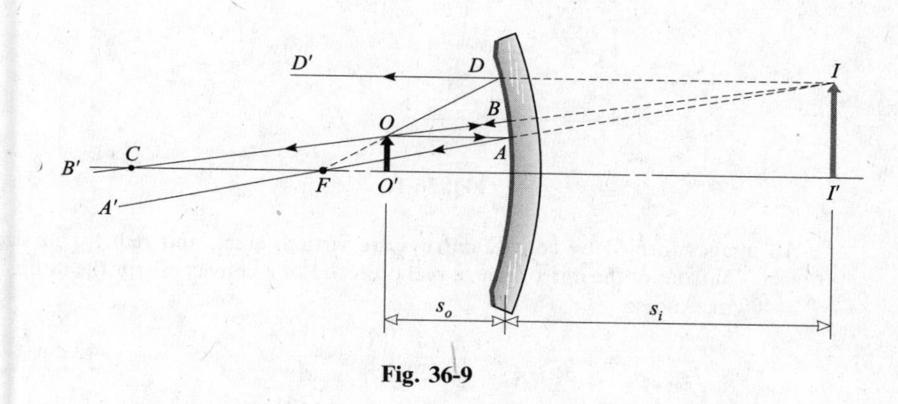

Fig. 36-9

(*a*) Two of the following three rays from O locate the image.

(1) A ray OA, parallel to the principal axis, is reflected through the focus F, 40 cm from the mirror.

(2) A ray OB, in the line of the radius COB, is normal to the mirror and is reflected back on itself through the center of curvature C.

(3) A ray OD, which (extended) passes through F, is reflected parallel to the axis. Because of the large curvature of the mirror from A to D, this ray is not as accurate as the other two.

The reflected rays (AA', BB', and DD') do not meet, but appear to originate from a point I behind the mirror. Thus II' represents the relative position and size of the image of OO'. The image is virtual (behind the mirror), erect, and magnified. Here the radius R is negative and so

(*b*)
$$\frac{1}{s_o}+\frac{1}{s_i}=-\frac{2}{R} \qquad \text{or} \qquad \frac{1}{25}+\frac{1}{s_i}=-\frac{2}{-80} \qquad \text{or} \qquad s_i=-67 \text{ cm}$$

The image is virtual (since s_i is negative) and 66.7 cm behind the mirror. Also,

$$M_T=-\frac{s_i}{s_o}=-\frac{-66.7 \text{ cm}}{25 \text{ cm}}=2.7 \text{ times}$$

Notice that M_T is positive and so the image is right-side-up.

36.6 [II] As shown in Fig. 36-10, an object 6 cm high is located 30 cm in front of a convex spherical mirror of radius 40 cm. Determine the position and height of its image, (*a*) by construction and (*b*) by use of the mirror equation.

(*a*) Choose two convenient rays coming from O at the top of the object:

(1) A ray OA, parallel to the principal axis, is reflected in the direction AA' as if it passed through the principal focus F.

(2) A ray OB, directed toward the center of curvature C, is normal to the mirror and is reflected back on itself.

The reflected rays, AA' and BO, never meet but appear to originate from a point I behind the mirror. Then II' represents the size and position of the image of OO'.

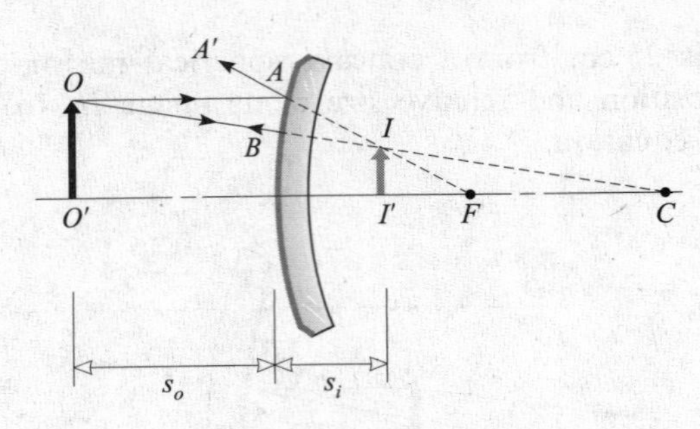

Fig. 36-10

All images formed by convex mirrors are virtual, erect, and reduced in size, provided the object is in front of the mirror (i.e., a real object). For a convex mirror the radius is positive; here $R = 40$ cm. And so

(*b*) $$\frac{1}{s_o} + \frac{1}{s_i} = -\frac{2}{R} \quad \text{or} \quad \frac{1}{30} + \frac{1}{s_i} = -\frac{2}{40} \quad \text{or} \quad s_i = -12 \text{ cm}$$

The image is virtual (s_i is negative) and 12 cm behind the mirror. Also,

$$M_T = -\frac{s_i}{s_o} = -\frac{-12 \text{ cm}}{30 \text{ cm}} = 0.40$$

Moreover, $M_T = y_i/y_o$ and so $y_i = M_T y_o = (0.40)(6.0 \text{ cm}) = 2.4$ cm

36.7 [II] Where should an object be placed, with reference to a concave spherical mirror of radius 180 cm, to form a real image that is half the size of the object?

All real images formed by the mirror are inverted and so the magnification is to be $-1/2$; hence $s_i = s_o/2$. Then, since $R = -180$ cm,

$$\frac{1}{s_o} + \frac{1}{s_i} = -\frac{2}{R} \quad \text{or} \quad \frac{1}{s_o} + \frac{2}{s_o} = -\frac{2}{-180} \quad \text{or} \quad s_o = 0.27 \text{ m from mirror}$$

36.8 [II] How far must a girl stand in front of a concave spherical mirror of radius 120 cm to see an erect image of her face four times its natural size?

The erect image must be virtual; hence s_i is negative. Since the magnification is +4 and $M_T = -s_i/s_o$, it follows that $s_i = -4s_o$. Then using $R = -120$ cm

$$\frac{1}{s_o} + \frac{1}{s_i} = -\frac{2}{R} \quad \text{or} \quad \frac{1}{s_o} - \frac{1}{4s_o} = \frac{2}{120} \quad \text{or} \quad s_o = 45 \text{ cm from mirror}$$

36.9 [II] What kind of spherical mirror must be used, and what must be its radius, in order to give an erect image one-fifth as large as an object placed 15 cm in front of it?

An erect image produced by a spherical mirror is virtual; hence s_i is negative. Moreover, since the magnification is $+1/5$, $s_i = -s_o/5 = -15/5 = -3$ cm. Because the virtual image is smaller than the object, a convex mirror is required. Its radius can be found using

$$\frac{1}{s_o} + \frac{1}{s_i} = -\frac{2}{R} \quad \text{or} \quad \frac{1}{15} - \frac{1}{3} = -\frac{2}{R} \quad \text{or} \quad R = +7.5 \text{ cm (convex mirror)}$$

36.10 [II] The diameter of the Sun subtends an angle of approximately 32 minutes (32′) at any point on the Earth. Determine the position and diameter of the solar image formed by a concave spherical mirror of radius 400 cm. Refer to Fig. 36-11.

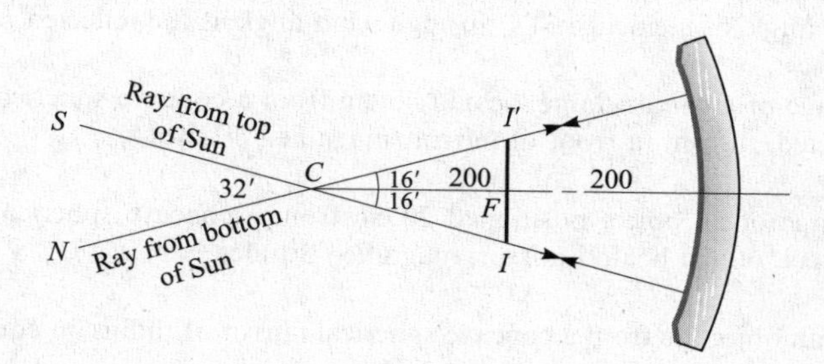

Fig. 36.11

Since the sun is very distant, s_o is very large and $1/s_o$ is practically zero. So with $R = -400$ cm

$$\frac{1}{s_o} + \frac{1}{s_i} = -\frac{2}{R} \qquad \text{or} \qquad 0 + \frac{1}{s_i} = \frac{2}{400}$$

and $s_i = 200$ cm. The image is at the principal focus F, 200 cm from the mirror.

The diameter of the Sun and its image II' subtend equal angles at the center of curvature C of the mirror. From the figure,

$$\tan 16' = \frac{\overline{II'}/2}{\overline{CF}} \qquad \text{or} \qquad \overline{II'} = 2\overline{CF}\,\tan 16' = (2)(2.00\text{ m})(0.00465) = 1.9\text{ cm}$$

36.11 [II] A dental technician uses a small mirror that gives a magnification of 4.0 when it is held 0.60 cm from a tooth. What is the radius of curvature of the mirror?

In order for the mirror to produce a right-side-up magnifieid image it must be concave. Accordingly R is negative.

Because the magnification is positive $-s_i/s_o = 4$ and with $s_o = 0.60$ cm it follows that $s_i = -2.4$ cm. The mirror equation becomes (in cm)

$$\frac{1}{0.60} + \frac{1}{-2.4} = -\frac{2}{R} \qquad \text{or} \qquad 1.667 - 0.417 = -\frac{2}{R}$$

and $R = -1.6$ cm. (This agrees with the fact that the image formed by a convex mirror is diminished, not magnified.)

Supplementary Problems

36.12 [I] If you wish to take a photo of yourself as you stand 3 m in front of a plane mirror, for what distance should you focus the camera you are holding? *Ans.* 6 m

36.13 [I] Two plane mirrors make an angle of 90° with each other. A point-like luminous object is placed between them. How many images are formed? *Ans.* 3

36.14 [I] Two plane mirrors are parallel to each other and spaced 20 cm apart. A luminous point is placed between them and 5.0 cm from one mirror. Determine the distance from each mirror of the three nearest images in each. *Ans.* 5.0, 35, 45 cm; 15, 25, 55 cm

36.15 [I] Two plane mirrors make an angle of 90° with each other. A beam of light is directed at one of the mirrors, reflects off it and the second mirror, and leaves the mirrors. What is the angle between the incident beam and the reflected beam? *Ans.* 180°

36.16 [I] A ray of light makes an angle of 25° with the normal to a plane mirror. If the mirror is turned through 6.0°, making the angle of incidence 31°, through what angle is the reflected ray rotated? *Ans.* 12°

36.17 [II] Describe the image of a candle flame located 40 cm from a concave spherical mirror of radius 64 cm. *Ans.* real, inverted, 0.16 m in front of mirror, magnified 4 times

36.18 [II] Describe the image of an object positioned 20 cm from a concave spherical mirror of radius 60 cm. *Ans.* virtual, erect, 60 cm behind mirror, magnified 3 times

36.19 [II] How far should an object be from a concave spherical mirror of radius 36 cm to form a real image one-ninth its size? *Ans.* 0.18 m

36.20 [II] An object 7.0 cm high is placed 15 cm from a convex spherical mirror of radius 45 cm. Describe its image. *Ans.* virtual, erect, 9.0 cm behind mirror, 4.2 cm high

36.21 [II] What is the focal length of a convex spherical mirror which produces an image one-sixth the size of an object located 12 cm from the mirror? *Ans.* −2.4 cm

36.22 [II] It is desired to cast the image of a lamp, magnified 5 times, upon a wall 12 m distant from the lamp. What kind of spherical mirror is required, and what is its position? *Ans.* concave, radius 5.0 m, 3.0 m from lamp

36.23 [II] Compute the position and diameter of the image of the Moon in a polished sphere of diameter 20 cm. The diameter of the Moon is 3500 km, and its distance from the Earth is 384 000 km, approximately. *Ans.* 5.0 cm inside sphere, 0.46 mm

Chapter 37

Refraction of Light

THE SPEED OF LIGHT (c) as ordinarily measured varies from material to material. Light (treated macroscopically) travels fastest in vacuum, where its speed is $c = 2.998 \times 10^8$ m/s. Its speed in air is c/1.000 3. In water its speed is c/1.33, and in ordinary glass it is about c/1.5. Nonetheless, on a microscopic level light is composed of photons, and photons exist only at the speed c. The apparent slowing down in material media arises from the absorption and re-emission as the light passes from atom to atom.

INDEX OF REFRACTION (n): The ***absolute index of refraction*** of a material is defined as

$$n = \frac{\text{speed of light in vacuum}}{\text{speed of light in the material}} = \frac{c}{v}$$

For any two materials, the *relative index of refraction* of material-1, with respect to material-2, is

$$\text{Relative index} = \frac{n_1}{n_2}$$

where n_1 and n_2 are the absolute refractive indices of the two materials.

REFRACTION: When a ray of light is transmitted obliquely through the boundary between two materials of unlike index of refraction, the ray bends. This phenomenon, called ***refraction***, is shown in Fig. 37-1. If $n_t > n_i$, the ray refracts as shown in the figure; it bends toward the normal as it enters the second material. If $n_t < n_i$, however, the ray refracts away from the normal. This would be the situation in Fig. 37-1 if the direction of the ray were reversed. In either case, the incident and refracted (or transmitted) rays and the normal all lie in the same plane. The angles θ_i and θ_t in Fig. 37-1 are called the ***angle of incidence*** and ***angle of transmission*** (or refraction), respectively.

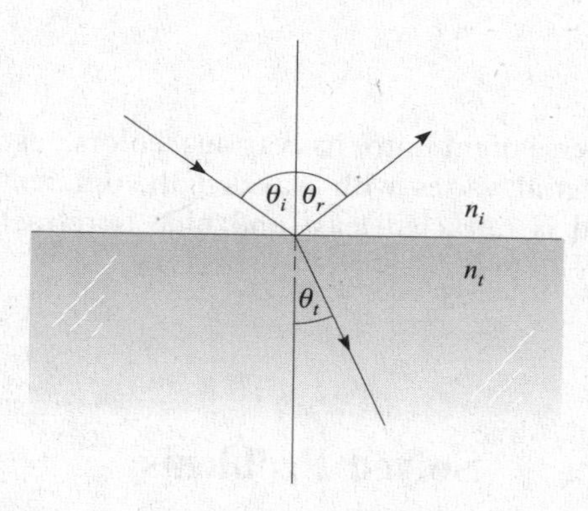

Fig. 37-1

SNELL'S LAW: The way in which a ray refracts at an interface between materials with indices of refraction n_i and n_t is given by **Snell's Law**:

$$n_i \sin \theta_i = n_t \sin \theta_t$$

where θ_i and θ_t are as shown in Fig. 37-1. Because this equation applies to light moving in either direction along the ray, a ray of light follows the same path when its direction is reversed.

CRITICAL ANGLE FOR TOTAL INTERNAL REFLECTION: When light reflects off an interface where $n_i < n_t$ the process is called ***external reflection***, when $n_i > n_t$ it's ***internal reflection***. Suppose that a ray of light passes from a material of higher index of refraction to one of lower index, as shown in Fig. 37-2. Part of the incident light is refracted and part is reflected at the interface. Because θ_t must be larger than θ_i, it is possible to make θ_i large enough so that $\theta_t = 90°$. This value for θ_i is called the ***critical angle*** θ_c. For θ_i larger than this, no refracted ray can exist; all the light is reflected.

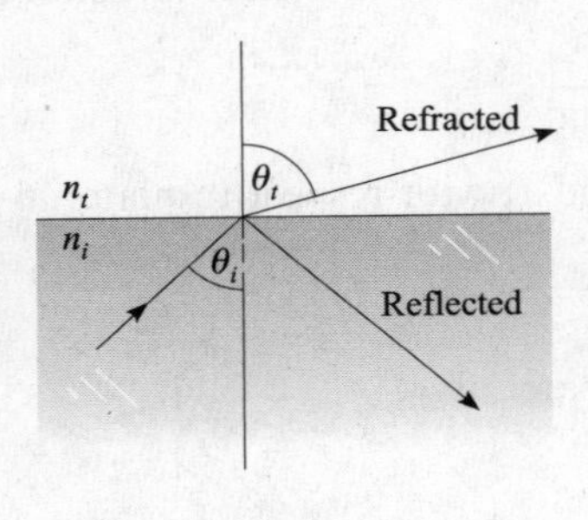

Fig. 37-2

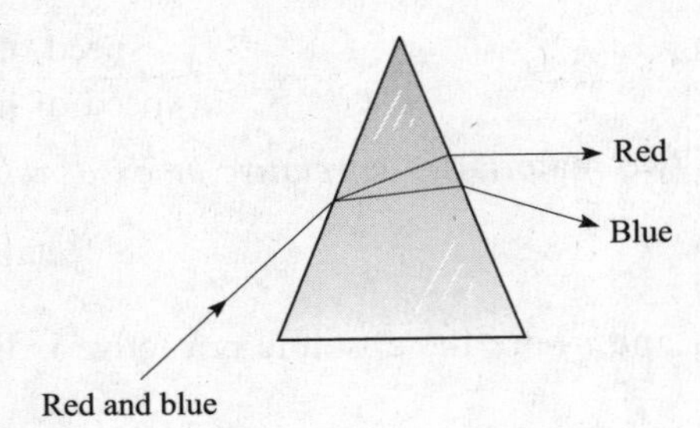

Fig. 37-3

The condition for ***total internal reflection*** is that θ_i exceed the critical angle θ_c where

$$n_i \sin \theta_c = n_t \sin 90° \qquad \text{or} \qquad \sin \theta_c = \frac{n_t}{n_i}$$

Because the sine of an angle can never be larger than unity, this relation confirms that total internal reflection can occur only if $n_i > n_t$.

A PRISM can be used to disperse light into its various colors, as shown in Fig. 37-3. Because the index of refraction of a material varies with wavelength, different colors of light refract differently. In nearly all materials, red is refracted least and blue is refracted most.

Solved Problems

37.1 [I] The speed of light in water is (3/4)c. What is the effect, on the frequency and wavelength of light, of passing from vacuum (or air, to good approximation) into water? Compute the refractive index of water.

The same number of wave peaks leave the air each second as enter into the water. Hence **the frequency is the same in the two materials**. But because wavelength = (speed)/(frequency), the wavelength in water is three-fourths that in air.

The (absolute) refractive index of water is

$$n = \frac{\text{speed in vacuum}}{\text{speed in water}} = \frac{c}{(3/4)c} = \frac{4}{3} = 1.33$$

37.2 [I] A glass plate is 0.60 cm thick and has a refractive index of 1.55. How long does it take for a pulse of light incident normally to pass through the plate?

$$t = \frac{x}{v} = \frac{0.0060 \text{ m}}{(2.998 \times 10^8/1.55) \text{ m/s}} = 3.1 \times 10^{-11} \text{ s}$$

37.3 [I] As is shown in Fig. 37-4, a ray of light in air strikes a glass plate ($n = 1.50$) at an incidence angle of 50°. Determine the angles of the reflected and transmitted rays.

The law of reflection applies to the reflected ray. Therefore, the angle of reflection is 50°, as shown.

For the refracted ray, $n_i \sin \theta_i = n_t \sin \theta_t$ becomes,

$$\sin \theta_t = \frac{n_i}{n_t} \sin \theta_i = \frac{1.0}{1.5} \sin 50° = 0.51$$

from which it follows that $\theta_t = 31°$.

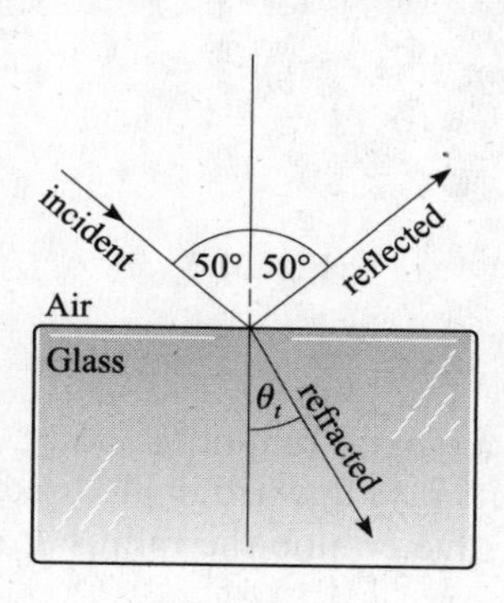

Fig. 37-4

37.4 [I] The refractive index of diamond is 2.42. What is the critical angle for light passing from diamond to air?

We use $n_i \sin \theta_i = n_t \sin \theta_t$ to obtain

$$(2.42) \sin \theta_c = (1) \sin 90.0°$$

from which it follows that $\sin \theta_c = 0.413$ and $\theta_c = 24.4°$.

37.5 [I] What is the critical angle for light passing from glass ($n = 1.54$) to water ($n = 1.33$)?

$$n_i \sin \theta_i = n_t \sin \theta_t \qquad \text{becomes} \qquad n_i \sin \theta_c = n_t \sin 90°$$

from which we get

$$\sin \theta_c = \frac{n_t}{n_i} = \frac{1.33}{1.54} = 0.864 \qquad \text{or} \qquad \theta_c = 59.7°$$

37.6 [II] A layer of oil ($n = 1.45$) floats on water ($n = 1.33$). A ray of light shines onto the oil with an incidence angle of 40.0°. Find the angle the ray makes in the water. (See Fig. 37-5.)

At the air–oil interface, Snell's Law gives

$$n_{\text{air}} \sin 40^\circ = n_{\text{oil}} \sin \theta_{\text{oil}}$$

At the oil–water interface, we have (using the equality of alternate angles)

$$n_{\text{oil}} \sin \theta_{\text{oil}} = n_{\text{water}} \sin \theta_{\text{water}}$$

Thus, $n_{\text{air}} \sin 40.0^\circ = n_{\text{water}} \sin \theta_{\text{water}}$; the overall refraction occurs just as though the oil layer were absent. Solving gives

$$\sin \theta_{\text{water}} = \frac{n_{\text{air}} \sin 40.0^\circ}{n_{\text{water}}} = \frac{(1)(0.643)}{1.33} \qquad \text{or} \qquad \theta_{\text{water}} = 28.9^\circ$$

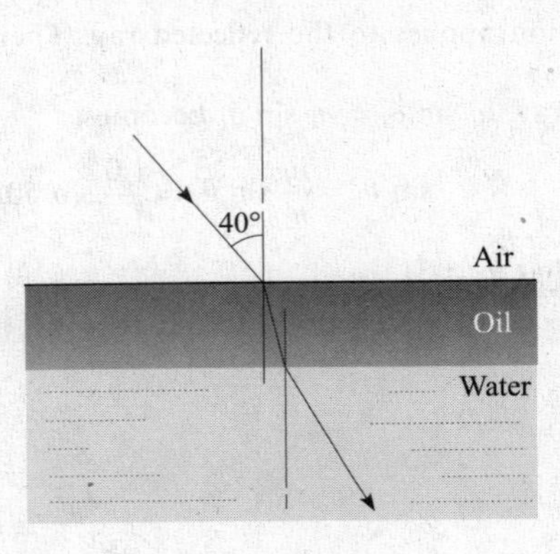

Fig. 37-5

37.7 [II] As shown in Fig. 37-6, a small luminous body, at the bottom of a pool of water ($n = 4/3$) 2.00 m deep, emits rays upward in all directions. A circular area of light is formed at the surface of the water. Determine the radius R of the circle of light.

The circular area is formed by rays refracted into the air. The angle θ_c must be the critical angle, because total internal reflection, and hence no refraction, occurs when the angle of incidence in the water is greater than the critical angle. We have, then,

$$\sin \theta_c = \frac{n_a}{n_w} = \frac{1}{4/3} \qquad \text{or} \qquad \theta_c = 48.6^\circ$$

From the figure,

$$R = (2.00 \text{ m}) \tan \theta_c = (2.00 \text{ m})(1.13) = 2.26 \text{ m}$$

37.8 [I] What is the minimum value of the refractive index for a 45.0° prism which is used to turn a beam of light by total internal reflection through a right angle? (See Fig. 37-7.)

The ray enters the prism without deviation, since it strikes side AB normally. It then makes an incidence angle of 45.0° with normal to side AC. The critical angle of the prism must be smaller than 45.0° if the ray is to be totally reflected at side AC and thus turned through 90°. From $n_i \sin \theta_c = n_t \sin 90^\circ$ with $n_t = 1.00$,

$$\text{Minimum } n_i = \frac{1}{\sin 45.0^\circ} = 1.41$$

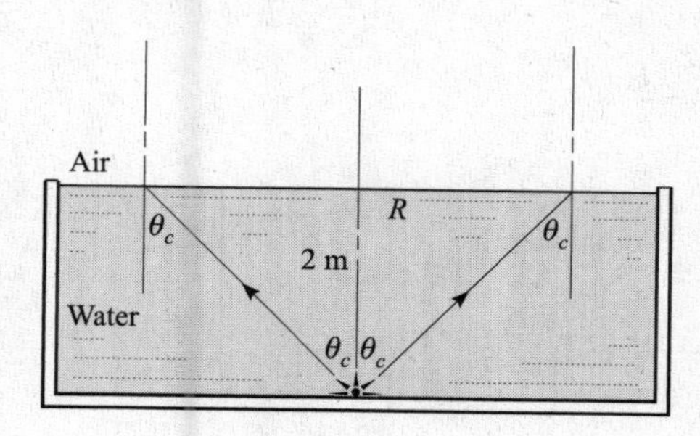

Fig. 37-6

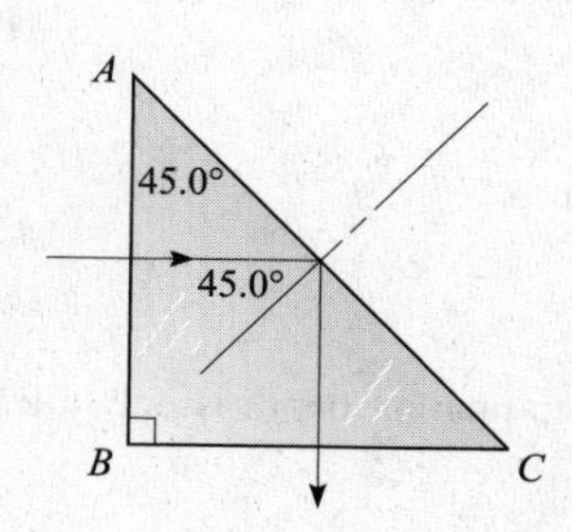

Fig. 37-7

37.9 [II] The glass prism shown in Fig. 37-8 has an index of refraction of 1.55. Find the angle of deviation D for the case shown.

No deflection occurs at the entering surface, because the incidence angle is zero. At the second surface, $\theta_i = 30°$ (because its sides are mutually perpendicular to the sides of the apex angle). Then, Snell's Law becomes

$$n_i \sin \theta_i = n_t \sin \theta_t \qquad \text{or} \qquad \sin \theta_t = \frac{1.55}{1} \sin 30°$$

from which $\theta_t = 50.8°$. But $D = \theta_t - \theta_i$ and so $D = 21°$.

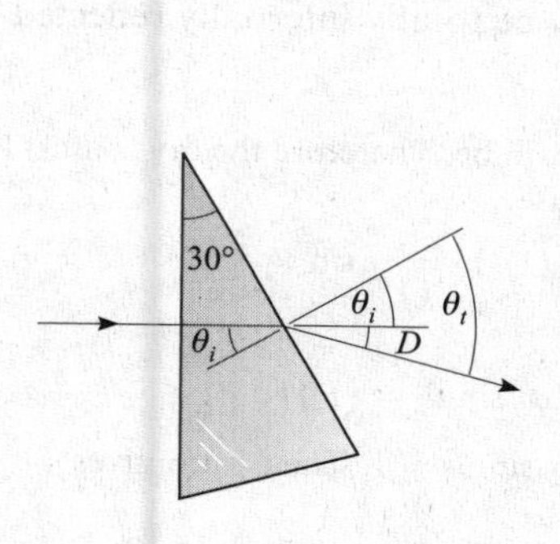

Fig. 37-8

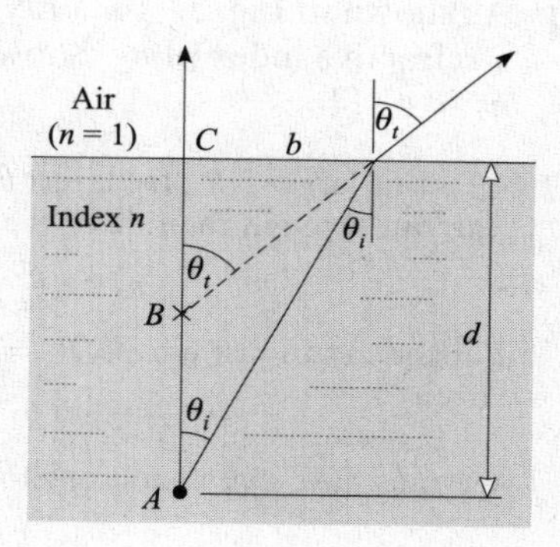

Fig. 37-9

37.10 [III] As shown in Fig. 37-9, an object is at a depth d beneath the surface of a transparent material of refractive index n. As viewed from a point almost directly above, how deep does the object appear to be?

The two rays from A that are shown emerging into the air both appear to come from point-B. Therefore, the apparent depth is CB. We have

$$\frac{b}{\overline{CB}} = \tan \theta_t \qquad \text{and} \qquad \frac{b}{\overline{CA}} = \tan \theta_i$$

If the object is viewed from nearly straight above, then angles θ_t and θ_i will be very small. For small angles, the sine and tangent are nearly equal. Therefore,

$$\frac{\overline{CB}}{\overline{CA}} = \frac{\tan \theta_i}{\tan \theta_t} \approx \frac{\sin \theta_i}{\sin \theta_t}$$

But $n \sin\theta_i = (1) \sin\theta_t$, from which

$$\frac{\sin\theta_i}{\sin\theta_t} = \frac{1}{n}$$

Hence,

$$\text{Apparent depth } \overline{CB} = \frac{\text{actual depth } \overline{CA}}{n}$$

The apparent depth is only a fraction $1/n$ of the actual depth d.

37.11 [I] A glass plate 4.00 mm thick is viewed from above through a microscope. The microscope must be lowered 2.58 mm as the operator shifts from viewing the top surface to viewing the bottom surface through the glass. What is the index of refraction of the glass? Use the results of Problem 37.10.

We found in Problem 37.10 that the apparent depth of the plate will be $1/n$ as large as its actual depth. Hence,

$$(\text{actual thickness})(1/n) = \text{apparent thickness}$$

or

$$(4.00 \text{ mm})(1/n) = 2.58 \text{ mm}$$

This yields $n = 1.55$ for the glass.

37.12 [III] As shown in Fig. 37-10, a ray enters the flat end of a long rectangular block of glass that has a refractive index of n_2. Show that all entering rays can be totally internally reflected only if $n_2 > 1.414$.

The larger θ_1 is, the larger θ_2 will be, and the smaller θ_3 will be. Therefore the ray is most likely to leak out through the side of the block if $\theta_1 = 90°$. In that case,

$$n_1 \sin\theta_1 = n_2 \sin\theta_2 \qquad \text{becomes} \qquad (1)(1) = n_2 \sin\theta_2$$

For the ray to just escape, $\theta_4 = 90°$. Then

$$n_2 \sin\theta_3 = n_1 \sin\theta_4 \qquad \text{becomes} \qquad n_2 \sin\theta_3 = (1)(1)$$

We thus have two conditions to satisfy: $n_2 \sin\theta_2 = 1$ and $n_2 \sin\theta_3 = 1$. Their ratio gives

$$\frac{\sin\theta_2}{\sin\theta_3} = 1$$

But we see from the figure that $\sin\theta_3 = \cos\theta_2$, and so this becomes

$$\tan\theta_2 = 1 \qquad \text{or} \qquad \theta_2 = 45.00°$$

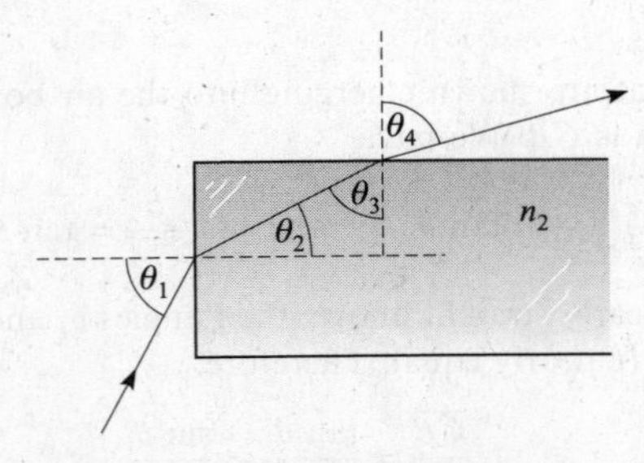

Fig. 37-10

Then, because $n_2 \sin \theta_2 = 1$, we have

$$n_2 = \frac{1}{\sin 45.00^\circ} = 1.414$$

This is the smallest possible value the index can have for total internal reflection of all rays that enter the end of the block. It is possible to obtain this answer by inspection. How?

Supplementary Problems

37.13 [I] The speed of light in a certain glass is 1.91×10^8 m/s. What is the refractive index of the glass? *Ans.* 1.57

37.14 [I] What is the frequency of light which has a wavelength in air of 546 nm? What is its frequency in water ($n = 1.33$)? What is its speed in water? What is its wavelength in water? *Ans.* 549 THz, 549 THz, 2.25×10^8 m/s, 411 nm

37.15 [I] A beam of light strikes the surface of water at an incidence angle of 60°. Determine the directions of the reflected and refracted rays. For water, $n = 1.33$. *Ans.* 60° reflected into air, 41° refracted into water

37.16 [I] The critical angle for light passing from rock salt into air is 40.5°. Calculate the index of refraction of rock salt. *Ans.* 1.54

37.17 [I] What is the critical angle when light passes from glass ($n = 1.50$) into air? *Ans.* 41.8°

37.18 [II] The absolute indices of refraction of diamond and crown glass are 5/2 and 3/2 respectively. Compute (*a*) the refractive index of diamond relative to crown glass and (*b*) the critical angle between diamond and crown glass. *Ans.* (*a*) 5/3; (*b*) 37°

37.19 [II] A pool of water ($n = 4/3$) is 60 cm deep. Find its apparent depth when viewed vertically through air. *Ans.* 45 cm

37.20 [III] In a vessel, a layer of benzene ($n = 1.50$) 6 cm deep floats on water ($n = 1.33$) 4 cm deep. Determine the apparent distance of the bottom of the vessel below the upper surface of the benzene when viewed vertically through air. *Ans.* 7 cm

37.21 [II] A mirror is made of plate glass ($n = 3/2$) 1.0 cm thick and silvered on the back. A man is 50.0 cm from the front face of the mirror. If he looks perpendicularly into it, at what distance behind the front face of the mirror will his image appear to be? *Ans.* 51.3 cm

37.22 [II] A straight rod is partially immersed in water ($n = 1.33$). Its submerged portion appears to be inclined 45° with the surface when viewed vertically through air. What is the actual inclination of the rod? *Ans.* $\arctan 1.33 = 53^\circ$

37.23 [II] The index of refraction for a certain type of glass is 1.640 for blue light and 1.605 for red light. When a beam of white light (one that contains all colors) enters a plate of this glass at an incidence angle of 40°, what is the angle in the glass between the blue and red parts of the refracted beam? *Ans.* 0.53°

Chapter 38

Thin Lenses

TYPES OF LENSES: As indicated in Fig. 38-1, **converging**, or **positive**, lenses are thicker at the center than at the rim and will converge a beam of parallel light to a real focus. **Diverging**, or **negative**, lenses are thinner at the center than at the rim and will diverge a beam of parallel light from a virtual focus.

The *principal focus* (or **focal point**) of a thin lens with spherical surfaces is the point F where rays parallel to and near the central or optical axis are brought to a focus; this focus is real for a converging lens and virtual for a diverging lens. The **focal length** f is the distance of the principal focus from the lens. Because each lens in Fig. 38-1 can be reversed without altering the rays, two symmetric focal points exist for each lens.

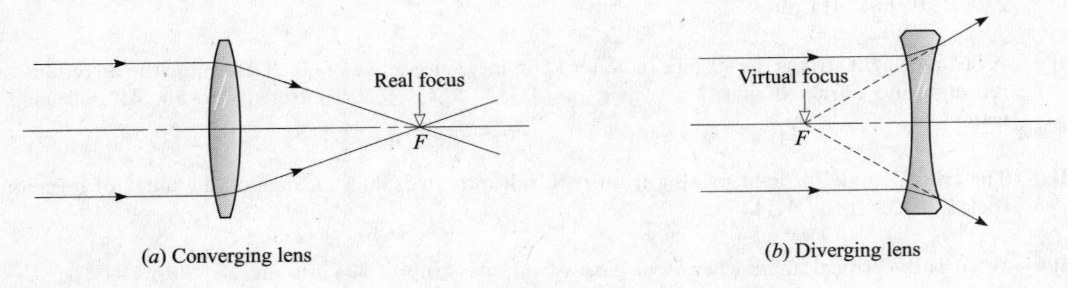

Fig. 38-1

RAY TRACING: When a ray passes through a lens it refracts or "bends" at each interface, as shown in Fig. 38-1. When dealing with thin lenses all of the bending can, for simplicity, be assumed to occur along a vertical plane running down the middle of the lens (see Fig. 38-2).

As in our previous treatment of mirrors (Chapter 36), any two rays originating from a point on the object, drawn through the system, will locate the image of that point. There are three especially convenient rays to use because we know, without making any calculations, exactly how they will pass through a lens. These rays are shown in Fig. 38-3 propagating through both a convex and a concave lens. Notice that a ray heading for the center (C) of a thin lens passes straight through it unbent.

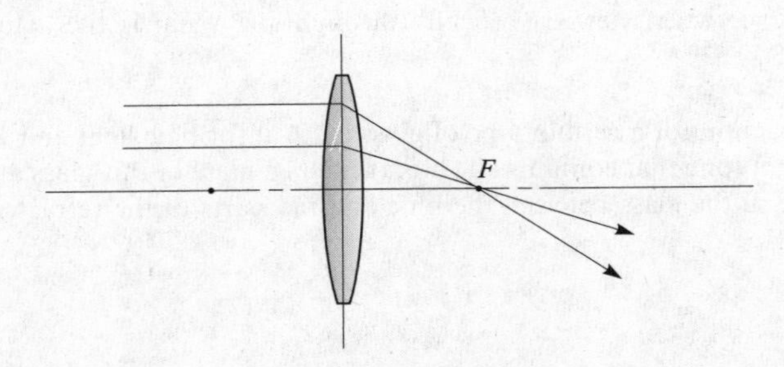

Fig. 38-2

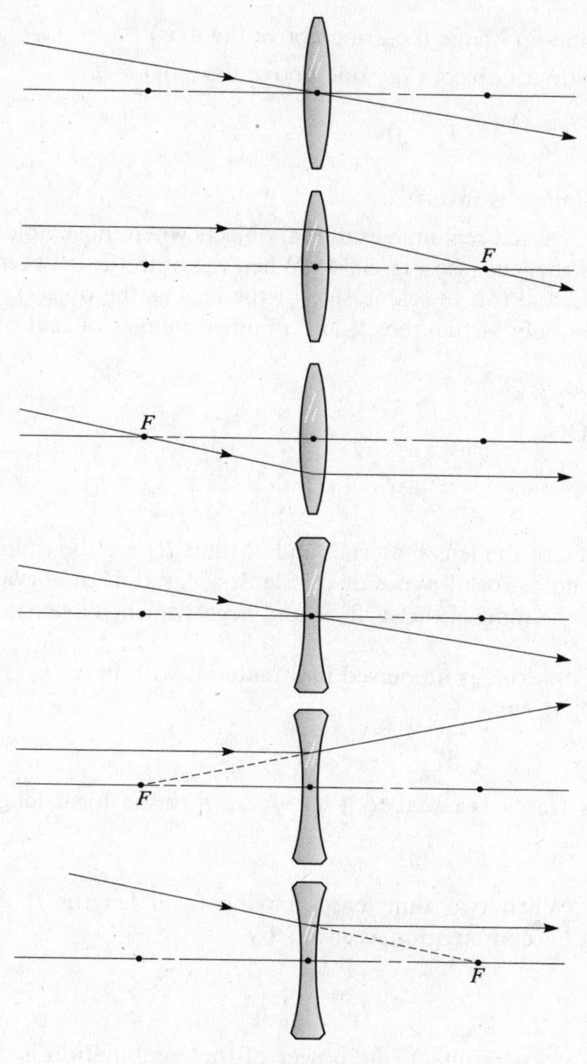

Fig. 38-3

OBJECT AND IMAGE RELATION for converging and diverging thin lenses:

$$\frac{1}{s_o} + \frac{1}{s_i} = \frac{1}{f}$$

where s_o is the object distance from the lens, s_i is the image distance from the lens, and f is the focal length of the lens. The lens is assumed to be thin, and the light rays **paraxial** (close to the principal axis). Then, with light entering from the left,

- s_o is positive when the object is to the left of the lens.
- s_o is positive for a real object, and negative for a virtual object (see Chapter 39).
- s_i is positive when the image is to the right of the lens.
- s_i is positive for a real image, and negative for a virtual image.
- f is positive for a converging lens, and negative for a diverging lens.

- y_i is positive for a right-side-up image (i.e., one above the axis).
- y_o is positive for a right-side-up object (i.e., one above the axis).

Also,
$$M_T = \frac{y_i}{y_o} = -\frac{s_i}{s_o}$$

- M_T is negative when the image is inverted.

Converging lenses form inverted real images of real objects when those objects are located to the left of the focal point, in front of the lens (see Fig. 38-4). When the object is between the focal point and the lens, the resulting image is virtual (on the same side of the lens as the object), erect, and enlarged.

Diverging lenses produce only virtual, erect, and minified images of real objects.

LENSMAKER'S EQUATION:

$$\frac{1}{f} = (n-1)\left(\frac{1}{R_1} - \frac{1}{R_2}\right)$$

where n is the refractive index of the lens material, and R_1 and R_2 are the radii of curvature of the two lens surfaces. This equation holds for all types of thin lenses. A radius of curvature, R, is positive when its center of curvature lies to the right of the surface, and negative when its center of curvature lies to the left of the surface.

If a lens with refractive index n_1 is immersed in a material with index n_2, then n in the lensmaker's equation is to be replaced by n_1/n_2.

LENS POWER in **diopters** (m^{-1}) is equal to $1/f$, where f is the focal length expressed in meters.

LENSES IN CONTACT: When two thin lenses having focal lengths f_1 and f_2 are in close contact, the focal length f of the combination is given by

$$\frac{1}{f} = \frac{1}{f_1} + \frac{1}{f_2}$$

Quite generally, for lenses in close contact, the power of the combination is equal to the sum of their individual powers.

Solved Problems

38.1 [II] An object OO', 4.0 cm high, is 20 cm in front of a thin convex lens of focal length +12 cm. Determine the position and height of its image II' (*a*) by construction and (*b*) by computation.

(*a*) The following two convenient rays from O will locate the images (see Fig. 38-4).

(1) A ray OP, parallel to the optical axis, must after refraction pass through the focus F.

(2) A ray passing through the optical center C of a thin lens is not appreciably deviated. Hence ray OCI may be drawn as a straight line.

The intersection I of these two rays is the image of O. Thus II' represents the position and size of the image of OO'. The image is real, inverted, enlarged, and at a greater distance from the lens than the object. (If the object were at II', the image at OO', would be real, inverted, and smaller.)

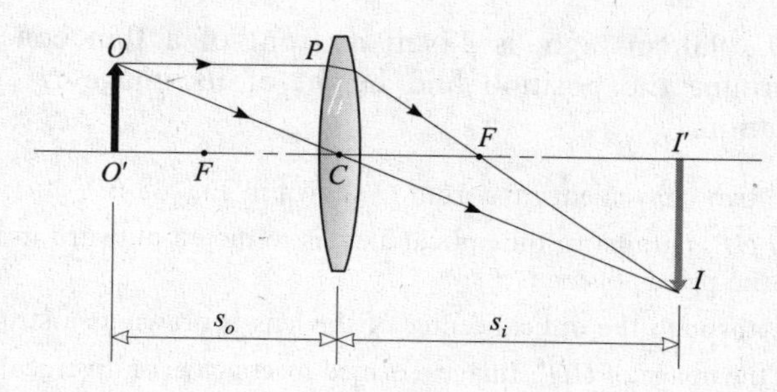

Fig. 38-4

(b) $$\frac{1}{s_o}+\frac{1}{s_i}=\frac{1}{f} \quad \text{or} \quad \frac{1}{20\text{ cm}}+\frac{1}{s_i}=\frac{1}{12\text{ cm}} \quad \text{or} \quad s_i = 30\text{ cm}$$

The image is real (since s_i is positive) and 30 cm behind the lens.

$$M_T=\frac{y_i}{y_o}=-\frac{s_i}{s_o}=-\frac{30\text{ cm}}{20\text{ cm}}=-1.5 \quad \text{or} \quad y_i = M_T y_o = (-1.5)(4.0\text{ cm}) = -6.0\text{ cm}$$

The negative magnification and image height both indicate an inverted image.

38.2 [II] An object OO' is 5.0 cm in front of a thin convex lens of focal length +7.5 cm. Determine the position and magnification of its image II' (*a*) by construction and (*b*) by computation.

(*a*) Choose two convenient rays from O, as in Fig. 38-5.

(1) A ray OP, parallel to the optical axis, is refracted so as to pass through the focus F.

(2) A ray OCN, through the optical center of the lens, is drawn as a straight line.

These two rays do not meet, but appear to originate from a point I. Thus II' represents the position and size of the image of OO'.

When the object is between F and C, the image is virtual, erect, and enlarged, as shown.

(b) $$\frac{1}{s_o}+\frac{1}{s_i}=\frac{1}{f} \quad \text{or} \quad \frac{1}{5.0\text{ cm}}+\frac{1}{s_i}=\frac{1}{7.5\text{ cm}} \quad \text{or} \quad s_i = -15\text{ cm}$$

Since s_i is negative, the image is virtual (on the same side of the lens as the object), and it is 15 cm in front of the lens. Also,

$$M_T=\frac{y_i}{y_o}=-\frac{s_i}{s_o}=-\frac{-15\text{ cm}}{5.0\text{ cm}}=3.0$$

Because the magnification is negative the image is right-side-up.

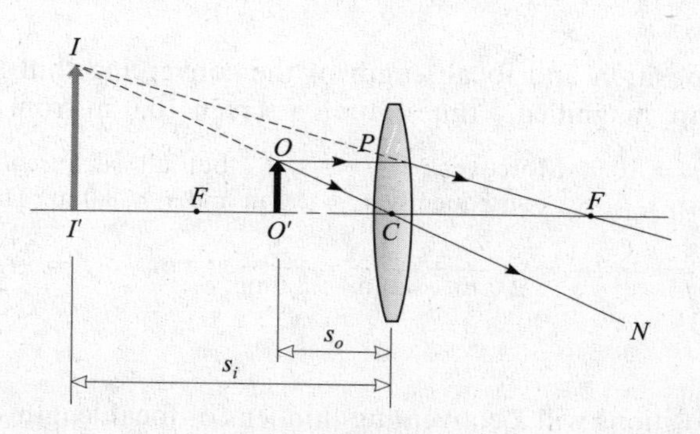

Fig. 38-5

38.3 [II] An object OO', 9.0 cm high, is 27 cm in front of a thin concave lens of focal length -18 cm. Determine the position and height of its image II' (*a*) by construction and (*b*) by computation.

(*a*) Choose the two convenient rays from O shown in Fig. 38-6.

(1) A ray OP, parallel to the optical axis, is refracted outward in the direction D as if it came from the principal focus F.

(2) A ray through the optical center of the lens is drawn as a straight line OC.

Then II' is the image of OO'. Images formed by concave or divergent lenses are virtual, erect, and smaller.

(*b*) $$\frac{1}{s_o}+\frac{1}{s_i}=\frac{1}{f} \qquad \text{or} \qquad \frac{1}{27\text{ cm}}+\frac{1}{s_i}=-\frac{1}{18\text{ cm}} \qquad \text{or} \qquad s_i=-10.8\text{ cm}=-11\text{ cm}$$

Since s_i is negative, the image is virtual, and it is 11 cm in front of the lens.

$$M_T=\frac{y_i}{y_o}=-\frac{s_i}{s_o}=-\frac{-10.8\text{ cm}}{27\text{ cm}}=0.40 \qquad \text{and so} \qquad y_i=y_oM_T=(0.40)(9.0\text{ cm})=3.6\text{ cm}$$

When $M_T>0$ the image is upright, and the same conclusion follows from the fact that $y_i>0$.

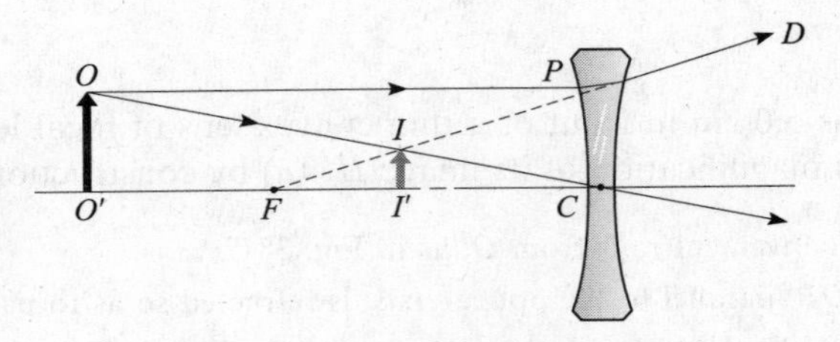

Fig. 38-6

38.4 [I] A converging thin lens ($f=20$ cm) is placed 37 cm in front of a screen. Where should the object be placed if its image is to appear on the screen?

We know that $s_i=+37$ cm and $f=+20$ cm. The lens equation gives

$$\frac{1}{s_o}+\frac{1}{37\text{ cm}}=\frac{1}{20\text{ cm}} \qquad \text{and} \qquad \frac{1}{s_o}=0.050\text{ cm}^{-1}-0.027\text{ cm}^{-1}=0.023\text{ cm}^{-1}$$

from which $s_o=43.5$ cm. The object should be placed 44 cm from the lens.

38.5 [II] Compute the position and focal length of the converging thin lens which will project the image of a lamp, magnified 4 times, upon a screen 10.0 m from the lamp.

Here $s_o+s_i=10.0$. Moreover, $M_T=-s_i/s_o$, but all such real images are inverted, hence $M_T=-4$. And so $s_i=4s_o$, consequently $s_o=2.0$ m and $s_i=8.0$ m. Then

$$\frac{1}{f}=\frac{1}{s_o}+\frac{1}{s_i}=\frac{1}{2.0\text{ m}}+\frac{1}{8.0\text{ m}}=\frac{5}{8.0\text{ m}} \qquad \text{or} \qquad f=\frac{8.0\text{ m}}{5}=+1.6\text{ m}$$

38.6 [II] In what two positions will a converging thin lens of focal length $+9.00$ cm form images of a luminous object on a screen located 40.0 cm from the object?

Given $s_o + s_i = 40.0\,\text{cm}$ and $f = +9.00\,\text{cm}$, we have

$$\frac{1}{s_o} + \frac{1}{40.0\text{ cm} - s_o} = \frac{1}{9.0\text{ cm}} \qquad \text{or} \qquad s_o^2 - 40.0s_o + 360 = 0$$

The use of the quadratic formula gives

$$s_o = \frac{40.0 \pm \sqrt{1600 - 1440}}{2}$$

from which $s_o = 13.7$ cm and $s_o = 26.3$ cm. The two lens positions are 13.7 cm and 26.3 cm from the object.

38.7 [II] A converging thin lens with 50 cm focal length forms a real image that is 2.5 times larger than the object. How far is the object from the image?

Real images formed by single converging lenses are all inverted. Accordingly, $M_T = -s_i/s_o = -2.5$ and so $s_i = 2.5s_o$. Therefore

$$\frac{1}{s_o} + \frac{1}{2.5s_o} = \frac{1}{50\text{ cm}} \qquad \text{or} \qquad s_o = 70\text{ cm}$$

This gives $s_i = (2.5)(70\text{ cm}) = 175$ cm. So the required distance is

$$s_i + s_o = 70\text{ cm} + 175\text{ cm} = 245\text{ cm} = 2.5\text{ m}$$

38.8 [II] A thin lens of focal length f projects upon a screen the image of a luminous object magnified N times. Show that the lens distance from the screen is $(N+1)f$.

The image is real, since it can be shown on a screen, and so $s_i > 0$. We then have

$$N = \left|-\frac{s_i}{s_o}\right| = s_i\left(\frac{1}{s_o}\right) = s_i\left(\frac{1}{f} - \frac{1}{s_i}\right) = \frac{s_i}{f} - 1 \qquad \text{or} \qquad s_i = (N+1)f$$

38.9 [II] A thin lens has a convex surface of radius 20 cm and a concave surface of radius 40 cm and is made of glass of refractive index 1.54. Compute the focal length of the lens, and state whether it is a converging or a diverging lens.

First, notice that $R_1 > 0$ and $R_2 > 0$ because both surfaces have their centers of curvature to the right. Consequently,

$$\frac{1}{f} = (n-1)\left(\frac{1}{R_1} - \frac{1}{R_2}\right) = (1.54 - 1)\left(\frac{1}{20\text{ cm}} - \frac{1}{40\text{ cm}}\right) = \frac{0.54}{40\text{ cm}} \qquad \text{or} \qquad f = +74\text{ cm}$$

Since f turns out to be positive, the lens is converging.

38.10 [II] A thin double convex lens has faces of radii 18 and 20 cm. When an object is 24 cm from the lens, a real image is formed 32 cm from the lens. Determine (*a*) the focal length of the lens and (*b*) the refractive index of the lens material.

Remember that a convex lens has a positive focal length.

(*a*) $$\frac{1}{f} = \frac{1}{s_o} + \frac{1}{s_i} = \frac{1}{24\text{ cm}} + \frac{1}{32\text{ cm}} = \frac{7}{96\text{ cm}} \qquad \text{or} \qquad f = \frac{96\text{ cm}}{7} = +13.7\text{ cm} = 14\text{ cm}$$

Here $R_1 > 0$ and $R_2 < 0$.

(*b*) $$\frac{1}{f} = (n-1)\left(\frac{1}{R_1} - \frac{1}{R_2}\right) \qquad \text{or} \qquad \frac{1}{13.7} = (n-1)\left(\frac{1}{18\text{ cm}} - \frac{1}{-20\text{ cm}}\right) \qquad \text{or} \qquad n = 1.7$$

38.11 [II] A thin glass lens ($n = 1.50$) has a focal length of $+10$ cm in air. Compute its focal length in water ($n = 1.33$).

Using

$$\frac{1}{f} = \left(\frac{n_1}{n_2} - 1\right)\left(\frac{1}{R_1} - \frac{1}{R_2}\right)$$

we get *For air*: $$\frac{1}{10} = (1.50 - 1)\left(\frac{1}{R_1} - \frac{1}{R_2}\right)$$

For water: $$\frac{1}{f} = \left(\frac{1.50}{1.33} - 1\right)\left(\frac{1}{R_1} - \frac{1}{R_2}\right)$$

Divide one equation by the other to obtain $f = 5.0/0.128 = 39$ cm.

38.12 [III] A double convex thin lens has radii of 20.0 cm. The index of refraction of the glass is 1.50. Compute the focal length of this lens (*a*) in air and (*b*) when it is immersed in carbon disulfide ($n = 1.63$).

For a thin lens with an index of n_1, immersed in a surrounding medium of index n_2,

$$\frac{1}{f} = \left(\frac{n_1}{n_2} - 1\right)\left(\frac{1}{R_1} - \frac{1}{R_2}\right)$$

Here $R_1 = +20.0$ cm and $R_2 = -20.0$ cm and so

(*a*) $$\frac{1}{f} = (1.50 - 1)\left(\frac{1}{20\text{ cm}} - \frac{1}{-20\text{ cm}}\right) \qquad \text{or} \qquad f = +20.0\text{ cm}$$

(*b*) $$\frac{1}{f} = \left(\frac{1.50}{1.63} - 1\right)\left(\frac{1}{20\text{ cm}} - \frac{1}{-20\text{ cm}}\right) \qquad \text{or} \qquad f = -125\text{ cm}$$

When $n_2 > n_1$ the focal length is negative and the lens is a diverging lens.

38.13 [I] Two thin lenses, of focal lengths $+9.0$ and -6.0 cm, are placed in contact. Calculate the focal length of the combination.

$$\frac{1}{f} = \frac{1}{f_1} + \frac{1}{f_2} = \frac{1}{9.0\text{ cm}} - \frac{1}{6.0\text{ cm}} = -\frac{1}{18\text{ cm}} \qquad \text{or} \qquad f = -18\text{ cm}$$

The combination lens is diverging.

38.14 An achromatic lens is formed from two thin lenses in contact, having powers of $+10.0$ diopters and -6.0 diopters. Determine the power and focal length of the combination.

Since reciprocal focal lengths add,

$$\text{Power} = +10.0 - 6.0 = +4.0\text{ diopters} \qquad \text{and} \qquad \text{focal length} = \frac{1}{\text{power}} = \frac{1}{+4.0\text{ m}^{-1}} = +25\text{ cm}$$

Supplementary Problems

38.15 [I] Draw diagrams to indicate qualitatively the position, nature, and size of the image formed by a converging lens of focal length f for the following object distances: (*a*) infinity, (*b*) greater than $2f$, (*c*) equal to $2f$, (*d*) between $2f$ and f, (*e*) equal to f, (*f*) less than f.

38.16 [I] Determine the nature, position, and transverse magnification of the image formed by a thin converging lens of focal length +100 cm when the object distance from the lens is (*a*) 150 cm, (*b*) 75.0 cm. *Ans.* (*a*) real, inverted, 300 cm beyond lens, 2 : 1; (*b*) virtual, erect, 300 cm in front of lens, 4 : 1

38.17 [II] Determine the two locations of an object such that its image will be enlarged 8.0 times by a thin lens of focal length +4.0 cm? *Ans.* 4.5 cm from lens (image is real and inverted), 3.5 cm from lens (image is virtual and erect)

38.18 [II] What are the nature and focal length of the thin lens that will form a real image having one-third the dimensions of an object located 9.0 cm from the lens? *Ans.* converging, +2.3 cm

38.19 [II] Describe fully the image of an object which is 10 cm high and 28 cm from a diverging lens of focal length −7.0 cm. *Ans.* virtual, erect, smaller, 5.6 cm in front of lens, 2.0 cm high

38.20 [II] Compute the focal length of a lens which will give an erect image 10 cm from the lens when the object distance from the lens is (*a*) 200 cm, (*b*) very great. *Ans.* (*a*) −11 cm; (*b*) −10 cm

38.21 [II] A luminous object and a screen are 12.5 m apart. What are the position and focal length of the lens which will throw upon the screen an image of the object magnified 24 times? *Ans.* 0.50 m from object, +0.48 m

38.22 [II] A plano-concave lens has a spherical surface of radius 12 cm, and its focal length is −22.2 cm. Compute the refractive index of the lens material. *Ans.* 1.5

38.23 [II] A convex-concave lens has faces of radii 3.0 and 4.0 cm, respectively, and is made of glass of refractive index 1.6. Determine (*a*) its focal length and (*b*) the linear magnification of the image when the object is 28 cm from the lens. *Ans.* (*a*) +20 cm; (*b*) 2.5 : 1

38.24 [II] A double convex glass lens ($n = 1.50$) has faces of radius 8 cm each. Compute its focal length in air and when immersed in water ($n = 1.33$). *Ans.* +8 cm, +0.3 m

38.25 [II] Two thin lenses, of focal lengths +12 and −30 cm, are in contact. Compute the focal length and power of the combination. *Ans.* +20 cm, +5.0 diopters

38.26 [II] What must be the focal length of a third thin lens, placed in close contact with two thin lenses of 16 cm and −23 cm focal length, to produce a lens with −12 cm focal length? *Ans.* −9.8 cm

Chapter 39

Optical Instruments

COMBINATION OF THIN LENSES: To locate the image produced by two lenses acting in combination, (1) compute the position of the image produced by the first lens alone, disregarding the second lens; (2) then consider this image as the object for the second lens, and locate its image as produced by the second lens alone. This latter image is the required image.

If the image formed by the first lens alone is computed to be behind the second lens, then that image is a virtual object for the second lens, and its distance from the second lens is considered negative.

THE EYE uses a variable-focus lens to form an image on the retina at the rear of the eye. The **near point** of the eye, represented by d_n, is the closest distance to the eye from which an object can be viewed clearly. For the normal eye, d_n is about 25 cm. *Farsighted* persons can see distinctly only objects that are far from the eye; *nearsighted* persons can see distinctly only objects that are close to the eye.

ANGULAR MAGNIFICATION (M_A), also sometimes called the ***magnifying power***, is the ratio of the respective angles subtended by the images on the retina with and without the instrument in place (see Fig. 39-1).

A MAGNIFYING GLASS is a converging lens used so that it forms an erect, enlarged, virtual image of an object placed inside its focal point. The angular magnification due to a magnifier with a focal length f (where the lens is close to the eye) is $(d_n/f) + 1$ if the image it casts is at the near point [Fig. 39-1(b)]. Alternatively, if the image is at infinity, for relaxed viewing, the angular magnification is d_n/f.

A MICROSCOPE that consists of two converging lenses, an objective lens (focal length f_O) and an eyepiece lens (f_E), has

$$M_A = M_{AE} M_{TO}$$

$$M_A = \left(\frac{d_n}{f_E} + 1\right)\left(\frac{s_{iO}}{f_O} - 1\right)$$

where s_{iO} is the distance from the objective lens to the image it forms. This equation holds when the final image is at the near point, $d_n = 25$ cm.

A TELESCOPE that has an objective lens (or mirror) with focal length f_O and an eyepiece with focal length f_E gives a magnification $M_A = -f_O/f_E$.

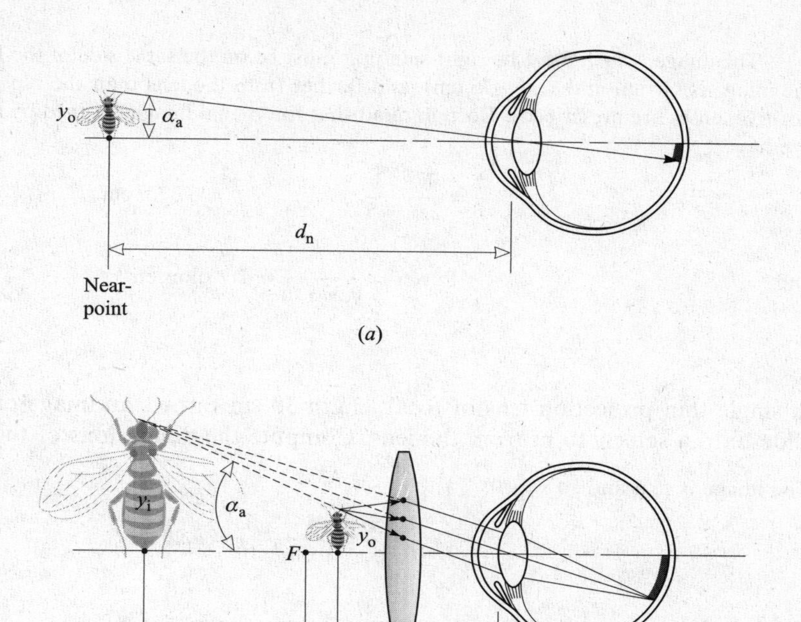

Fig. 39-1

Solved Problems

39.1 [II] A certain nearsighted person cannot see distinctly objects beyond 80 cm from the eye. What is the power in diopters of the spectacle lenses that will enable him to see distant objects clearly?

The image, which must be right-side-up, must be on the same side of the lens as the distant object (hence the image is virtual and $s_i = -80$ cm), and nearer to the lens than the object (hence diverging or negative lenses are indicated). Keep in mind that for virtual images formed by a concave lens $s_o > |s_i|$. As the object is at a great distance, s_o is very large and $1/s_o$ is practically zero. Then

$$\frac{1}{s_o}+\frac{1}{s_i}=\frac{1}{f} \qquad \text{or} \qquad 0-\frac{1}{80}=\frac{1}{f} \qquad \text{or} \qquad f=-80 \text{ cm (diverging)}$$

and

$$\text{Power in diopters} = \frac{1}{f \text{ in meters}} = \frac{1}{-0.80 \text{ m}} = -1.3 \text{ diopters}$$

39.2 [II] A certain farsighted person cannot see clearly objects closer to the eye than 75 cm. Determine the power of the spectacle lenses which will enable her to read type at a distance of 25 cm.

The image, which must be right-side-up, must be on the same side of the lens as the type (hence the image is virtual and $s_i = -75$ cm), and farther from the lens than the type (hence converging or positive lenses are prescribed). Keep in mind that for virtual images formed by a convex lens $|s_i| > s_o$. We have

$$\frac{1}{f} = \frac{1}{25} - \frac{1}{75} \qquad \text{or} \qquad f = +37.5 \text{ cm}$$

and

$$\text{Power} = \frac{1}{0.375 \text{ m}} = 2.7 \text{ diopters}$$

39.3 [II] A single thin projection lens of focal length 30 cm throws an image of a 2.0 cm × 3.0 cm slide onto a screen 10 m from the lens. Compute the dimensions of the image.

The image is real and so $s_i > 0$:

$$\frac{1}{s_o} = \frac{1}{f} - \frac{1}{s_i} = \frac{1}{0.30} - \frac{1}{10} = 3.23 \text{ m}^{-1}$$

and so

$$M_T = -\frac{s_i}{s_o} = -\frac{10 \text{ m}}{(1/3.23) \text{ m}} = -32$$

The magnification is negative because the image is inverted. The length and width of the slide are each magnified 32 times, so

$$\text{Size of image} = (32 \times 2.0 \text{ cm}) \times (32 \times 3.0 \text{ cm}) = 64 \text{ cm} \times 96 \text{ cm}$$

39.4 [II] A camera produces a clear image of a distant landscape when the thin lens is 8 cm from the film. What adjustment is required to get a good photograph of a map placed 72 cm from the lens?

When the camera is focused for distant objects (for parallel rays), the distance between lens and film is the focal length of the lens, namely, 8 cm. For an object 72 cm distant:

$$\frac{1}{s_i} = \frac{1}{f} - \frac{1}{s_o} = \frac{1}{8} - \frac{1}{72} \qquad \text{or} \qquad s_i = 9 \text{ cm}$$

The lens should be moved farther away from the film a distance of (9 – 8) cm = 1 cm.

39.5 [II] With a given illumination and film, the correct exposure for a camera lens set at $f/12$ is (1/5) s. What is the proper exposure time with the lens working at $f/4$?

A setting of $f/12$ means that the diameter of the opening, or stop, of the lens is 1/12 of the focal length; $f/4$ means that it is 1/4 of the focal length.

The amount of light passing through the opening is proportional to its area, and therefore to the square of its diameter. The diameter of the stop at $f/4$ is three times that at $f/12$, so $3^2 = 9$ times as much light will pass through the lens at $f/4$, and the correct exposure at $f/4$ is

$$(1/9)(\text{exposure time at } f/12) = (1.45) \text{ s}$$

39.6 [II] An engraver who has normal eyesight uses a converging lens of focal length 8.0 cm which he holds very close to his eye. At what distance from the work should the lens be placed, and what is the magnification of the lens?

Method 1

When a converging lens is used as a magnifying glass, the object is between the lens and the focal point. The virtual erect, and enlarged image forms at the distance of distinct vision, 25 cm from the eye. For a virtual image $s_i < 0$. Thus

$$\frac{1}{s_o}+\frac{1}{s_i}=\frac{1}{f} \quad \text{or} \quad \frac{1}{s_o}+\frac{1}{-25\text{ cm}}=\frac{1}{8.0\text{ cm}} \quad \text{or} \quad s_o=\frac{200}{33}=6.06\text{ cm}=6.1\text{ cm}$$

and

$$M_T=-\frac{s_i}{s_o}=-\frac{25\text{ cm}}{6.06\text{ cm}}=4.1$$

Method 2

By the formula,

$$M_A=\frac{d_n}{f}+1=\frac{25}{8.0}+1=4.1$$

Note that in this simple case $M_T = M_A$.

39.7 [III] Two positive lenses, having focal lengths of +2.0 cm and +5.0 cm, are 14 cm apart as shown in Fig. 39-2. An object AB is placed 3.0 cm in front of the +2.0 lens. Determine the position and magnification of the final image $A''B''$ formed by this combination of lenses.

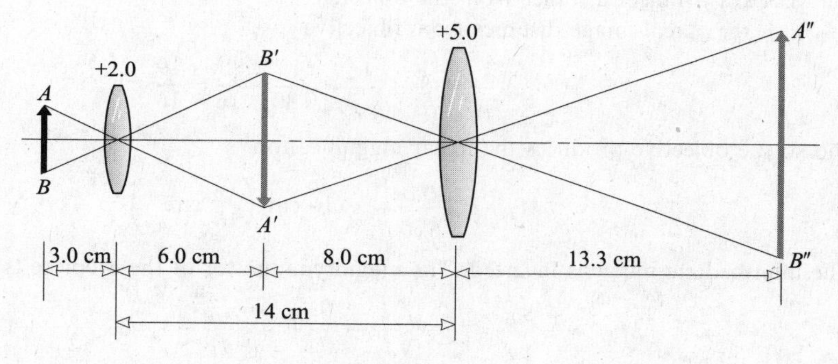

Fig. 39-2

To locate image $A'B'$ formed by the +2.0 lens alone:

$$\frac{1}{s_i}=\frac{1}{f}-\frac{1}{s_o}=\frac{1}{2.0}-\frac{1}{3.0}=\frac{1}{6.0} \quad \text{or} \quad s_i=6.0\text{ cm}$$

The image $A'B'$ is real, inverted, and 6.0 cm beyond the +2.0 lens.

To locate the final image $A''B''$: The image $A'B'$ is (14 – 6.0) cm = 8.0 cm in front of the +5.0 lens and is taken as a real object for the +5.0 lens.

$$\frac{1}{s_i}=\frac{1}{5.0}-\frac{1}{8.0} \quad \text{or} \quad s_i=13.3\text{ cm}$$

$A''B''$ is real, erect, and 13 cm from the +5 lens. Then,

$$M_T=\frac{\overline{A''B''}}{\overline{AB}}=\frac{\overline{A'B'}}{\overline{AB}}\times\frac{\overline{A''B''}}{\overline{A'B'}}=\frac{6.0}{3.0}\times\frac{13.3}{8.0}=3.3$$

Note that the magnification produced by a combination of lenses is the product of the individual magnifications.

39.8 [II] In the compound microscope shown in Fig. 39-3, the objective and eyepiece have focal lengths of +0.80 and +2.5 cm, respectively. The real image $A'B'$ formed by the objective is 16 cm from the objective. Determine the total magnification if the eye is held close to the eyepiece and views the virtual image $A''B''$ at a distance of 25 cm.

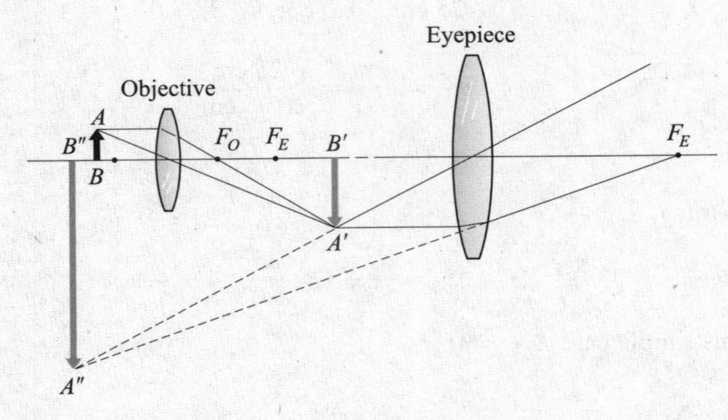

Fig. 39-3

Method 1

Let s_{oO} = object distance from the objective
s_{iO} = real-image distance from objective

$$\frac{1}{s_{oO}} = \frac{1}{f_O} - \frac{1}{s_{iO}} = \frac{1}{0.80} - \frac{1}{16} = \frac{19}{16}\ \text{cm}^{-1}$$

and so the objective produces the linear magnification

$$M_{TO} = -\frac{s_{iO}}{s_{oO}} = -(16\ \text{cm})\left(\frac{19}{16}\ \text{cm}^{-1}\right) = -19$$

The intermediate image is inverted. The magnifying power of the eyepiece is

$$M_{TE} = -\frac{s_{iE}}{s_{oE}} = -s_{iE}\left(\frac{1}{f_E} - \frac{1}{s_{iE}}\right) = -\frac{s_{iE}}{f_E} + 1 = -\frac{-25}{+2.5} + 1 = 11$$

The eyepiece does not flip the image: the intermediate image is inverted and the final image is inverted. Therefore, the magnifying power of the instrument is $-19 \times 11 = -2.1 \times 10^2$.

Alternatively, under the conditions stated, the magnifying power of the eyepiece can be found as

$$\frac{25}{f_E} + 1 = \frac{25}{2.5} + 1 = 11$$

Method 2

By the formula on p.372, with $s_{iO} = 16$ cm,

$$\text{Magnification} = \left(\frac{d_n}{f_E} + 1\right)\left(\frac{s_{iO}}{f_O} - 1\right) = \left(\frac{25}{2.5} + 1\right)\left(\frac{16}{0.8} - 1\right) = 2.1 \times 10^2$$

39.9 [III] The telephoto lens shown in Fig. 39-4 consists of a converging lens of focal length +6.0 cm placed 4.0 cm in front of a diverging lens of focal length −2.5 cm. (*a*) Locate the image of a very distant object. (*b*) Compare the size of the image formed by this lens combination with the size of the image that could be produced by the positive lens alone.

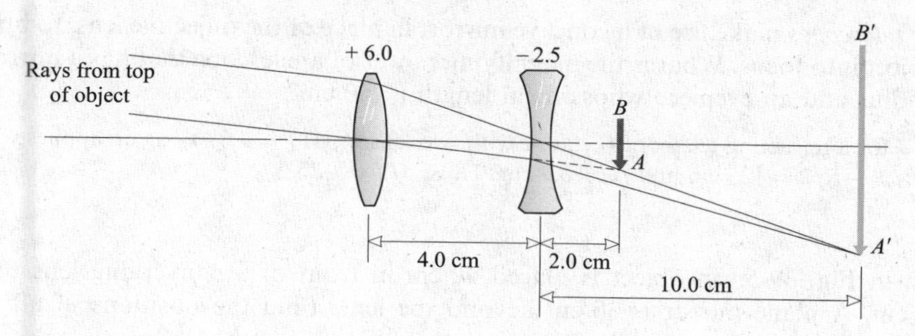

Fig. 39-4

(*a*) If the negative lens were not employed, the image AB would be formed at the focal point of the $+6.0$ lens, 6.0 cm distant from the $+6.0$ lens. The negative lens decreases the convergence of the rays refracted by the positive lens and causes them to focus at $A'B'$ instead of AB.

The image AB (that would have been formed by the $+6.0$ lens alone) is $6.0 - 4.0 = 2.0$ cm beyond the -2.5 lens and is taken as the (virtual) object for the -2.5 lens. Then $s_o = -2.0$ cm (negative because AB is virtual), and

$$\frac{1}{s_i} = \frac{1}{f} - \frac{1}{s_o} = \frac{1}{-2.5\text{ cm}} - \frac{1}{-2.0\text{ cm}} = \frac{1}{10\text{ cm}} \qquad \text{or} \qquad s_i = +10\text{ cm}$$

The final image $A'B'$ is real and 10 cm beyond the negative lens.

(*b*)

$$\text{Magnification by negative lens} = \frac{\overline{A'B'}}{\overline{AB}} = -\frac{s_i}{s_o} = -\frac{10\text{ cm}}{-2.0\text{ cm}} = 5.0$$

so the diverging lens increases the magnification by a factor of 5.0.

Notice that the magnification produced by the convex lens is negative and so the net magnification of both lenses is negative: the final image is inverted.

39.10 [II] A microscope has two interchangeable objective lenses (3.0 mm and 7.0 mm) and two interchangeable eyepieces (3.0 cm and 5.0 cm). What magnifications can be obtained with the microscope if it is adjusted so that the image formed by the objective is 17 cm from that lens?

Because $s_{iO} = 17$ cm the magnification formula for a microscope, with $d_n = 25$ cm, gives the following possibilities for M_A:

$$\begin{aligned} &\text{For } f_E = 3\text{ cm},\ f_O = 0.3\text{ cm}: && M_A = (9.33)(55.6) = 518 = 5.2 \times 10^2 \\ &\text{For } f_E = 3\text{ cm},\ f_O = 0.7\text{ cm}: && M_A = (9.33)(23.2) = 216 = 2.2 \times 10^2 \\ &\text{For } f_E = 5\text{ cm},\ f_O = 0.3\text{ cm}: && M_A = (5)(55.6) = 278 = 2.8 \times 10^2 \\ &\text{For } f_E = 5\text{ cm},\ f_O = 0.7\text{ cm}: && M_A = (5)(23.2) = 116 = 1.2 \times 10^2 \end{aligned}$$

39.11 [I] Compute the magnifying power of a telescope, having objective and eyepiece lenses of focal lengths $+60$ and $+3.0$ cm respectively, when it is focused for parallel rays.

$$\text{Magnifying power} = -\frac{\text{focal length of objective}}{\text{focal length of eyepiece}} = -\frac{60\text{ cm}}{3.0\text{ cm}} = -20$$

The image is inverted.

39.12 [II] *Reflecting telescopes* make use of a concave mirror, in place of the objective lens, to bring the distant object into focus. What is the magnifying power of a telescope that has a mirror with 250 cm radius and an eyepiece whose focal length is 5.0 cm?

As it is for a refracting telescope (i.e., one with two lenses), $M_A = -f_O/f_E$ again applies where, in this case, $f_O = -R/2 = 125$ cm and $f_E = 5.0$ cm. Thus, $M_A = -25$.

39.13 [III] As shown in Fig. 39-5, an object is placed 40 cm in front of a converging lens that has $f = +8.0$ cm. A plane mirror is 30 cm beyond the lens. Find the positions of all images formed by this system.

For the lens

$$\frac{1}{s_i} = \frac{1}{f} - \frac{1}{s_o} = \frac{1}{8.0} - \frac{1}{40} = \frac{4}{40} \qquad \text{or} \qquad s_i = 10 \text{ cm}$$

This is image $A'B'$ in the figure. It is real and inverted.

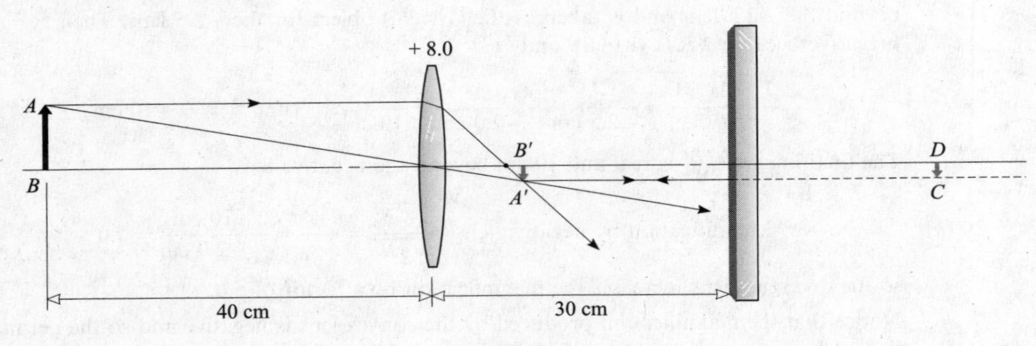

Fig. 39-5

$A'B'$ acts as an object for the plane mirror, 20 cm away. A virtual image CD is formed 20 cm behind the mirror.

Light reflected by the mirror appears to come from the image at CD. With CD as object, the lens forms an image of it to the left of the lens. The distance s_i from the lens to this latter image is given by

$$\frac{1}{s_i} = \frac{1}{f} - \frac{1}{s_o} = \frac{1}{8} - \frac{1}{50} = 0.105 \qquad \text{or} \qquad s_i = 9.5 \text{ cm}$$

The real images are therefore located 10 cm to the right of the lens and 9.5 cm to the left of the lens. (This latter image is upright.) A virtual inverted image is found 20 cm behind the mirror.

Supplementary Problems

39.14 [II] A farsighted woman cannot see objects clearly that are closer to her eye than 60.0 cm. Determine the focal length and power of the spectacle lenses that will enable her to read a book at a distance of 25.0 cm. *Ans.* +42.9 cm, +2.33 diopters

39.15 [II] A nearsighted man cannot see objects clearly that are beyond 50 cm from his eye. Determine the focal length and power of the glasses that will enable him to see distant objects clearly. *Ans.* −50 cm, −2.0 diopters

39.16 [II] A projection lens is employed to produce 2.4 m × 3.2 m pictures from 3.0 cm × 4.0 cm slides on a screen that is 25 cm from the lens. Compute its focal length. *Ans.* 31 cm

39.17 [II] A camera gives a life-size picture of a flower when the thin lens is 20 cm from the film. What should be the distance between lens and film to photograph a flock of birds high overhead? *Ans.* 10 cm

39.18 [II] What is the maximum stop rating of a camera lens having a focal length of +10 cm and a diameter of 2.0 cm? If the correct exposure at $f/6$ is (1/90) s, what exposure is needed when the diaphragm setting is changed to $f/9$? *Ans.* $f/5$, (1/40) s

39.19 [I] What is the magnifying power of a lens of focal length +2.0 cm when it used as a magnifying glass (or simple microscope)? The lens is held close to the eye, and the virtual image forms at the distance of distinct vision, 25 cm from the eye. *Ans.* 14

39.20 [II] When the object distance from a converging lens is 5.0 cm, a real image is formed 20 cm from the lens. What magnification is produced by this lens when it is used as a magnifying glass, the distance of most distinct vision being 25 cm? *Ans.* 7.3

39.21 [II] In a compound microscope, the focal lengths of the objective and eyepiece are +0.50 cm and +2.0 cm respectively. The instrument is focused on an object 0.52 cm from the objective lens. Compute the magnifying power of the microscope if the virtual image is viewed by the eye at a distance of 25 cm. *Ans.* 3.4×10^2

39.22 [II] A refracting astronomical telescope has a magnifying power of 150 when adjusted for minimum eye-strain. Its eyepiece has a focal length of +1.20 cm. (*a*) Determine the focal length of the objective lens. (*b*) How far apart must the two lenses be so as to project a real image of a distant object on a screen 12.0 cm from the eyepeice? *Ans.* (*a*) +180 cm; (*b*) 181 cm

39.23 [III] The large telescope at Mt Palomar has a concave objective mirror diameter of 5.0 m and radius of curvature 46 m. What is the magnifying power of the instrument when it is used with an eyepiece of focal length 1.25 cm? *Ans.* 1.8×10^3

39.24 [II] An astronomical telescope with an objective lens of focal length +80 cm is focused on the moon. By how much must the eyepiece be moved to focus the telescope on an object 40 meters distant? *Ans.* 1.6 cm

39.25 [II] A lens combination consists of two lenses with focal lengths of +4.0 cm and +8.0 cm, which are spaced 16 cm apart. Locate and describe the image of an object placed 12 cm in front of the +4.0-cm lens. *Ans.* 40 cm beyond +8.0 lens, real, erect

39.26 [II] Two lenses, of focal lengths +6.0 cm and −10 cm, are spaced 1.5 cm apart. Locate and describe the image of an object 30 cm in front of the +6.0-cm lens. *Ans.* 15 cm beyond negative lens, real, inverted, 5/8 as large as the object.

39.27 [II] A telephoto lens consists of a positive lens of focal length +3.5 cm placed 2.0 cm in front of a negative lens of focal length −1.8 cm. (*a*) Locate the image of a very distant object. (*b*) Determine the focal length of the single lens that would form as large an image of a distant object as is formed by this lens combination. *Ans.* (*a*) real image 9.0 cm in back of negative lens; (*b*) +21 cm

39.28 [II] An opera glass has an objective lens of focal length +3.60 cm and a negative eyepiece of focal length −1.20 cm. How far apart must the two lenses be for the viewer to see a distant object at 25.0 cm from the eye? *Ans.* 2.34 cm

39.29 [II] Repeat Problem 39.13 if the distance between the plane mirror and the lens is 8.0 cm. *Ans.* at 6.0 cm (real) and 24 cm (virtual) to the right of the lens

39.30 [II] Solve Problem 39.13 if the plane mirror is replaced by a concave mirror with a 20 cm radius of curvature. *Ans.* at 10 cm (real, inverted), 10 cm (real, upright), −40 cm (real, inverted) to the right of the lens

Chapter 40

Interference and Diffraction of Light

A PROPAGATING WAVE is a self-sustaining disturbance of a medium that carries energy and momentum from one location to another. All such waves are ultimately associated with the motion of an underlying distribution of particles.

COHERENT WAVES (be they light, sound, or disturbances on a string) are waves that have the same form, the same frequency, and a fixed phase difference (i.e., the amount by which the peaks of one wave lead or lag those of the other wave does not change with time).

THE RELATIVE PHASE of two coherent waves traveling along the same line specifies their relative positions on the line. If the crests of one wave fall on the crests of the other, the waves are completely **in-phase**. If the crests of one fall on the troughs of the other, the waves are 180° (or one-half wavelength) **out-of-phase**. Two waves can be out of phase by any amount greater than zero up to and including 180°.

INTERFERENCE EFFECTS occur when two or more coherent waves overlap. If two coherent waves of the same amplitude are superposed, **total destructive interference** (cancellation, or in the case of light, darkness) occurs when the waves are 180° out-of-phase. **Total constructive interference** (reinforcement, or in the case of light, brightness) occurs when they are in-phase.

DIFFRACTION refers to the deviation from straight-line propagation that occurs when a wave passes beyond a partial obstruction. It usually corresponds to the bending or spreading of waves around the edges of apertures and obstacles. The simplest form of the diffraction of light is *far-field* or ***Fraunhofer diffraction***. It is observed on a screen that is far away from the aperture or obstacle which is obstructing an incident stream of plane waves. Diffraction places a limit on the size of details that can be observed optically.

SINGLE-SLIT FRAUNHOFER DIFFRACTION: When parallel rays of light of wavelength λ are incident normally upon a slit of width D, a diffraction pattern is observed beyond the slit. On a far-away screen complete darkness is observed at angles $\theta_{m'}$ to the straight-through beam, where

$$m'\lambda = D \sin \theta_{m'}$$

Here, $m' = \pm 1, \pm 2, \pm 3, \ldots$, is the *order number* of the diffraction dark band (or minimum). The pattern consists of a broad central bright band flanked on both sides by an alternating succession of faint narrow light and dark bands ($m' = \pm 1, \pm 2$ etc.).

LIMIT OF RESOLUTION of two objects due to diffraction: If two objects are viewed through an optical instrument, the diffraction patterns caused by the aperture of the instrument limit our ability to distinguish the objects from each other. For distinguishability, the angle θ subtended at the aperture by the objects must be larger than a critical value θ_{cr}, given by

$$\sin \theta_{cr} = (1.22)\frac{\lambda}{D}$$

where D is the diameter of the circular aperture of the instrument (be it an eye, telescope, or camera).

DIFFRACTION GRATING EQUATION: A **diffraction grating** is a repetitive array of apertures or obstacles that alters the amplitude or phase of a wave. It usually consists of a large number of equally spaced, parallel slits or ridges; the distance between slits is the grating spacing a. When waves of wavelength λ are incident normally upon a grating with spacing a, maxima are observed beyond the grating at angles θ_m to the normal, where

$$m\lambda = a \sin \theta_m$$

Here, $m = 0, \pm 1, \pm 2, \pm 3, \ldots$, is the *order number* of the diffracted image. Usually there will be a bright central undeviated band of colored light ($m = 0$) flanked on either side by blackness and then another band of colored light ($m = \pm 1$), and so on. These are known as the zeroth order spectrum, the first order spectrum, and so forth.

This same relation applies to the major maxima in the interference patterns of even two and three slits. In these cases, however, the maxima are not nearly so sharply defined as for a grating consisting of hundreds or thousands of slits. The pattern may become quite complex if the slits are wide enough so that the single-slit diffraction pattern from each slit shows several minima.

THE DIFFRACTION OF X-RAYS of wavelength λ by reflection from a crystal is described by the *Bragg equation*. Strong reflections are observed at grazing angles ϕ_m (where ϕ is the angle between the face of the crystal and the reflected beam) given by

$$m\lambda = 2d \sin \phi_m$$

where d is the distance between reflecting planes in the crystal, and $m = 1, 2, 3, \ldots$, is the *order* of reflection.

OPTICAL PATH LENGTH: In the same time that it takes a beam of light to travel a distance d in a material of index of refraction n, the beam would travel a distance nd in vacuum. For this reason, nd is defined as the **optical path length** of the material.

Solved Problems

40.1 [II] Figure 40.1 shows a thin film of a transparent material of thickness d and index n_f where $n_2 > n_f > n_1$. For what three smallest gap thicknesses will reflected light rays-1 and -2 interfere totally (*a*) constructively and (*b*) destructively? Assume the monochromatic light has a wavelength in the film of 600 nm.

Because $n_2 > n_f > n_1$ each reflection is at the interface with a more optically dense medium and so each is an *external reflection*. Accordingly, the two rays will not experience a relative phase shift due to the reflections.

(*a*) Ray-2 travels a distance of roughly $2d$ farther than ray-1. The rays reinforce if this distance is $0, \lambda, 2\lambda, 3\lambda, \ldots, m\lambda$, where m is an integer. Hence for reinforcement,

$$m\lambda = 2d \quad \text{or} \quad d = (\tfrac{1}{2}m)(600 \text{ nm}) = 300m \text{ nm}$$

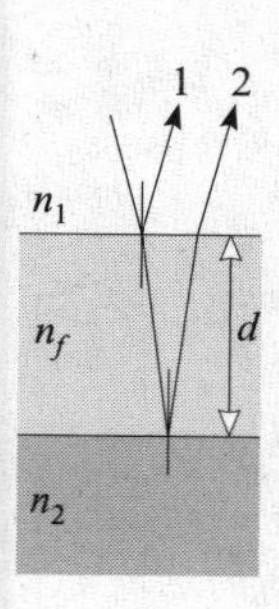

Fig. 40-1

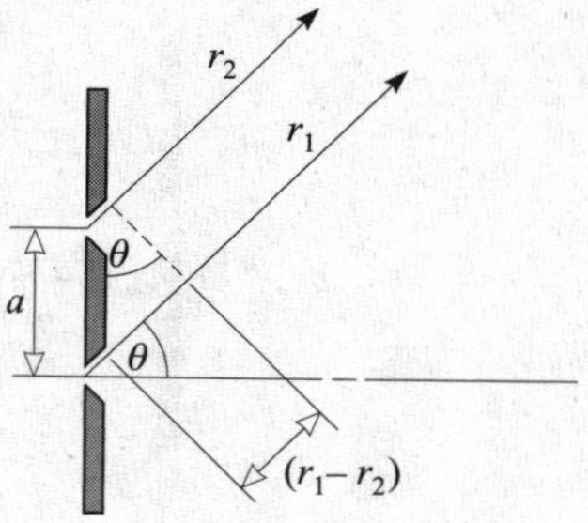

Fig. 40-2

The three smallest values for d are 0, 300 nm, and 600 nm.

(*b*) The waves cancel if they are 180° out-of-phase. This occurs when $2d$ is $\frac{1}{2}\lambda$, $(\lambda + \frac{1}{2}\lambda)$, $(2\lambda + \frac{1}{2}\lambda), \ldots,$ $(m\lambda + \frac{1}{2}\lambda), \ldots,$ with m an integer. Therefore, for cancellation,

$$2d = m\lambda + \tfrac{1}{2}\lambda \qquad \text{or} \qquad d = \tfrac{1}{2}(m + \tfrac{1}{2})\lambda = (m + \tfrac{1}{2})(300)\ \text{nm}$$

The three smallest values for d, that is, the ones corresponding to $m = 0, 1,$ and 2, are 150 nm, 450 nm, and 750 nm.

40.2 [III] Two narrow horizontal parallel slits (a distance $a = 0.60$ mm apart) are illuminated by a beam of 500-nm light as shown in Fig. 40-2. Light that is diffracted at certain angles θ reinforces; at others, it cancels. Find the three smallest values for θ at which (*a*) reinforcement occurs and (*b*) cancellation occurs. (See Fig. 40-3.)

The difference in path lengths for the two beams is $(r_1 - r_2)$. From Fig. 40-2.

$$\sin\theta = \frac{(r_1 - r_2)}{a}$$

(*a*) For reinforcement, $(r_1 - r_2) = 0, \pm\lambda, \pm 2\lambda, \ldots$ and so $\sin\theta_m = m\lambda/a$ where $m = 0, \pm 1, \pm 2, \ldots.$ The corresponding three smallest values for θ_m are found using

$$m = 0 \qquad \sin\theta_0 = 0 \qquad \text{or} \quad \theta_0 = 0$$

$$m = \pm 1 \qquad \sin\theta_1 = \pm\frac{500 \times 10^{-9}\ \text{m}}{6 \times 10^{-4}\ \text{m}} = \pm 8.33 \times 10^{-4} \qquad \text{or} \quad \theta_1 = \pm 0.048°$$

$$m = \pm 2 \qquad \sin\theta_2 = \pm\frac{2(500 \times 10^{-9}\ \text{m})}{6 \times 10^{-4}\ \text{m}} = \pm 16.7 \times 10^{-4} \qquad \text{or} \quad \theta_2 = \pm 0.095°$$

(*b*) For cancellation, $(r_1 - r_2) = \pm\frac{1}{2}\lambda$, $\pm(\lambda + \frac{1}{2}\lambda)$, $\pm(2\lambda + \frac{1}{2}\lambda), \ldots$ and so $\sin\theta_{m'} = \frac{1}{2}m'\lambda/a$ where $m' = \pm 1, \pm 3, \pm 5, \ldots.$ The corresponding three smallest values for $\theta_{m'}$ are found using

$$m' = \pm 1 \qquad \sin\theta_1 = \pm\frac{250\ \text{nm}}{600\,000\ \text{nm}} = \pm 4.17 \times 10^{-4} \qquad \text{or} \quad \theta_1 = \pm 0.024°$$

$$m' = \pm 3 \qquad \sin\theta_3 = \pm\frac{750\ \text{nm}}{600\,000\ \text{nm}} = \pm 0.001\,25 \qquad \text{or} \quad \theta_3 = \pm 0.072°$$

$$m' = \pm 5 \qquad \sin\theta_5 = \pm\frac{1250\ \text{nm}}{600\,000\ \text{nm}} = \pm 0.002\,08 \qquad \text{or} \quad \theta_5 = \pm 0.12°$$

40.3 [II] Monochromatic light from a point source illuminates two narrow, horizontal parallel slits. The centers of the two slits are $a = 0.80$ mm apart, as shown in Fig. 40-3. An interference pattern forms on the screen, 50 cm away. In the pattern, the bright and dark fringes are evenly spaced. The distance y_1 shown is 0.304 mm. Compute the wavelength λ of the light.

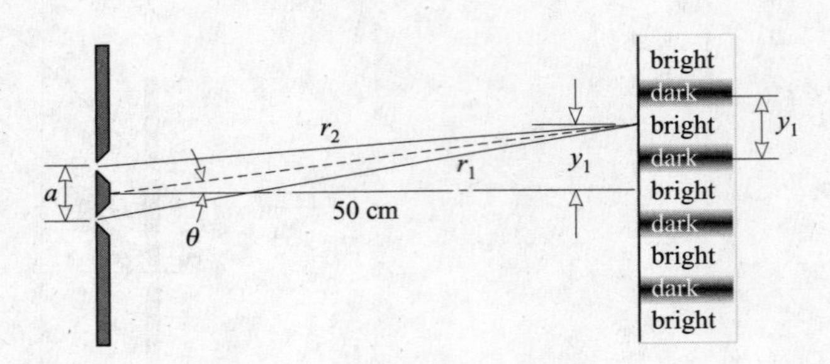

Fig. 40-3

Notice first that Fig. 40-3 is not to scale. The rays from the slits would actually be nearly parallel. We can therefore use the result of Problem 40.2 with $(r_1 - r_2) = m\lambda$ at the maxima (bright spots), where $m = 0, \pm 1, \pm 2, \ldots$. Then

$$\sin\theta = \frac{(r_1 - r_2)}{a} \qquad \text{becomes} \qquad m\lambda = a\sin\theta_m$$

Or, alternatively, we could use the grating equation, since a double slit is simply a grating with two lines. Both approaches result in $m\lambda = a\sin\theta_m$.

We know that the distance from the central maximum to the first maximum on either side is 0.304 mm. Therefore, from Fig. 40-3,

$$\sin\theta_1 = \frac{0.0304\ \text{cm}}{50\ \text{cm}} = 0.000\,608$$

Then, for $m = 1$,

$$m\lambda = a\sin\theta_m \qquad \text{becomes} \qquad (1)\lambda = (0.80 \times 10^{-3}\ \text{m})(6.08 \times 10^{-4})$$

from which $\lambda = 486$ nm, or to two significant figures 0.49×10^3 nm.

40.4 [III] Repeat Problem 40.1 for the case in which $n_1 < n_f > n_2$ or $n_1 > n_f < n_2$.

Experiment shows that, in this situation, cancellation occurs when d is near zero. This is due to the fact that light generally undergoes a phase shift upon reflection. The process is rather complicated, but for incident angles less than about 30° it's fairly straightforward. Then there will be a net phase difference of 180° introduced between the internally and externally reflected beams. Thus when the film is very thin compared to λ and $d \approx 0$, there will be an apparent path difference for the two beams of $\frac{1}{2}\lambda$ and cancellation will occur. (This was not the situation in Problem 40.1, because there both beams were externally reflected.)

Destructive interference occurs for $d \approx 0$, as we have just seen. When $d = \frac{1}{2}\lambda$ cancellation again occurs. The same thing happens at $d = \frac{1}{2}\lambda + \frac{1}{2}\lambda$. Therefore in this problem cancellation occurs at $d = 0$, 300 nm, and 600 nm.

Reinforcement occurs when $d = \frac{1}{4}\lambda$, because then beam-2 acts as though it had traveled an additional $\frac{1}{2}\lambda + (2)(\frac{1}{4}\lambda) = \lambda$. Reinforcement again occurs when d is increased by $\frac{1}{2}\lambda$ and by λ. Hence, for reinforcement, $d = 150$ nm, 450 nm, and 750 nm.

40.5 [III] When one leg of a Michelson interferometer is lengthened slightly, 150 dark fringes sweep through the field of view. If the light used has a wavelength of $\lambda = 480$ nm, how far was the mirror in that leg moved?

Darkness is observed when the light beams from the two legs are 180° out of phase. As the length of one leg is increased by $\frac{1}{2}\lambda$, the path length (down and back) increases by λ and the field of view changes from dark to bright to dark. When 150 fringes pass, the leg is lengthened by an amount

$$(150)(\tfrac{1}{2}\lambda) = (150)(240\ \text{nm}) = 36\,000\ \text{nm} = 0.0360\ \text{mm}$$

40.6 [III] As shown in Fig. 40-4, two flat glass plates touch along the leftmost edge and are separated at the other end by a spacer. Using vertical viewing and light with $\lambda = 589.0$ nm, five dark fringes (indicated by a D in the diagram) are obtained from edge to edge. What is the thickness of the spacer?

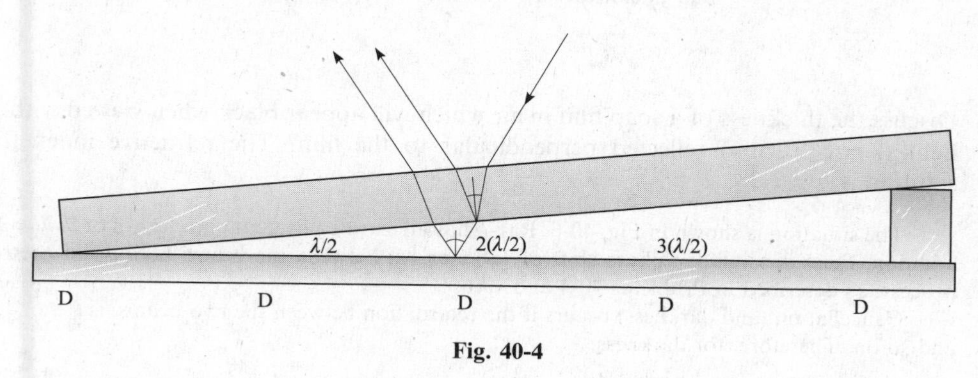

Fig. 40-4

The pattern is caused by interference between a beam reflected from the upper surface of the air wedge and a beam reflected from the lower surface of the wedge. The two reflections are of different nature in that reflection at the upper surface takes place at the boundary of a medium (air) of lower refractive index, while reflection at the lower surface occurs at the boundary of a medium (glass) of higher refractive index. In such cases, the act of reflection by itself involves a phase displacement of 180° between the two reflected beams. This explains the presence of a dark fringe at the left-hand edge.

As we move from a dark fringe to the next dark fringe, the beam that traverses the wedge must be held back by a path-length difference of λ. Because the beam travels twice through the wedge (down and back up), the wedge thickness changes by only $\frac{1}{2}\lambda$ as we move from fringe to fringe. Thus,

$$\text{Spacer thickness} = 4(\tfrac{1}{2}\lambda) = 2(589.0 \text{ nm}) = 1178 \text{ nm}$$

40.7 [III] In an experiment used to show *Newton's rings*, a plano-convex lens is placed on a flat glass plate, as in Fig. 40-5. When the lens is illuminated from directly above, a top-side viewer sees a series of bright and dark rings centered on the contact point, which is dark. Find the air-gap thickness at (*a*) the third dark ring and (*b*) the second bright ring. Assume 500 nm light is being used.

Because one reflection is internal and the other external, there will be a relative phase shift of 180°.

(*a*) The gap thickness is zero at the central dark spot. It increases by $\frac{1}{2}\lambda$ as we move from a position of darkness to the next position of darkness. (Why $\frac{1}{2}\lambda$?) Therefore, at the third dark ring,

$$\text{Gap thickness} = 3(\tfrac{1}{2}\lambda) = 3(250 \text{ nm}) = 750 \text{ nm}$$

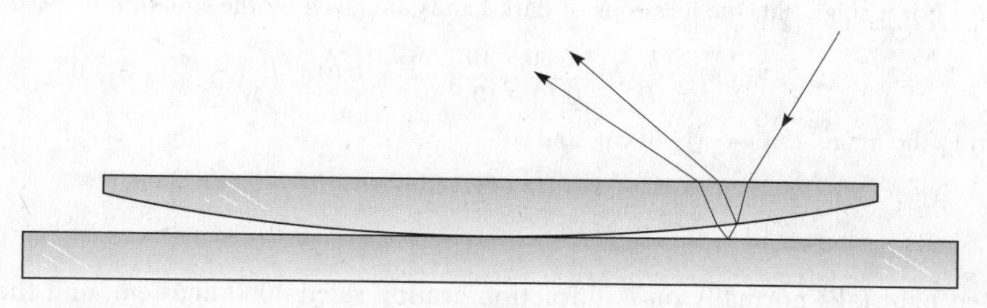

Fig. 40-5

(*b*) The gap thickness at the first bright ring must be large enough to increase the path length by $\frac{1}{2}\lambda$. Since the ray traverses the gap twice, the thickness there is $\frac{1}{4}\lambda$. As we go from one bright ring to the next, the gap thickness increases by $\frac{1}{2}\lambda$. Therefore, at the second bright ring,

$$\text{Gap thickness} = \tfrac{1}{4}\lambda + \tfrac{1}{2}\lambda = (0.750)(500\ \text{nm}) = 375\ \text{nm}$$

40.8 [II] Discuss the thickness of a soap film in air which will appear black when viewed with sodium light ($\lambda = 589.3$ nm) reflected perpendicular to the film? The refractive index for soap solution is $n = 1.38$.

The situation is shown in Fig. 40-6. Ray-*b* has an extra equivalent path length of $2nd = 2.76d$. In addition, there is a relative phase shift of 180°, or $\frac{1}{2}\lambda$, between the beams because of the reflection process, as described in Problems 40-4 and 40-6.

Cancellation (and darkness) occurs if the retardation between the two beams, is $\frac{1}{2}\lambda$, or $\frac{3}{2}\lambda$, or $\frac{5}{2}\lambda$, and so on. Therefore, for darkness,

$$2.76d + \tfrac{1}{2}\lambda = m(\tfrac{1}{2}\lambda) \qquad \text{where} \qquad m = 1, 3, 5, \ldots$$

When $m = 1$, this gives $d = 0$. For $m = 3$, we have

$$d = \frac{\lambda}{2.76} = \frac{589.3\ \text{nm}}{2.76} = 214\ \text{nm}$$

as the thinnest possible film other than zero. In practice, the film will become black when $d << \lambda/4$.

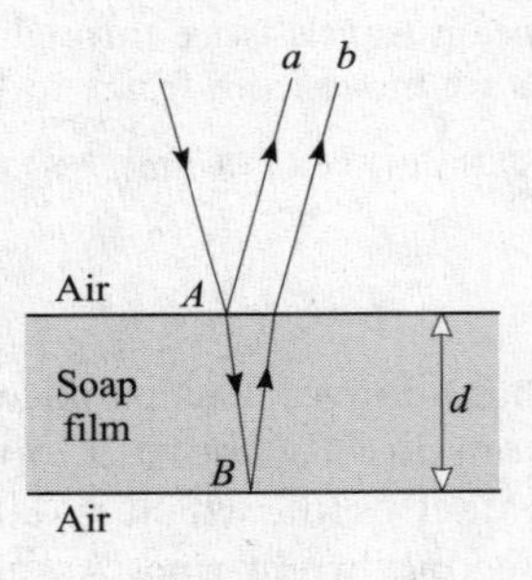

Fig. 40-6

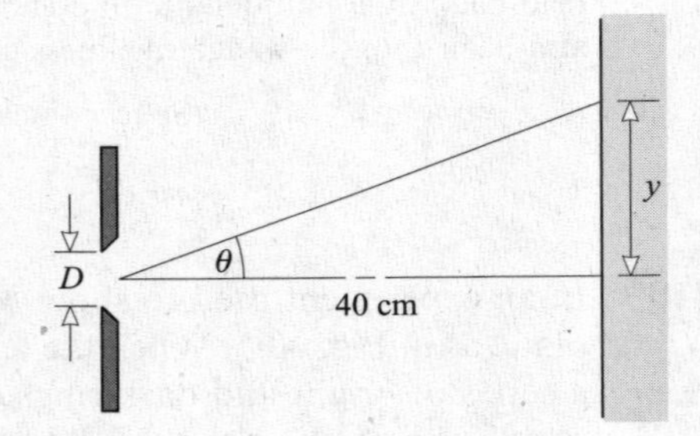

Fig. 40-7

40.9 [II] A single slit of width $D = 0.10$ mm is illuminated by parallel light of wavelength 600 nm, and diffraction bands are observed on a screen 40 cm from the slit. How far is the third dark band from the central bright band? (Refer to Fig. 40-7.)

For a single slit, the locations of dark bands are given by the equation $m'\lambda = D \sin \theta_{m'}$. Then

$$\sin \theta_3 = \frac{3\lambda}{D} = \frac{3(6.00 \times 10^{-7}\ \text{m})}{0.10 \times 10^{-3}\ \text{m}} = 0.018 \qquad \text{or} \qquad \theta_3 = 1.0^\circ$$

From the figure, $\tan \theta_3 = y/40$ cm, and so

$$y = (40\ \text{cm})(\tan \theta_3) = (40\ \text{cm})(0.018) = 0.72\ \text{cm}$$

40.10 [I] Red light falls normally on a diffraction grating ruled 4000 lines/cm, and the second-order image is diffracted 34.0° from the normal. Compute the wavelength of the light.

From the grating equation $m\lambda = a \sin \theta_m$,

$$\lambda = \frac{a \sin \theta_2}{2} = \frac{\left(\dfrac{1}{4000} \text{ cm}\right)(0.559)}{2} = 6.99 \times 10^{-5} \text{ cm} = 699 \text{ nm}$$

40.11 [I] Figure 40-8 shows a laboratory setup for grating experiments. The diffraction grating has 5000 lines/cm and is 1.00 m from the slit, which is illuminated with sodium light. On either side of the slit, and parallel to the grating, is a meterstick. The eye, placed close to the grating, sees virtual images of the slit along the metersticks. Determine the wavelength of the light if each first-order image is 31.0 cm from the slit.

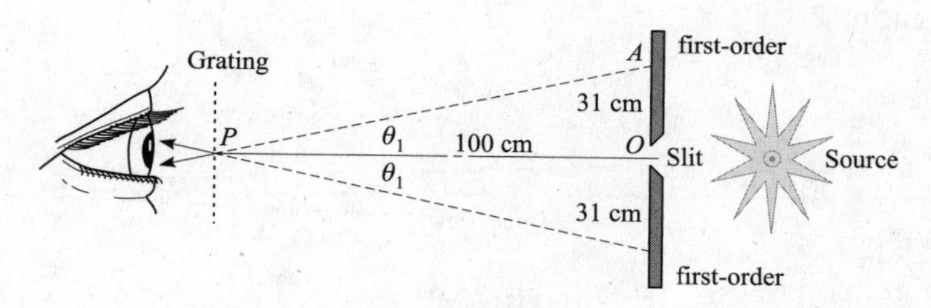

Fig. 40-8

$$\tan \theta_1 = 31.0/100 \qquad \text{or} \qquad \theta_1 = 17.2°$$

so

$$\lambda = \frac{a \sin \theta_1}{1} = \frac{(0.000\,200 \text{ cm})(0.296)}{1} = 592 \times 10^{-7} \text{ cm} = 592 \text{ nm}$$

40.12 [I] Green light of wavelength 540 nm is diffracted by a grating ruled with 2000 lines/cm. (*a*) Compute the angular deviation of the third-order image. (*b*) Is a 10th-order image possible?

(*a*)

$$\sin \theta_3 = \frac{3\lambda}{a} = \frac{3(5.40 \times 10^{-5} \text{ cm})}{5.00 \times 10^{-4} \text{ cm}} = 0.324 \qquad \text{or} \qquad \theta = 18.9°$$

(*b*)

$$\sin \theta_{10} = \frac{10\lambda}{a} = \frac{10(5.40 \times 10^{-5} \text{ cm})}{5.00 \times 10^{-4} \text{ cm}} = 1.08 \quad \text{(impossible)}$$

Since the value of $\sin \theta_{10}$ cannot exceed 1, a 10th-order image is impossible.

40.13 [II] Show that, in a spectrum of white light obtained with a grating, the red ($\lambda_r = 700$ nm) of the second order overlaps the violet ($\lambda_v = 400$ nm) of the third order.

For the red: $$\sin \theta_2 = \frac{2\lambda_r}{a} = \frac{2(700)}{a} = \frac{1400}{a} \quad (a \text{ in nm})$$

For the violet: $$\sin \theta_3 = \frac{3\lambda_v}{a} = \frac{3(400)}{a} = \frac{1200}{a}$$

As $\sin \theta_2 > \sin \theta_3$, $\theta_2 > \theta_3$. Thus the angle of diffraction of red in the second order is greater than that of violet in the third order.

40.14 [I] A parallel beam of X-rays is diffracted by a rock salt crystal. The first-order strong reflection is obtained when the glancing angle (the angle between the crystal face and the beam) is $6°50'$. The distance between reflection planes in the crystal is 2.8 Å. What is the wavelength of the X-rays? (1 angstrom $= 1\text{ Å} = 0.1$ nm.)

Note that the Bragg equation involves the glancing angle, not the angle of incidence.

$$\lambda = \frac{2d \sin \phi_1}{1} = \frac{(2)(2.8\text{ Å})(0.119)}{1} = 0.67\text{ Å}$$

40.15 [II] Two point sources of light are 50 cm apart, as shown in Fig. 40-9. They are viewed by the eye at a distance L. The entrance opening (pupil) of the viewer's eye has a diameter of 3.0 mm. If the eye were perfect, the limiting factor for resolution of the two sources would be diffraction. In that limit, how large could we make L and still have the sources seen as separate entities?

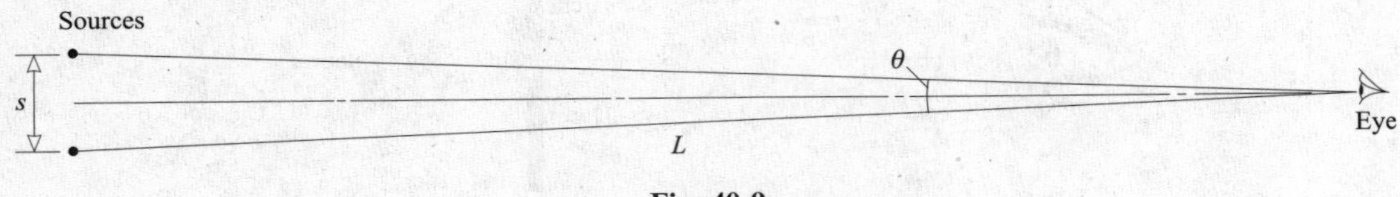

Fig. 40-9

This problem is about the *limit of resolution* as defined on p. 381. In the limiting case, $\theta = \theta_{cr}$, where $\sin \theta_{cr} = (1.22)(\lambda/D)$. But, we see from the figure that $\sin \theta_{cr}$ is nearly equal to s/L, because s is so much smaller than L. Substitution of this value gives

$$L \approx \frac{sD}{1.22\lambda} \approx \frac{(0.50\text{ m})(3.0 \times 10^{-3}\text{ m})}{(1.22)(5.0 \times 10^{-7}\text{ m})} = 2.5\text{ km}$$

We have taken $\lambda = 500$ nm, about the middle of the visible range.

Supplementary Problems

40.16 [II] Two sound sources send identical waves of 20 cm wavelength out along the $+x$-axis. At what separations of the sources will a listener on the axis beyond them hear (*a*) the loudest sound and (*b*) the weakest sound? *Ans.* (*a*) $m(20\text{ cm})$, where $m = 0, 1, 2, \ldots$; (*b*) $10\text{ cm} + m(20\text{ cm})$

40.17 [II] In an experiment such as that described in Problem 40.1, brightness is observed for the following film thicknesses: 2.90×10^{-7} m, 5.80×10^{-7} m, and 8.70×10^{-7} m. (*a*) What is the wavelength of the light being used? (*b*) At what thicknesses would darkness be observed? *Ans.* (*a*) 580 nm; (*b*) $145(1 + 2m)$ nm

40.18 [I] A double-slit experiment is done in the usual way with 480-nm light and narrow slits that are 0.050 cm apart. At what angle to the central axis will one observe (*a*) the third-order bright spot and (*b*) the second minimum from the central maximum? *Ans.* (*a*) 0.17°; (*b*) 0.083°

40.19 [I] In Problem 40.18, if the slit-to-screen distance is 200 cm, how far from the central maximum are (*a*) the third-order bright spot and (*b*) the second minimum? *Ans.* (*a*) 0.58 cm; (*b*) 0.29 cm

40.20 [I] Red light of wavelength 644 nm, from a point source, passes through two parallel and narrow slits which are 1.00 mm apart. Determine the distance between the central bright fringe and the third dark interference fringe formed on a screen parallel to the plane of the slits and 1.00 m away. *Ans.* 1.61 mm

40.21 [I] Two flat glass plates are pressed together at the top edge and separated at the bottom edge by a strip of tinfoil. The air wedge is examined in yellow sodium light (589 nm) reflected normally from its two surfaces, and 42 dark interference fringes are observed. Compute the thickness of the tinfoil. *Ans.* 12.4 μm

40.22 [I] A mixture of yellow light of wavelength 580 nm and blue light of wavelength 450 nm is incident normally on an air film 290 nm thick. What is the color of the reflected light? *Ans.* blue

40.23 [II] Repeat Problem 40.1 if the film has a refractive index of 1.40 and the vacuum wavelength of the incident light is 600 nm. *Ans.* (*a*) 0, 214 nm, 429 nm; (*b*) 107 nm, 321 nm, 536 nm

40.24 [II] Repeat Problem 40.6 if the wedge is filled with a fluid that has a refractive index of 1.50 instead of air. *Ans.* 785 nm

40.25 [II] A single slit of width 0.140 mm is illuminated by monochromatic light, and diffraction bands are observed on a screen 2.00 m away. If the second dark band is 16.0 mm from the central bright band, what is the wavelength of the light? *Ans.* 560 nm

40.26 [II] Green light of wavelength 500 nm is incident normally on a grating, and the second-order image is diffracted 32.0° from the normal. How many lines/cm are marked on the grating? *Ans.* 5.30×10^3 lines/cm

40.27 [II] A narrow beam of yellow light of wavelength 600 nm is incident normally on a diffraction grating ruled 2000 lines/cm, and images are formed on a screen parallel to the grating and 1.00 m distant. Compute the distance along the screen from the central bright line to the first-order lines. *Ans.* 12.1 cm

40.28 [II] Blue light of wavelength 4.7×10^{-7} m is diffracted by a grating ruled 5000 lines/cm. (*a*) Compute the angular deviation of the second-order image. (*b*) What is the highest-order image theoretically possible with this wavelength and grating? *Ans.* (*a*) 28°; (*b*) fourth

40.29 [II] Determine the ratio of the wavelengths of two spectral lines if the second-order image of one line coincides with the third-order image of the other line, both lines being examined by means of the same grating. *Ans.* 3 : 2

40.30 [II] A spectrum of white light is obtained with a grating ruled with 2500 lines/cm. Compute the angular separation between the violet ($\lambda_v = 400$ nm) and red ($\lambda_r = 700$ nm) in the (*a*) first order and (*b*) second order. (*c*) Does yellow ($\lambda_y = 600$ nm) in the third order overlap the violet in the fourth order? *Ans.* (*a*) 4°20′; (*b*) 8°57′; (*c*) yes

40.31 [II] A spectrum of the Sun's radiation in the infrared region is produced by a grating. What is the wavelength being studied, if the infrared line in the first order occurs at an angle of 25.0° with the normal, and the fourth-order image of the hydrogen line of wavelength 656.3 nm occurs at 30.0°? *Ans.* 2.22×10^{-6} m

40.32 [III] How far apart are the diffracting planes in a NaCl crystal for which X-rays of wavelength 1.54 Å make a glancing angle of 15°54′ in the first order? *Ans.* 2.81 Å

Chapter 41

Relativity

A REFERENCE FRAME is a coordinate system relative to which physical measurements are taken. An ***inertial reference frame*** is one which moves with constant velocity, i.e., one which is not accelerating.

THE SPECIAL THEORY OF RELATIVITY was proposed by Albert Einstein (1905) and is concerned with bodies that are moving with constant velocity. The theory is predicated on two postulates:

(1) The laws of physics are the same in all inertial reference frames. Therefore, all motion is relative. The velocity of an object can only be given relative to some other object.
(2) The speed of light in free space, c, has the same value for all observers, independent of the motion of the source (or the motion of the observer).

These postulates lead to the following conclusions.

THE RELATIVISTIC LINEAR MOMENTUM ($\vec{p}$) of a body of mass m and speed v is

$$\vec{p} = \frac{m\vec{v}}{\sqrt{1-(v/c)^2}} = \gamma m\vec{v}$$

where $\gamma = 1/\sqrt{1-(v/c)^2}$ and $\gamma > 1$. Some physicists prefer to associate the γ with the mass and introduce a relativistic mass $m_R = \gamma m$. That allows you to write the momentum as $p = m_R v$, but m_R is then speed dependent. Here we will use only one mass, m, which is independent of its speed, just like the two other fundamental properties of particles of matter, charge and spin.

LIMITING SPEED: When $v = c$, the momentum of an object becomes infinite. We conclude that no object can be accelerated to the speed of light c, and so c is an upper limit for speed.

RELATIVISTIC ENERGY (E): The total energy of a body of mass m is given by

$$E = \gamma mc^2$$

where

$$\text{total energy} = \text{kinetic energy} + \text{rest energy}$$

or

$$E = KE + E_0$$

When a body is at rest $\gamma = 1$, KE = 0 and the **rest energy** (E_0) is given by

$$E_0 = mc^2$$

The rest energy includes all forms of energy internal to the system.

The **kinetic energy** of a body of mass m is

$$KE = \gamma mc^2 - mc^2$$

If the speed of the object is not too large, this reduces to the usual expression

$$\mathrm{KE} = \tfrac{1}{2}mv^2 \qquad (v \ll c)$$

Using the expression $p = \gamma mv$, the total energy of a body can be written as

$$\mathrm{E}^2 = m^2\mathrm{c}^4 + p^2\mathrm{c}^2$$

TIME DILATION: *Time is relative*, it "flows" at different rates for differently moving observers. Suppose a spaceship and a planet are moving with respect to one another at a relative speed v and each carries an identical clock. The ship's pilot will see an interval of time Δt_S pass on her clock, with respect to which she is *stationary*. An observer on the ground will also notice a time interval Δt_S pass on the ship's clock, which is *moving* with respect to him. He, however, will notice that interval to take a time (measured via his own clock) of Δt_M where $\Delta t_M > \Delta t_S$. The observer on the ground will see time running more slowly on board the ship. For example, he might see 10 min (i.e., Δt_S) go by on the clock in the spaceship while his own clock shows that perhaps 20 min (i.e., Δt_M) went by. Accordingly,

$$\Delta t_M = \gamma \Delta t_S$$

Remember that $\gamma > 1$. Similarly the pilot will see time running more slowly on the ground.

The time taken for an event to occur, as recorded by a stationary observer at the site of the event, is called the **proper time**, Δt_S. All observers moving past the site record a longer time for the event to occur. Hence the proper time for the duration of an event is the smallest measured time for the event.

SIMULTANEITY: Suppose that for an observer two events occur at *different locations*, but at the same time. The events are simultaneous for this observer, but in general they are not simultaneous for a second observer moving relative to the first.

LENGTH CONTRACTION: Suppose an object is measured to have an x-component length L_S when at rest (L_S is called the **proper length**). The object is then given an x-directed speed v, so that it is moving with respect to an observer. That observer will see the object to have been shortened in the x-direction (but not in the y- and z-directions). Its x-length as measured by the observer with respect to whom it is moving (L_M) will then be

$$L_M = L_S\sqrt{1-(v/c)^2}$$

where $L_S > L_M$.

VELOCITY ADDITION FORMULA: Figure 41-1 shows a coordinate system S' moving at a speed $v_{O'O}$ with respect to a coordinate system S. Now consider an object at point P moving in the x-direction at a speed $v_{PO'}$ relative to point O'. Special Relativity establishes that the speed of the object with respect to O is not the classical value of $v_{PO'} + v_{O'O}$, but instead

$$v_{PO} = \frac{v_{PO'} + v_{O'O}}{1 + \dfrac{v_{PO'}v_{O'O}}{\mathrm{c}^2}}$$

Notice that even when $v_{PO'} = v_{O'O} = \mathrm{c}$ the value of $v_{PO} = \mathrm{c}$.

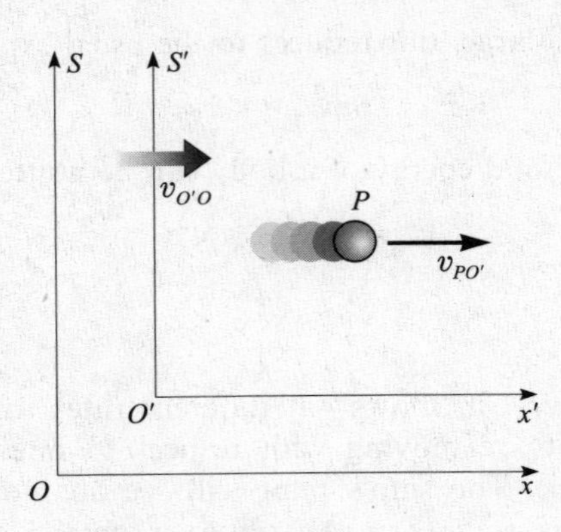

Fig. 41-1

Solved Problems

41.1 [I] How fast must an object be moving if its corresponding value of γ is to be 1.0 percent larger than γ is when the object is at rest? Give your answer to two significant figures.

Use the definition $\gamma = 1/\sqrt{1-(v/c)^2}$ to find that at $v = 0$, $\gamma = 1.0$. Hence the new value of $\gamma = 1.01(1.0)$, and so

$$1 - \left(\frac{v}{c}\right)^2 = \left(\frac{1}{1.01}\right)^2 = 0.980$$

Solving yields $v = 0.14c = 4.2 \times 10^7$ m/s.

41.2 [I] Compute the value of γ for a particle traveling at half the speed of light. Give your answer to three significant figures.

$$\gamma = \frac{1}{\sqrt{1-(v/c)^2}} = \frac{1}{\sqrt{1-(0.500)^2}} = \frac{1}{\sqrt{0.750}} = \frac{1}{0.866} = 1.15$$

41.3 [II] If 1.00 g of matter could be converted entirely into energy, what would be the value of the energy so produced, at 10.0 cents per kW·h?

We make use of $\Delta E_0 = (\Delta m)c^2$ to find

$$\text{Energy gained} = (\text{mass lost})\,c^2 = (1.00 \times 10^{-3}\ \text{kg})(2.998 \times 10^8\ \text{m/s})^2 = 8.99 \times 10^{13}\ \text{J}$$

$$\text{Value of energy} = (8.99 \times 10^{13}\ \text{J})\left(\frac{1\ \text{kW}\cdot\text{h}}{3.600 \times 10^6\ \text{J}}\right)\left(\frac{\$\ 0.10}{\text{kW}\cdot\text{h}}\right) = \$\ 2.50 \times 10^6$$

41.4 [II] A 2.0 kg object is lifted from the floor to a tabletop 30 cm above the floor. By how much did the mass of the system consisting of the Earth and the object increase because of this increased PE_G?

We use $\Delta E_0 = (\Delta m)c^2$, with $\Delta E_0 = mgh$. Therefore,

$$\Delta m = \frac{\Delta E_0}{c^2} = \frac{mgh}{c^2} = \frac{(2.0\ \text{kg})(9.81\ \text{m/s}^2)(0.30\ \text{m})}{(2.998 \times 10^8\ \text{m/s})^2} = 6.5 \times 10^{-17}\ \text{kg}$$

41.5 [III] An electron is accelerated from rest through a potential difference of 1.5 MV and thereby acquires 1.5 MeV of energy. Find its final speed.

Using $KE = \gamma mc^2 - mc^2$ and the fact that $KE = \Delta PE_E$ we have

$$KE = (1.5 \times 10^6\ \text{eV})(1.6 \times 10^{-19}\ \text{J/eV}) = 2.4 \times 10^{-13}\ \text{J}$$

Then $$(\gamma m - m) = \frac{KE}{c^2} = \frac{2.4 \times 10^{-13}\ \text{J}}{(2.998 \times 10^8\ \text{m/s})^2} = 2.67 \times 10^{-30}\ \text{kg}$$

But $m = 9.11 \times 10^{-31}$ kg and so $\gamma m = 3.58 \times 10^{-30}$ kg.

To find its speed, we use $\gamma = 1/\sqrt{1 - (v/c)^2}$, which gives us

$$\frac{1}{\gamma^2} = 1 - \left(\frac{v}{c}\right)^2 = \left(\frac{m}{\gamma m}\right)^2 = \left(\frac{0.91}{3.58}\right)^2 = 0.0646$$

from which $$v = c\sqrt{1 - 0.0646} = 0.967c = 2.9 \times 10^8\ \text{m/s}$$

41.6 [II] Determine the energy required to give an electron a speed equal to 0.90 that of light, starting from rest.

$$KE = (\gamma m - m)c^2 = \left[\frac{m}{\sqrt{1 - (v/c)^2}} - m\right]c^2 = mc^2\left[\frac{1}{\sqrt{1 - (v/c)^2}} - 1\right]$$

$$= (9.11 \times 10^{-31}\ \text{kg})(2.998 \times 10^8\ \text{m/s})^2\left[\frac{1}{\sqrt{1 - (0.90)^2}} - 1\right] = 1.06 \times 10^{-13}\ \text{J} = 0.66\ \text{MeV}$$

41.7 [III] Show that $KE = (\gamma m - m)c^2$ reduces to $KE = \frac{1}{2}mv^2$ when v is very much smaller than c.

$$KE = (\gamma m - m)c^2 = \left[\frac{m}{\sqrt{1 - (v/c)^2}} - m\right]c^2 = mc^2\left[\left(1 - \frac{v^2}{c^2}\right)^{-1/2} - 1\right]$$

Let $b = -v^2/c^2$ and expand $(1 + b)^{-1/2}$ by the binomial theorem:

$$(1 + b)^{-1/2} = 1 + (-1/2)b + \frac{(-1/2)(-3/2)}{2!}b^2 + \cdots = 1 + \frac{1}{2}\frac{v^2}{c^2} + \frac{3}{8}\frac{v^4}{c^4} + \cdots$$

Then $$KE = mc^2\left[\left(1 + \frac{1}{2}\frac{v^2}{c^2} + \frac{3}{8}\frac{v^4}{c^4} + \cdots\right) - 1\right] = \frac{1}{2}mv^2 + \frac{3}{8}mv^2\frac{v^2}{c^2} + \cdots$$

If v is very much smaller than c, the terms after $\frac{1}{2}mv^2$ are negligibly small.

41.8 [III] An electron traveling at high (or relativistic) speed moves perpendicularly to a magnetic field of 0.20 T. Its path is circular, with a radius of 15 m. Find (*a*) the momentum, (*b*) the speed, and (*c*) the kinetic energy of the electron. Recall that, in nonrelativistic situations, the magnetic force qvB furnishes the centripetal force mv^2/r. Thus, since $p = mv$ it follows that

$$p = qBr$$

and this relation holds even when relativistic effects are important.

First find the momentum using $p = qBr$

(*a*)
$$p = (1.60 \times 10^{-19}\ \text{C})(0.20\ \text{T})(15\ \text{m}) = 4.8 \times 10^{-19}\ \text{kg·m/s}$$

(*b*) Because $p = mv/\sqrt{1-(v^2/c^2)}$ with $m = 9.11 \times 10^{-31}$ kg, we have

$$4.8 \times 10^{-19}\ \text{kg·m/s} = \frac{(mc)(v/c)}{\sqrt{1-(v^2/c^2)}}$$

Squaring both sides and solving for $(v/c)^2$ give

$$\frac{v^2}{c^2} = \frac{1}{1 + 3.23 \times 10^{-7}} \qquad \text{or} \qquad \frac{v}{c} = \frac{1}{\sqrt{1 + 3.23 \times 10^{-7}}}$$

Most hand calculators cannot handle this. Accordingly, we make use of the fact that $1/\sqrt{1+x} \approx 1 - \frac{1}{2}x$ for $x \ll 1$. Then

$$v/c \approx 1 - 1.61 \times 10^{-7} = 0.999\,999\,84$$

(*c*)
$$\text{KE} = (\gamma m - m)c^2 = mc^2\left[\frac{1}{\sqrt{1-(v^2/c^2)}} - 1\right]$$

But we already found $(v/c)^2 = 1/(1 + 3.23 \times 10^{-7})$. If we use the approximation $1/(1+x) \approx 1 - x$ for $x \ll 1$, we have $(v/c)^2 \approx 1 - 3.23 \times 10^{-7}$. Then

$$\text{KE} = mc^2\left(\frac{1}{\sqrt{3.23 \times 10^{-7}}} - 1\right) = (mc^2)(1.76 \times 10^3)$$

Evaluating the above expression yields

$$\text{KE} = 1.4 \times 10^{-10}\ \text{J} = 9.0 \times 10^8\ \text{eV}$$

An alternative solution method would be to use $E^2 = p^2c^2 + m^2c^4$ and recall that $\text{KE} = E - mc^2$

41.9 [II] The Sun radiates energy equally in all directions. At the position of the Earth ($r = 1.50 \times 10^{11}$ m), the irradiance of the Sun's radiation is 1.4 kW/m^2. How much mass does the Sun lose per day because of the radiation?

The area of a spherical shell centered on the Sun and passing through the Earth is

$$\text{Area} = 4\pi r^2 = 4\pi(1.50 \times 10^{11}\ \text{m})^2 = 2.83 \times 10^{23}\ \text{m}^2$$

Through each square meter of this area, the Sun radiates an energy per second of 1.4 kW/m^2. Therefore the Sun's total radiation per second is

$$\text{Energy/s} = (\text{area})(1400\ \text{W/m}^2) = 3.96 \times 10^{26}\ \text{W}$$

The energy radiated in one day (86 400 s) is

$$\text{Energy/day} = (3.96 \times 10^{26}\ \text{W})(86\,400\ \text{s/day}) = 3.42 \times 10^{31}\ \text{J/day}$$

Because mass and energy are related through $\Delta E_0 = \Delta mc^2$, the mass loss per day is

$$\Delta m = \frac{\Delta E_0}{c^2} = \frac{3.42 \times 10^{31}\ \text{J}}{(2.998 \times 10^8\ \text{m/s})^2} = 3.8 \times 10^{14}\ \text{kg}$$

For comparison, the Sun's mass is 2×10^{30} kg.

41.10 [I] A beam of radioactive particles is measured as it shoots through the laboratory. It is found that, on the average, each particle "lives" for a time of 2.0×10^{-8} s; after that time, the particle changes to a new form. When at rest in the laboratory, the same particles "live" 0.75×10^{-8} s on the average. How fast are the particles in the beam moving?

Some sort of timing mechanism within the particle determines how long it "lives". This internal clock, which gives the proper lifetime, must obey the time-dilation relation. We have $\Delta t_M = \gamma \, \Delta t_S$ where the observer with respect to whom the particle (clock) is moving sees a time interval of $\Delta t_M = 2.0 \times 10^{-8}$ s. Hence

$$2.0 \times 10^{-8}\ \text{s} = \gamma(0.75 \times 10^{-8}\ \text{s}) \qquad \text{or} \qquad 0.75 \times 10^{-8} = (2.0 \times 10^{-8})\sqrt{1 - (v/c)^2}$$

Squaring both sides of the equation and solving for v leads to $v = 0.927c = 2.8 \times 10^8$ m/s.

41.11 [II] Two twins are 25.0 years old when one of them sets out on a journey through space at nearly constant speed. The twin in the spaceship measures time with an accurate watch. When he returns to Earth, he claims to be 31.0 years old, while the twin left on Earth knows that she is 43.0 years old. What was the speed of the spaceship?

The spaceship clock as seen by the space-twin reads the trip time to be Δt_S which is 6.0 years long. The Earth bound twin sees her brother age 6.0 years but her clocks tell her that a time $\Delta t_M = 18.0$ years has actually passed. Hence $\Delta t_M = \gamma \Delta t_S$ becomes $\Delta t_S = \Delta t_M \sqrt{1 - (v/c)^2)}$ and so

$$6 = 18\sqrt{1 - (v/c)^2}$$

from which $$(v/c)^2 = 1 - 0.111 \qquad \text{or} \qquad v = 0.943c = 2.83 \times 10^8\ \text{m/s}$$

41.12 [II] Two cells that subdivide on Earth every 10.0 s start from the Earth on a journey to the Sun (1.50×10^{11} m away) in a spacecraft moving at 0.850c. How many cells will exist when the spacecraft crashes into the Sun?

According to Earth observers, with respect to whom the cells are moving, the time taken for the trip to the Sun is the distance traveled (x) over the speed (v),

$$\Delta t_M = \frac{x}{v} = \frac{1.50 \times 10^{11}\ \text{m}}{(0.850)(2.998 \times 10^8\ \text{m/s})} = 588\ \text{s}$$

Because spacecraft clocks are moving with respect to the planet, they appear from Earth to run more slowly. The time these clocks read is

$$\Delta t_S = \Delta t_M/\gamma = \Delta t_M\sqrt{1 - (v/c)^2}$$

and so $$\Delta t_S = 310\ \text{s}$$

The cells divide according to the spacecraft clock, a clock that is at rest relative to them. They therefore undergo 31 divisions in this time, since they divide each 10.0 s. Therefore the total number of cells present on crashing is

$$(2)^{31} = 2.1 \times 10^9\ \text{cells}$$

41.13 [I] A person in a spaceship holds a meterstick as the ship shoots past the Earth with a speed v parallel to the Earth's surface. What does the person in the ship notice as the stick is rotated from parallel to perpendicular to the ship's motion?

The stick behaves normally; it does not change its length, because it has no translational motion relative to the observer in the spaceship. However, an observer on Earth would measure the stick to be $(1\ \text{m})\sqrt{1-(v/c)^2}$ long when it is parallel to the ship's motion, and 1 m long when it is perpendicular to the ship's motion.

41.14 [II] A spacecraft moving at 0.95c travels from the Earth to the star Alpha Centauri, which is 4.5 light years away. How long will the trip take according to (*a*) Earth clocks and (*b*) spacecraft clocks? (*c*) How far is it from Earth to the star according to spacecraft occupants? (*d*) What do they compute their speed to be?

A light year is the distance light travels in 1 year, namely

$$1 \text{ light year} = (2.998 \times 10^8 \text{ m/s})(3.16 \times 10^7 \text{ s}) = 9.47 \times 10^{15} \text{ m}$$

Hence the distance to the star (according to earthlings) is

$$d_e = (4.5)(9.47 \times 10^{15} \text{ m}) = 4.3 \times 10^{16} \text{ m}$$

(*a*)
$$\Delta t_e = \frac{d_e}{v} = \frac{4.3 \times 10^{16} \text{ m}}{(0.95)(2.998 \times 10^8 \text{ m/s})} = 1.5 \times 10^8 \text{ s}$$

(*b*) Because clocks on the moving spacecraft run slower,

$$\Delta t_{\text{craft}} = \Delta t_e \sqrt{1-(v/c)^2} = (1.51 \times 10^8 \text{ s})(0.312) = 4.7 \times 10^7 \text{ s}$$

(*c*) For the spacecraft occupants, the Earth–star distance is moving past them with speed 0.95c. Therefore that distance is shortened for them; they find it to be

$$d_{\text{craft}} = (4.3 \times 10^{16} \text{ m})\sqrt{1-(0.95)^2} = 1.3 \times 10^{16} \text{ m}$$

(*d*) For the spacecraft occupants, their relative speed is

$$v = \frac{d_{\text{craft}}}{\Delta t_{\text{craft}}} = \frac{1.34 \times 10^{16} \text{ m}}{4.71 \times 10^7 \text{ s}} = 2.8 \times 10^8 \text{ m/s}$$

which is 0.95c. Both Earth and spacecraft observers measure the same relative speed.

41.15 [II] As a rocket ship sweeps past the Earth with speed v, it sends out a pulse of light ahead of it. How fast does the light pulse move according to people on the Earth?

Method 1

With speed c (by the second postulate of Special Relativity).

Method 2

Here $v_{O'O} = v$ and $v_{PO'} = \text{c}$. According to the velocity addition formula, the observed speed will be (since $u = \text{c}$ in this case)

$$v_{PO} = \frac{v_{PO'} + v_{O'O}}{1 + \dfrac{v_{PO'}v_{O'O}}{\text{c}^2}} = \frac{v + \text{c}}{1 + (v/\text{c})} = \frac{(v + \text{c})\text{c}}{\text{c} + v} = \text{c}$$

Supplementary Problems

41.16 [I] At what speed must a particle move for γ to be 2.0? *Ans.* 2.6×10^8 m/s

41.17 [I] A particle is traveling at a speed v such that $v/c = 0.99$. Find γ for the particle. *Ans.* 7.1

41.18 [I] Compute the *rest energy* of an electron, i.e., the energy equivalent of its mass, 9.11×10^{-31} kg. *Ans.* 0.512 MeV = 820 pJ

41.19 [I] Determine the speed of an electron having a kinetic energy of 1.0×10^5 eV (or equivalently} 1.6×10^{-14} J). *Ans.* 1.6×10^8 m/s

41.20 [II] A proton ($m = 1.67 \times 10^{-27}$ kg) is accelerated to a kinetic energy of 200 MeV. What is its speed at this energy? *Ans.* 1.70×10^8 m/s

41.21 [II] Starting with the definition of linear momentum and the relation between mass and energy, prove that $E^2 = p^2c^2 + m^2c^4$. Use this relation to show that the translational KE of a particle is $\sqrt{m^2c^4 + p^2c^2} - mc^2$.

41.22 [II] A certain strain of bacteria doubles in number each 20 days. Two of these bacteria are placed on a spaceship and sent away from the Earth for 1000 Earth-days. During this time, the speed of the ship is 0.9950c. How many bacteria are aboard when the ship lands on the Earth? *Ans.* 64

41.23 [II] A certain light source sends out 2×10^{15} pulses each second. As a spaceship travels parallel to the Earth's surface with a speed of 0.90 c, it uses this source to send pulses to the Earth. The pulses are sent perpendicular to the path of the ship. How many pulses are recorded on Earth each second? *Ans.* 8.7×10^{14} pulses/s

41.24 [II] The insignia painted on the side of a spaceship is a circle with a line across it at 45° to the vertical. As the ship shoots past another ship in space, with a relative speed of 0.95c, the second ship observes the insignia. What angle does the observed line make to the vertical? *Ans.* $\tan\theta = 0.31$ and $\theta = 17°$

41.25 [II] As a spacecraft moving at 0.92c travels past an observer on Earth, the Earthbound observer and the occupants of the craft each start identical alarm clocks that are set to ring after 6.0 h have passed. According to the Earthling, what does the Earth clock read when the spacecraft clock rings? *Ans.* 15 h

41.26 [III] Find the speed and momentum of a proton ($m = 1.67 \times 10^{-27}$ kg) that has been accelerated through a potential difference of 2000 MV. (We call this a 2 GeV proton.) Give your answers to three significant figures. *Ans.* 0.948c, 1.49×10^{-18} kg·m/s

Chapter 42

Quantum Physics and Wave Mechanics

QUANTA OF RADIATION: All the various forms of electromagnetic radiation, including light, have a dual nature. When traveling through space, they act like waves and give rise to interference and diffraction effects. But when electromagnetic radiation interacts with atoms and molecules, the beam acts like a stream of energy corpuscles called **photons** or *light-quanta.*

The energy (E) of each photon depends upon the frequency f (or wavelength λ) of the radiation:

$$\mathrm{E} = hf = \frac{hc}{\lambda}$$

where $h = 6.626 \times 10^{-34}\ \mathrm{J\cdot s}$ is a constant of nature called **Planck's constant.**

PHOTOELECTRIC EFFECT: When electromagnetic radiation is incident on the surface of certain metals electrons may be ejected. A photon of energy hf penetrates the material and is absorbed by an electron. If enough energy is available, the electron will be raised to the surface and ejected with some kinetic energy, $\frac{1}{2}mv^2$. Depending on how deep in the material they are, electrons having a range of values of KE will be emitted. Let ϕ be the energy required for an electron to break free of the surface, the so-called **work function**. For electrons up near the surface to begin with, an amount of energy $(hf - \phi)$ will be available and this is the maximum kinetic energy that can be imparted to any electron.

Accordingly, *Einstein's* **photoelectric equation** is

$$\tfrac{1}{2}mv_{\mathrm{max}}^2 = hf - \phi$$

The energy of the ejected electron may be found by determining what potential difference must be applied to stop its motion; then $\frac{1}{2}mv^2 = V_s e$. For the most energetic electron,

$$hf - \phi = V_s e$$

where V_s is called the **stopping potential.**

For any surface, the radiation must be of short enough wavelength so that the photon energy hf is large enough to eject the electron. At the **threshold wavelength** (or *frequency*), the photon's energy just equals the work function. For ordinary metals the threshold wavelength lies in the visible or ultraviolet range. X-rays will eject photoelectrons readily; far-infrared photons will not.

THE MOMENTUM OF A PHOTON: Because $\mathrm{E}^2 = m^2\mathrm{c}^4 + p^2\mathrm{c}^2$, when $m = 0$, $\mathrm{E} = p\mathrm{c}$. Hence, since $\mathrm{E} = hf$

$$\mathrm{E} = p\mathrm{c} = hf \qquad \text{and} \qquad p = \frac{hf}{\mathrm{c}} = \frac{h}{\lambda}$$

The momentum of a photon is $p = h/\lambda$.

COMPTON EFFECT: A photon can collide with a particle having mass, such as an electron. When it does so, the scattered photon can have a new energy and momentum. If a photon of initial wavelength λ_i collides with a free, stationary electron of mass m_e and is deflected through an angle θ, then its scattered wavelength is increased to λ_s, where

$$\lambda_s = \lambda_i + \frac{h}{m_e \mathrm{c}}(1 - \cos\theta)$$

The fractional change in wavelength is very small except for high-energy radiation such as X-rays or γ-rays.

DE BROGLIE WAVELENGTH (λ): A particle of mass m moving with momentum p has associated with it a **de Broglie wavelength**

$$\lambda = \frac{h}{p} = \frac{h}{mv}$$

A beam of particles can be diffracted and can undergo interference phenomena. These wavelike properties of particles can be computed by assuming the particles to behave like waves (*de Broglie waves*) having the de Broglie wavelength.

RESONANCE OF DE BROGLIE WAVES: A particle that is confined to a finite region of space is said to be a *bound* particle. Typical examples of bound-particle systems are a gas molecule in a closed container and an electron in an atom. The de Broglie wave that represents a bound particle will undergo resonance within the confinement region if the wavelength fits properly into the region. We call each possible resonance form a (stationary) *state* of the system. The particle is most likely to be found at the positions of the antinodes of the resonating wave; it is never found at the positions of the nodes.

QUANTIZED ENERGIES for bound particles arise because each resonance situation has a discrete energy associated with it. Since the particle is likely to be found only in a resonance state, its observed energies are discrete (*quantized*). Only in atomic (and smaller) particle systems are the energy differences between resonance states large enough to be easily observable.

Solved Problems

42.1 [I] Show that the photons in a 1240 nm infrared light beam have energies of 1.00 eV.

$$\mathrm{E} = hf = \frac{hc}{\lambda} = \frac{(6.63 \times 10^{-34}\ \mathrm{J \cdot s})(2.998 \times 10^{8}\ \mathrm{m/s})}{1240 \times 10^{-9}\ \mathrm{m}} = 1.602 \times 10^{-19}\ \mathrm{J} = 1.00\ \mathrm{eV}$$

42.2 [I] Compute the energy of a photon of blue light of wavelength 450 nm.

$$\mathrm{E} = \frac{hc}{\lambda} = \frac{(6.63 \times 10^{-34}\ \mathrm{J \cdot s})(2.998 \times 10^{8}\ \mathrm{m/s})}{450 \times 10^{-9}\ \mathrm{m}} = 4.42 \times 10^{-19}\ \mathrm{J} = 2.76\ \mathrm{eV}$$

42.3 [I] To break a chemical bond in the molecules of human skin and thus cause sunburn, a photon energy of about 3.50 eV is required. To what wavelength does this correspond?

$$\lambda = \frac{hc}{\mathrm{E}} = \frac{(6.63 \times 10^{-34}\ \mathrm{J \cdot s})(2.998 \times 10^{8}\ \mathrm{m/s})}{(3.50\ \mathrm{eV})(1.602 \times 10^{-19}\ \mathrm{J/eV})} = 354\ \mathrm{nm}$$

Ultraviolet radiation causes sunburn.

42.4 [II] The work function of sodium metal is 2.3 eV. What is the longest-wavelength light that can cause photoelectron emission from sodium?

At threshold, the photon energy just equals the energy required to tear the electron loose from the metal. In other words, the electron's KE is zero and so $hf = \phi$. Since $f = c/\lambda$,

$$\phi = \frac{hc}{\lambda}$$

$$(2.3\ \text{eV})\left(\frac{1.602 \times 10^{-19}\ \text{J}}{1.00\ \text{eV}}\right) = \frac{(6.63 \times 10^{-34}\ \text{J}\cdot\text{s})(2.998 \times 10^{8}\ \text{m/s})}{\lambda}$$

$$\lambda = 5.4 \times 10^{-7}\ \text{m}$$

42.5 [II] What potential difference must be applied to stop the fastest photoelectrons emitted by a nickel surface under the action of ultraviolet light of wavelength 200 nm? The work function of nickel is 5.01 eV.

$$E = \frac{hc}{\lambda} = \frac{(6.63 \times 10^{-34}\ \text{J}\cdot\text{s})(2.998 \times 10^{8}\ \text{m/s})}{2000 \times 10^{-10}\ \text{m}} = 9.95 \times 10^{-19}\ \text{J} = 6.21\ \text{eV}$$

Then, from the photoelectric equation, the energy of the fastest emitted electron is

$$6.21\ \text{eV} - 5.01\ \text{eV} = 1.20\ \text{eV}$$

Hence a negative retarding potential of 1.20 V is required. This is the stopping potential.

42.6 [II] Will photoelectrons be emitted by a copper surface, of work function 4.4 eV, when illuminated by visible light?

As in Problem 42.4, the released-electron's KE = 0 and so

$$\text{Threshold } \lambda = \frac{hc}{\phi} = \frac{(6.63 \times 10^{-34}\ \text{J}\cdot\text{s})(2.998 \times 10^{8}\ \text{m/s})}{4.4(1.602 \times 10^{-19})\ \text{J}} = 282\ \text{nm}$$

Hence visible light (400 nm to 700 nm) cannot eject photoelectrons from copper.

42.7 [II] A beam ($\lambda = 633$ nm) from a typical laser designed for student use has an intensity of 3.0 mW. How many photons pass a given point in the beam each second?

The energy that is carried past the point each second is 0.0030 J. Because the energy per photon is hc/λ, which works out to be 3.14×10^{-19} J, the number of photons passing the point per second is

$$\text{Number/s} = \frac{0.0030\ \text{J/s}}{3.14 \times 10^{-19}\ \text{J/photon}} = 9.5 \times 10^{15}\ \text{photon/s}$$

42.8 [III] In a process called *pair production*, a photon is transformed into an electron and a positron. A positron has the same mass (m_e) as the electron, but its charge is $+e$. To three significant figures, what is the minimum energy a photon can have if this process is to occur? What is the corresponding wavelength?

The electron-positron pair will come into existence moving with some minimum amount of KE. The particles will separate, and as they do they will slow down. When far apart each will have a mass of 9.11×10^{-31} kg. In effect, KE goes into PE, which is manifested as mass.

Thus the minimum energy photon at the start of the process must have the energy equivalent of the free-particle mass of the pair at the end of the process. Hence,

$$E = 2m_e c^2 = (2)(9.11 \times 10^{-31}\ \text{kg})(2.998 \times 10^8\ \text{m/s})^2 = 1.64 \times 10^{-13}\ \text{J} = 1.02\ \text{MeV}$$

Because this energy must equal hc/λ, the photon's energy,

$$\lambda = \frac{hc}{1.64 \times 10^{-13}\ \text{J}} = 1.21 \times 10^{-12}\ \text{m}$$

This wavelength is in the very short X-ray region, the region of γ rays.

42.9 [II] What wavelength must electromagnetic radiation have if a photon in the beam is to have the same momentum as an electron moving with a speed of 2.00×10^5 m/s?

The requirement is that $(mv)_{\text{electron}} = (h/\lambda)_{\text{photon}}$. From this,

$$\lambda = \frac{h}{mv} = \frac{6.63 \times 10^{-34}\ \text{J}\cdot\text{s}}{(9.11 \times 10^{-31}\ \text{kg})(2.00 \times 10^5\ \text{m/s})} = 3.64\ \text{nm}$$

This wavelength is in the X-ray region.

42.10 [II] Suppose that a 3.64-nm photon moving in the $+x$-direction collides head-on with a 2×10^5 m/s electron moving in the $-x$-direction. If the collision is perfectly elastic, find the conditions after collision.

From the law of conservation of momentum,

$$\text{momentum before} = \text{momentum after}$$

$$\frac{h}{\lambda_0} - mv_0 = \frac{h}{\lambda} - mv$$

But, from Problem 42.9, $h/\lambda_0 = mv_0$ in this case. Hence, $h/\lambda = mv$. Also, for a perfectly elastic collision,

$$\text{KE before} = \text{KE after}$$

$$\frac{hc}{\lambda_0} + \frac{1}{2}mv_0^2 = \frac{hc}{\lambda} + \frac{1}{2}mv^2$$

Using the facts that $h/\lambda_0 = mv_0$ and $h/\lambda = mv$, we find

$$v_0(c + \tfrac{1}{2}v_0) = v(c + \tfrac{1}{2}v)$$

Therefore $v = v_0$ and the electron moves in the $+x$-direction with its original speed. Because $h/\lambda = mv = mv_0$, the photon also "rebounds," and with its original wavelength.

42.11 [I] A photon ($\lambda = 0.400$ nm) strikes an electron at rest and rebounds at an angle of 150° to its original direction. Find the speed and wavelength of the photon after the collision.

The speed of a photon is always the speed of light in vacuum, c. To obtain the wavelength after collision, we use the equation for the Compton effect:

$$\lambda_s = \lambda_i + \frac{h}{m_e c}(1 - \cos\theta)$$

$$\lambda_s = 4.00 \times 10^{-10}\ \text{m} + \frac{6.63 \times 10^{-34}\ \text{J}\cdot\text{s}}{(9.11 \times 10^{-31}\ \text{kg})(2.998 \times 10^8\ \text{m/s})}(1 - \cos 150^\circ)$$

$$\lambda_s = 4.00 \times 10^{-10}\ \text{m} + (2.43 \times 10^{-12}\ \text{m})(1 + 0.866) = 0.405\ \text{nm}$$

42.12 [I] What is the de Broglie wavelength for a particle moving with speed 2.0×10^6 m/s if the particle is (*a*) an electron, (*b*) a proton, and (*c*) a 0.20 kg ball?

We make use of the definition of the de Broglie wavelength:

$$\lambda = \frac{h}{mv} = \frac{6.63 \times 10^{-34}\ \text{J}\cdot\text{s}}{m(2.0 \times 10^6\ \text{m/s})} = \frac{3.31 \times 10^{-40}\ \text{m}\cdot\text{kg}}{m}$$

Substituting the required values for m, one finds that the wavelength is 3.6×10^{-10} m for the electron, 2.0×10^{-13} m for the proton, and 1.7×10^{-39} m for the 0.20-kg ball.

42.13 [II] An electron falls from rest through a potential difference of 100 V. What is its de Broglie wavelength?

Its speed will still be far below c, so relativistic effects can be ignored. The KE gained, $\frac{1}{2}mv^2$, equals the electrical PE lost, Vq. Therefore,

$$v = \sqrt{\frac{2Vq}{m}} = \sqrt{\frac{2(100\ \text{V})(1.60 \times 10^{-19}\ \text{C})}{9.11 \times 10^{-31}\ \text{kg}}} = 5.927 \times 10^6\ \text{m/s}$$

and

$$\lambda = \frac{h}{mv} = \frac{6.626 \times 10^{-34}\ \text{J}\cdot\text{s}}{(9.11 \times 10^{-31}\ \text{kg})(5.927 \times 10^6\ \text{m/s})} = 0.123\ \text{nm}$$

42.14 [II] What potential difference is required in an electron microscope to give electrons a wavelength of 0.500 Å?

$$\text{KE of electron} = \frac{1}{2}mv^2 = \frac{1}{2}m\left(\frac{h}{m\lambda}\right)^2 = \frac{h^2}{2m\lambda^2}$$

where use has been made of the de Broglie relation, $\lambda = h/mv$. Substitution of the known values gives the KE as 9.66×10^{-17} J. But KE $= Vq$, and so

$$V = \frac{\text{KE}}{q} = \frac{9.66 \times 10^{-17}\ \text{J}}{1.60 \times 10^{-19}\ \text{C}} = 600\ \text{V}$$

42.15 [II] By definition, a thermal neutron is a free neutron in a neutron gas at about 20 °C (293 K). What are the KE and wavelength of such a neutron?

From Chapter 17, the thermal energy of a gas molecule is $3kT/2$, where k is Boltzmann's constant $(1.38 \times 10^{-23}\ \text{J/K})$. Then

$$\text{KE} = \tfrac{3}{2}kT = 6.07 \times 10^{-21}\ \text{J}$$

This is a nonrelativistic situation for which we can write

$$\text{KE} = \frac{1}{2}mv^2 = \frac{m^2v^2}{2m} = \frac{p^2}{2m} \qquad \text{or} \qquad p^2 = (2m)(\text{KE})$$

Then

$$\lambda = \frac{h}{p} = \frac{h}{\sqrt{(2m)(\text{KE})}} = \frac{6.63 \times 10^{-34}\ \text{J}\cdot\text{s}}{\sqrt{(2)(1.67 \times 10^{-27}\ \text{kg})(6.07 \times 10^{-21}\ \text{J})}} = 0.147\ \text{nm}$$

42.16 [III] Find the pressure exerted on a surface by the photon beam of Problem 42.7 if the cross-sectional area of the beam is 3.0 mm^2. Assume perfect reflection at normal incidence.

Each photon has a momentum

$$p = \frac{h}{\lambda} = \frac{6.63 \times 10^{-34}\ \text{J}\cdot\text{s}}{633 \times 10^{-9}\ \text{m}} = 1.05 \times 10^{-27}\ \text{kg}\cdot\text{m/s}$$

When a photon reflects, it changes momentum from $+p$ to $-p$, a total change of $2p$. Since (from Problem 42.7) 9.5×10^{15} photons strike the surface each second, we have

$$\text{Momentum change/s} = (9.5 \times 10^{15}/\text{s})(2)(1.05 \times 10^{-27}\ \text{kg}\cdot\text{m/s}) = 2.0 \times 10^{-11}\ \text{kg}\cdot\text{m/s}^2$$

From the impulse equation (Chapter 8),

$$\text{Impulse} = Ft = \text{change in momentum}$$

we have

$$F = \text{momentum change/s} = 1.99 \times 10^{-11}\ \text{kg}\cdot\text{m/s}^2$$

Then

$$\text{Pressure} = \frac{F}{A} = \frac{1.99 \times 10^{-11}\ \text{kg}\cdot\text{m/s}^2}{3.0 \times 10^{-6}\ \text{m}^2} = 6.6 \times 10^{-6}\ \text{N/m}^2$$

42.17 [III] A particle of mass m is confined to a narrow tube of length L. Find (*a*) the wavelengths of the de Broglie waves which will resonate in the tube, (*b*) the corresponding particle momenta, and (*c*) the corresponding energies. (*d*) Evaluate the energies for an electron in a tube with $L = 0.50$ nm.

(*a*) The de Broglie waves will resonate with a node at each end of the tube because the ends are impervious. A few of the possible resonance forms are shown in Fig. 42-1. They indicate that, for resonance, $L = \frac{1}{2}\lambda_1, 2(\frac{1}{2}\lambda_2), 3(\frac{1}{2}\lambda_3), \ldots, n(\frac{1}{2}\lambda_n), \ldots$ or

$$\lambda_n = \frac{2L}{n} \qquad n = 1, 2, 3, \ldots$$

(*b*) Because the de Broglie wavelengths are $\lambda_n = h/p_n$, the resonance momenta are

$$p_n = \frac{nh}{2L} \qquad n = 1, 2, 3, \ldots$$

(*c*) As shown in Problem 42.15, $p^2 = (2m)(\text{KE})$, and so

$$(\text{KE})_n = \frac{n^2h^2}{8L^2m} \qquad n = 1, 2, 3, \ldots$$

Notice that the particle can assume only certain discrete energies. The energies are quantized.

(*d*) With $m = 9.1 \times 10^{-31}$ kg and $L = 5.0 \times 10^{-10}$ m, substitution gives

$$(\text{KE})_n = 2.4 \times 10^{-19}\ n^2\ \text{J} = 1.5n^2\ \text{eV}$$

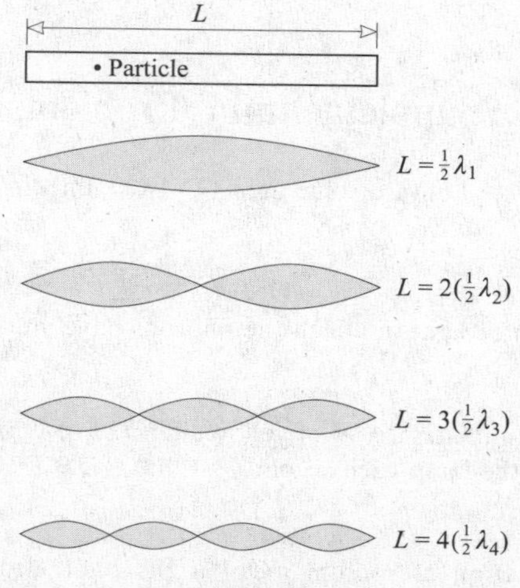

Fig. 42-1

42.18 [III] A particle of mass m is confined to a circular orbit with radius R. For resonance of its de Broglie wave on this orbit, what energies can the particle have? Determine the KE for an electron with $R = 0.50$ nm.

To resonate on a circular orbit, a wave must circle back on itself in such a way that crest falls upon crest and trough falls upon trough. One resonance possibility (for an orbit circumference that is four wavelengths long) is shown in Fig. 42-2. In general, resonance occurs when the circumference is n wavelengths long, where $n = 1, 2, 3, \ldots$. For such a de Broglie wave we have

$$n\lambda_n = 2\pi R \qquad \text{and} \qquad p_n = \frac{h}{\lambda_n} = \frac{nh}{2\pi R}$$

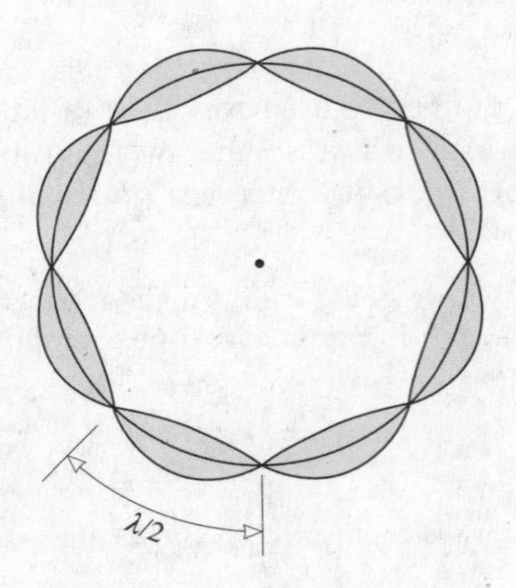

Fig. 42-2

As in Problem 42.17,

$$(\text{KE})_n = \frac{p_n^2}{2m} = \frac{n^2h^2}{8\pi^2R^2m}$$

The energies are obviously quantized. Placing in the values requested gives

$$(\text{KE})_n = 2.4 \times 10^{-20}\ n^2\ \text{J} = 0.15n^2\ \text{eV}$$

Supplementary Problems

42.19 [I] Compute the energy of a photon of blue light ($\lambda = 450$ nm), in joules and in eV.
Ans. 4.41×10^{-19} J = 2.76 eV

42.20 [I] What is the wavelength of light in which the photons have an energy of 600 eV? *Ans.* 2.07 nm

42.21 [I] A certain sodium lamp radiates 20 W of yellow light ($\lambda = 589$ nm). How many photons of the yellow light are emitted from the lamp each second? *Ans.* 5.9×10^{19}

42.22 [I] What is the work function of sodium metal if the photoelectric threshold wavelength is 680 nm?
Ans. 1.82 eV

42.23 [II] Determine the maximum KE of photoelectrons ejected from a potassium surface by ultraviolet radiation of wavelength 200 nm. What retarding potential difference is required to stop the emission of electrons? The photoelectric threshold wavelength for potassium is 440 nm. *Ans.* 3.38 eV, 3.38 V

42.24 [II] With what speed will the fastest photoelectrons be emitted from a surface whose threshold wavelength is 600 nm, when the surface is illuminated with light of wavelength 4×10^{-7} m? *Ans.* 6×10^{5} m/s

42.25 [II] Electrons with a maximum KE of 3.00 eV are ejected from a metal surface by ultraviolet radiation of wavelength 150 nm. Determine the work function of the metal, the threshold wavelength of the metal, and the retarding potential difference required to stop the emission of electrons. *Ans.* 5.27 eV, 235 nm, 3.00 V

42.26 [I] What are the speed and momentum of a 500-nm photon? *Ans.* 2.998×10^{8} m/s, 133×10^{-27} kg·m/s

42.27 [II] An X-ray beam with a wavelength of exactly 5.00×10^{-14} m strikes a proton that is at rest ($m = 1.67 \times 10^{-27}$ kg). If the X-rays are scattered through an angle of 110°, what is the wavelength of the scattered X-rays? *Ans.* 5.18×10^{-14} m

42.28 [III] A photon produces an electron and a positron which each have a kinetic energy of 220 keV even when they are separated by a great distance. Find the energy and wavelength of the photon. *Ans.* 1.46 MeV, 8.49×10^{-13} m

42.29 [II] Show that the de Broglie wavelength of an electron accelerated from rest through a potential difference of V volts is $1.228/\sqrt{V}$ nm. Ignore relativistic effects and take a look at Problem 42.13.

42.30 [II] Compute the de Broglie wavelength of an electron that has been accelerated through a potential difference of 9.0 kV. Ignore relativistic effects. *Ans.* 1.3×10^{-11} m

42.31 [III] What is the de Broglie wavelength of an electron that has been accelerated through a potential difference of 1.0 MV? (You must use the relativistic mass and energy expressions at this high energy.) *Ans.* 8.7×10^{-13} m

42.32 [II] It is proposed to send a beam of electrons through a diffraction grating. The electrons have a speed of 400 m/s. How large must the distance between slits be if a strong beam of electrons is to emerge at an angle of 25° to the straight-through beam? *Ans.* $n(4.3 \times 10^{-6}$ m), where $n = 1, 2, 3, \ldots$.

Chapter 43

The Hydrogen Atom

THE HYDROGEN ATOM has a diameter of about 0.1 nm; it consists of a proton as the nucleus (with a radius of about 10^{-15} m) and a single electron.

ELECTRON ORBITS: The first effective model of the atom was introduced by Niels Bohr in 1913. Although it has been surpassed by quantum mechanics, many of its simple results are still valid. The earliest version of the ***Bohr model*** pictured electrons in circular orbits around the nucleus. The hydrogen atom was then one electron circulating around a single proton. For the electron's de Broglie wave to resonate or "fit" (see Fig. 42-2) in an orbit of radius r, the following must be true (see Problem 42.18):

$$mv_n r_n = \frac{nh}{2\pi}$$

where n is an integer. The quantity $mv_n r_n$ is the angular momentum of the electron in its nth orbit. The speed of the electron is v, its mass is m, and h is Planck's constant, 6.63×10^{-34} J·s.

The centripetal force that holds the electron in orbit is supplied by Coulomb attraction between the nucleus and the electron. Hence, $F = ke^2/r^2 = ma = mv_n^2/r_n$ and

$$\frac{mv_n^2}{r_n} = k\frac{e^2}{r^2}$$

Simultaneous solution of these equations gives the radii of stable orbits as $r_n = (0.053\text{ nm})n^2$. The energy of the atom when it is in the nth state (i.e., with its electron in the nth orbit configuration) is

$$E_n = -\frac{13.6}{n^2}\text{ eV}$$

As in Problems 42.17 and 42.18, the energy is quantized because a stable configuration corresponds to a resonance form of the bound system. For a nucleus with charge Ze orbited by a single electron, the corresponding relations are

$$r_n = (0.053\text{ nm})\left(\frac{n^2}{Z}\right) \qquad \text{and} \qquad E_n = -\frac{13.6Z^2}{n^2}\text{ eV}$$

where Z is called the **atomic number** of the nucleus.

ENERGY-LEVEL DIAGRAMS summarize the allowed energies of a system. On a vertical energy scale, the allowed energies are shown by horizontal lines. The energy-level diagram for hydrogen is shown in Fig. 43-1. Each horizontal line represents the energy of a resonance state of the atom. The zero of energy is taken to be the ionized atom, i.e., the state in which the atom has an infinite orbital radius. As the electron falls closer to the nucleus, its potential energy decreases from the zero level, and thus the energy of the atom is negative as indicated. The lowest possible state, $n = 1$, corresponds to the electron in its smallest possible orbit; it is called the **ground state.**

EMISSION OF LIGHT: When an isolated atom falls from one energy level to a lower one, a photon is emitted. This photon carries away the energy lost by the atom in its transition to the lower energy state. The wavelength and frequency of the photon are given by

$$hf = \frac{hc}{\lambda} = \text{energy lost by the system}$$

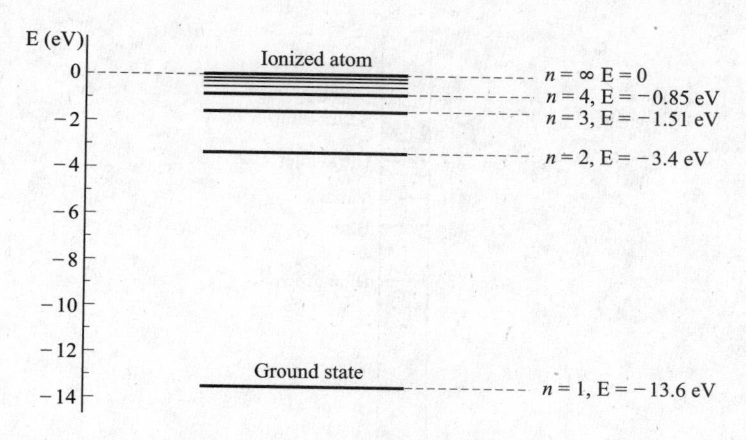

Fig. 43-1

The emitted radiation has a precise wavelength and gives rise to a single *spectral line* in the emission spectrum of the atom. It is convenient to remember that a 1240 nm photon has an energy of 1 eV, and that photon energy varies *inversely* with wavelength.

THE SPECTRAL LINES emitted by excited isolated hydrogen atoms occur in series. Typical is the series that appears at visible wavelengths, the ***Balmer series*** shown in Fig. 43-2. Other series exist; one, in the ultraviolet, is called the ***Lyman series***; there are others in the infrared, the one closest to the visible portion of the spectrum being the ***Paschen series***. Their wavelengths are given by simple formulas:

$$\text{Lyman:} \quad \frac{1}{\lambda} = R\left(\frac{1}{1^2} - \frac{1}{n^2}\right) \qquad n = 2, 3, \ldots$$

$$\text{Balmer:} \quad \frac{1}{\lambda} = R\left(\frac{1}{2^2} - \frac{1}{n^2}\right) \qquad n = 3, 4, \ldots$$

$$\text{Paschen:} \quad \frac{1}{\lambda} = R\left(\frac{1}{3^2} - \frac{1}{n^2}\right) \qquad n = 4, 5, \ldots$$

where $R = 1.0974 \times 10^7\ \text{m}^{-1}$ is called the **Rydberg constant.**

ORIGIN OF SPECTRAL SERIES: The Balmer series of lines in Fig. 43-2 arises when an electron in the atom falls from higher states to the $n = 2$ state. The transition from $n = 3$ to $n = 2$ gives rise to a photon energy $\Delta E_{3,2} = 1.89$ eV, which is equivalent to a wavelength of 656 nm, the first line of the series. The second line originates in the transition from $n = 4$ to $n = 2$. The series limit line represents the transition from $n = \infty$ to $n = 2$. Similarly, transitions ending in the $n = 1$ state give rise to the Lyman series; transitions that end in the $n = 3$ state give lines in the Paschen series.

ABSORPTION OF LIGHT: An atom in its ground state can absorb a photon in a process called *resonance absorption* only if that photon will raise the atom to one of its allowed energy levels.

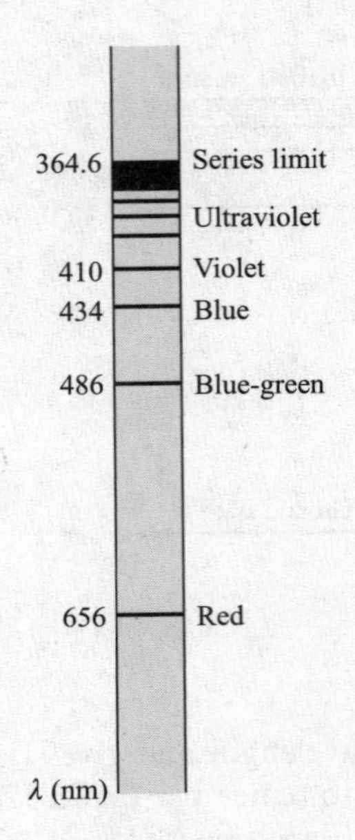

Fig. 43-2

Solved Problems

43.1 [II] What wavelength does a hydrogen atom emit as its excited electron falls from the $n = 5$ state to the $n = 2$ state? Give your answer to three significant figures.

Since $E_n = -13.6/n^2$ eV, we have

$$E_5 = -0.54 \text{ eV} \qquad \text{and} \qquad E_2 = -3.40 \text{ eV}$$

The energy difference between these states is $3.40 - 0.54 = 2.86$ eV. Because 1240 nm corresponds to 1.00 eV in an inverse proportion, we have, for the wavelength of the emitted photon,

$$\lambda = \left(\frac{1.00 \text{ eV}}{2.86 \text{ eV}}\right)(1240 \text{ nm}) = 434 \text{ nm}$$

43.2 [II] When a hydrogen atom is bombarded, the atom may be raised into a higher energy state. As the excited electron falls back to the lower energy levels, light is emitted. What are the three longest-wavelength spectral lines emitted by the hydrogen atom as it returns to the $n = 1$ state from higher energy states? Give your answers to three significant figures.

We are interested in the following transitions (see Fig. 43-1):

$$n = 2 \rightarrow n = 1: \qquad \Delta E_{2,1} = -3.4 - (-13.6) = 10.2 \text{ eV}$$
$$n = 3 \rightarrow n = 1: \qquad \Delta E_{3,1} = -1.5 - (-13.6) = 12.1 \text{ eV}$$
$$n = 4 \rightarrow n = 1: \qquad \Delta E_{4,1} = -0.85 - (-13.6) = 12.8 \text{ eV}$$

To find the corresponding wavelengths we can proceed as in Problem 43.1, or we can use $\Delta E = hf = hc/\lambda$. For example, for the $n = 2$ to $n = 1$ transition,

$$\lambda = \frac{hc}{\Delta E_{2,1}} = \frac{(6.63 \times 10^{-34} \text{ J}\cdot\text{s})(2.998 \times 10^8 \text{ m/s})}{(10.2 \text{ eV})(1.60 \times 10^{-19} \text{ J/eV})} = 1.22 \text{ nm}$$

The other lines are found in the same way to be 102 nm and 96.9 nm. These are the first three lines of the Lyman series.

43.3 [I] The ***series limit*** wavelength of the Balmer series is emitted as the electron in the hydrogen atom falls from the $n = \infty$ state to the $n = 2$ state. What is the wavelength of this line (to three significant figures)?

From Fig. 43-1, $\Delta E = 3.40 - 0 = 3.40$ eV. We find the corresponding wavelength in the usual way from $\Delta E = hc/\lambda$. The result is 365 nm.

43.4 [I] What is the greatest wavelength of radiation that will ionize unexcited hydrogen atoms?

The incident photons must have enough energy to raise the atom from the $n = 1$ level to the $n = \infty$ level when absorbed by the atom. Because $E_\infty - E_1 = 13.6$ eV, we can use $E_\infty - E_1 = hc/\lambda$ to find the wave-length as 91.2 nm. Wavelengths shorter than this would not only remove the electron from the atom, but would add KE to the removed electron.

43.5 [I] The energy levels for singly ionized helium atoms (atoms from which one of the two electrons has been removed) are given by $E_n = (-54.4/n^2)$ eV. Construct the energy-level diagram for this system.

See Fig. 43-3.

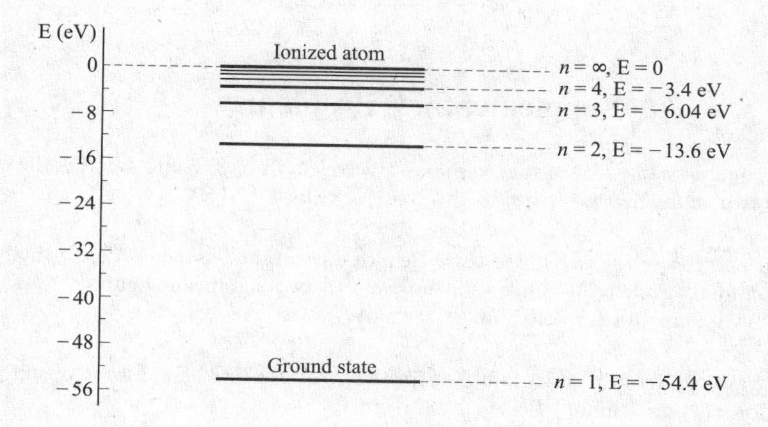

Fig. 43-3

43.6 [I] What are the two longest wavelengths of the Balmer series for singly ionized helium atoms?

The pertinent energy-level diagram is shown in Fig. 43-3. Recall that the Balmer series corresponds to transitions from higher states to the $n = 2$ state. From the diagram, the two smallest-energy transitions to the $n = 2$ states are

$$n = 3 \rightarrow n = 2 \qquad \Delta E_{3,2} = 13.6 - 6.04 = 7.6 \text{ eV}$$

$$n = 4 \rightarrow n = 2 \qquad \Delta E_{4,2} = 13.6 - 3.4 = 10.2 \text{ eV}$$

Using the fact that 1 eV corresponds to 1240 nm, we find the corresponding wavelengths to be 163 nm and 122 nm; both wavelengths are in the far ultraviolet or long X-ray region.

43.7 [II] Unexcited hydrogen atoms are bombarded with electrons that have been accelerated through 12.0 V. What wavelengths will the atoms emit?

When an atom in the ground state is given 12.0 eV of energy, the most these electrons can supply, the atom can be excited no higher than 12.0 eV above the ground state. Only one state exists in this energy region, the $n = 2$ state. Hence the only transition possible is

$$n = 2 \rightarrow n = 1: \qquad \Delta E_{2,1} = 13.6 - 3.4 = 10.2 \text{ eV}$$

The only emitted wavelength will be

$$\lambda = (1240 \text{ nm})\left(\frac{1.00 \text{ eV}}{10.2 \text{ eV}}\right) = 122 \text{ nm}$$

which is the longest-wavelength line in the Lyman series.

43.8 [II] Unexcited hydrogen gas is an electrical insulator because it contains no free electrons. What maximum-wavelength photon beam incident on the gas can cause the gas to conduct electricity?

The photons in the beam must ionize the atom so as to produce free electrons. (This is called the *atomic photoelectric effect*.) To do this, the photon energy must be at least 13.6 eV, and so the maximum wavelength is

$$\lambda = (1240 \text{ nm})\left(\frac{1.00 \text{ eV}}{13.6 \text{ eV}}\right) = 91.2 \text{ nm}$$

which is the series limit for the Lyman series.

Supplementary Problems

43.9 [I] One spectral line in the hydrogen spectrum has a wavelength of 821 nm. What is the energy difference between the two states that gives rise to this line? *Ans.* 1.51 eV

43.10 [II] What are the energies of the two longest-wavelength lines in the Paschen series for hydrogen? What are the corresponding wavelengths? Give your answers to two significant figures. *Ans.* 0.66 eV and 0.97 eV, 1.9×10^{-6} m and 1.3×10^{-6} m

43.11 [I] What is the wavelength of the series limit line for the hydrogen Paschen series? Consult Problem 43.3 for an explanation of "series limit." *Ans.* 821 nm

43.12 [II] The lithium atom has a nuclear charge of $+3e$. Find the energy required to remove the third electron from a lithium atom that has already lost two of its electrons. Assume the third electron to be initially in the ground state. *Ans.* 122 eV

43.13 [II] Electrons in an electron beam are accelerated through a potential difference V and are incident on hydrogen atoms in their ground state. What is the maximum value for V if the collisions are to be perfectly elastic? *Ans.* < 10.2 V

43.14 [II] What are the three longest photon wavelengths that singly ionized helium atoms (in their ground state) will absorb strongly? (See Fig. 43-3.) *Ans.* 30.4 nm, 25.6 nm, 24.3 nm

43.15 [II] How much energy is required to remove the second electron from a singly ionized helium atom? What is the maximum wavelength of an incident photon that could tear this electron from the ion? *Ans.* 54.4 eV, 22.8 nm

43.16 [II] In the spectrum of singly ionized helium, what is the series limit for its Balmer series? *Ans.* 91 nm

Chapter 44

Multielectron Atoms

A NEUTRAL ATOM whose nucleus carries a positive charge of Ze has Z electrons. When the electrons have the least energy possible, the atom is in its *ground state*. The state of an atom is specified by the *quantum numbers* for its individual electrons.

THE QUANTUM NUMBERS that are used to specify the parameters of an atomic electron are as follows:

- The **principal quantum number n** specifies the orbit, or shell, in which the electron is to be found. In the hydrogen atom, it specifies the electron's energy via $E_n = -13.6/n^2$ eV.
- The **orbital quantum number** ℓ specifies the angular momentum L of the electron in its orbit:

$$L = \left(\frac{h}{2\pi}\right)\sqrt{\ell(\ell+1)}$$

  where h is Planck's constant, and $\ell = 0, 1, 2, \ldots, n-1$.
- The **magnetic quantum number** m_ℓ describes the orientation of the orbital angular momentum vector relative to the z direction, the direction of an impressed magnetic field:

$$L_z = \left(\frac{h}{2\pi}\right)(m_\ell)$$

  where $m_\ell = 0, \pm 1, \pm 2, \ldots, \pm \ell$.
- The **spin quantum number** m_s has allowed values of $\pm\frac{1}{2}$.

THE PAULI EXCLUSION PRINCIPLE maintains that no two electrons in the same atom can have the same set of quantum numbers. In other words, no two electrons can be in the same state.

Solved Problems

44.1 [II] Estimate the energy required to remove an $n = 1$ (i.e., inner-shell) electron from a gold atom ($Z = 79$).

Because an electron in the innermost shell of the atom is not much influenced by distant electrons in outer shells, we can consider it to be the only electron present. Then its energy is given approximately by an appropriately modified version of the energy formula of Chapter 43 that takes into consideration the charge (Ze) of the nucleus. With $n = 1$, that formula, $E_n = -13.6Z^2/n^2$, gives

$$E_1 = -13.6(79)^2 = -84\,900 \text{ eV} = -84.9 \text{ keV}$$

To tear the electron loose (i.e., remove it to the $E_\infty = 0$ level), we must give it an energy of about 84.9 keV.

44.2 [II] What are the quantum numbers for the electrons in the lithium atom ($Z = 3$) when the atom is in its ground state?

Keeping in mind that $\ell = 0, 1, 2, \ldots, (n-1)$ and $m_\ell = 0, \pm 1, \pm 2, \ldots, \pm \ell$ while $m_s = \pm\frac{1}{2}$, the Pauli Exclusion Principle tells us that the lithium atom's three electrons can take on the following quantum numbers:

$$\begin{array}{llll} \text{Electron 1:} & n=1, \quad \ell=0, & m_\ell=0, & m_s=+\frac{1}{2} \\ \text{Electron 2:} & n=1, \quad \ell=0, & m_\ell=0, & m_s=-\frac{1}{2} \\ \text{Electron 3:} & n=2, \quad \ell=0, & m_\ell=0, & m_s=+\frac{1}{2} \end{array}$$

Notice that, when $n = 1$, ℓ must be zero and m_ℓ must be zero (why?). Then there are only two $n = 1$ possibilities, and the third electron has to go into the $n = 2$ level. Since it is in the second Bohr orbit, it is more easily removed from the atom than an $n = 1$ electron. That is why lithium ionizes easily to Li^+.

44.3 [II] Why is sodium ($Z = 11$) the next univalent atom after lithium?

Sodium has a single electron in the $n = 3$ shell. To see why this is necessarily so, notice that the Pauli Exclusion Principle allows only two electrons in the $n = 1$ shell. The next eight electrons can fit in the $n = 2$ shell, as follows:

$$\begin{array}{llll} n=2, & \ell=0, & m_\ell=0, & m_s=\pm\frac{1}{2} \\ n=2, & \ell=1, & m_\ell=0, & m_s=\pm\frac{1}{2} \\ n=2, & \ell=1, & m_\ell=1, & m_s=\pm\frac{1}{2} \\ n=2, & \ell=1, & m_\ell=-1, & m_s=\pm\frac{1}{2} \end{array}$$

The eleventh electron must go into the $n = 3$ shell, from which it is easily removed to yield Na^+.

44.4 [II] (*a*) Estimate the wavelength of the photon emitted as an electron falls from the $n = 2$ shell to the $n = 1$ shell in the gold atom ($Z = 79$). (*b*) About how much energy must bombarding electrons have to excite gold to emit this emission line?

(*a*) As noted in Problem 44.1, to a first approximation the energies of the innermost electrons of a large-Z atom are given by $E_n = -13.6Z^2/n^2$ eV. Thus, we have

$$\Delta E_{2,1} = 13.6(79)^2\left(\tfrac{1}{1} - \tfrac{1}{4}\right) = 63\,700 \text{ eV}$$

This corresponds to a photon with

$$\lambda = (1240 \text{ nm})\left(\frac{1 \text{ eV}}{63\,700 \text{ eV}}\right) = 0.019\,5 \text{ nm}$$

It is clear from this result that inner-shell transitions in high-Z atoms give rise to the emission of X-rays.

(*b*) Before an $n = 2$ electron can fall to the $n = 1$ shell, an $n = 1$ electron must be thrown to an empty state of large n, which we approximate as $n = \infty$ (with $E_\infty = 0$). This requires an energy

$$\Delta E_{1,\infty} = 0 - \frac{-13.6Z^2}{n^2} = \frac{13.6(79)^2}{1} = 84.9 \text{ keV}$$

The bombarding electrons must thus have an energy of about 84.9 keV.

44.5 [II] Suppose electrons had no spin, so that the spin quantum number did not exist. If the Exclusion Principle still applied to the remaining quantum numbers, what would be the first three univalent atoms?

The electrons would take on the following quantum numbers:

Electron 1: $n = 1,\quad \ell = 0,\quad m_\ell = 0$ (univalent)
Electron 2: $n = 2,\quad \ell = 0,\quad m_\ell = 0$ (univalent)
Electron 3: $n = 2,\quad \ell = 1,\quad m_\ell = 0$
Electron 4: $n = 2,\quad \ell = 1,\quad m_\ell = +1$
Electron 5: $n = 2,\quad \ell = 1,\quad m_\ell = -1$
Electron 6: $n = 3,\quad \ell = 0,\quad m_\ell = 0$ (univalent)

Each electron marked "univalent" is the first electron in a new shell. Since an electron is easily removed if it is the outermost electron in the atom, atoms with that number of electrons are univalent. They are the atoms with $Z = 1$ (hydrogen), $Z = 2$ (helium), and $Z = 6$ (carbon). Can you show that $Z = 15$ (phosphorus) would also be univalent?

44.6 [II] Electrons in an atom that have the same value for ℓ but different values for m_ℓ and m_s are said to be in the same *subshell*. How many electrons exist in the $\ell = 3$ subshell?

Because m_ℓ is restricted to the values 0, ± 1, ± 2, ± 3, and $m_s = \pm\frac{1}{2}$ only, the possibilities for $\ell = 3$ are

$$(m_\ell,\ m_s) = (0,\ \pm\tfrac{1}{2}),\ (1,\ \pm\tfrac{1}{2}),\ (-1,\ \pm\tfrac{1}{2}),\ (2,\ \pm\tfrac{1}{2}),\ (-2,\ \pm\tfrac{1}{2}),\ (3,\ \pm\tfrac{1}{2}),\ (-3,\ \pm\tfrac{1}{2})$$

which gives 14 possibilities. Therefore, 14 electrons can exist in this subshell.

44.7 [II] An electron beam in an X-ray tube is accelerated through 40 kV and is incident on a tungsten target. What is the shortest wavelength emitted by the tube?

When an electron in the beam is stopped by the target, the photons emitted have an upper limit for their energy, namely, the energy of the incident electron. In this case, that energy is 40 keV. The corresponding photon has a wavelength given by

$$\lambda = (1240\ \text{nm})\left(\frac{1.0\ \text{eV}}{40\,000\ \text{eV}}\right) = 0.031\ \text{nm}$$

Supplementary Problems

44.8 [II] If there were no m_ℓ quantum number, what would be the first four univalent atoms? *Ans.* H, Li, N, Al

44.9 [II] Helium has a closed (completely filled) outer shell and is nonreactive because the atom does not easily lose an electron. Show why neon ($Z = 10$) is the next nonreactive element.

44.10 [II] It is desired to eject an electron from the $n = 1$ shell of a uranium atom ($Z = 92$) by means of the atomic photoelectric effect. Approximately what is the longest-wavelength photon capable of doing this? *Ans.* 0.0108 nm

44.11 [II] Show that the maximum number of electrons that can exist in the ℓth subshell is $2(2\ell + 1)$.

Chapter 45

Nuclei and Radioactivity

THE NUCLEUS of an atom is a positively charged entity at the atom's center. Its radius is roughly 10^{-15} m, which is about 10^{-5} as large as the radius of the atom. Hydrogen is the lightest and simplest of all the atoms. Its nucleus is a single proton. All other nuclei contain both protons and neutrons. Protons and neutrons are collectively called ***nucleons***. Although the positively charged protons repel each other, the much stronger, short-range *nuclear force* (which is a manifestation of the more fundamental *strong force*) holds the nucleus together. The nuclear attractive force between nucleons decreases rapidly with particle separation and is essentially zero for nucleons more than 5×10^{-15} m apart.

NUCLEAR CHARGE AND ATOMIC NUMBER: Each proton within the nucleus carries a charge $+e$, whereas the neutrons carry no electromagnetic charge. If there are Z protons in a nucleus, then the charge on the nucleus is $+Ze$. We call Z the **atomic number** of that nucleus.

Because normal atoms are neutral electrically, the atom has Z electrons outside the nucleus. These Z electrons determine the chemical behavior of the atom. As a result, all atoms of the same chemical element have the same value of Z. For example, all hydrogen atoms have $Z = 1$, while all carbon atoms have $Z = 6$.

ATOMIC MASS UNIT (u): A convenient mass unit used in nuclear calculations is the **atomic mass unit** (u). By definition, 1 u is exactly 1/12 of the mass of the common form of carbon atom found on the Earth. It turns out that

$$1\text{ u} = 1.660\,5 \times 10^{-27}\text{ kg} = 931.494\text{ MeV}/c^2$$

Table 45-1 lists the masses of some common particles and nuclei, as well as their charges.

Table 45-1

Particle	Symbol	Mass, u	Charge
Proton	p, ${}^{1}_{1}\text{H}$	1.007 276	$+e$
Neutron	n, ${}^{1}_{0}n$	1.008 665	0
Electron	e^-, β^-, ${}^{\ 0}_{-1}e$	0.000 548 6	$-e$
Positron	e^+, β^+, ${}^{\ 0}_{+1}e$	0.000 548 6	$+e$
Deuteron	d, ${}^{2}_{1}\text{H}$	2.013 55	$+e$
Alpha particle	α, ${}^{4}_{2}\text{He}$	4.001 5	$+2e$

THE MASS NUMBER (A) of an atom is equal to the number of nucleons (neutrons plus protons) in the nucleus of the atom. Because each nucleon has a mass close to 1 u, the mass number A is nearly equal to the nuclear mass in atomic mass units. In addition, because the atomic electrons have such small mass, A is nearly equal to the mass of the atom in atomic mass units.

ISOTOPES: The number of neutrons in the nucleus has very little effect on the chemical behavior of all but the lightest atoms. In nature, atoms of the same element (same Z) often exist that have unlike numbers of neutrons in their nuclei. Such atoms are called **isotopes** of each other. For example, ordinary oxygen consists of three isotopes that have mass numbers 16, 17, and 18. Each of the isotopes has $Z = 8$, or eight protons in the nucleus. Hence these isotopes have the following numbers of neutrons in their nuclei: $16 - 8 = 8$, $17 - 8 = 9$, and $18 - 8 = 10$. It is customary to represent the isotopes in the following way: ${}^{16}_{8}\text{O}$, ${}^{17}_{8}\text{O}$, ${}^{18}_{8}\text{O}$, or simply as ${}^{16}\text{O}$, ${}^{17}\text{O}$, and ${}^{18}\text{O}$, where it is understood that oxygen always has $Z = 8$.

In keeping with this notation, we designate the nucleus having mass number A and atomic number Z by the symbolism

$${}^{A}_{Z}(\text{CHEMICAL SYMBOL})$$

BINDING ENERGIES: The mass of an atom is not equal to the sum of the masses of its component protons, neutrons, and electrons. Imagine a reaction in which free electrons, protons, and neutrons combine to form an atom; in such a reaction, you would find that the mass of the atom is *slightly less* than the combined masses of the component parts, and that a tremendous amount of energy is released when the reaction occurs. The loss in mass is exactly equal to the mass equivalent of the released energy, according to Einstein's equation $\Delta E_0 = (\Delta m)c^2$. Conversely, this same amount of energy, ΔE_0 would have to be given to the atom to separate it completely into its component particles. We call ΔE_0 the **binding energy** of the atom. A mass loss of $\Delta m = 1$ u is equivalent to

$$(1.66 \times 10^{-27}\ \text{kg})(2.99 \times 10^{8}\ \text{m/s})^2 = 1.49 \times 10^{-10}\ \text{J} = 931\ \text{MeV}$$

of binding energy.

The percentage "loss" of mass is different for each isotope of any element. The atomic masses of some of the lighter isotopes are given in Table 45-2. These masses are for neutral atoms and include the orbital electrons.

Table 45-2

Neutral atom	Atomic mass, u	Neutral atom	Atomic mass, u
${}^{1}_{1}\text{H}$	1.007 83	${}^{7}_{4}\text{Be}$	7.016 93
${}^{2}_{1}\text{H}$	2.014 10	${}^{9}_{4}\text{Be}$	9.012 19
${}^{3}_{1}\text{H}$	3.016 04	${}^{12}_{6}\text{C}$	12.000 00
${}^{4}_{2}\text{He}$	4.002 60	${}^{14}_{7}\text{N}$	14.003 07
${}^{6}_{3}\text{Li}$	6.015 13	${}^{16}_{8}\text{O}$	15.994 91
${}^{7}_{3}\text{Li}$	7.016 00		

RADIOACTIVITY: Nuclei found in nature with Z greater than that of lead, 82, are unstable or **radioactive**. Many artificially produced elements with smaller Z are also radioactive. A radioactive nucleus spontaneously ejects one or more particles in the process of transforming into a different nucleus.

The stability of a radioactive nucleus against spontaneous decay is measured by its **half-life** $t_{1/2}$. The half-life is defined as the time in which half of any large sample of identical nuclei will undergo decomposition. The half-life is a fixed number for each isotope.

Radioactive decay is a random process. No matter when one begins to observe a material, only half the material will remain unchanged after a time $t_{1/2}$; after an additional time of $t_{1/2}$ only $\frac{1}{2} \times \frac{1}{2} = \frac{1}{4}$ of the material will remain unchanged. After n half-lives have passed, only $(\frac{1}{2})^n$ of the material will remain unchanged.

A simple relation exists between the number N of atoms of radioactive material present and the number ΔN that will decay in a short time Δt. It is

$$\Delta N = \lambda N \, \Delta t$$

where λ, the **decay constant**, is related to the half-life $t_{1/2}$ through

$$\lambda t_{1/2} = 0.693$$

The decay constant has the unit of s^{-1}, and can be thought of as the fractional disintegration rate. The quantity $\Delta N/\Delta t$, which is the rate of disintegrations, is called the **activity** of the sample. It is equal to λN, and therefore it steadily decreases with time. The SI unit for activity is the **becquerel** (Bq), where 1 Bq = 1 decay/s.

NUCLEAR EQUATIONS: In a balanced equation the sum of the subscripts (atomic numbers) must be the same on the two sides of the equation. The sum of the superscripts (mass numbers) must also be the same on the two sides of the equation. Thus the equation for the primary radioactivity of radium is

$${}^{226}_{88}\text{Ra} \rightarrow {}^{222}_{86}\text{Rn} + {}^{4}_{2}\text{He}$$

Many nuclear processes may be indicated by a condensed notation, in which a light bombarding particle and a light product particle are represented by symbols in parentheses between the symbols for the initial target nucleus and the final product nucleus. The symbols n, p, d, α, e^-, and γ are used to represent neutron, proton, deuteron (${}^{2}_{1}\text{H}$), alpha, particle, electron, and gamma rays (photons), respectively. Here are three examples of corresponding long and condensed notations:

$${}^{14}_{7}\text{N} + {}^{1}_{1}\text{H} \rightarrow {}^{11}_{6}\text{C} + {}^{4}_{2}\text{He} \qquad {}^{14}\text{N}(p, a){}^{11}\text{C}$$

$${}^{27}_{13}\text{Al} + {}^{1}_{0}n \rightarrow {}^{27}_{12}\text{Mg} + {}^{1}_{1}\text{H} \qquad {}^{27}\text{Al}(n,\ p){}^{27}\text{Mg}$$

$${}^{55}_{25}\text{Mn} + {}^{2}_{1}\text{H} \rightarrow {}^{55}_{26}\text{Fe} + 2{}^{1}_{0}n \qquad {}^{55}\text{Mn}(d,\ 2n){}^{55}\text{Fe}$$

The slow neutron is a very efficient agent in causing transmutations, since it has no positive charge and hence can approach the nucleus without being repelled. By contrast, a positively charged particle such as a proton must have a high energy to cause a transformation. Because of their small masses, even very high-energy electrons are relatively inefficient in causing nuclear transmutations.

Solved Problems

45.1 [II] The radius of a carbon nucleus is about 3×10^{-15} m and its mass is 12 u. Find the average density of the nuclear material. How many more times dense than water is this?

$$\rho = \frac{m}{V} = \frac{m}{4\pi r^3/3} = \frac{(12\text{ u})(1.66 \times 10^{-27}\text{ kg/u})}{4\pi(3 \times 10^{-15}\text{ m})^3/3} = 1.8 \times 10^{17}\text{ kg/m}^3$$

$$\frac{\rho}{\rho_{\text{water}}} = \frac{1.8 \times 10^{17}}{1000} = 2 \times 10^{14}$$

45.2 [II] In a *mass spectrograph*, the masses of ions are determined from their deflections in a magnetic field. Suppose that singly charged ions of chlorine are shot perpendicularly into a magnetic field $B = 0.15$ T with a speed of 5.0×10^4 m/s. (The speed could be measured by use of a velocity selector.) Chlorine has two major isotopes, of masses 34.97 u and 36.97 u. What would be the radii of the circular paths described by the two isotopes in the magnetic field? (See Fig. 45-1.)

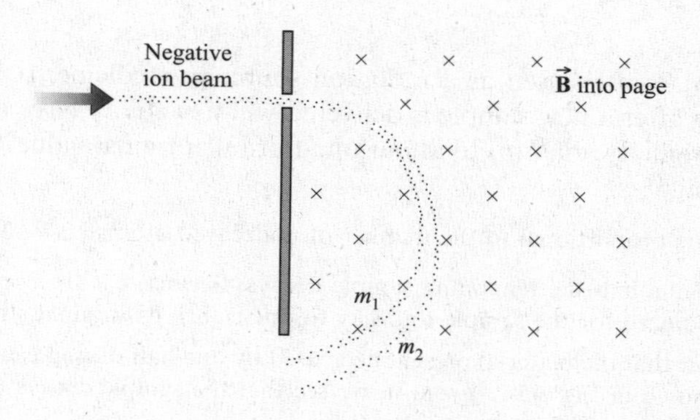

Fig. 45-1

The masses of the two isotopes are

$$m_1 = (34.97 \text{ u})(1.66 \times 10^{-27} \text{ kg/u}) = 5.81 \times 10^{-26} \text{ kg}$$

$$m_2 = (36.97 \text{ u})(1.66 \times 10^{-27} \text{ kg/u}) = 6.14 \times 10^{-26} \text{ kg}$$

Because the magnetic force qvB must provide the centripetal force mv^2/r, we have

$$r = \frac{mv}{qB} = \frac{m(5.0 \times 10^4 \text{ m/s})}{(1.6 \times 10^{-19} \text{ C})(0.105 \text{ T})} = m(2.98 \times 10^{24} \text{ m/kg})$$

Substituting the values for m found above gives the radii as 0.17 m and 0.18 m.

45.3 [I] How many protons, neutrons, and electrons are there in (*a*) ^{3}He, (*b*) ^{12}C, and (*c*) ^{206}Pb?

(*a*) The atomic number of He is 2; therefore the nucleus must contain 2 protons. Since the mass number of this isotope is 3, the sum of the protons and neutrons in the nucleus must equal 3; therefore there is 1 neutron. The number of electrons in the atom is the same as the atomic number, 2.

(*b*) The atomic number of carbon is 6; hence the nucleus must contain 6 protons. The number of neutrons in the nucleus is equal to $12 - 6 = 6$. The number of electrons is the same as the atomic number, 6.

(*c*) The atomic number of lead is 82; hence there are 82 protons in the nucleus and 82 electrons in the atom. The number of neutrons is $206 - 82 = 124$.

45.4 [II] What is the binding energy of the atom ^{12}C?

One atom of ^{12}C consists of 6 protons, 6 electrons, and 6 neutrons. The mass of the uncombined protons and electrons is the same as that of six ^{1}H atoms (if we ignore the very small binding energy of each proton-electron pair). The component particles may thus be considered as six ^{1}H atoms and six neutrons. A mass balance may be computed as follows.

$$\begin{aligned}
\text{Mass of six } {}^1\text{H atoms} = 6 \times 1.0078\ \text{u} &= 6.0468\ \text{u} \\
\text{Mass of six neutrons} = 6 \times 1.0087\ \text{u} &= \underline{6.0522\ \text{u}} \\
\text{Total mass of component particles} &= 12.0990\ \text{u} \\
\text{Mass of } {}^{12}\text{C atom} &= \underline{12.0000\ \text{u}} \\
\text{Loss in mass on forming } {}^{12}\text{C} &= 0.0990\ \text{u} \\
\text{Binding energy} = (931 \times 0.0990)\ \text{MeV} &= 92\ \text{MeV}
\end{aligned}$$

45.5 [II] Cobalt-60 (^{60}Co) is often used as a radiation source in medicine. It has a half-life of 5.25 years. How long after a new sample is delivered will its activity have decreased (*a*) to about one-eighth its original value? (*b*) to about one-third its original value? Give your answers to two significant figures.

The activity is proportional to the number of undecayed atoms ($\Delta N/\Delta t = \lambda N$).

(*a*) In each half-life, half the remaining sample decays. Because $\frac{1}{2} \times \frac{1}{2} \times \frac{1}{2} = \frac{1}{8}$, three half-lives, or 16 years, are required for the sample to decay to one-eighth its original strength.

(*b*) Using the fact that the material present decreased by one-half during each 5.25 years, we can plot the graph shown in Fig. 45-2. From it, we see that the sample decays to 0.33 its original value after a time of about 8.3 years.

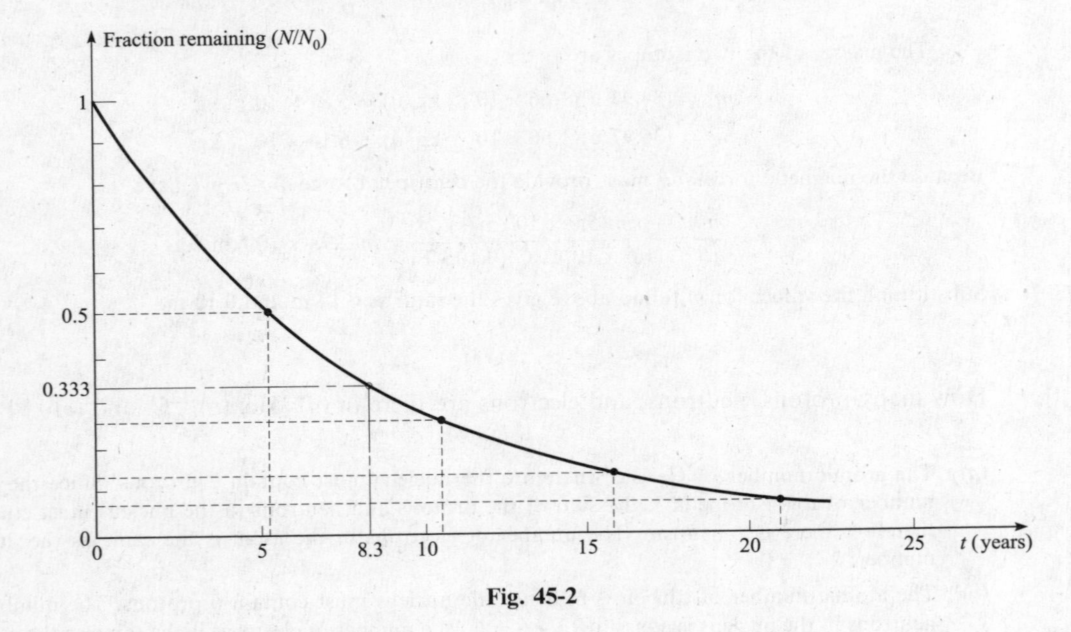

Fig. 45-2

45.6 [II] Solve Problem 45.5(*b*) by using the exponential function.

The curve in Fig. 45-2 is an *exponential decay curve* and it is expressed by the equation

$$\frac{N}{N_0} = e^{-\lambda t}$$

where λ is the decay constant, and N/N_0 is the fraction of the original N_0 particles that remain undecayed after a time t. Inasmuch as $\lambda t_{1/2} = 0.693$, $\lambda = 0.693/t_{1/2} = 0.132/\text{year}$ and $N/N_0 = 0.333$. Thus,

$$0.333 = e^{-0.132t/\text{year}}$$

Take the natural logarithm of each side to find

$$\ln (0.333) = -0.132t/\text{year}$$

from which $t = 8.3$ years.

45.7 [II] For the situation described in Problems 45.5 and 45.6, what is N/N_0 after 20 years?

As in the previous problem, where now $\lambda = 0.132$/year

$$\frac{N}{N_0} = e^{-\lambda t} = e^{-(0.132)(20)} = e^{-2.64}$$

from which $N/N_0 = 0.071$.

In this and the previous problem, we used t in years because λ was expressed in $(\text{years})^{-1}$. More often, λ would be expressed in s^{-1} and t would be in seconds. Be careful that the same time units are used for t and λ.

45.8 [II] Potassium found in nature contains two isotopes. One isotope constitutes 93.4 percent of the whole and has an atomic mass of 38.975 u; the other 6.6 percent has a mass of 40.974 u. Compute the atomic mass of potassium as found in nature.

The atomic mass of the material found in nature is obtained by combining the individual atomic masses in proportion to their abundances. The 38.975 u material is 93.4%, while the 40.974 u material is 6.6%, hence in combination

$$\text{Atomic mass} = (0.934)(38.975 \text{ u}) + (0.066)(40.974 \text{ u}) = 39.1 \text{ u}$$

45.9 [III] The half-life of radium is 1.62×10^3 years. How many radium atoms decay in 1.00 s in a 1.00 g sample of radium? The atomic weight of radium is 226 kg/kmol.

A 1.00 g sample is 0.001 00 kg, which for radium of atomic number 226 is (0.001 00/226) kmol. Since each kilomole contains 6.02×10^{26} atoms,

$$N = \left(\frac{0.001\,00}{226} \text{ kmol}\right)\left(6.02 \times 10^{26} \frac{\text{atoms}}{\text{kmol}}\right) = 2.66 \times 10^{21} \text{ atoms}$$

The decay constant is

$$\lambda = \frac{0.693}{t_{1/2}} = \frac{0.693}{(1620 \text{ y})(3.156 \times 10^7 \text{ s/y})} = 1.36 \times 10^{-11} \text{ s}^{-1}$$

Then

$$\frac{\Delta N}{\Delta t} = \lambda N = (1.36 \times 10^{-11} \text{ s}^{-1})(2.66 \times 10^{21}) = 3.61 \times 10^{10} \text{ s}^{-1}$$

is the number of disintegrations per second in 1.00 g of radium.

The above result leads to the definition of the *curie* (Ci) as a unit of activity:

$$1 \text{ Ci} = 3.7 \times 10^{10} \text{ disintegrations/s}$$

Because of its convenient size, we shall sometimes use the curie in subsequent problems, even though the official SI unit of activity is the becquerel.

45.10 [III] Technetium-99 ($^{99}_{43}\text{Tc}$) has an excited state that decays by emission of a gamma ray. The half-life of the excited state is 360 min. What is the activity, in curies, of 1.00 mg of this excited isotope?

Because we have the half-life ($t_{1/2}$) we can determine the decay constant since $\lambda t_{1/2} = 0.693$. The activity of a sample is λN. In this case,

$$\lambda = \frac{0.693}{t_{1/2}} = \frac{0.693}{21\,600\ \text{s}} = 3.21 \times 10^{-5}\ \text{s}^{-1}$$

We also know that 99.0 kg of Tc contains 6.02×10^{26} atoms. A mass m will therefore contain $[m/(99.0\ \text{kg})](6.02 \times 10^{26})$ atoms. In our case, $m = 1.00 \times 10^{-6}$ kg, and so

$$\text{Activity} = \lambda N = (3.21 \times 10^{-5}\ \text{s}^{-1})\left(\frac{1.00 \times 10^{-6}\ \text{kg}}{99.0\ \text{kg}}\right)(6.02 \times 10^{26})$$

$$= 1.95 \times 10^{14}\ \text{s}^{-1} = 1.95 \times 10^{14}\ \text{Bq}$$

45.11 [III] How much energy must a bombarding proton possess to cause the reaction ${}^{7}\text{Li}(p, n){}^{7}\text{Be}$? Give your answer to three significant figures.

The reaction is as follows:

$${}^{7}_{3}\text{Li} + {}^{1}_{1}\text{H} \rightarrow {}^{7}_{4}\text{Be} + {}^{1}_{0}n$$

where the symbols represent the *nuclei* of the atoms indicated. Because the masses listed in Table 45-2 include the masses of the atomic electrons, the appropriate number of electron masses (m_e) must be subtracted from the values given.

	Reactant mass		Product mass
${}^{7}_{3}\text{Li}$	$7.016\,00 - 3m_e$	${}^{7}_{4}\text{Be}$	$7.016\,93 - 4m_e$
${}^{1}_{1}\text{H}$	$1.007\,83 - 1m_e$	${}^{1}_{0}n$	$1.008\,66$
TOTAL	$8.023\,83 - 4m_e$	TOTAL	$8.025\,59 - 4m_e$

Subtracting the total reactant mass from the total product mass gives the increase in mass as 0.001 76 u. (Notice that the electron masses cancel out. This happens frequently, but not always.)

To create this mass in the reaction, energy must have been supplied to the reactants. The energy corresponding to 0.001 76 u is $(931 \times 0.001\,76)$ MeV = 1.65 MeV. This energy is supplied as KE of the bombarding proton. The incident proton must have more than this energy because the system must possess some KE even after the reaction, so that momentum is conserved. With momentum conservation taken into account, the minimum KE that the incident particle must have can be found with the formula

$$\left(1 + \frac{m}{M}\right)(1.65)\ \text{MeV}$$

where M is the mass of the target particle, and m that of the incident particle. Therefore, the incident particle must have an energy of at least

$$(1 + \tfrac{1}{7})(1.65)\ \text{MeV} = 1.89\ \text{MeV}$$

45.12 [II] Complete the following nuclear equations:

(*a*) ${}^{14}_{7}\text{N} + {}^{4}_{2}\text{He} \rightarrow {}^{17}_{8}\text{O} + ?$ (*d*) ${}^{30}_{15}\text{P} \rightarrow {}^{30}_{14}\text{Si} + ?$

(*b*) ${}^{9}_{4}\text{Be} + {}^{4}_{2}\text{He} \rightarrow {}^{12}_{6}\text{C} + ?$ (*e*) ${}^{3}_{1}\text{H} \rightarrow {}^{3}_{2}\text{He} + ?$

(*c*) ${}^{9}_{4}\text{Be}(p, \alpha)?$ (*f*) ${}^{43}_{20}\text{Ca}(\alpha, ?){}^{46}_{21}\text{Sc}$

(*a*) The sum of the subscripts on the left is $7 + 2 = 9$. The subscript of the first product on the right is 8. Hence the second product on the right must have a subscript (net charge) of 1. Also, the sum of the superscripts on the left is $14 + 4 = 18$. The superscript of the first product is 17. Hence the second product on the right must have a superscript (mass number) of 1. The particle with nuclear charge 1 and mass number 1 is the proton, ${}^{1}_{1}\text{H}$.

(*b*) The nuclear charge of the second product particle (its subscript) is $(4 + 2) - 6 = 0$. The mass number of the particle (its superscript) is $(9 + 4) - 12 = 1$. Hence the particle must be the neutron, ${}^{1}_{0}n$.

(c) The reactants ^9_4Be and ^1_1H have a combined nuclear charge of 5 and a mass number of 10. In addition to the alpha particle, a product will be formed of charge $5 - 2 = 3$ and mass number $10 - 4 = 6$. This is ^6_3Li.

(d) The nuclear charge of the second product particle is $15 - 14 = +1$. Its mass number is $30 - 30 = 0$. Hence the particle must be a positron, $^{0}_{+1}e$.

(e) The nuclear charge of the second product particle is $1 - 2 = -1$. Its mass number is $3 - 3 = 0$. Hence the particle must be a beta particle (an electron), $^{0}_{-1}e$.

(f) The reactants, $^{43}_{20}Ca$ and ^4_2He, have a combined nuclear charge of 22 and mass number of 47. The ejected product will have charge $22 - 21 = 1$, and mass number $47 - 46 = 1$. This is a proton and should be represented in the parentheses by p.

In some of these reactions a neutrino and/or a photon are emitted. We ignore them for this discussion since the mass and charge for both are zero.

45.13 [II] Uranium-238 ($^{238}_{92}U$) is radioactive and decays into a succession of different elements. The following particles are emitted before the nucleus reaches a stable form: α, β, β, α, α, α, α, α, β, β, α, β, β, and α (β stands for "beta particle," e^-). What is the final stable nucleus?

The original nucleus emitted 8 alpha particles and 6 beta particles. When an alpha particle is emitted, Z decreases by 2, since the alpha particle carries away a charge of $+2e$. A beta particle carries away a charge of $-1e$, and so as a result the charge on the nucleus must increase to $(Z + 1)e$. We then have, for the final nucleus,

$$\text{Final } Z = 92 + 6 - (2)(8) = 82$$
$$\text{Final } A = 238 - (6)(0) - (8)(4) = 206$$

The final stable nucleus is $^{206}_{82}Pb$.

45.14 [I] The half-life of uranium-238 is about 4.5×10^9 years, and its end product is lead-206. We notice that the oldest uranium-bearing rocks on Earth contain about a 50:50 mixture of ^{238}U and ^{206}Pb. Roughly what is the age of these rocks?

Apparently about half the ^{238}U has decayed to ^{206}Pb during the existence of the rock. Hence the rock must have been formed about 4.5 billion years ago.

45.15 [II] A 5.6-MeV alpha particle is shot directly at a uranium atom ($Z = 92$). About how close will it get to the center of the uranium nucleus?

At such high energies the alpha particle will easily penetrate the electron cloud and the effects of the atomic electrons can be ignored. We also assume the uranium atom to be so massive that it does not move appreciably. Then the original KE of the alpha particle will be changed into electrostatic potential energy. This energy, for a charge q' at a distance r from a point charge q, is (Chapter 25)

$$\text{Potential energy} = q'V = k\frac{qq'}{r}$$

Equating the KE of the alpha particle to this potential energy, we find that

$$(5.6 \times 10^6 \text{ eV})(1.60 \times 10^{-19} \text{ J/eV}) = (8.99 \times 10^9)\frac{(2e)(92e)}{r}$$

where $e = 1.60 \times 10^{-19}$ C. We find from this that $r = 4.7 \times 10^{-14}$ m.

45.16 [II] Neon-23 beta-decays in the following way:

$$^{23}_{10}\text{Ne} \rightarrow {}^{23}_{11}\text{Na} + {}^{0}_{-1}e + {}^{0}_{0}\bar{\nu}$$

where $\bar{\nu}$ is an antineutrino, a particle with no charge and no mass. Depending on circumstances, the energy carried away by the antineutrino can range from zero to the maximum energy available from the reaction. Find the minimum and maximum KE that the beta particle ${}_{-1}^{0}e$ can have. Pertinent atomic masses are 22.994 5 u for ^{23}Ne, and 22.989 8 u for ^{23}Na. The mass of the beta particle is 0.000 55 u.

Before we begin, note that the given reaction is a *nuclear* reaction, while the masses provided are those of neutral *atoms*. To calculate the mass lost in the reaction, we must subtract the mass of the atomic electrons from the atomic masses given. We have the following nuclear masses:

	Reactant mass		Product mass
${}_{10}^{23}\text{Ne}$	$22.9945 - 10m_e$	${}_{11}^{23}\text{Na}$	$22.9898 - 11m_e$
		${}_{-1}^{0}e$	m_e
		${}_{0}^{0}\bar{\nu}$	0
TOTAL	$22.9945 - 10m_e$	TOTAL	$22.9898 - 10m_e$

which gives a mass loss of $22.9945 - 22.9898 = 0.0047$ u. Since 1.00 u corresponds to 931 MeV, this mass loss corresponds to an energy of 4.4 MeV. The beta particle and antineutrino share this energy. Hence the energy of the beta particle can range from zero to 4.4 MeV.

45.17 [II] A nucleus ${}_{n}^{M}\text{P}$, the *parent* nucleus, decays to a *daughter* nucleus D by positron decay:

$$ {}_{n}^{M}\text{P} \rightarrow \text{D} + {}_{+1}^{0}e + {}_{0}^{0}\nu $$

where ν is a neutrino, a particle that has zero mass and charge. (*a*) What are the subscript and superscript for D? (*b*) Prove that the mass loss in the reaction is $M_p - M_d - 2m_e$, where M_p and M_d are the *atomic* masses of the parent and daughter.

(*a*) To balance the subscripts and superscripts, we must have ${}_{n-1}^{M}\text{D}$.

(*b*) The table of masses for the *nuclei* involved is

	Reactant mass		Product mass
${}_{n}^{M}\text{P}$	$M_p - nm_e$	${}_{n-1}^{M}\text{D}$	$M_d - (n-1)m_e$
		${}_{1}^{0}e$	m_e
		${}_{0}^{0}\nu$	0
TOTAL	$M_p - nm_e$	TOTAL	$M_d - nm_e + 2m_e$

Subtraction gives the mass loss:

$$ (M_p - nm_e) - (M_d - nm_e + 2m_e) = M_p - M_d - 2m_e $$

Notice how important it is to keep track of the electron masses in this and the previous problem.

Supplementary Problems

45.18 [I] How many protons, neutrons, and electrons does an atom of ${}_{92}^{235}\text{U}$ possess? *Ans.* 92, 143, 92

45.19 [I] By how much does the mass of a heavy nucleus change when it emits a 4.8-MeV gamma ray?
Ans. 5.2×10^{-3} u $= 8.6 \times 10^{-30}$ kg

45.20 [II] Find the binding energy of ${}_{47}^{107}\text{Ag}$, which has an atomic mass of 106.905 u. Give your answer to three significant figures. *Ans.* 915 eV

45.21 [II] The binding energy per nucleon for elements near iron in the periodic table is about 8.90 MeV per nucleon. What is the atomic mass, including electrons, of ${}^{56}_{26}Fe$? *Ans.* 55.9 u

45.22 [II] What mass of ${}^{60}_{27}Co$ has an activity of 1.0 Ci? The half-life of cobalt-60 is 5.25 years. *Ans.* 8.8×10^{-7} kg

45.23 [II] An experiment is done to determine the half-life of a radioactive substance that emits one beta particle for each decay process. Measurements show that an average of 8.4 beta particles are emitted each second by 2.5 mg of the substance. The atomic mass of the substance is 230. Find the half-life of the substance. *Ans.* 1.7×10^{10} years

45.24 [II] The half-life of carbon-14 is 5.7×10^3 years. What fraction of a sample of ${}^{14}C$ will remain unchanged after a period of five half-lives? *Ans.* 0.031

45.25 [II] Cesium-124 has a half-life of 31 s. What fraction of a cesium-124 sample will remain after 0.10 h? *Ans.* 0.000 32

45.26 [II] A certain isotope has a half-life of 7.0 h. How many seconds does it take for 10 percent of the sample to decay? *Ans.* 3.8×10^3 s

45.27 [II] By natural radioactivity ${}^{238}U$ emits an α-particle. The heavy residual nucleus is called UX_1. UX_1 in turn emits a beta particle. The resultant nucleus is called UX_2. Determine the atomic number and mass number for (*a*) UX_1 and (*b*) UX_2. *Ans.* (*a*) 90, 234; (*b*) 91, 234

45.28 [I] Upon decaying ${}^{239}_{93}Np$ emits a beta particle. The residual heavy nucleus is also radioactive, and gives rise to ${}^{235}U$ by the radioactive process. What small particle is emitted simultaneously with the formation of uranium-235? *Ans.* alpha particle

45.29 [II] Complete the following equations. (See Appendix H for a table of the elements.)

(*a*) ${}^{23}_{11}Na + {}^{4}_{2}He \rightarrow {}^{26}_{12}Mg + ?$
(*b*) ${}^{64}_{29}Cu \rightarrow {}^{0}_{+1}e + ?$
(*c*) ${}^{106}Ag \rightarrow {}^{106}Cd + ?$
(*d*) ${}^{10}_{5}B + {}^{4}_{2}He \rightarrow {}^{13}_{6}N + ?$
(*e*) ${}^{105}_{48}Cd + {}^{0}_{-1}e \rightarrow ?$
(*f*) ${}^{238}_{92}U \rightarrow {}^{234}_{90}Th + ?$

Ans. (*a*) ${}^{1}_{1}H$; (*b*) ${}^{64}_{28}Ni$; (*c*) ${}^{0}_{-1}e$; (*d*) ${}^{1}_{0}n$; (*e*) ${}^{105}_{47}Ag$; (*f*) ${}^{4}_{2}He$

45.30 [I] Complete the notations for the following processes.

(*a*) ${}^{24}Mg(d, \alpha)?$
(*b*) ${}^{26}Mg(d, p)?$
(*c*) ${}^{40}Ar(\alpha, p)?$
(*d*) ${}^{12}C(d, n)?$
(*e*) ${}^{130}Te(d, 2n)?$
(*f*) ${}^{55}Mn(n, \gamma)?$
(*g*) ${}^{59}Co(n, \alpha)?$

Ans. (*a*) ${}^{22}Na$; (*b*) ${}^{27}Mg$; (*c*) ${}^{43}K$; (*d*) ${}^{13}N$; (*e*) ${}^{130}I$; (*f*) ${}^{56}Mn$; (*g*) ${}^{56}Mn$

45.31 [II] How much energy is released during reactions (*a*) ${}^{1}_{1}H + {}^{7}_{3}Li \rightarrow 2{}^{4}_{2}He$ and (*b*) ${}^{3}_{1}H + {}^{2}_{1}H \rightarrow {}^{4}_{2}He + {}^{1}_{0}n$? *Ans.* (*a*) 17.4 MeV; (*b*) 17.6 MeV

45.32 [II] In the ${}^{14}N(n, p){}^{14}C$ reaction, the proton is ejected with an energy of 0.600 MeV. Very slow neutrons are used. Calculate the mass of the ${}^{14}C$ atom. *Ans.* 14.003 u

Chapter 46

Applied Nuclear Physics

NUCLEAR BINDING ENERGIES differ from the atomic binding energies discussed in Chapter 45 by the relatively small amount of energy that binds the electrons to the nucleus. The **binding energy per nucleon** (the total energy liberated on assembling the nucleus, divided by the number of protons and neutrons) turns out to be largest for nuclei near $Z = 30$ ($A = 60$). Hence the nuclei at the two ends of the table of elements can liberate energy if they are in some way transformed into middle-sized nuclei.

FISSION REACTION: A very large nucleus, such as the nucleus of the uranium atom, liberates energy as it is split into two or three middle-sized nuclei. Such a **fission reaction** can be induced by striking a large nucleus with a low- or moderate-energy neutron. The fission reaction produces additional neutrons, which, in turn, can cause further fission reactions and more neutrons. If the number of neutrons remains constant or increases in time, the process is a self-perpetuating *chain reaction*.

FUSION REACTION: In a ***fusion reaction***, small nuclei, such as those of hydrogen or helium, are joined together to form more massive nuclei, thereby liberating energy.

This reaction is usually difficult to initiate and sustain because the nuclei must be fused together even though they repel each other with the Coulomb force. Only when the particles move toward each other with very high energies do they come close enough for the strong force to bind them together. The fusion reaction can occur in stars because of the high densities and high thermal energies of the particles in these extremely hot objects.

RADIATION DOSE (D) is defined as the amount of energy imparted to a unit mass of substance via the absorption of ionizing radiation. A material receives a dose of 1 **gray** (Gy) when 1 J of radiation is absorbed in each kilogram of the material:

$$D = \frac{\text{energy absorbed in J}}{\text{mass of absorber in kg}}$$

so a gray is 1 J/kg. Although the gray is the SI unit for radiation dose, another unit is widely used. It is the **rad** (rd), where 1 rd = 0.01 Gy.

RADIATION DAMAGE POTENTIAL: Each type (and energy) of radiation causes its own characteristic degree of damage to living tissue. The damage also varies among types of tissue. The potential damaging effects of a specific type of radiation are expressed as the **quality factor** Q of that radiation. Arbitrarily, the damage potential is determined relative to the damage caused by 200-keV X-rays:

$$Q = \frac{\text{biological effect of 1 Gy of the radiation}}{\text{biological effect of 1 Gy of 200-keV X-rays}}$$

For example, if 10 Gy of a particular radiation will cause 7 times more damage than 10 Gy of 200-keV X-rays, then the Q for that radiation is 7. Quite often, the unit RBE (relative biological effectiveness) is used in place of quality factor. The two are equivalent.

EFFECTIVE RADIATION DOSE (H), also called the ***equivalent dose***, is the radiation dose modified to express radiation damage to living tissue. The SI unit od H is the sievert (Sv). It is defined as the product of the dose in grays and the quality factor of the radiation:

$$H = (Q)(\mathrm{D})$$

For example, suppose a certain type of tissue is subjected to a dose of 5 Gy of a radiation for which the quality factor is 3. Then the dose in sieverts is $3 \times 5 = 15$ Sv. Note that the units of Q are Sv/Gy.

While the sievert is the SI unit, another unit, the *rem* (radiation equivalent, man), is very widely used. The two are related through 1 rem = 0.01 Sv.

HIGH-ENERGY ACCELERATORS: Charged particles can be accelerated to high energies by causing them to follow a circular path repeatedly. Each time a particle (of charge q) circles the path, it is caused to fall through a potential difference V. After n trips around the path, its energy is $q(nV)$.

Magnetic fields are used to supply the centripetal force required to keep the particle moving in a circle. Equating magnetic force qvB to centripetal force mv^2/r gives

$$mv = qBr$$

In this expression, m is the mass of the particle that is traveling with speed v on a circle of radius r perpendicular to a magnetic field B.

THE MOMENTUM OF A PARTICLE is related to its KE. From Chapter 41, since the total energy of a particle is the sum of its kinetic energy plus its rest energy, $\mathrm{E} = \mathrm{KE} + mc^2$, and with $\mathrm{E}^2 = m^2c^4 + p^2c^2$ it follows that

$$\mathrm{KE} = \sqrt{p^2c^2 + m^2c^4} - mc^2$$

Solved Problems

46.1 [I] The binding energy per nucleon for ^{238}U is about 7.6 MeV, while it is about 8.6 MeV for nuclei of half that mass. If a ^{238}U nucleus were to split into two equal-size nuclei, about how much energy would be released in the process?

There are 238 nucleons involved. Each nucleon will release about $8.6 - 7.6 = 1.0$ MeV of energy when the nucleus undergoes fission. The total energy liberated is therefore about 238 MeV or 2.4×10^2 MeV.

46.2 [II] What is the binding energy per nucleon for the $^{238}_{92}$U nucleus? The *atomic* mass of ^{238}U is 238.050 79 u; also $m_p = 1.007\,276$ u and $m_n = 1.008\,665$ u.

The mass of 92 free protons plus $238 - 92 = 146$ free neutrons is

$$(92)(1.007\,276\ \mathrm{u}) + (146)(1.008\,665\ \mathrm{u}) = 239.934\,48\ \mathrm{u}$$

The mass of the ^{238}U *nucleus* is

$$238.050\,79 - 92m_e = 238.050\,79 - (92)(0.000\,549) = 238.000\,28\ \mathrm{u}$$

The mass lost in assembling the nucleus is then

$$\Delta m = 239.934\,48 - 238.000\,28 = 1.934\,2 \text{ u}$$

Since 1.00 u corresponds to 931 MeV, we have

$$\text{Binding energy} = (1.934\,2 \text{ u})(931 \text{ MeV/u}) = 1800 \text{ MeV}$$

and
$$\text{Binding energy per nucleon} = \frac{1800 \text{ MeV}}{238} = 7.57 \text{ MeV}$$

46.3 [III] When an atom of ^{235}U undergoes fission in a reactor, about 200 MeV of energy is liberated. Suppose that a reactor using uranium-235 has an output of 700 MW and is 20 percent efficient. (*a*) How many uranium atoms does it consume in one day? (*b*) What mass of uranium does it consume each day?

(*a*) Each fission yields

$$200 \text{ MeV} = (200 \times 10^6)(1.6 \times 10^{-19}) \text{ J}$$

of energy. Only 20 percent of this is utilized efficiently, and so

$$\text{Usable energy per fission} = (200 \times 10^6)(1.6 \times 10^{-19})(0.20) = 6.4 \times 10^{-12} \text{ J}$$

Because the reactor's usable output is 700×10^6 J/s, the number of fissions required per second is

$$\text{Fissions/s} = \frac{7 \times 10^8 \text{ J/s}}{6.4 \times 10^{-12} \text{ J}} = 1.1 \times 10^{20} \text{ s}^{-1}$$

and
$$\text{Fissions/day} = (86\,400 \text{ s/d})(1.1 \times 10^{20} \text{ s}^{-1}) = 9.5 \times 10^{24} \text{ d}^{-1}$$

(*b*) There are 6.02×10^{26} atoms in 235 kg of uranium-235. Therefore the mass of uranium-235 consumed in one day is

$$\text{Mass} = \left(\frac{9.5 \times 10^{24}}{6.02 \times 10^{26}}\right)(235 \text{ kg}) = 3.7 \text{ kg}$$

46.4 [III] Neutrons produced by fission reactions must be slowed by collisions with moderator nuclei before they are effective in causing further fissions. Suppose an 800-keV neutron loses 40 percent of its energy on each collision. How many collisions are required to decrease its energy to 0.040 eV? (This is the average thermal energy of a gas particle at 35°C.)

After one collision, the neutron energy is down to (0.6)(800 keV). After two, it is (0.6)(0.6)(800 keV); after three, it is $(0.6)^3$(800 keV). Therefore, after n collisions, the neutron energy is $(0.6)^n$(800 keV). We want n large enough so that

$$(0.6)^n(8 \times 10^5 \text{ eV}) = 0.040 \text{ eV}$$

Taking the logarithms of both sides of this equation gives

$$n \log_{10} 0.6 + \log_{10}(8 \times 10^5) = \log_{10} 0.04$$
$$(n)(-0.222) + 5.903 = -1.398$$

from which we find n to be 32.9. So 33 collisions are required.

46.5 [II] To examine the structure of a nucleus, pointlike particles with de Broglie wavelengths below about 10^{-16} m must be used. Through how large a potential difference must an electron fall to have this wavelength? Assume the electron is moving in a relativistic way.

The KE and momentum of the electron are related through

$$\text{KE} = \sqrt{p^2c^2 + m^2c^4} - mc^2$$

Because the de Broglie wavelength is $\lambda = h/p$, this equation becomes

$$\text{KE} = \sqrt{\left(\frac{hc}{\lambda}\right)^2 + m^2c^4} - mc^2$$

Using $\lambda = 10^{-16}$ m, $h = 6.63 \times 10^{-34}$ J·s, and $m = 9.1 \times 10^{-31}$ kg, we find that

$$\text{KE} = 1.99 \times 10^{-9}\ \text{J} = 1.24 \times 10^{10}\ \text{eV}$$

The electron must be accelerated through a potential difference of about 10^{10} eV.

46.6 [III] The following fusion reaction takes place in the Sun and furnishes much of its energy:

$$4\,{}_1^1\text{H} \rightarrow {}_2^4\text{He} + 2\,{}_{+1}^{\ 0}e + \text{energy}$$

where ${}_{+1}^{\ 0}e$ is a positron electron. How much energy is released as 1.00 kg of hydrogen is consumed? The masses of ^{1}H, ^{4}He, and ${}_{+1}^{\ 0}e$ are, respectively, 1.007 825, 4.002 604, and 0.000 549 u, where atomic electrons are included in the first two values.

Ignoring the electron binding energy, the mass of the reactants, 4 protons, is 4 times the atomic mass of hydrogen (^{1}H), less the mass of 4 electrons:

$$\begin{aligned}\text{Reactant mass} &= (4)(1.007\,825\ \text{u}) - 4m_e \\ &= 4.031\,300\ \text{u} - 4m_e\end{aligned}$$

where m_e is the mass of the electron (or positron). The reaction products have a combined mass

$$\begin{aligned}\text{Product mass} &= (\text{mass of } {}_2^4\text{He nucleus}) + 2m_e \\ &= (4.002\,604\ \text{u} - 2m_e) + 2m_e \\ &= 4.002\,604\ \text{u}\end{aligned}$$

The mass loss is therefore

$$(\text{reactant mass}) - (\text{product mass}) = (4.031\,3\ \text{u} - 4m_e) - 4.002\,6\ \text{u}$$

Substituting $m_e = 0.000\,549$ u gives the mass loss as 0.026 5 u.

But 1.00 kg of ^{1}H contains 6.02×10^{26} atoms. For each four atoms that undergo fusion, 0.026 5 u is lost. The mass lost when 1.00 kg undergoes fusion is therefore

$$\begin{aligned}\text{Mass loss/kg} &= (0.026\,5\ \text{u})(6.02 \times 10^{26}/4) = 3.99 \times 10^{24}\ \text{u} \\ &= (3.99 \times 10^{24}\ \text{u})(1.66 \times 10^{-27}\ \text{kg/u}) = 0.006\,63\ \text{kg}\end{aligned}$$

Then, from the Einstein relation,

$$\Delta E = (\Delta m)c^2 = (0.006\,63\ \text{kg})(2.998 \times 10^8\ \text{m/s})^2 = 5.96 \times 10^{14}\ \text{J}$$

46.7 [III] Lithium hydride, LiH, has been proposed as a possible nuclear fuel. The nuclei to be used and the reaction involved are as follows:

$$\begin{matrix} {}_3^6\text{Li} & + & {}_1^2\text{H} & \rightarrow & 2\,{}_2^4\text{He} \\ 6.015\,13 & & 2.014\,10 & & 4.002\,60 \end{matrix}$$

the listed masses being those of the neutral atoms. Calculate the expected power production, in megawatts, associated with the consumption of 1.00 g of LiH per day. Assume 100 percent efficiency.

Ignoring the electron binding energies, the change in mass for the reaction must be computed first:

	Reactant mass		Product mass
^6_3Li	$6.015\,13\ \text{u} - 3m_e$	$2\,^4_2\text{He}$	$2(4.002\,60\ \text{u} - 2m_e)$
^2_1H	$2.014\,10\ \text{u} - 1m_e$		
TOTAL	$8.029\,23\ \text{u} - 4m_e$	TOTAL	$8.005\,20\ \text{u} - 4m_e$

We find the loss in mass by subtracting the product mass from the reactant mass. In the process, the electron masses drop out and the mass loss is found to be 0.024 03 u.

The fractional loss in mass is $0.0240/8.029 = 2.99 \times 10^{-3}$. Therefore, when 1.00 g reacts, the mass loss is

$$(2.99 \times 10^{-3})(1.00 \times 10^{-3}\ \text{kg}) = 2.99 \times 10^{-6}\ \text{kg}$$

This corresponds to an energy of

$$\Delta E = (\Delta m)c^2 = (2.99 \times 10^{-6}\ \text{kg})(2.998 \times 10^8\ \text{m/s})^2 = 2.687 \times 10^{11}\ \text{J}$$

Then $$\text{Power} = \frac{\text{energy}}{\text{time}} = \frac{2.687 \times 10^{11}\ \text{J}}{86\,400\ \text{s}} = 3.11\ \text{MW}$$

46.8 [II] Cosmic rays bombard the CO_2 in the atmosphere and, by nuclear reaction, cause the formation of the radioactive carbon isotope $^{14}_6\text{C}$. This isotope has a half-life of 5730 years. It mixes into the atmosphere uniformly and is taken up in plants as they grow. After a plant dies, the ^{14}C decays over the ensuing years. How old is a piece of wood that has a ^{14}C content which is only 9 percent as large as the average ^{14}C content of new-grown wood?

During the years, the ^{14}C has decayed to 0.090 its original value. Hence (see Problem 45.6),

$$\frac{N}{N_0} = e^{-\lambda t} \qquad \text{becomes} \qquad 0.090 = e^{-0.693t/(5730\ \text{years})}$$

After taking the natural logarithms of both sides, we have

$$\ln 0.090 = \frac{-0.693t}{5730\ \text{years}}$$

from which $$t = \left(\frac{5730\ \text{years}}{-0.693}\right)(-2.41) = 1.99 \times 10^4\ \text{years}$$

The piece of wood is about 20 000 years old.

46.9 [III] Iodine-131 has a half-life of about 8.0 days. When consumed in food, it localizes in the thyroid. Suppose 7.0 percent of the ^{131}I localizes in the thyroid and that 20 percent of its disintegrations are detected by counting the emitted gamma rays. How much ^{131}I must be ingested to yield a thyroid count rate of 50 counts per second?

Because only 20 percent of the disintegrations are counted, there must be a total of 50/20% or $50/0.20 = 250$ disintegrations per second, which is what $\Delta N/\Delta t$ is. From Chapter 45,

$$\frac{\Delta N}{\Delta t} = \lambda N = \frac{0.693N}{t_{1/2}} \qquad \text{and so} \qquad 250\ \text{s}^{-1} = \frac{0.693N}{(8.0\ \text{d})(3600\ \text{s/h})(24\ \text{h/d})}$$

from which $N = 2.49 \times 10^8$.

However, this is only 7.0 percent of the ingested ^{131}I. Hence the number of ingested atoms is $N/0.070 = 3.56 \times 10^9$. And, since 1.00 kmol of ^{131}I is approximately 131 kg, this number of atoms represents

$$\left(\frac{3.56 \times 10^9 \text{ atoms}}{6.02 \times 10^{26} \text{ atoms/kmol}}\right)(131 \text{ kg/kmol}) = 7.8 \times 10^{-16} \text{ kg}$$

which is the mass of ^{131}I that must be ingested.

46.10 [II] A beam of gamma rays has a cross-sectional area of 2.0 cm^2 and carries 7.0×10^8 photons through the cross-section each second. Each photon has an energy of 1.25 MeV. The beam passes through a 0.75 cm thickness of flesh ($\rho = 0.95$ g/cm^3) and loses 5.0 percent of its intensity in the process. What is the average dose (in Gy and in rd) applied to the flesh each second?

The dose in this case is the energy absorbed per kilogram of flesh. Since 5.0% of the intensity is absorbed we have

$$\text{Number of photons absorbed/s} = (7.0 \times 10^8 \text{ s}^{-1})(0.050) = 3.5 \times 10^7 \text{ s}^{-1}$$

and each such photon carries an energy of 1.25 MeV. Hence,

$$\text{Energy absorbed/s} = (3.5 \times 10^7 \text{ s}^{-1})(1.25 \text{ MeV}) = 4.4 \times 10^7 \text{ MeV/s}$$

We need the mass of flesh in which this energy was absorbed. The beam was delivered to a region of area 2.0 cm^2 and thickness 0.75 cm. Thus

$$\text{Mass} = \rho V = (0.95 \text{ g/cm}^3)[(2.0 \text{ cm}^2)(0.75 \text{ cm})] = 1.43 \text{ g}$$

Keeping in mind that 1rd = 0.01 Gy, we than have

$$\text{Dose/s} = \frac{\text{energy/s}}{\text{mass}} = \frac{(4.4 \times 10^7 \text{ MeV/s})(1.6 \times 10^{-13} \text{ J/MeV})}{1.43 \times 10^{-3} \text{ kg}} = 4.9 \text{ mGy/s} = 0.49 \text{ rd/s}$$

46.11 [II] A beam of alpha particles passes through flesh and deposits 0.20 J of energy in each kilogram of flesh. The Q for these particles is 12 Sv/Gy. Find the dose in Gy and rd, as well as the effective dose in Sv and rem.

Keeping in mind the $H = QD$ where

$$D = \text{Dose} = \frac{\text{absorbed energy}}{\text{mass}} = 0.20 \text{ J/kg} = 0.20 \text{ Gy} = 20 \text{ rd}$$

Hence $\quad H = \text{Effective dose} = (Q)(\text{dose}) = (12 \text{ Sv/Gy})(0.20 \text{ Gy}) = 2.4 \text{ Sv} = 2.4 \times 10^2 \text{ rem}$

46.12 [III] A tumor on a person's leg has a mass of 3.0 g. What is the minimum activity a radiation source can have if it is to furnish a dose of 10 Gy to the tumor in 14 min? Assume each disintegration within the source, on the average, provides an energy 0.70 MeV to the tumor.

A dose of 10 Gy corresponds to 10 J of radiation energy being deposited per kilogram. Since the tumor has a mass of 0.0030 kg, the energy required for a 10 Gy dose is (0.0030 kg)(10 J/kg) = 0.030 J.

Each disintegration provides 0.70 MeV, which in joules is

$$(0.70 \times 10^6 \text{ eV})(1.60 \times 10^{-19} \text{ J/eV}) = 1.12 \times 10^{-13} \text{ J}$$

A dose of 10 Gy requires that an energy of 0.030 J be delivered. That total energy divided by the energy per disintegration, yields the number of disintegrations:

$$\frac{0.030 \text{ J}}{1.12 \times 10^{-13} \text{ J/disintegration}} = 2.68 \times 10^{11} \text{ disintegrations}$$

They are to occur in 14 min (or 840 s), and so the disintegration rate is

$$\frac{2.68 \times 10^{11}}{840 \text{ s}} \text{disintegrations} = 3.2 \times 10^8 \text{ disintegrations/s}.$$

Hence the source activity must be at least 3.2×10^8 Bq. Since 1 Ci = 3.70×10^{10} Bq, the source activity must be at least 8.6 mCi.

46.13 [II] A beam of 5.0 MeV alpha particles ($q = 2e$) has a cross-sectional area of 1.50 cm^2. It is incident on flesh ($\rho = 950$ kg/m^3) and penetrates to a depth of 0.70 mm. (*a*) What dose (in Gy) does the beam provide to the flesh in a time of 3.0 s? (*b*) What effective dose does it provide? Assume the beam to carry a current of 2.50×10^{-9} A and to have $Q = 14$.

Using the current we can find the number of particles deposited in the flesh in 3.0 s, keeping in mind that for each particle $q = 2e$:

$$\text{Number in 3.0 s} = \frac{It}{q} = \frac{(2.50 \times 10^{-9}\ \text{C/s})(3.0\ \text{s})}{3.2 \times 10^{-19}\ \text{C}} = 2.34 \times 10^{10}\ \text{particles}$$

Each 5.0-MeV alpha particle deposits an energy of $(5.0 \times 10^6\ \text{eV})(1.60 \times 10^{-19}\ \text{J/eV}) = 8.0 \times 10^{-13}$ J. In 3.0 s a total energy of 2.34×10^{10} particles) (8.0×10^{-13} J/particle) is deposited. And it is delivered to a volume of area 1.50 cm^2 and thickness 0.70 mm. Therefore,

$$\text{Dose} = \frac{\text{energy}}{\text{mass}} = \frac{(2.34 \times 10^{10})(8.0 \times 10^{-13}\ \text{J})}{(950\ \text{kg/m}^3)(0.070 \times 1.5 \times 10^{-6}\ \text{m}^3)} = 188\ \text{Gy} = 1.9 \times 10^2\ \text{Gy}$$

$$\text{Effective dose} = (Q)(\text{dose}) = (14)(188) = 2.6 \times 10^3\ \text{Sv}$$

Supplementary Problems

46.14 [II] Consider the following fission reaction:

$$\begin{array}{ccccccccccc} {}^{1}_{0}n & + & {}^{235}_{92}\text{U} & \rightarrow & {}^{138}_{56}\text{Ba} & + & {}^{93}_{41}\text{Nb} & + & 5\,{}^{1}_{0}n & + & 5\,{}^{0}_{-1}e \\ 1.0087 & & 235.0439 & & 137.9050 & & 92.9060 & & 1.0087 & & 0.00055 \end{array}$$

where the neutral atomic masses are given. How much energy is released when (*a*) 1 atom undergoes this type of fission, and (*b*) 1.0 kg of atoms undergoes fission? *Ans.* (*a*) 182 MeV; (*b*) 7.5×10^{13} J

46.15 [II] It is proposed to use the nuclear fusion reaction

$$\begin{array}{ccc} 2\,{}^{2}_{1}\text{H} & \rightarrow & {}^{4}_{2}\text{He} \\ 2.014102 & & 4.002604 \end{array}$$

to produce industrial power (neutral atomic masses are given). If the output is to be 150 MW and the energy of the reaction will be used with 30 percent efficiency, how many grams of deuterium fuel will be needed per day? *Ans.* 75 g/day

46.16 [II] One of the most promising fusion reactions for power generation involves deuterium (^{2}H) and tritium (^{3}H):

$$\begin{array}{ccccccc} {}^{2}_{1}\text{H} & + & {}^{3}_{1}\text{H} & \rightarrow & {}^{4}_{2}\text{He} & + & {}^{1}_{0}n \\ 2.01410 & & 3.01605 & & 4.00260 & & 1.00867 \end{array}$$

where the atomic masses including electrons are as given. How much energy is produced when 2.0 kg of ^{2}H fuses with 3.0 kg of ^{3}H to form ^{4}He? *Ans.* 1.7×10^{15} J

46.17 [I] What is the average KE of a neutron at the center of the Sun, where the temperature is about 10^7 K? Give your answer to two significant figures. *Ans.* 1.3 keV

46.18 [II] Find the energy released when two deuterons ($^2_1$H, atomic mass = 2.014 10 u) fuse to form $^3_2$He (atomic mass = 3.016 03 u) with the release of a neutron. Give your answer to three significant figures.
Ans. 3.27 MeV

46.19 [II] The tar in an ancient tar pit has a ^{14}C activity that is only about 4.00 percent of that found for new wood of the same density. What is the approximate age of the tar? *Ans.* 26.6×10^3 years

46.20 [II] Rubidium-87 has a half-life of 4.9×10^{10} years and decays to strontium-87, which is stable. In an ancient rock, the ratio of ^{87}Sr to ^{87}Rb is 0.005 0. If we assume all the strontium came from rubidium decay, about how old is the rock? Repeat if the ratio is 0.210. *Ans.* 3.5×10^8 years, 1.35×10^{10} years

46.21 [II] The luminous dial of an old watch gives off 130 fast electrons each minute. Assume that each electron has an energy of 0.50 MeV and deposits that energy in a volume of skin that is 2.0 cm^2 in area and 0.20 cm thick. Find the dose (in both Gy and rd) that the volume experiences in 1.0 day. Take the density of skin to be 900 kg/m^3. *Ans.* 42 μGy, 4.2 mrd

46.22 [II] An alpha-particle beam enters a charge collector and is measured to carry 2.0×10^{-14} C of charge into the collector each second. The beam has a cross-sectional area of 150 mm^2, and it penetrates human skin to a depth of 0.14 mm. Each particle has an initial energy of 4.0 MeV. The Q for such particles is about 15. What effective dose, in Sv and in rem, does a person's skin receive when exposed to this beam for 20 s? Take $\rho = 900$ kg/m^3 for skin. *Ans.* 0.63 Sv, 63 rem

Appendix A

Significant Figures

INTRODUCTION: The numerical value of every measurement is an approximation. Consider that the length of an object is recorded as 15.7 cm. By convention, this means that the length was measured to the *nearest* tenth of a centimeter and that its exact value lies between 15.65 and 15.75 cm. If this measurement were exact to the nearest hundredth of a centimeter, it would have been recorded as 15.70 cm. The value 15.7 cm represents *three significant figures* (1, 5, 7), while the value 15.70 represents *four significant figures* (1, 5, 7, 0). A significant figure is one that is known to be reasonably reliable.

Similarly, a recorded mass of 3.406 2 kg means that the mass was determined to the nearest tenth of a gram and represents five significant figures (3, 4, 0, 6, 2), the last figure (2) being reasonably correct and guaranteeing the certainty of the preceding four figures.

ZEROS may be significant or they may merely serve to locate the decimal point. We will take zeros to the left of the normal position of the decimal point (in numbers like 100, 2500, 40, etc.) to be significant. For instance the statement that a body of ore weighs 9800 N will be understood to mean that we know the weight to the nearest newton: there are four significant figures here. Alternatively, if it was weighed to the nearest hundred newtons, the weight contains only two significant figures (9, 8) and may be written exponentially as 9.8×10^3 N. If it was weighed to the nearest ten newtons, it should be written as 9.80×10^3 N, displaying three significant figures. If the object was weighed to the nearest newton, the weight can also be written as 9.800×10^3 N (four significant figures). Of course, if a zero stands between two significant figures, it is itself significant. Zeros to the immediate right of the decimal are significant only when there is a nonzero figure to the left of the decimal. Thus the numbers 0.001, 0.001 0, 0.001 00, and 1.001 have one, two, three, and four significant figures, respectively.

ROUNDING OFF: A number is rounded off to the desired number of significant figures by dropping one or more digits to the right. When the first digit dropped is less than 5, the last digit retained should remain unchanged; when it is 5 or more, 1 is added to the last digit retained.

ADDITION AND SUBTRACTION: The result of adding or subtracting should be rounded off, so as to retain digits only as far as the first column containing estimated figures. (Remember that the last significant figure is estimated.) In other words, the answer should have the same number of figures to the right of the decimal point as does the least precisely known number being added or subtracted.

Examples: Add the following quantities expressed in meters.

(*a*)	(*b*)	(*c*)	(*d*)
25.340	58.0	4.20	415.5
5.465	0.003 8	1.652 3	3.64
0.322	0.000 01	0.015	0.238
31.127 m (*Ans.*)	58.003 81	5.867 3	419.378
	= 58.0 m (*Ans.*)	= 5.87 m (*Ans.*)	= 419.4 m (*Ans.*)

MULTIPLICATION AND DIVISION: Here the result should be rounded off to contain only as many significant figures as are contained in the least exact factor.

There are some exceptional cases, however. Consider the division $9.84 \div 9.3 = 1.06$, to three places. By the rule given above, the answer should be 1.1 (two significant figures). However, a difference of 1 in in the last place of 9.3 (9.3 ± 0.1) results in an error of about 1 percent, while a difference of 1 in the last place of 1.1 (1.1 ± 0.1) yields an error of roughly 10 percent. Thus the answer 1.1 is of much lower percentage accuracy than 9.3. Hence in this case the answer should be 1.06, since a difference of 1 in the last place of the least exact factor used in the calculation (9.3) yields a percentage of error about the same (about 1 percent) as a difference of 1 in the last place of 1.06 (1.06 ± 0.01). Similarly, $0.92 \times 1.13 = 1.04$. We shall not worry about such exceptions.

TRIGONOMETRIC FUNCTIONS: As a rule, the values of sines, cosines, tangents, and so forth, should have the same number of significant figures as their arguments. For example, $\sin 35° = 0.57$ whereas $\sin 35.0° = 0.574$.

Exercises

1 [I] How many significant figures are given in the following quantities?

(*a*) 454 g	(*e*) 0.035 3 m	(*i*) 1.118×10^{-3} V
(*b*) 2.2 N	(*f*) 1.008 0 hr	(*j*) 1030 kg/m^3
(*c*) 2.205 N	(*g*) 14.0 A	(*k*) 125 000 N
(*d*) 0.393 7 s	(*h*) 9.3×10^7 km	

Ans.

(*a*) 3	(*e*) 3	(*i*) 4
(*b*) 2	(*f*) 5	(*j*) 4
(*c*) 4	(*g*) 3	(*k*) 6
(*d*) 4	(*h*) 2	

2 [I] Add:

(*a*)	(*b*)	(*c*)	(*d*)
703 h	18.425 cm	0.003 5 s	4.0 N
7 h	7.21 cm	0.097 s	0.632 N
0.66 h	5.0 cm	0.225 s	0.148 N

Ans. (*a*) 711 h, (*b*) 30.6 cm, (*c*) 0.326 s, (*d*) 4.8 N

3 [I] Subtract:

(*a*)	(*b*)	(*c*)
7.26 J	562.4 m	34 kg
0.2 J	16.8 m	0.2 kg

Ans. (*a*) 7.1 J, (*b*) 545.6 m, (*c*) 34 kg

4 [I] Multiply:

(*a*) 2.21×0.3	(*d*) 107.88×0.610
(*b*) 72.4×0.084	(*e*) 12.4×84.0
(*c*) 2.02×4.113	(*f*) 72.4×8.6

Ans.

(*a*) 0.7	(*d*) 65.8
(*b*) 6.1	(*e*) 1.04×10^3
(*c*) 8.31	(*f*) 6.2×10^2

5 [I] Divide: (*a*) $\frac{97.52}{2.54}$ (*b*) $\frac{14.28}{0.714}$ (*c*) $\frac{0.032}{0.004}$ (*d*) $\frac{9.80}{9.30}$

Ans. (*a*) 38.4, (*b*) 20.0, (*c*) 8, (*d*) 1.05

Appendix B

Trigonometry Needed for College Physics

FUNCTIONS OF AN ACUTE ANGLE: The trigonometric functions most often used are the sine, cosine, and tangent. It is convenient to put the definitions of the functions of an acute angle in terms of the sides of a right triangle.

In any right triangle: The **sine** of either acute angle is equal to the length of the side opposite that angle divided by the length of the hypotenuse. The **cosine** of either acute angle is equal to the length of the side adjacent to that angle divided by the length of the hypotenuse. The **tangent** of either acute angle is equal to the length of the side opposite that angle divided by the length of the side adjacent to that angle.

If θ and ϕ are the acute angles of any right triangle and A, B, and C are the sides, as shown in the diagram, then

$$\sin\theta = \frac{\text{opposite}}{\text{hypotenuse}} = \frac{B}{C} \qquad \sin\phi = \frac{\text{opposite}}{\text{hypotenuse}} = \frac{A}{C}$$

$$\cos\theta = \frac{\text{adjacent}}{\text{hypotenuse}} = \frac{A}{C} \qquad \cos\phi = \frac{\text{adjacent}}{\text{hypotenuse}} = \frac{B}{C}$$

$$\tan\theta = \frac{\text{opposite}}{\text{adjacent}} = \frac{B}{A} \qquad \tan\phi = \frac{\text{opposite}}{\text{adjacent}} = \frac{A}{B}$$

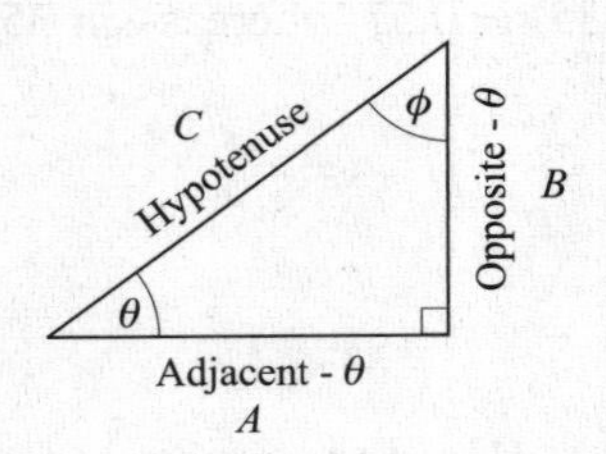

Note that $\sin\theta = \cos\phi$; thus the sine of any angle equals the cosine of its complementary angle. For example,

$$\sin 30^\circ = \cos(90^\circ - 30^\circ) = \cos 60^\circ \qquad \cos 50^\circ = \sin(90^\circ - 50^\circ) = \sin 40^\circ$$

As an angle increases from 0° to 90°, its sine increases from 0 to 1, its tangent increases from 0 to infinity, and its cosine decreases from 1 to 0.

LAW OF SINES AND OF COSINES: These two laws give the relations between the sides and angles of *any* plane triangle. In any plane triangle with angles α, β, and γ and sides opposite A, B, and C, respectively, the following relations apply:

Law of Sines

$$\frac{A}{\sin\alpha} = \frac{B}{\sin\beta} = \frac{C}{\sin\gamma}$$

or

$$\frac{A}{B} = \frac{\sin\alpha}{\sin\beta} \qquad \frac{B}{C} = \frac{\sin\beta}{\sin\gamma} \qquad \frac{C}{A} = \frac{\sin\gamma}{\sin\alpha}$$

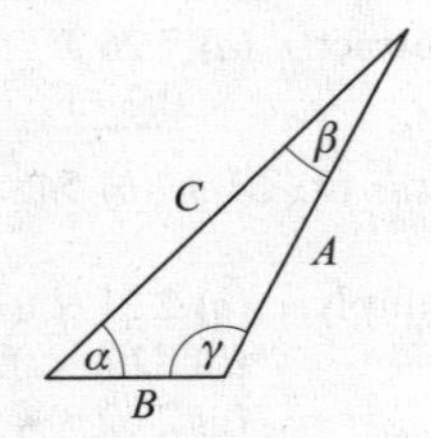

Law of Cosines

$$A^2 = B^2 + C^2 - 2BC\ \cos\alpha$$

$$B^2 = A^2 + C^2 - 2AC\ \cos\beta$$

$$C^2 = A^2 + B^2 - 2AB\ \cos\gamma$$

If the angle θ is between 90° and 180°, as in the case of angle C in the above diagram, then

$$\sin\theta = \sin(180^\circ - \theta) \qquad \text{and} \qquad \cos\theta = -\cos(180^\circ - \theta)$$

Thus
$$\sin 120° = \sin(180° - 120°) = \sin 60° = 0.866$$
$$\cos 120° = -\cos(180° - 120°) = -\cos 60° = -0.500$$

Solved Problems

1 [I] In right triangle ABC, given $A = 8$, $B = 6$, $\gamma = 90°$. Find the values of the sine, cosine, and tangent of angle α and of angle β.

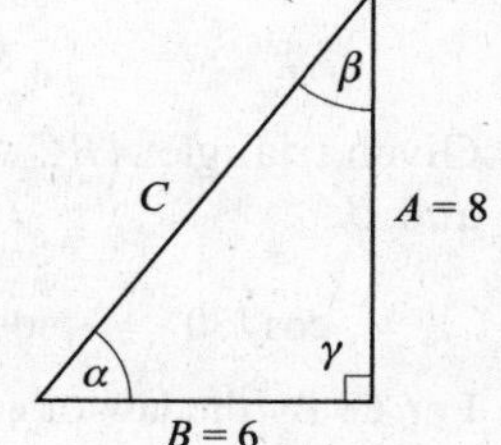

$$C = \sqrt{8.0^2 + 6.0^2} = \sqrt{100} = 10$$

$\sin\alpha = A/C = 8.0/10 = 0.80$ $\quad\sin\beta = B/C = 6.0/10 = 0.60$

$\cos\alpha = B/C = 6.0/10 = 0.60$ $\quad\cos\beta = A/C = 8.0/10 = 0.80$

$\tan\alpha = A/B = 8.0/6.0 = 1.3$ $\quad\tan\beta = B/A = 6.0/8.0 = 0.75$

2 [I] Given a right triangle with one acute angle 40.0° and hypotenuse 400. Find the other sides and angles.

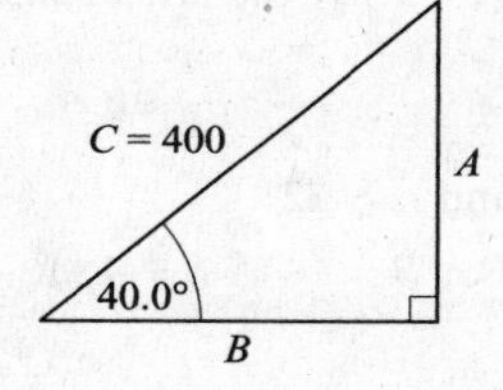

$$\sin 40.0° = \frac{A}{400} \quad \text{and} \quad \cos 40.0° = \frac{B}{400}$$

Using a calculator, we find that $\sin 40.0° = 0.642\,8$ and $\cos 40.0° = 0.766\,0$. Then

$$a = 400 \sin 40.0° = 400(0.642\,8) = 257$$
$$b = 400 \cos 40.0° = 400(0.766\,0) = 306$$
$$B = 90.0° - 40.0° = 50.0°$$

3 [II] Given triangle ABC with $\alpha = 64.0°$, $\beta = 71.0°$, $B = 40.0°$. Find A and C.

$$\gamma = 180.0° - (\alpha + \beta) = 180.0° - (64.0° + 71.0°) = 45.0°$$

By the law of sines,

$$\frac{A}{\sin\alpha} = \frac{B}{\sin\beta} \quad \text{and} \quad \frac{C}{\sin\gamma} = \frac{B}{\sin\beta}$$

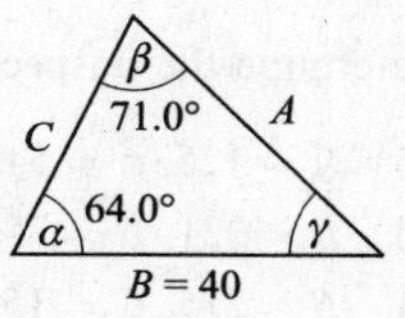

so
$$A = \frac{B\sin\alpha}{\sin\beta} = \frac{40.0 \sin 64.0°}{\sin 71.0°} = \frac{40.0(0.898\,8)}{0.945\,5} = 38.0$$

and
$$C = \frac{B\sin\gamma}{\sin\beta} = \frac{40.0 \sin 45.0°}{\sin 71.0°} = \frac{40.0(0.707\,1)}{0.945\,5} = 29.9$$

4 [I] (*a*) If $\cos\alpha = 0.438$, find α to the nearest degree. (*b*) If $\sin\beta = 0.800\,0$, find β to the nearest tenth of a degree. (*c*) If $\cos\gamma = 0.712\,0$, find γ to the nearest tenth of a degree.

(*a*) On your calculator use the inverse and cosine keys to get $\alpha = 64°$; or if you have a $\cos^{-1}$ key use it.

(*b*) Enter 0.800 0 on your calculator and use the inverse and sine keys to get $\beta = 53.1°$.

(*c*) Use your calculator as in (*a*) to get 44.6°.

5 [II] Given triangle ABC with $\alpha = 130.8°$, $A = 525$, $C = 421$. Find B, β, and γ.

$$\sin 130.8° = \sin(180° - 130.8°) = \sin 49.2° = 0.757$$

Most hand calculators give sin 130.8° directly.

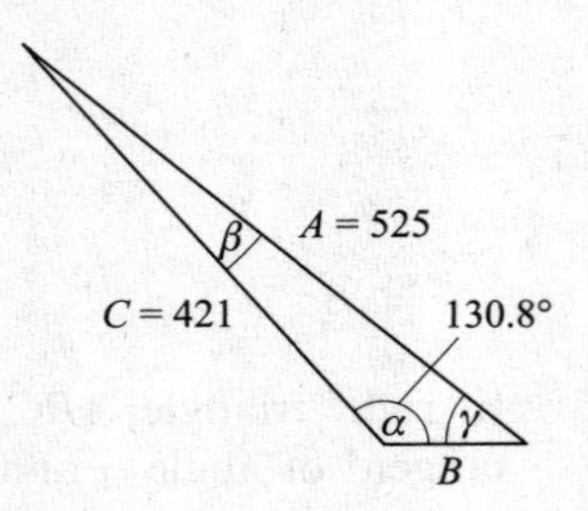

For γ: $\quad \sin\gamma = \dfrac{C\sin\alpha}{A} = \dfrac{421\sin 30.8°}{525} = \dfrac{421(0.757)}{525} = 0.607$

from which $\gamma = 37.4°$.

For β: $\quad \beta = 180° - (\gamma + \alpha) = 180° - (37.4° + 130.8°) = 11.8°$

For B: $\quad B = \dfrac{A\sin\beta}{\sin\alpha} = \dfrac{525\sin 11.8°}{\sin 130.8°} = \dfrac{525(0.204)}{0.757} = 142$

6 [II] Given triangle ABC with $A = 14$, $B = 8.0$, $\gamma = 130°$. Find C, α, and β.

$$\cos 130° = -\cos(180° - 130°) = -\cos 50° = -0.64$$

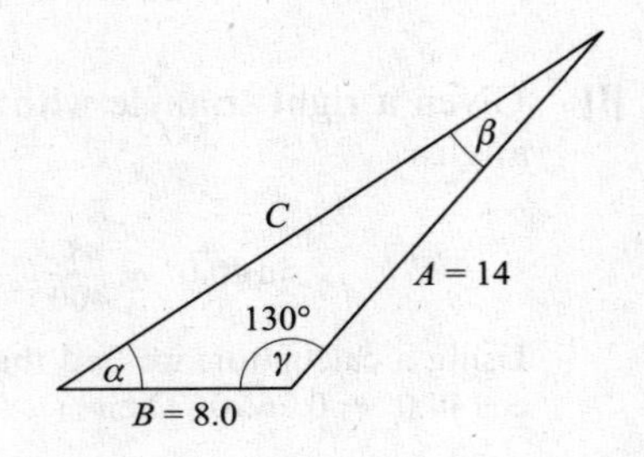

For C: By the law of cosines,

$$C^2 = A^2 + B^2 - 2AB\cos 130°$$
$$= 14^2 + 8.0^2 - 2(14)(8.0)(-0.643) = 404$$

and $C = \sqrt{404} = 20$.

For α: By the law of sines,

$$\sin\alpha = \frac{A\sin\gamma}{C} = \frac{14(0.766)}{20.1} = 0.533$$

and $\alpha = 32°$.

For β: $\quad \beta = 180° - (\alpha + \gamma) = 180° - (32° + 130°) = 18°$

7 Determine the unspecified sides and angles of the following right triangles ABC, with $\gamma = 90°$.

(*a*) $\alpha = 23.3°$, $C = 346$ (*d*) $A = 25.4$, $B = 38.2$
(*b*) $\beta = 49.2°$, $B = 222$ (*e*) $B = 673$, $C = 888$
(*c*) $\alpha = 66.6°$, $A = 113$

Ans. (*a*) $\beta = 66.7°$, $A = 137$, $B = 318$ (*d*) $\alpha = 33.6°$, $\beta = 56.4°$, $C = 45.9$
(*b*) $\alpha = 40.8°$, $A = 192$, $C = 293$ (*e*) $\alpha = 40.7°$, $\beta = 49.3°$, $A = 579$
(*c*) $\beta = 23.4°$, $B = 48.9$, $C = 123$

8 Determine the unspecified sides and angles of the following oblique triangles ABC.

(*a*) $A = 125$, $\alpha = 54.6°$, $\beta = 65.2°$ (*e*) $B = 50.4$, $C = 33.3$, $\beta = 118.5°$
(*b*) $B = 321$, $\alpha = 75.3°$, $\gamma = 38.5°$ (*f*) $B = 120$, $C = 270$, $\alpha = 118.7°$
(*c*) $B = 215$, $C = 150$, $\beta = 42.7°$ (*g*) $A = 24.5$, $B = 18.6$, $C = 26.4$
(*d*) $A = 512$, $B = 426$, $\alpha = 48.8°$ (*h*) $A = 6.34$, $B = 7.30$, $C = 9.98$

Ans. (*a*) $B = 139$, $C = 133$, $\gamma = 60.2°$ (*e*) $A = 25.1$, $\alpha = 26.0°$, $\gamma = 35.5°$
(*b*) $A = 339$, $C = 218$, $\beta = 66.2°$ (*f*) $A = 344$, $\beta = 17.8°$, $\gamma = 43.5°$
(*c*) $A = 300$, $\alpha = 109.1°$, $\gamma = 28.2°$ (*g*) $\alpha = 63.2°$, $\beta = 42.7°$, $\gamma = 74.1°$
(*d*) $C = 680$, $\beta = 38.8°$, $\gamma = 92.4°$ (*h*) $\alpha = 39.3°$, $\beta = 46.9°$, $\gamma = 93.8°$

Appendix C

Exponents

POWERS OF 10: The following is a partial list of powers of 10. (See also Appendix E.)

$$10^0 = 1 \qquad 10^{-1} = \frac{1}{10} = 0.1$$

$$10^1 = 10$$

$$10^2 = 10 \times 10 = 100 \qquad 10^{-2} = \frac{1}{10^2} = \frac{1}{100} = 0.01$$

$$10^3 = 10 \times 10 \times 10 = 1000$$

$$10^4 = 10 \times 10 \times 10 \times 10 = 10\,000 \qquad 10^{-3} = \frac{1}{10^3} = \frac{1}{1000} = 0.001$$

$$10^5 = 10 \times 10 \times 10 \times 10 \times 10 = 100\,000$$

$$10^6 = 10 \times 10 \times 10 \times 10 \times 10 \times 10 = 1\,000\,000 \qquad 10^{-4} = \frac{1}{10^4} = \frac{1}{10\,000} = 0.000\,1$$

In the expression 10^5, the *base* is 10 and the *exponent* is 5.

MULTIPLICATION AND DIVISION: In multiplication, exponents of like bases are added:

$$a^3 \times a^5 = a^{3+5} = a^8 \qquad 10^7 \times 10^{-3} = 10^{7-3} = 10^4$$

$$10^2 \times 10^3 = 10^{2+3} = 10^5 \qquad (4 \times 10^4)(2 \times 10^{-6}) = 8 \times 10^{4-6} = 8 \times 10^{-2}$$

$$10 \times 10 = 10^{1+1} = 10^2 \qquad (2 \times 10^5)(3 \times 10^{-2}) = 6 \times 10^{5-2} = 6 \times 10^3$$

In division, exponents of like bases are subtracted:

$$\frac{a^5}{a^3} = a^{5-3} = a^2 \qquad \frac{8 \times 10^2}{2 \times 10^{-6}} = \frac{8}{2} \times 10^{2+6} = 4 \times 10^8$$

$$\frac{10^2}{10^5} = 10^{2-5} = 10^{-3} \qquad \frac{5.6 \times 10^{-2}}{1.6 \times 10^4} = \frac{5.6}{1.6} \times 10^{-2-4} = 3.5 \times 10^{-6}$$

SCIENTIFIC NOTATION: Any number may be expressed as an integral power of 10, or as the product of two numbers one of which is an integral power of 10. For example,

$$2806 = 2.806 \times 10^3 \qquad 0.045\,4 = 4.54 \times 10^{-2}$$

$$22\,406 = 2.240\,6 \times 10^4 \qquad 0.000\,06 = 6 \times 10^{-5}$$

$$454 = 4.54 \times 10^2 \qquad 0.003\,06 = 3.06 \times 10^{-3}$$

$$0.454 = 4.54 \times 10^{-1} \qquad 0.000\,000\,5 = 5 \times 10^{-7}$$

OTHER OPERATIONS: A nonzero expression with an exponent of zero is equal to 1. Thus,

$$a^0 = 1 \qquad 10^0 = 1 \qquad (3 \times 10)^0 = 1 \qquad 8.2 \times 10^0 = 8.2$$

A power may be transferred from the numerator to the denominator of a fraction, or vice versa, by changing the sign of the exponent. For example,

$$10^{-4} = \frac{1}{10^4} \qquad 5 \times 10^{-3} = \frac{5}{10^3} \qquad \frac{7}{10^{-2}} = 7 \times 10^2 \qquad -5a^{-2} = -\frac{5}{a^2}$$

The meaning of the fractional exponent is illustrated by the following:

$$10^{2/3} = \sqrt[3]{10^2} \qquad 10^{3/2} = \sqrt{10^3} \qquad 10^{1/2} = \sqrt{10} \qquad 4^{3/2} = \sqrt{4^3} = \sqrt{64} = 8$$

To take a power to a power, multiply exponents:

$$(10^3)^2 = 10^{3\times 2} = 10^6 \qquad (10^{-2})^3 = 10^{-2\times 3} = 10^{-6} \qquad (a^3)^{-2} = a^{-6}$$

To extract the square root, divide the exponent by 2. If the exponent is an odd number it should first be increased or decreased by 1, and the coefficient adjusted accordingly. To extract the cube root, divide the exponent by 3. The coefficients are treated independently. Thus,

$$\sqrt{9 \times 10^4} = 3 \times 10^2 \qquad \sqrt{4.9 \times 10^{-5}} = \sqrt{49 \times 10^{-6}} = 7.0 \times 10^{-3}$$

$$\sqrt{3.6 \times 10^7} = \sqrt{36 \times 10^6} = 6.0 \times 10^3 \qquad \sqrt[3]{1.25 \times 10^8} = \sqrt[3]{125 \times 10^6} = 5.00 \times 10^2$$

Most hand calculators give square roots directly. Cube roots and other roots are easily found using the y^x key.

Exercises

1 [I] Express the following in powers of 10.

(*a*) 326 (*b*) 32 608 (*c*) 1006 (*d*) 36 000 008 (*e*) 0.831 (*f*) 0.03 (*g*) 0.000 002 (*h*) 0.000 706 (*i*) $\sqrt{0.000\,081}$ (*j*) $\sqrt[3]{0.000\,027}$

Ans. (*a*) 3.26×10^2 (*b*) $3.260\,8 \times 10^4$ (*c*) 1.006×10^3 (*d*) $3.600\,000\,8 \times 10^7$ (*e*) 8.31×10^{-1} (*f*) 3×10^{-2} (*g*) 2×10^{-6} (*h*) 7.06×10^{-4} (*i*) 9.0×10^{-3} (*j*) 3.0×10^{-2}

2 [I] Evaluate the following and express the results in powers of 10.

(*a*) 1500×260

(*b*) $220 \times 35\,000$

(*c*) $40 \div 20\,000$

(*d*) $82\,800 \div 0.12$

(*e*) $\dfrac{1.728 \times 17.28}{0.000\,172\,8}$

(*f*) $\dfrac{(16\,000)(0.000\,2)(1.2)}{(2000)(0.006)(0.000\,32)}$

(*g*) $\dfrac{0.004 \times 32\,000 \times 0.6}{6400 \times 3000 \times 0.08}$

(*h*) $(\sqrt{14\,400})(\sqrt{0.000\,025})$

(*i*) $(\sqrt[3]{2.7 \times 10^7})(\sqrt[3]{1.25 \times 10^{-4}})$

(*j*) $(1 \times 10^{-3})(2 \times 10^5)^2$

(*k*) $\dfrac{(3 \times 10^2)^3(2 \times 10^{-5})^2}{3.6 \times 10^{-8}}$

(*l*) $8(2 \times 10^{-2})^{-3}$

Ans. (*a*) 3.90×10^5 (*b*) 7.70×10^6 (*c*) 2.0×10^{-3} (*d*) 6.9×10^5 (*e*) 1.728×10^5 (*f*) 1×10^3 (*g*) 5×10^{-5} (*h*) 6.0×10^{-1} (*i*) 1.5×10^1 (*j*) 4×10^7 (*k*) 3×10^5 (*l*) 1×10^6

Appendix D

Logarithms

THE LOGARITHM TO BASE 10 of a number is the exponent or power to which 10 must be raised to yield that number. Since 1000 is 10^3, the logarithm to base 10 of 1000 (written log 1000) is 3. Similarly, $\log 10\,000 = 4$, $\log 10 = 1$, $\log 0.1 = -1$, and $\log 0.001 = -3$.

Most hand calculators have a log key. When a number is entered into the calculator, its logarithm to base 10 can be found by pressing the log key. In this way we find that $\log 50 = 1.698\,97$ and $\log 0.035 = -1.455\,93$. Also, $\log 1 = 0$, which reflects the fact that $10^0 = 1$.

NATURAL LOGARITHMS are taken to the base $e = 2.718$, rather than 10. They can be found on most hand calculators by pressing the ln key. Since $e^0 = 1$, we have $\ln 1 = 0$.

Examples:

$$\log 971 = 2.987\,2 \qquad \ln 971 = 6.878\,3$$
$$\log 9.71 = 0.987\,2 \qquad \ln 9.71 = 2.273\,2$$
$$\log 0.097\,1 = -1.012\,8 \qquad \ln 0.097\,1 = -2.332\,0$$

Exercises: Find the logarithms to base 10 of the following numbers.

(*a*) 454 (*f*) 0.621
(*b*) 5280 (*g*) 0.946 3
(*c*) 96 500 (*h*) 0.035 3
(*d*) 30.48 (*i*) 0.002 2
(*e*) 1.057 (*j*) 0.000 264 5

Ans. (*a*) 2.657 1 (*f*) −0.206 9
(*b*) 3.722 6 (*g*) −0.023 97
(*c*) 4.984 5 (*h*) −1.452 2
(*d*) 1.484 0 (*i*) −2.657 6
(*e*) 0.024 1 (*j*) −3.577 6

ANTILOGARITHMS: Suppose we have an equation such as $3.5 = 10^{0.544}$; then we know that 0.544 is the log to base 10 of 3.5. Or, inversely, we can say that 3.5 is the *antilogarithm* (or *inverse logarithm*) of 0.544. Finding the antilogarithm of a number is simple with most hand calculators: Enter the number; then press first the inverse key and then the log key. Or, if the base is e rather than 10, press the inverse and ln keys.

Exercises: Find the numbers corresponding to the following logarithms.

(*a*) 3.156 8 (*f*) 0.914 2
(*b*) 1.693 4 (*g*) 0.000 8
(*c*) 5.693 4 (*h*) −0.249 3
(*d*) 2.500 0 (*i*) −1.996 5
(*e*) 2.043 6 (*j*) −2.799 4

Ans. (*a*) 1435 (*f*) 8.208
(*b*) 49.37 (*g*) 1.002
(*c*) 4.937×10^5 (*h*) 0.563 2
(*d*) 316.2 (*i*) 0.010 08
(*e*) 110.6 (*j*) 0.001 587

BASIC PROPERTIES OF LOGARITHMS: Since logarithms are exponents, all properties of exponents are also properties of logarithms.

(1) The logarithm of the product of two numbers is the sum of their logarithms. Thus,

$$\log ab = \log a + \log b \qquad \log(5280 \times 48) = \log 5280 + \log 48$$

(2) The logarithm of the quotient of two numbers is the logarithm of the numerator minus the logarithm of the denominator. For example,

$$\log \frac{a}{b} = \log a - \log b \qquad \log \frac{536}{24.5} = \log 536 - \log 24.5$$

(3) The logarithm of the *n*th power of a number is *n* times the logarithm of the number. Thus,

$$\log a^n = n \log a \qquad \log(4.28)^3 = 3 \log 4.28$$

(4) The logarithm of the *n*th root of a number is $1/n$ times the logarithm of the number. Thus,

$$\log \sqrt[n]{a} = \frac{1}{n}\log a \qquad \log\sqrt{32} = \frac{1}{2}\log 32 \qquad \log\sqrt[3]{792} = \frac{1}{3}\log 792$$

Solved Problem

1 [I] Use a hand calculator to evaluate (*a*) $(5.2)^{0.4}$, (*b*) $(6.138)^3$, (*c*) $\sqrt[3]{5}$, (*d*) $(7.25 \times 10^{-11})^{0.25}$.

(*a*) Enter 5.2; press y^x key; enter 0.4; press = key. The displayed answer is 1.934.

(*b*) Enter 6.138; press y^x key; enter 3; press = key. The displayed answer is 231.2.

(*c*) Enter 5; press y^x key; enter 0.333 3; press = key. The displayed answer is 1.710.

(*d*) Enter 7.25×10^{-11}; press y^x key; enter 0.25; press = key. The displayed answer is 2.918×10^{-3}.

Exercises

2 [I] Evaluate each of the following.

(1) $28.32 \times 0.082\,54$

(2) $573 \times 6.96 \times 0.004\,81$

(3) $\dfrac{79.28}{63.57}$

(4) $\dfrac{65.38}{225.2}$

(5) $\dfrac{1}{239}$

(6) $\dfrac{0.572 \times 31.8}{96.2}$

(7) $47.5 \times \dfrac{779}{760} \times \dfrac{273}{300}$

(8) $(8.642)^2$

(9) $(0.086\,42)^2$

(10) $(11.72)^3$

(11) $(0.052\,3)^3$

(12) $\sqrt{9463}$

(13) $\sqrt{946.3}$

(14) $\sqrt{0.006\,61}$

(15) $\sqrt[3]{1.79}$

(16) $\sqrt[4]{0.182}$

(17) $\sqrt{643} \times (1.91)^3$

(18) $(8.73 \times 10^{-2})(7.49 \times 10^6)$

(19) $(3.8 \times 10^{-5})^2(1.9 \times 10^{-5})$

(20) $\dfrac{8.5 \times 10^{-45}}{1.6 \times 10^{-22}}$

(21) $\sqrt{2.54 \times 10^6}$

(22) $\sqrt{9.44 \times 10^5}$

(23) $\sqrt{7.2 \times 10^{-13}}$

(24) $\sqrt[3]{7.3 \times 10^{-14}}$

(25) $\sqrt{\dfrac{(1.1 \times 10^{-23})(6.8 \times 10^{-2})}{1.4 \times 10^{-24}}}$

(26) $2.04 \log 97.2$

(27) $37 \log 0.029\,8$

(28) $6.30 \log (2.95 \times 10^3)$

(29) $8.09 \log (5.68 \times 10^{-16})$

(30) $(2.00)^{0.714}$

Ans.

(1) 2.337	(9) 0.007 467	(17) 177	(25) 0.73
(2) 19.2	(10) 1611	(18) 6.54×10^5	(26) 4.05
(3) 1.247	(11) 0.000 143	(19) 2.7×10^{-14}	(27) −56
(4) 0.290 2	(12) 97.27	(20) 5.3×10^{-23}	(28) 21.9
(5) 0.004 18	(13) 30.76	(21) 1.59×10^3	(29) −123
(6) 0.189	(14) 0.081 3	(22) 9.72×10^2	(30) 1.64
(7) 44.3	(15) 1.21	(23) 8.5×10^{-7}	
(8) 74.67	(16) 0.653	(24) 4.2×10^{-5}	

Appendix E

Prefixes for Multiples of SI Units

Multiplication Factor	Prefix	Symbol
10^{12}	tera	T
10^{9}	giga	G
10^{6}	mega	M
10^{3}	kilo	k
10^{2}	hecto	h
10	deka	da
10^{-1}	deci	d
10^{-2}	centi	c
10^{-3}	milli	m
10^{-6}	micro	μ
10^{-9}	nano	n
10^{-12}	pico	p
10^{-15}	femto	f
10^{-18}	atto	a

The Greek Alphabet

A	α	alpha	H	η	eta	N	ν	nu	T	τ	tau
B	β	beta	Θ	θ	theta	Ξ	ξ	xi	Y	υ	upsilon
Γ	γ	gamma	I	ι	iota	O	o	omicron	Φ	ϕ	phi
Δ	δ	delta	K	κ	kappa	Π	π	pi	X	χ	chi
E	ϵ	epsilon	Λ	λ	lambda	P	ρ	rho	Ψ	ψ	psi
Z	ζ	zeta	M	μ	mu	Σ	σ	sigma	Ω	ω	omega

Appendix F

Factors for Conversions to SI Units

Acceleration	$1\ \text{ft/s}^2 = 0.3048\ \text{m/s}^2$
	$g = 9.807\ \text{m/s}^2$
Area	$1\ \text{acre} = 4047\ \text{m}^2$
	$1\ \text{ft}^2 = 9.290 \times 10^{-2}\ \text{m}^2$
	$1\ \text{in.}^2 = 6.45 \times 10^{-4}\ \text{m}^2$
	$1\ \text{mi}^2 = 2.59 \times 10^{6}\ \text{m}^2$
Density	$1\ \text{g/cm}^3 = 10^3\ \text{kg/m}^3$
Energy	$1\ \text{Btu} = 1054\ \text{J}$
	1 calorie (cal) $= 4.184\ \text{J}$
	1 electron volt (eV) $= 1.602 \times 10^{-19}\ \text{J}$
	1 foot pound (ft·lb) $= 1.356\ \text{J}$
	1 kilowatt hour (kW·h) $= 3.60 \times 10^{6}\ \text{J}$
Force	$1\ \text{dyne} = 10^{-5}\ \text{N}$
	$1\ \text{lb} = 4.448\ \text{N}$
Length	1 angstrom (Å) $= 10^{-10}\ \text{m}$
	$1\ \text{ft} = 0.3048\ \text{m}$
	$1\ \text{in.} = 2.54 \times 10^{-2}\ \text{m}$
	1 light year $= 9.461 \times 10^{15}\ \text{m}$
	$1\ \text{mile} = 1069\ \text{m}$
Mass	1 atomic mass unit (u) $= 1.6606 \times 10^{-27}\ \text{kg}$
	$1\ \text{gram} = 10^{-3}\ \text{kg}$
Power	$1\ \text{Btu/s} = 1054\ \text{W}$
	$1\ \text{cal/s} = 4.184\ \text{W}$
	$1\ \text{ft·lb/s} = 1.356\ \text{W}$
	1 horsepower (hp) $= 746\ \text{W}$
Pressure	1 atmosphere (atm) $= 1.013 \times 10^{5}\ \text{Pa}$
	$1\ \text{bar} = 10^{5}\ \text{Pa}$
	$1\ \text{cmHg} = 1333\ \text{Pa}$
	$1\ \text{lb/ft}^2 = 47.88\ \text{Pa}$
	$1\ \text{lb/in.}^2$ (psi) $= 6895\ \text{Pa}$
	$1\ \text{N/m}^2 = 1$ pascal (Pa)
	$1\ \text{torr} = 133.3\ \text{Pa}$
Speed	1 ft/s (fps) $= 0.3048\ \text{m/s}$
	$1\ \text{km/h} = 0.2778\ \text{m/s}$
	1 mi/h (mph) $= 0.44704\ \text{m/s}$
Temperature	$T_{\text{Kelvin}} = T_{\text{Celsius}} + 273.15$
	$T_{\text{Kelvin}} = \frac{5}{9}(T_{\text{Fahrenheit}} + 459.67)$
	$T_{\text{Celsius}} = \frac{5}{9}(T_{\text{Fahrenheit}} - 32)$
	$T_{\text{Kelvin}} = \frac{5}{9} T_{\text{Rankine}}$
Time	$1\ \text{day} = 86\,400\ \text{s}$
	$1\ \text{year} = 3.16 \times 10^{7}\ \text{s}$
Volume	$1\ \text{ft}^3 = 2.832 \times 10^{-2}\ \text{m}^3$
	$1\ \text{gallon} = 3.785 \times 10^{-3}\ \text{m}^3$
	$1\ \text{in.}^3 = 1.639 \times 10^{-5}\ \text{m}^3$
	$1\ \text{liter} = 10^{-3}\ \text{m}^3$

Appendix G

Physical Constants

Constant	Symbol	Value
Speed of light in free space	c	$= 2.997\,924\,58 \times 10^8$ m/s
Acceleration due to gravity (normal)	g	$= 9.807$ m/s^2
Gravitational constant	G	$= 6.672\,59 \times 10^{-11}$ N·m^2/kg^2
Coulomb constant	k_0	$= 8.988 \times 10^9$ N·m^2/C^2
Density of water (maximum)		$= 0.999\,972 \times 10^3$ kg/m^3
Density of mercury (S.T.P.)		$= 13.595 \times 10^3$ kg/m^3
Standard atmosphere		$= 1.013\,2 \times 10^5$ N/m^2
Volume of ideal gas at S.T.P.		$= 22.4$ m^3/kmol
Avogadro's number	N_A	$= 6.022 \times 10^{26}$ kmol^{-1}
Universal gas constant	R	$= 8314$ J/kmol·K
Ice point		$= 273.15$ K
Mechanical equivalent of heat		$= 4.184$ J/cal
Stefan–Boltzmann constant	σ	$= 5.67 \times 10^{-8}$ W/m^2·K^4
Planck's constant	h	$= 6.626 \times 10^{-34}$ J·s
Faraday	F	$= 9.648\,5 \times 10^4$ C/mol
Electronic charge	e	$= 1.602\,2 \times 10^{-19}$ C
Boltzmann's constant	k_B	$= 1.38 \times 10^{-23}$ J/K
Ratio of electron charge to mass	e/m_e	$= 1.758\,8 \times 10^{11}$ C/kg
Electron mass	m_e	$= 9.109 \times 10^{-31}$ kg
Proton mass	m_p	$= 1.672\,6 \times 10^{-27}$ kg
Neutron mass	m_n	$= 1.674\,9 \times 10^{-27}$ kg
Alpha particle mass		$= 6.645 \times 10^{-27}$ kg
Atomic mass unit (1/12 mass of ^{12}C)	u	$= 1.660\,6 \times 10^{-27}$ kg
Rest energy of 1 u		$= 931.5$ MeV

Appendix H

Table of the Elements

The masses listed are based on $^{12}_{6}C = 12$ u. A value in parentheses is the mass number of the most stable (long-lived) of the known isotopes.

Element	Symbol	Atomic Number Z	Average Atomic Mass, u
Actinium	Ac	89	(227)
Aluminum	Al	13	26.981 5
Americium	Am	95	(243)
Antimony	Sb	51	121.75
Argon	Ar	18	39.948
Arsenic	As	33	74.921 6
Astatine	At	85	(210)
Barium	Ba	56	137.34
Berkelium	Bk	97	(247)
Beryllium	Be	4	9.012 2
Bismuth	Bi	83	208.980
Boron	B	5	10.811
Bromine	Br	35	79.904
Cadmium	Cd	48	112.40
Calcium	Ca	20	40.08
Californium	Cf	98	(251)
Carbon	C	6	12.011 2
Cerium	Ce	58	140.12
Cesium	Cs	55	132.905
Chlorine	Cl	17	35.453
Chromium	Cr	24	51.996
Cobalt	Co	27	58.933 2
Copper	Cu	29	63.546
Curium	Cm	96	(247)
Dysprosium	Dy	66	162.50
Einsteinium	Es	99	(254)
Erbium	Er	68	167.26
Europium	Eu	63	151.96
Fermium	Fm	100	(257)
Fluorine	F	9	18.998 4
Francium	Fr	87	(223)
Gadolinium	Gd	64	157.25
Gallium	Ga	31	69.72
Germanium	Ge	32	72.59
Gold	Au	79	196.967
Hafnium	Hf	72	178.49
Helium	He	2	4.002 6
Holmium	Ho	67	164.930
Hydrogen	H	1	1.008 0
Indium	In	49	114.82
Iodine	I	53	126.904 4
Iridium	Ir	77	192.2
Iron	Fe	26	55.847
Krypton	Kr	36	83.80
Lanthanum	La	57	138.91
Lawrencium	Lr	103	(257)
Lead	Pb	82	207.19
Lithium	Li	3	6.939
Lutetium	Lu	71	174.97

Table of the Elements (*Continued*)

Element	Symbol	Atomic Number Z	Average Atomic Mass, u
Magnesium	Mg	12	24.312
Manganese	Mn	25	54.938 0
Mendelevium	Md	101	(256)
Mercury	Hg	80	200.59
Molybdenum	Mo	42	95.94
Neodymium	Nd	60	144.24
Neon	Ne	10	20.183
Neptunium	Np	93	(237)
Nickel	Ni	28	58.71
Niobium	Nb	41	92.906
Nitrogen	N	7	14.006 7
Nobelium	No	102	(254)
Osmium	Os	76	190.2
Oxygen	O	8	15.999 4
Palladium	Pd	46	106.4
Phosphorus	P	15	30.973 8
Platinum	Pt	78	195.09
Plutonium	Pu	94	(244)
Polonium	Po	84	(209)
Potassium	K	19	39.102
Praseodymium	Pr	59	140.907
Promethium	Pm	61	(145)
Protactinium	Pa	91	(231)
Radium	Ra	88	(226)
Radon	Rn	86	222
Rhenium	Re	75	186.2
Rhodium	Rh	45	102.905
Rubidium	Rb	37	85.47
Ruthenium	Ru	44	101.07
Samarium	Sm	62	150.35
Scandium	Sc	21	44.956
Selenium	Se	34	78.96
Silicon	Si	14	28.086
Silver	Ag	47	107.868
Sodium	Na	11	22.989 8
Strontium	Sr	38	87.62
Sulfur	S	16	32.064
Tantalum	Ta	73	180.948
Technetium	Tc	43	(97)
Tellurium	Te	52	127.60
Terbium	Tb	65	158.924
Thallium	Tl	81	204.37
Thorium	Th	90	232.038 1
Thulium	Tm	69	168.934
Tin	Sn	50	118.69
Titanium	Ti	22	47.90
Tungsten	W	74	183.85
Uranium	U	92	238.03
Vanadium	V	23	50.942
Xenon	Xe	54	131.30
Ytterbium	Yb	70	173.04
Yttrium	Y	39	88.905
Zinc	Zn	30	65.37
Zirconium	Zr	40	91.22

Index